P9-CCI-185

# Modeling and Simulation of Dynamic Systems

# Modeling and Simulation of Dynamic Systems

Robert L. Woods
*Mechanical and Aerospace Engineering*
*University of Texas at Arlington*

Kent L. Lawrence
*Mechanical and Aerospace Engineering*
*University of Texas at Arlington*

PRENTICE HALL, Upper Saddle River, New Jersey 07458

**Library of Congress Cataloging-in-Publication Data**

Woods, Robert L.
Modeling and simulation of dynamic systems / Robert L. Woods, Kent L. Lawrence.
p. cm.
Includes bibliographical references and index.
ISBN (invalid) 0-13-337379-1
1. Engineering—Mathematical models. 2. Systems engineering —Mathematical models. I. Lawrence, Kent L. II. Title.
TA342.W66 1997
620'.001'1—dc21 96-49673
CIP

Editor-in-chief: Marcia Horton
Acquisitions editor: Bill Stenquist
Managing editor: Bayani Mendoza DeLeon
Project manager: Jennifer Wenzel
Cover director: Amy Rosen
Manufacturing buyer: Julia Meehan
Editorial assistant: Meg Weist

Simon & Schuster/A Viacom Company
Upper Saddle River, New Jersey 07458



Printed in the United States of America

10 9 8 7 6 5 4 3 2 1

ISBN 0-13-337379-1

Prentice-Hall International (UK) Limited, *London*
Prentice-Hall of Australia Pty. Limited, *Sydney*
Prentice-Hall Canada Inc., *Toronto*
Prentice-Hall Hispanoamericana, S.A., *Mexico*
Prentice-Hall of India Private Limited, *New Delhi*
Prentice-Hall of Japan, Inc., *Tokyo*
Simon & Schuster Asia Pte. Ltd., *Singapore*
Editora Prentice-Hall do Brasil, Ltda., *Rio de Janeiro*

*To my parents, Aldon Woods and Gwen Woods;*
*my wife, Brenda; and my daughter, Lori.*

RLW

*To my mother, Doris W. Lawrence; the loving memory*
*of my father, James H. Lawrence, Sr.; my wife Carol;*
*and my children, Dory, Kent, Jr., and Jamie.*

KLL

# Contents

# Preface

## Approach

This text is intended to provide a comprehensive treatment of the modeling techniques of the major types of engineering systems, the methods for solving the resulting differential equations, and the attendant mathematical procedures related to the representation of dynamic systems and the determination of their response characteristics.

The material is designed for a one-semester course in engineering at the undergraduate or graduate level. The reader is expected to have completed the basic lower division courses in chemistry, physics, mathematics, and engineering. The appendices supplement this background as necessary.

The introductory material on system dynamics gives a broad overview of the concepts of dynamic systems and the systems approach to the analysis and design of engineering systems. This should give the reader a feel for the similarity of the different engineering disciplines.

The modeling portions of the book begin with a discussion of the basic system components of each engineering discipline and then show how to combine these components into systems and how to obtain the appropriate governing differential equation models. Mechanical, electrical, fluid (hydraulic and pneumatic), and thermal systems are treated in detail, and emphasis is placed on the similarity of the response characteristics embodied in each of these physically dissimilar systems.

We stress the importance of matching the system mathematical model to the solution technique. Thus, classical differential equation models are recommended for low-order linear systems that are to be solved using analytical differential equation methods or transfer function techniques, and state-space formulations are emphasized for use with digital and analog simulation solutions. Both frequency-domain and time-domain analysis techniques are presented.

## Contents

Chapters 1 and 2 present the necessary background and preliminaries that tie together the concepts of system dynamics for the major engineering disciplines. Chapters 3–6 present the techniques of modeling using conservation laws and properties of each of the technical disciplines of mechanical, electrical, fluid, and thermal systems analysis. Mixed systems, i.e., systems composed of combinations of mechanical, electrical, fluid, and/or thermal components, are discussed in Chapter 7.

The frequency response of dynamic systems, or how dynamic systems respond to a steady-state sinusoidal input, is presented in Chapter 8. Chapter 9 is devoted to methods of time response and digital simulation.

In Chapter 10, the methods of the preceding chapters are brought to bear on the design and selection of components of representive sample systems, using linear and nonlinear sytem analysis. These systems serve to tie the previous modeling and analysis develpments together and to unify the reader's understanding of the modeling and simulation process as it might be employed by the practicing engineer.

International System (SI) units and British units are used alternatively throughout the text, and Appendix A presents conversions between the two. We emphasize the use of units and the compatibility of equations in terms of units. Appendix B provides often-used information on the properties of the components of mechanical systems.

Reference material on vector and matrix algebra and the solution of algebraic equations is presented in a complete form in Appendices C and D. For the reader not familiar with these subjects, the appendices are intended to provide substantial background coverage.

The solution of ordinary differential equations by classical and by Laplace transform methods is discussed in Appendices E and F. Methods for converting linear differential equations to and from the state-space format are presented in Appendix G.

Illustrative examples, practice problems, and suitable references are provided to supplement the major ideas that are discussed in the text, and simulation software tools are emphasized throughout. The example problems in the text were solved using the Student Edition of **MATLAB®,** and in Appendix H the reader is provided guidance on the use of MATLAB for the solution of typical dynamic simulation problems.[1]

Analog simulation techniques are covered in Appendix I.

System model representations by classical equation, transfer function, or state-space differential equations are interspersed in each of the chapters to give the reader experience with these approaches. The techniques for solving the three representations of differential equations are presented in enough detail that the reader

[1]MATLAB is a registered trademark of The MathWorks, Inc.

can solve typical equations. Classical solutions to differential equations and the Laplace transform solution technique should be a review from a course in differential equations for most students; however, enough material is presented in Appendices E and F to allow the reader, if necessary, to obtain a first-time knowledge of these methods.

### "Road maps"

The chapters, as well as the appendices, are more or less independent of one another, and a number of "road maps" may be used to move through the material.

One approach is to work straight through the text material covering modeling first, frequency and time response simulation second, and applications last. The appendix may be used as required for review or to fill in any deficiencies.

Another way to use the book effectively is to study Chapters 1 through 3 (or 4), followed by Chapter 8 (on the frequency response) and/or Chapter 9 (on the time response). The reader is then in a position to evaluate system behavior through frequency and/or time simulation methods. Using a computational and visualizational tool such as MATLAB with real-world example systems usually promotes greater understanding of the basic principles. Covering the material in Appendix H on the use of MATLAB early in the course will assist students in preparing for subsequent chapters. One may then take up Chapter 7 (on combined technology systems), Chapter 10 (on systems design), and the additional disciplines covered in Chapters 5 and 6 in any convenient order.

### Acknowledgments

We wish to thank our colleagues Kliff Black, Raul Fernandez, A. Haji-Sheikh, David Hullender, Frank Lewis, Panos Shiakolas, Dan Tuckness, and B. P. Wang, who read portions of the manuscript and suggested improvements or suggested material for inclusion as homework problems.

Our thanks also go to Prentice Hall Executive Editor Bill Stenquist for his invaluable support in bringing this project to completion and to Ms. Margaret McGranahan for encouraging us to complete the work. We appreciate the expert assistance of Project Manager Jennifer Wenzel and Copy Editor Brian Baker in bringing the manuscript to final form.

We are deeply indebted to our wives, Brenda Woods and Carol Lawrence, for their patience during the preparation of the text.

Your comments regarding the material and its presentation are welcome. You can contact us by phone or by surface or electronic mail (woods@mae.uta.edu, lawrence@mae.uta.edu).

*Robert L. Woods*
*Kent L. Lawrence*

# Introduction

The launch of Apollo 13. (Photo courtesy of NASA)

## "UNLUCKY 13"*

A dramatic chapter in America's space exploration occurred during the flight of Apollo 13.

> Apollo 13 took off on 11 April 1970. The initial part of the journey proceeded with accustomed smoothness. But two days later, 55 hours 55 minutes after lift-off, the mission became the unlucky 13$^{th}$. The spacecraft was over 200,000 miles (320,000 km) from home, and accelerating towards the Moon. The crew had just completed a telecast to Earth and were clearing up. John Swigert was in the command module (code-named Odyssey), Fred Haise was in the lunar module (Aquarius), with James Lovell in between. Suddenly there was a loud bang, the spacecraft vibrated, and within seconds the master alarm sounded.
>
> Mission control back on Earth were stunned when Swigert radioed through: "Okay Houston, we've had a problem here." It was a masterly understatement. What had happened was a potential disaster. A liquid oxygen tank in the service module had exploded, destroying the fuel cells that supplied power to the spacecraft and cutting off the oxygen supply. The service module, including its propulsion motor, was dead. The command module had a back-up battery pack, but that would be needed for re-entry. In any case it had a life of only 10 hours, and Apollo 13 was 87 hours from home.

NASA engineers scrambled to understand the problem that had occurred, develop a solution, and assist in bringing the astronauts safely back to earth. Mathematical models of the spacecraft and its systems had, of course, been developed long ago to predict the in-flight behavior of the vehicle before any Apollo missions were launched. These dynamic simulation models had been refined and updated as data were obtained from previous Apollo flights. The digital computer implementation of the models would now be invaluable as engineers attempted to convert an exploratory mission into a rescue mission. Numerous proposed changes in life support, propulsion, and trajectory parameters could be examined through dynamic simulation of dozens "what if" scenarios, and the various outcomes could be evaluated before selecting an appropriate rescue approach.

> The crew's salvation rested with the lunar module Aquarius. For the next three days they had to rely on its limited power supply, its oxygen, and its engines to get them back home. They fired the LM's descent engine to change Apollo's trajectory into one that would swing them around the Moon and direct them back to Earth....
>
> On 17 April Apollo 13 was within sight of home. The crew transferred to the command module and cut loose the lifeless service module. As the structure drifted away they could see the extent of the damage. A whole side had been blasted away. Next, they jettisoned their lunar module "life raft" and prepared to hit the Earth's atmosphere traveling at thirty-five times the speed of sound. Within the quarter of an hour, they were

*Robin Kerrod, *The Illustrated History of NASA,* Gallery Books, New York, 1986.

> rocking gently in a Pacific swell. The agony was over.... "You did not reach the Moon, but you reached the hearts of millions of people on Earth," the president remarked upon welcoming the astronauts back home.

Neither the United States' successful reach for the moon nor the associated rescue just described were feasible prior to the advent of advanced dynamic system modeling and simulation together with its contemporary computer implementation. The whole fabric of engineering practice has changed from a cut-and-try experimental art to a process in which informed decisions can be made using system simulation very early in the design process before any prototype hardware is constructed and tested. This in turn has led to improved engineering designs, more intimate integration of design with manufacturing, shortened time to production, and lower system costs. Indeed, advances in modeling and simulation methods for dynamic systems have contributed enormously to the technological progress of the late twentieth century.

# Part 1

# Overview of Dynamic Systems

The Bell Boeing V22 Osprey Tiltrotor Aircraft. (Photo courtesy of Bell Helicopter Textron, Inc.)

"If I have seen further [than you and Descartes] it is by standing upon the shoulders of Giants"

Sir Isaac Newton in a 1675 letter to Robert Hooke

# CHAPTER 1

# Introduction to Modeling and Simulation

## 1.1 DYNAMIC SYSTEMS

This text discusses both the modeling of dynamic systems as found in the major engineering disciplines and solutions of the resulting differential equations by analytical and computational means. The book is intended for use at the introductory level, but serves the practicing engineer as a reference source as well. It presents modeling of the engineering disciplines using a unified approach. Since the equations that represent the dynamics of a physical system can take several different forms, we emphasize selecting the form most compatible with the mathematical method or numerical process that is ultimately to be used in solving the equations. This chapter presents general considerations that will be used throughout the text.

### 1.1.1 Examples of Dynamic Systems

**Static systems** have an output response to an input that does not change with time; i.e., the input is held constant. This means that the output always has the same instantaneous relationship with the input. **Dynamic systems** have a response to an input that is not instantaneously proportional to the input or disturbance and that may continue after the input is held constant. Dynamic systems can respond to input signals, disturbance signals, or initial conditions.

Examples of dynamic systems are all around us. They may be observed in common devices employed in everyday living, Figure 1.1, as well as in sophisticated engineering systems such as those in spacecraft that took astronauts to the moon. Dynamic systems are found in all major engineering disciplines and include mechanical,

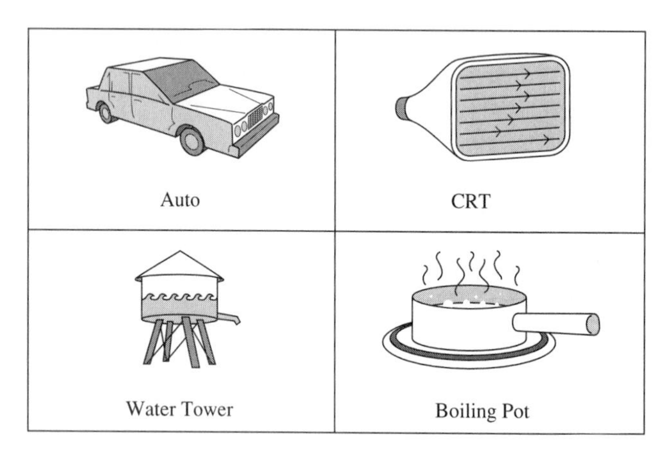


**Figure 1.1** Examples of dynamic systems.

electrical, fluid, and thermal systems. They can be observed as well in natural systems (ecological, biological, economic, traffic, etc.); but while these have a dynamic behavior that is similar to that of engineering systems, they are not treated here.

**Mechanical Systems.** Systems that possess significant mass, inertia, and spring and energy dissipation components driven by forces, torques, specified displacements are considered to be **mechanical systems**. An automobile is a good example of a dynamic mechanical system. It has a dynamic response as it speeds up, slows down, or rounds a curve in the road. The body and the suspension system of the car have a dynamic response of the position of the vehicle as it goes over a bump. An airplane in flight has a dynamic response of its speed and altitude as it maneuvers in the air. A paint shaker at the hardware store, with its unbalanced motor suspended on springs, provides a dynamic response of the position of the frame when the device is in use. A musical drum has a dynamic response or vibration of the position of the membrane. The structural frame of a building may have a dynamic response or vibration due to external loading, such as wind forces or ground motions.

**Electrical Systems.** **Electrical systems** include circuits with resistive, capacitive, or inductive components excited by voltage or current. Electronic circuits can include transistors or amplifiers. We need not look far to find good examples of electrical systems with important dynamic response characteristics. A television receiver has a dynamic response of the beam that traces the picture on the screen of the set. The TV tuning circuit, which allows you to select the desired channel, also has a dynamic response, and a simpler, though no less important, example is the dynamic voltage and current responses that occur when you switch a light on or off.

**Fluid Systems.** **Fluid systems** employ orifices, restrictions, control valves, accumulators (capacitors), long tubes (inductors), and actuators excited by pressure or fluid flow. A city water tower has a dynamic response of the height of the water as a function of the amount of water pumped into the tower and the amount being used by the citizens. If a garden hose is suddenly blocked at its end when water is flowing through it, the pressure in the hose will have a dynamic response. Airflow across a cavity in a tube will cause a dynamic response (generate an acoustic tone) in an organ pipe. A water pump with an accumulator to damp out pulsations will have a dynamic response of the output pressure when in use.

**Thermal Systems.** **Thermal systems** have components that provide resistance (conduction, convection, or radiation) and capacitance (mass and specific heat) when excited by temperature or heat flow. A heating system warming a house has a dynamic response as the temperature rises to meet the set point on the thermostat. Placing a pot of water over a burner to boil has a dynamic response of the temperature. The size of the pot, the material it is made of, the amount of water in the pot, and the size of the burner all play a role in how quickly the water comes to a boil.

**Mixed Systems.** Some of the more interesting dynamic systems use two or more of the previously mentioned engineering disciplines, with energy conversion between the various components. Figure 1.2 shows several examples.

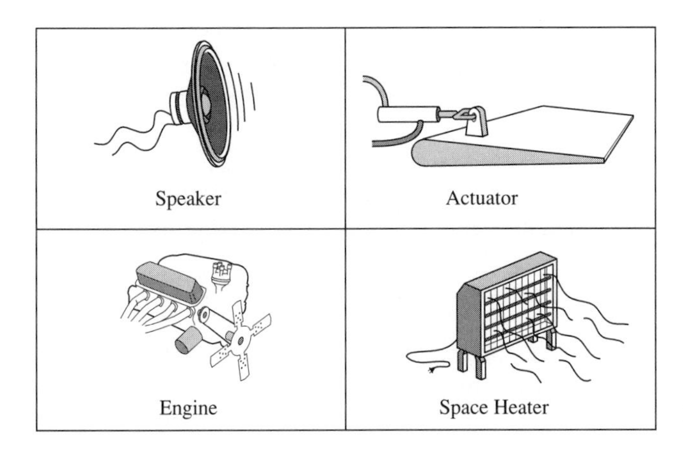


**Figure 1.2** Examples of mixed systems.

***Electro-Mechanical.*** Systems employing an electromagnetic component that converts a current into a force generally have a dynamic response. Examples are a loudspeaker in a stereo system, a solenoid actuator, and electric motors. In a loudspeaker, electrical current from the amplifier is transformed into movement of the speaker cone and the subsequent air pressure fluctuations that cause us to hear the amplified sound.

***Fluid-Mechanical.*** Hydraulic or pneumatic systems with fluid-mechanical conversion components exhibit dynamic behavior. Examples are a hydraulic pump, a valve-controlled actuator, and a hydraulic motor drive. A hydraulic servo system used for flight control in an airplane is a good example of a common electro-fluid-mechanical dynamic system.

***Thermo-Mechanical.*** A combustion engine used in a car, truck, ship, or airplane is a thermo-fluid-mechanical (or simply, thermomechanical) device, since it converts thermal energy into fluid power and then into mechanical power. Thermal dynamics, fluid dynamics, and mechanical dynamics are all involved in the process.

***Electro-Thermal.*** A space heater that uses electric current to heat a filament, which in turn warms the air, has a dynamic response to the surrounding environment. An electric water heater is another common example of an electrothermal dynamic system.

### 1.1.2 Definitions Related to Dynamic Systems

**Modeling** is the process of identifying the principal physical dynamic effects to be considered in analyzing a system, writing the differential and algebraic equations from the conservation laws and property laws of the relevant discipline, and reducing the equations to a convenient differential equation form.

A **system** is a set of interacting components connected together in such a way that the variation or response in the state of one component affects the state of the others. In this text, "system" refers to a collection of components from the major engineering disciplines.

The **major disciplines** of engineering systems are mechanics, electricity and electronics, fluid mechanics and fluid controls (including hydraulics and pneumatics), and thermodynamics. Magnetism and optics also involve dynamic systems, but are not covered here.

The behavior of a system is characterized by its **response** to external inputs, disturbances, and initial conditions. Figure 1.3 shows this relationship. By **outputs,** we mean the dependent variables of the differential equation that represent the response of the system. By **inputs,** we mean functions of the independent variable of the differential equation, the excitation, or the forcing function to the system. By **external disturbances** or **perturbations,** we mean those external environmental effects

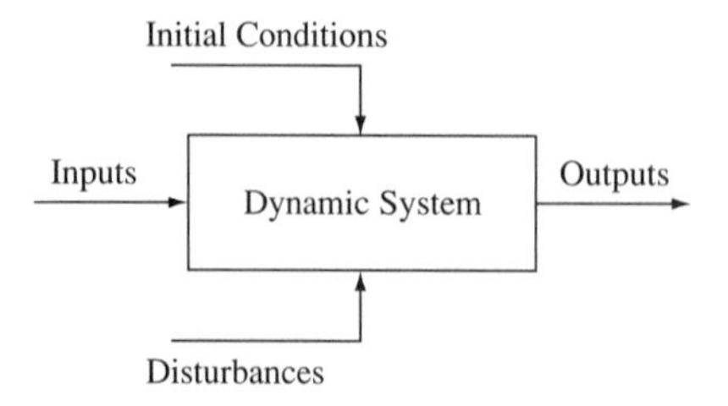


**Figure 1.3** Excitation and response of a system.

that may occur randomly or unexpectedly. The **initial conditions** are the initial values of the dynamic variables of the system. The **dynamic variables** of a system are those variables whose time derivatives appear in the governing equations.

As an example of a system and its response, consider a vehicle traveling down the road and passing over a bump. The positions of the wheel and the body of the vehicle relative to the ground could be the system variables. Their differential equations could be written from a knowledge of the spring rates, mass, and damping values of the vehicle's components. The initial conditions would be the values of these variables just before the vehicle hits the bump. The bump would be the input to the system, and any aerodynamic turbulence could be considered a disturbance. From these considerations, it is possible to develop equations and solve them for the outputs, i.e., the displacements of the wheel and body. The maximum stresses in the springs could then be found, as could other critical performance parameters necessary in the design of the vehicle.

A **dynamic system** is described by time-differential equations; therefore, the future response of the system is determined by the present state of the system (the initial conditions) and the present input. Thus, a dynamic system may continue to have a time-varying response after the inputs are held constant. In contrast, a **static system** is described by algebraic equations, so that the present response of the system is totally determined by the present value of the input, and there will be no change in the response in the future if the input is held constant.

The **transient response** of a dynamic system to an external input refers to the behavior of the system as it makes a transition from the initial condition to the final condition. The transient response is expressed as a function of time. Figure 1.4 shows a typical dynamic response of a system. A dynamic system will reach a **steady state** after all of the transients have died out. The time it takes to reach the steady state is

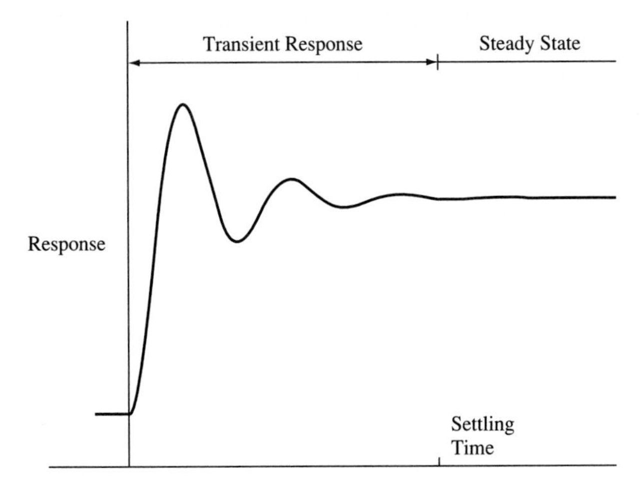


**Figure 1.4** Typical dynamic response and settling of a system.

called the **settling time**. We normally think of the steady state as the response that occurs after the input is held constant; however, we can also have a steady-state response that occurs due to a steady periodic input.

A **classical differential equation** involves terms in the dependent variable (the response) and some of its derivatives, summed or differenced together, with the result equated to the input function. The response of a system can be due to the initial conditions of the dependent variable or an excitation by a variation in the inputs. An example of the classical differential equation form is shown in Eq. (1.1):

$$\frac{d^2x}{dt^2} + a_1 \frac{dx}{dt} + a_0 x = f(t)$$
$$x(0^-) = x_0 \tag{1.1}$$
$$\frac{dx}{dt}(0^-) = \dot{x}_0$$

Here $x(t)$ is the response variable, the *a's* are constants that depend upon the system parameters, $f(t)$ includes external inputs, and $x_0$ and $\dot{x}_0$ represent the initial state and initial rate of the system at the instant just before $t = 0$ (designated as $0^-$). We are interested in calculating the function $x(t)$ that describes the response of the system. Note that a dot above the variable represents a differentiation with respect to time, so that this system of equations can be written in the following familiar form:

$$\ddot{x} + a_1 \dot{x} + a_0 x = f(t)$$
$$x(0^-) = x_0 \tag{1.2}$$
$$\dot{x}(0^-) = \dot{x}_0$$

The $0^-$ notation is not used frequently in this text; however, it is always important to note that for initial conditions and inputs, we should consider the values just before and just after the time $t = 0$.

We can also use the **differential operator *D*** to express differentiation with respect to time (see Ref. 1.6):

$$D = \frac{d}{dt} \tag{1.3}$$

Thus,

$$Dx = \frac{dx}{dt}, \qquad D^2x = \frac{d^2x}{dt^2}, \text{ etc.}$$

Using $D$-operator notation, we can write Eq. (1.2) as

$$[D^2 + a_1 D + a_0]\, x(t) = f(t) \tag{1.4}$$

It is often desirable to express a system response variable normalized by the input variable as an **output-input ratio**. For linear systems, the **transfer function** is defined as the ratio of the output to the input for a system with zero initial conditions, as determined from the **Laplace transform** of the system equation. The Laplace transform (see Appendix F) of Eq. (1.2) with zero initial conditions gives

$$[s^2 + a_1 s + a_0]X(s) = F(s) \tag{1.5}$$

The Laplace transform converts the system equation from one in which time is the independent variable to one in which $s$ is the independent variable. The result is an **$s$-domain** expression rather than a **time-domain** expression.

Solving Eq. (1.5) for the output-input ratio gives the **transfer function**:

$$\frac{X(s)}{F(s)} = \frac{1}{s^2 + a_1 s + a_0} \tag{1.6}$$

The output-input ratio can also be written symbolically in the time domain using the $D$-operator notation, Ref. 1.6. Rearranging Eq. (1.4) gives

$$\frac{x(t)}{f(t)} = \frac{1}{D^2 + a_1 D + a_0} \tag{1.7}$$

While the Laplace transform ($s$-domain) expression, Eq. (1.6), is the usual definition of the transfer function, in this text we will also find it convenient to refer to the time-domain form, Eq. (1.7), as a transfer function. If derivatives of $f(t)$ occur in the right-hand side of the differential equation, the numerator of the transfer function also contains $s$ or $D$ terms, and we frequently refer to the transfer function as the ratio of two polynomials in $s$ (or $D$).

**State-space differential equations** form a set of simultaneous first-order differential equations. The state variables are the dependent variables of each first-order differential equation and represent the dynamic response variables of the system. An example of equations put into the state-space format is as follows:

$$\dot{x} = a_1 x + a_2 y + f(t) \tag{1.8a}$$

$$\dot{y} = a_3 x y \sin(x) + g(t) \tag{1.8b}$$

$$\dot{z} = a_4 x z + a_5 e^{-yt} + h(t) \tag{1.8c}$$

$$x(0) = x_0$$

$$y(0) = y_0$$

$$z(0) = z_0$$

The **order** of a differential equation or dynamic system is given by the number of independent derivatives in the system. In an ordinary differential equation, the order is the difference of the highest and lowest derivatives in the equation. Equation (1.2) is a second-order equation.

In a system of simultaneous equations, the order is the number of independent derivatives found in all of the equations. Equations (1.8a) through (1.8c) represent a third-order system.

A **linear function** has the following two properties: (1) If you multiply the argument of the function by a constant factor, the value of the function is multiplied by that same factor, and (2) the sum of the values of the function for first one value of the argument and then for a second value of the argument, is the same as the function evaluated with the sum of the two arguments. Mathematically, the first condition is expressed as

$$f(ax) = a\,f(x) \tag{1.9}$$

In other words, a multiple of the input results in a multiple of the output.

The second condition is

$$f(x_1 + x_2) = f(x_1) + f(x_2) \tag{1.10}$$

In other words, a sum of the input results in a sum of the output.

We can combine these two properties as follows:

$$f(a_1x_1 + a_2x_2) = a_1\,f(x_1) + a_2\,f(x_2) \tag{1.11}$$

Thus, the only linear function is a straight line through the origin; the form $y = mx + b$ is not a linear function according to these definitions!

A **linear combination** of variables results from the sum of a multiple of the variables. For example, $L = ax + by$, where $a$ and $b$ are constants, is a linear combination of the variables $x$ and $y$. A **linear equation** is formed by a linear combination of the variables and their derivatives. A **linear differential equation** is an equation formed by a linear combination of the derivatives of the system variables. For example, $L = ax + b\dot{x} + c\ddot{x}$, where $a$, $b$, and $c$ are constants, is a linear differential equation, and Eqs. (1.1), (1.2), and (1.8a) are linear, constant-coefficient differential equations.

A **linear system** is described by linear algebraic and differential equations. By contrast, a **nonlinear system** has nonlinear combinations of the variables and their derivatives. Examples of nonlinear functions are the product of two variables, the square of a variable, a trigonometric function of a variable, and so on. Equations (1.8b) and (1.8c) are examples of nonlinear differential equations.

The **analytic solution** of a differential equation is the mathematical expression of the dependent variable as a function of time and may include exponentials, sinusoids, and/or other mathematical functions. An analytic solution of a given differential equation requires knowledge of the initial conditions and the inputs as explicit functions of time. Analytic solutions are found by employing techniques for solving classical differential equations or by using Laplace transform techniques. (See Appendices E and F for a discussion of these methods.)

Linear differential equations are well understood, and their analytical solutions usually can be obtained by applying the widely accepted methods that are discussed in a course in elementary differential equations. Nonlinear systems, on the other hand, with the exception of a few first-order systems and a limited number of second- or higher order systems, do not have known analytic solutions. If an analytic solution is not possible for a nonlinear system, a numerical approximation to the solution of the nonlinear differential equation might be found by using appropriate simulation methods. We call such an approximation a computational solution.

**Computational solutions** to differential equations can be found by numerical integration, using a digital computer. Numerical integration is the process of computing an approximate solution to the integral of a derivative function by a numerical algorithm. The algorithm propagates the solution of the differential equation by using small increments in time. Thus, the solution of the differential equation is known only at certain discrete times. Computational methods commonly employ the state-space representation of differential equations. Calculation of the response of a dynamic system in this way is commonly called **digital simulation**.

In **analog computation,** the differential equation is represented by an interconnection of linear and/or nonlinear electrical components and electronic integrators (operational amplifiers with capacitive feedback). Since the equations that govern the electrical system are the same as the equations that govern the dynamic system under consideration, an analogy between the two systems is formed. The electronic integrators then "solve" the differential equation by executing an electrical dynamic behavior corresponding to that of the system being studied. Many systems can be simulated by making an "analogy" between the voltages displayed by an analog computer and the variables of the equations being solved.

## 1.2 MODELING OF DYNAMIC SYSTEMS

### 1.2.1 Steps in Modeling and Representing Dynamic Systems

Figure 1.5 illustrates the several stages involved in the modeling of dynamic systems. Shown first is the **actual dynamic system** of interest. It has all of the dynamic response characteristics that correspond to the exact linear or nonlinear behavior of the system and the dynamic terms that naturally occur in the system. The actual system, of course, possesses the true response that we wish to determine.

Second is the **engineer's perception** of the system—its linearity (or lack of it) and its dynamic characteristics. The modeler may neglect some nonlinearities or higher order dynamic characteristics for simplicity; however, the actual system does include all of these effects and characteristics. Therefore, the engineer's perception might not truly represent the actual dynamic system.

Third is the **mathematical model** of the system represented by the differential equations derived from the conservation and property laws of the appropriate disciplines. If the actual system is linear, the development of a suitable mathematical model is quite straightforward. However, if the system is nonlinear,

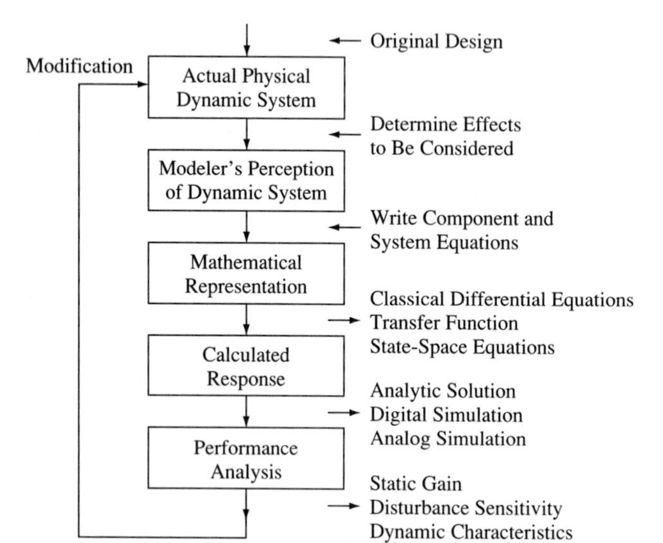


**Figure 1.5** Modeling sequence and levels of representation.

the mathematical representation could include some approximations to simplify the analysis; therefore, the equations may not exactly represent the engineer's perception of the real system.

Fourth is the **calculated response** of the system. The system response expressed by analytic solutions to differential equations is an exact solution of those equations. Some slight error might exist, however, between the response computed by a numerical or analog computer method and the actual solution of the differential equation.

The fifth stage involves the **analysis of the performance** of the dynamic system, as expressed by specific response measures. This text discusses both frequency-domain and time-domain analysis methods for evaluating the system's performance.

One of the most important responsibilities of the engineer is the process of examining a certain combination of values for the components, determining the performance of the system, and adjusting the values of the components until an acceptable performance is achieved. This cut-and-try procedure can be done by assembling hardware or by mathematical modeling. The advantage of mathematical modeling and analysis is that it is usually faster and much less expensive than experimenting with hardware! Thus, the iteration loop shown in Fig. 1.5 in which the system is modified is an important part of the engineering process.

It is important to note that, because the model may be simplified (e.g., by removing nonlinearities and higher order dynamics) or approximated, the calculated response from the analytic solution of the differential equation may be *one* step removed from the response of the actual system. If the mathematical model contains some approximations, then the analytic solution is *two* steps removed from the actual response.

When the response is found by digital or analog simulation, it is *three* steps removed from the response of the actual system since there might be some inaccuracies in the numerical integration or electronic response and the mathematical model might not include all of the nonlinearities or higher order dynamics.

When we say "the system," we could be referring to the actual system or to the mathematical model of the system. Similarly, when we refer to the "system response," we could mean the actual system response or the computed or simulated response of the system. It is important to note these differences.

Modeling of the system, calculation of the response, analysis of the performance of the system, and system redesign naturally fit together in a complete cycle as an engineering tool. No single step is valuable, except as an academic exercise, unless the cycle is complete. If, after the performance has been evaluated, the system does not display the desired response, the system and its model should be modified or the components of the existing model adjusted to obtain the required output.

### 1.2.2 Utility of Modeling and Simulation

The major advantage of modeling a system and analyzing its response is that it allows us to predict the behavior of the system before it is built. This is sometimes called "virtual prototyping". We also can analyze the performance of an existing system with the intent of improving the static or dynamic behavior of the system, or we can determine what might happen to a system with an unusual input or condition without exposing the actual system to danger.

For example, if we were designing a new system and wanted to have a certain level of performance, we could select components of the correct size to give us the desired result. On the other hand, if the system already exists, and we want to improve its performance, modeling and analysis could help us determine which components to change and the amount of change necessary. If we wanted to change the system damping characteristics, modeling could tell us not only which components affect the damping, but how much to change them so as to get the desired result. If we were curious about what would happen to a system with an unusual input or condition, such as the failure of a specific hydraulic servoactuator on the flight control system of an aircraft, then modeling and analysis could be used, and we would not put a multimillion-dollar aircraft at risk in determining the behavior of the system.

A good mathematical model of a dynamic system can immediately tell us the performance characteristics of the system and, with a little analysis, the dynamic characteristics. If we are familiar with the dynamic characteristics, this will give us a feel for the system response, which might be enough to convey to us what we want to know about the response. Or we might want to compute the frequency and/or time response of the system to provide additional insight.

Of course, our solutions to differential equations might not be an exact representation of the actual system under consideration; however, the simulation could show the trends of the response, so we would know which components to change and approximately how much to change them to get a desired response.

In many cases, simulation by digital computation is favored over the use of analytic solutions to differential equations for several reasons. In the first place, given the right software, it does not require a lot of effort, experience, or time on the part of the engineer to set up a simulation. Second, we can make changes in the model (differential equations) of the system and quickly obtain the response using the same computational technique as before, whereas by analytical methods, a change in the differential equations might require a completely different analytic solution to be developed. Third, and perhaps most important, nonlinear systems can be handled by simulation with relative ease, whereas analytic solutions of specific nonlinear systems may not be possible.

It is important to note that a numerical solution of a set of system equations gives only a numerical result for that particular circumstance; a large number of simulations may be required to detect performance trends corresponding to variations in the values of parameters. So while it may sometimes be difficult to find, an analytical solution contains the complete representation of the behavior that is modeled and may be evaluated over the entire range of performance as easily as it is evaluated for a specific point of the system's operation. Much more information is always given by a complete analytical solution, provided that it can be found.

## 1.3 SCOPE OF THE TEXT

The text is organized into four parts:

**Part I, "Overview of Dynamic Systems,"** consisting of Chapters 1 and 2, presents the necessary background and preliminaries that tie together the concepts of system dynamics in the major engineering disciplines.

**Part II, "Modeling of Engineering Systems,"** includes Chapters 3–7. Chapters 3–6 present the techniques of modeling using conservation laws and properties of each of the technical disciplines of mechanical systems, electrical systems, fluid systems, and thermal systems. Mixed systems, i.e., systems composed of combinations of mechanical, electrical, fluid, and/or thermal components, are considered in Chapter 7.

**Part III, "System Dynamic Response Analysis,"** comprises Chapters 8 and 9. Chapter 8 is concerned with frequency response behavior and the stability of dynamic systems, while Chapter 9 is devoted to methods of time response and digital simulation.

**Part IV, "Engineering Applications,"** consists of Chapter 10 alone, in which the concepts of the preceding chapters are brought to bear on the sizing and selection of components of a number of representative homogeneous and mixed, linear and nonlinear systems. These examples serve to tie the previous developments together and to unify the reader's understanding of the modeling and simulation process.

The International System of units (SI) and British units are used alternately throughout the text. Appendix A presents basic information on conversions between the two. We emphasize the consistent use of units and the importance of the compatibility of equations in terms of units. One of the best ways to detect errors in an equation that has been derived is to check the dimensions of each term; a correct expression of physical principles must be homogeneous in its dimensions.

Appendix B provides often-used information on geometrical, mass, and material properties of components of mechanical systems.

Reference material on vector and matrix algebra and on the solution of algebraic equations is presented in a complete form in Appendices C and D, respectively. For the reader who is not familiar with these subjects, these appendices provide much more than just a brief review.

A discussion of the solution of ordinary differential equations by classical methods is given in Appendix E, and the solution of linear differential equations by Laplace transform techniques is presented in Appendix F. The definition and format of the state-space representation for dynamic systems are presented in Appendix G as are methods for converting linear differential equations to and from the state-space form. In Appendix H the reader is provided instruction on the use of the MATLAB software system for analysis and design of dynamic systems. Appendix I sets forth the fundamentals of analog simulation.

Modeling techniques resulting in the derivation of classical equations, transfer functions, and state-space differential equations are included throughout to give the reader experience with each approach. The techniques for solving each of these three kinds of differential equations are presented in enough detail that the student can master typical situations. Solutions of classical differential equations and the Laplace transform solution technique should be a review from a course in differential equations; however, enough material is presented in Appendices E and F to allow the student, if necessary, to obtain a first-time knowledge of these solution techniques.

The derivation of the final mathematical representation of a dynamic system depends upon the intended solution or analysis technique as indicated in Table 1.1. We would express the system in a classical differential equation if we intended to solve the equation by classical analytic techniques, in a transfer function if we were going to use Laplace transform techniques, or in the state-space format if we were going to use digital or analog simulation. Accordingly, examples alternate in their presentation among these three representations.

**TABLE 1.1** DIFFERENTIAL EQUATION SOLUTION METHODS

| System Equation Format | Solution Method |
|---|---|
| Classical | Analytical |
| Transfer Function | Laplace Transform |
| State Space | Digital Simulation |
| | Analog Simulation |

While it may be necessary to omit coverage of one of the chapters on modeling in a particular discipline (Chapter 3, 4, 5, or 6), the interdependence of modeling and solution techniques must be emphasized. First of all, what good is a model of a dynamic system if we do not analyze it to determine the performance of the system? And second, what good is a differential equation solution technique if we cannot model a system and derive the associated differential equations in a form fit for solving? Therefore, we must be able to model dynamic systems and to arrange the resulting equations into the form necessary for the customary differential equation solution technique—classical, Laplace transform, or digital or analog simulation.

This text is intended to give a complete presentation of the modeling techniques of major types of engineering systems, the solution techniques for the resulting differential equations, and the attendant mathematical procedures related to the representation of dynamic systems and the determination of their response characteristics. Illustrative examples, practice problems, and references are provided to complement the principal ideas discussed.

## REFERENCES

1.1 Close, Charles M., and Frederick, Dean K. *Modeling and Analysis of Dynamic Systems,* 2d ed. Houghton Mifflin Co., Boston, 1993.

1.2 Dorny, C. Nelson. *Understanding Dynamic Systems: Approaches to Modeling, Analysis, and Design*. Prentice Hall, Englewood Cliffs, NJ, 1993.

1.3 Ogata, Katsuhiko. *System Dynamics,* 2d ed. Prentice Hall, Englewood Cliffs, NJ, 1992.

1.4 Palm, William J., III. *Modeling, Analysis, and Control of Dynamic Systems*. John Wiley & Sons, New York, 1983.

1.5 Shearer, J. Lowen, and Kulakowski, Bohdan T. *Dynamic Modeling and Control of Engineering Systems*. Macmillan Publishing Company, New York, 1990.

1.6 Takahashi, Yasundo, Rabins, Michael J., and Auslander, David M. *Control and Dynamic Systems*. Addison-Wesley Publishing Co., Reading, MA, 1970.

## NOMENCLATURE

| | |
|---|---|
| $a_0, a_1, a_2$ | coefficients |
| $D$ | differential operator $= d/dt$ |
| $f$ | external input |
| $s$ | Laplace transform variable |
| $t$ | time (independent variable) |
| $x$ | response variable |
| $X(s), F(s)$ | Laplace transform of $x(t)$ and $f(t)$ |

## PROBLEMS

**1.1** *Make* a list of five systems or subsystems that you come in contact with on a regular basis, and indicate how their dynamic response is important to their function. *Identify* each system or subsystem as mechanical, electrical, fluid, electromechanical, etc.

**1.2** For two of the systems in the previous problem, *draw* a diagram that identifies the physical quantities that constitute the input, the output, and any disturbances. *Make* a sketch of the expected response as a function of time.

**1.3** *Determine* the response time of a dynamic system by performing an experiment to evaluate the reaction of the system to inputs. For example, you could turn your thermostat up or down two degrees and determine how long it takes the heating or air conditioning system to reach the temperature set point. Then repeat the procedure for four degrees, and remark on any differences. Or you could hold a ruler or yardstick flat against a tabletop, with most of it projecting over the edge. Then pull down the tip a specified amount, and let the measuring rod vibrate. How long does it take to stop? How does the amount of overhang affect things? Heat a pan of tap water. How long is it until the water boils? How does the amount of water influence things? You can think of other possibilities as well.

*Submit* a description of the experiment you performed and how the results varied with changes in the parameters. *Draw* any conclusions you think are appropriate.

**1.4** Suppose you accidentally leave a flashlight turned on after using it, and the battery runs down. *Describe* this electrical system and its inputs, outputs, and response. *Make* a sketch of the response as a function of time. *Gather* the relevant physical data on a typical flashlight, and see whether you can estimate the time required to run the batteries down.

**1.5** *Determine* the approximate circumference of a circle by dividing the circle into a number of straight lines of equal length. Use several increasingly shorter chord lengths, and show how this numerical process approaches the exact result.

**1.6** *Repeat* the steps in the previous problem, but now use the triangular wedge to compute successive approximations to the area of the circle.

**1.7** *Identify* one or two system processes with which you are familiar in which (a) the input and output are linearly related, (that is, if you double the input value, the output doubles) and (b) the input-output relationship is nonlinear.

**1.8** Using the definitions of linearity (Eqs. 1.8 and 1.9), *prove* that $f(x) = mx + b$ is not a linear function.

**1.9** Using the definitions of linearity (Eqs. 1.8 and 1.9), *prove* that $f(x) = ax^2$ is not a linear function.

**1.10** Using the definitions of linearity (Eqs. 1.8 and 1.9), *prove* that $f(x) = a \sin(x)$ is not a linear function.

**1.11** Using the definitions of linearity (Eqs. 1.8 and 1.9), *prove* that $f(x) = ae^{-x}$ is not a linear function.

# CHAPTER 2

# Models for Dynamic Systems and Systems Similarity

## 2.1 FORMULATION OF MODELS FOR ENGINEERING SYSTEMS

Mathematical models for dynamic systems are derived from the conservation laws of physics and the engineering properties of each system component. These basic formulations are then combined or simplified to achieve the desired differential equation form. The preferred forms are:

**(1)** The classical representation of a single $n$th-order differential equation.

**(2)** The transfer function format, which gives the output in terms of the input.

**(3)** The state-space format of $n$ simultaneous first-order differential equations.

Analytic solutions in the form of the sum of exponential terms are found from the classical format of linear differential equations. Laplace transform solutions to the system dynamic response are found from the transfer function format. Computational solutions using a digital or analog computer are found from the state-space format. The equations from the modeling exercises in this textbook may be expressed in any of the three formats.

### 2.1.1 Conservation Laws of Engineering Systems

The development of models for dynamic systems is based upon the conservation laws of physics and other special properties of a specific discipline. Each of the conservation laws is discussed in greater detail in the separate chapters on each subject; however, it is interesting to introduce them here to see the similarity among them.

In general, the conservation law for a given quantity simply states that at a given location in space (or in a volume selected for consideration—a "control volume"), whatever comes in must either go out or be stored. It is interesting to note that even though the basic conservation laws of engineering are stated in terms of the quantity conserved, we almost always *use* the law in the time-derivative format; for example, the conservation of charge in electronic systems is expressed as the law of conservation of current (current being the time derivative of charge). Thus, the time derivative of the quantity stored provides the framework on which the differential equations used in each discipline are based.

The storage term can be treated directly or can be treated as a separate component that is connected to the location in question. In the subsequent discussion of each law, notice that the time-differential conservation laws are set equal either to zero (if dynamic storage elements are considered) or to the derivative of the quantity being conserved (which gives us the differential equation directly).

**Linear Momentum.** The principle of **linear impulse-momentum** states that the translational momentum is equal to the applied impulse. The momentum equation can be applied to the motion of a rigid body or a fluid. The momentum of a mass particle is the product of the mass $m$ and the translational velocity $v$.

Stated in the time-differential format, the momentum law becomes the familiar second law of motion established by Isaac Newton (English mathematician, 1642–1727) and can be stated as follows: The sum of all forces acting on a mass particle causes a rate of change in the momentum of the particle with respect to time. Mathematically,

$$\sum F_{net} = \frac{d}{dt}[mv] \tag{2.1}$$

If we employ the D'Alembert principle for fixed-mass systems (J. D'Alembert, French mathematician, 1717–83), we can think of the rate of change of momentum with respect to time as another component of the forces acting on the system. Then the preceding equation is considered a statement about the balance of forces on the system and can be written as

$$\sum F_{net} - m\frac{dv}{dt} = 0 \tag{2.2}$$

**Angular Momentum.** The principle of **angular impulse-momentum** states that the angular impulse is equal to the rate of change of angular momentum with respect to time. The angular momentum of a rigid body in planar motion is the product of the mass moment of inertia $J$ and the rotational velocity $\omega$, both taken with respect to an axis through the center of mass of the body or with respect to a fixed axis of rotation for the body.

Stated in the time-differential format, the angular momentum law becomes Newton's second law, which states that the sum of all torques and moments about an axis through the center of mass of a body or about a fixed axis of rotation for the body is equal to the rate of change, with respect to time, of the angular momentum for that axis:

$$\sum T_{net} = \frac{d}{dt}[J\omega] \tag{2.3}$$

D'Alembert's principle can also be applied to problems involving rotational motion, in which case we obtain the following statement about the balance of torques on the body:

$$\sum T_{net} - J\frac{d\omega}{dt} = 0 \tag{2.4}$$

**Conservation of Charge.** The law of **conservation of charge** states that the charge in an electrical system is constant. Stated in time-differential format, the law becomes Kirchhoff's current law (G. R. Kirchhoff, German physicist, 1824–87) and states that the sum of all currents at a node in an electrical circuit is equal to the rate at which charge is being stored at the node:

$$\sum i_{node} = \frac{dQ}{dt} = C\frac{de}{dt} \tag{2.5a}$$

If we interpret this to mean that charge can be stored only in a capacitor $C$, then we can simply connect a capacitor to the node and state that the sum of all currents is equal to zero:

$$\sum i_{node} - C\frac{de}{dt} = 0 \tag{2.5b}$$

**Conservation of Mass.** The law of **conservation of mass** states that the mass of a fluid system is constant. Written in time-differential format, the law becomes the law of conservation of mass flow rate. The net mass flow rate at a location is equal to the rate of change, with respect to time, of the mass at that location. With the mass of a fluid at a location equal to the product of the density $\rho$ and the volume $V$ of the fluid, at that location, we can write

$$\sum \dot{m}_{net} = \frac{d}{dt}[\rho V] = \rho\dot{V} + V\dot{\rho} \tag{2.6a}$$

Again, we could use the concept that mass is stored in a "fluid capacitor" (or equivalently, the fluid is compressible or in a "compliant" container) at that location and, considering the capacitor as a dynamic component, state that the sum of all mass flow rates is zero:

$$\sum \dot{m}_{net} - \frac{d}{dt}[\rho V] = 0 \tag{2.6b}$$

Chapter 5 explains in detail that in order to use this relationship, the rate of change of the density with respect to time must be expressed as a function of the compressibility of the fluid. When this is done, the mass flow rate equation becomes the so-called continuity equation relating volume flow rates.

**Conservation of Energy.** The law of **conservation of energy** states that the energy in a system is constant. The system can be mechanical, electrical, thermal, fluid, or of combined types. Stated in time-differential format, the law becomes the law of conservation of power, or the first law of thermodynamics, which states that the sum of all power (heat transfer, mechanical power, and thermal power) in and out of a system is equal to the rate at which energy is being stored in a control volume of the system. Mathematically,

$$\sum Q_h - \dot{W} + \dot{m}_{net}\left[h + \frac{v^2}{2} + z\right] = \frac{d}{dt}\left[mu + \frac{mv^2}{2} + mz\right]_{cv} \tag{2.7}$$

Again, if we consider the storage term as a separate dynamic storage component, then the sum of all power at a control volume must be equal to zero.

When there is no significant energy exchange (heat transfer or work) with the environment, and no energy is being stored, the law of conservation of power reduces to the Bernoulli equation (D. Bernoulli, Swiss physicist and mathematician, 1700–82), for fluid flow:

$$\frac{P}{\rho} + \frac{v^2}{2} + zg = \text{constant along a streamline} \tag{2.8}$$

### 2.1.2 Property Laws for Engineering Systems

The property laws are derived from the various special properties of each discipline in engineering and perhaps geometric or mathematical properties as well.

In **mechanical systems,** the **viscous friction** of a piston and cylinder filled with fluid gives rise to **damping**. This type of friction can exhibit a linear or nonlinear relationship between force and velocity. It is actually the combination of a fluid system with a mechanical system. **Coulomb friction** is a purely mechanical effect also called dry friction (friction without lubrication). This type of friction is always nonlinear and may be relatively difficult to handle analytically when modeling dynamic systems. The relations between stress and strain (Hooke's law; R. Hooke, English scientist, 1635–1703) determine the value of **spring stiffness** for a given size and diameter of a wire. The mass or rotational inertia property of an object represents **mechanical inductance** (Newton's second law).

In **electrical systems,** the resistivity property of a material causes electrical **resistance** in a given size and shape of material. The dielectric properties of materials and their spacing give rise to **capacitance,** while the magnetic properties of wire wrapped in a coil cause **inductance**.

In **fluid systems,** fluid flow causes pressure losses due to viscous shear caused by Hagen-Poiseuille flow (Hagen, German engineer, published in 1839, Poiseuille, French scientist, published in 1940) or velocity head losses in orifices (Bernoulli equation). These represent **fluid resistance**. **Fluid capacitance** is caused by the compressibility of the fluid itself, due to the bulk modulus property of the fluid, or by a compliant container, due to the mechanical flexibility of the container. **Fluid inductance** is caused by the inertial properties of the fluid as it accelerates in a pipe. (See Eq. 5.53.)

In **thermal systems,** the processes of thermal conduction, convection, and radiation represent **thermal resistance,** and the specific heat property of a mass constitutes **thermal capacitance**. Oddly enough, there is no thermal inductance.

### 2.1.3 Reduction of the Modeling Equations

Once the conservation laws and the property laws of the discipline are written for a particular system, we arrive at a group of simultaneous differential and algebraic equations. These equations must be simplified or reduced to the desired differential equation form.

If the system is linear, the equations can be combined and reduced to a single $n$th-order differential equation. If the equations are nonlinear, it may or may not be possible to reduce the system to a single equation, depending upon the nature of the nonlinearity. If it is not possible to reduce the nonlinear system of equations to a single differential equation, the system is best treated using the state-space representation and employing the solution techniques discussed for state space.

The **classical form** of a differential equation is developed by reducing all of the system equations to a single $n$th-order differential equation. This form will be an equation having a summation of the 0th through the $n$th derivatives of the dependent (or output) variable set equal to the input function and, possibly, derivatives of the input function.

The **transfer function** notation for a linear differential equation is found from the classical equation by solving explicitly for the dependent variable and expressing it as a ratio of the output to the input. This form is required for the Laplace transform solution technique and for control system analysis.

The **state-space** form for differential equations consists of $n$ simultaneous first-order differential equations. They may be found by taking the differential equations from the conservation law equations and substituting algebraic equations from the property law equations, to eliminate the algebraic equations. The equations are then arranged in first-order form.

The **state variables** are defined as the system variables that appear as first derivatives and can be thought of as the minimum number of variables needed to define the state of the system at any instant. By proper simplification of equations, there will be $n$ state variables. These first-order equations are the derivatives of the state variables, expressed as functions of the sum of other state variables and inputs. If the equations are linear, they can be written in matrix form.

## 2.2 SOLUTION OF THE DIFFERENTIAL EQUATIONS

Linear differential equations have **classical analytic solutions** that can be expressed with exponentials, sinusoids, and other explicit mathematical functions of time. For low-order systems, it is convenient to express the solution as the summation of exponential terms in time, plus the solution of the system that is due to the particular form of the input function. This method, the subject of most elementary courses on differential equations, is discussed in Appendix E.

The **Laplace transform** technique provides a powerful approach for higher order systems and for systems with complicated inputs. The differential equation is transformed to the $s$-domain ($s$ is a complex number) using the Laplace integral transform procedure, and the output is expressed as a transfer function that includes the system characteristic dynamics in the denominator and the inputs and initial conditions in the numerator. After factoring the transfer function according to the roots of its denominator polynomial, the explicit time-domain analytic response of each factor of the system response may be determined using inverse Laplace transforms. Laplace transform techniques are reviewed in Appendix F.

**Digital simulation,** which is the solution of differential equations using a digital computer and numerical integration, is quite popular for a variety of reasons. First, numerical integration is easy, since the user has only to write the differential equations in computer code and does not have to solve them by analytic techniques. Second, digital simulation can handle time-varying inputs that are not analytic and that cannot easily be expressed by a mathematical function. Third, and perhaps most important, is that digital simulation can easily handle nonlinear systems. Since most nonlinear systems do not have an analytical solution, we *must* resort to digital simulation in these cases.

### 2.2.1 Classical Differential Equations

A classical differential equation has a dependent variable (or response of the system) $x(t)$ that varies with time according to the initial conditions $x(0)$ and the input $u(t)$. The input is a specified function of time. Therefore, three pieces of information must be known to find the future response of the system: the initial conditions, the differential equation, and the input as an explicit function of time.

**Forms for Differential Equations.** A **classical linear differential equation** is a summation of terms in the dependent (or response) variable $x(t)$ and terms in the successive derivatives of the dependent variable, equated to terms in the input function $u(t)$ and, possibly, its derivatives.

***First-Order Differential Equations.*** A linear first-order differential equation with constant coefficients $a_1$, $a_0$, and $b_0$ is expressed as

$$a_1 \frac{dx}{dt} + a_0 x = b_0 u \tag{2.9}$$

The differential operator $D = d/dt$ may be substituted to indicate a derivative with respect to time:

$$a_1 Dx + a_0 x = b_0 u \tag{2.10}$$

The Laplace operator $s$ can be used when the differential equation is transformed for solution by the Laplace transform technique. With a zero initial condition on the dependent variable, this results in

$$a_1 sX(s) + a_0 X(s) = b_0 U(s) \tag{2.11}$$

Either form can be simplified further for linear systems by having the coefficients and the operators multiply the dependent variable:

$$[a_1 D + a_0]x = b_0 u \tag{2.12}$$

$$[a_1 s + a_0]\, X(s) = b_0 U(s) \tag{2.13}$$

In this form, the coefficients, $a_1$, $a_0$, and $b_0$, are derived from the dynamic system and do not by themselves reveal much about the system's inherent characteristics. A more revealing form is obtained when we normalize the differential equation with respect to the coefficient of the lowest order dependent variable, $a_0$:

$$[\tau D + 1]x = Gu \tag{2.14}$$

When the coefficients are normalized in this manner, it is easy to identify terms that we call the time constant $\tau$ and the static gain $G$ of the system. The **time constant** of a first-order system relates to the responsiveness of the system, or the amount of time required to settle to steady-state operation from a disturbance or initial condition. Figure 2.1 illustrates that the time constant is the amount of time required for the first-order system to achieve 63% of the response from the initial condition to the final condition when the input is held constant. In four time constants ($4\tau$), the response is 98% complete. In this example, $\tau = a_1/a_0$. (See Appendix E for a further discussion of the time constant.)

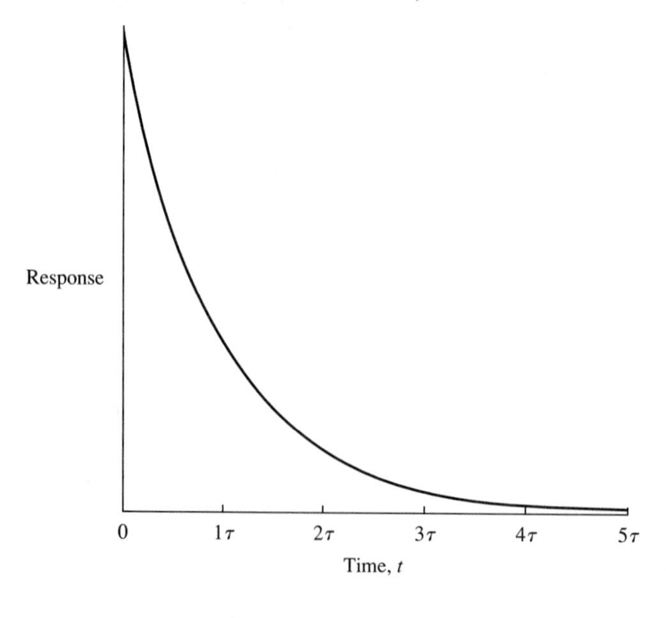


**Figure 2.1** Time response of a first-order system.

The **static gain** of the system is the change in the output as the input is varied very slowly relative to the time constant of the system. In other words, the static gain is the partial derivative of the output with respect to the input in the steady state; therefore, when $D = 0$,

$$G = \left.\frac{\partial x}{\partial u}\right|_{D=0} = \frac{b_0}{a_0} \tag{2.15}$$

Figure 2.2 illustrates the static gain of a system.

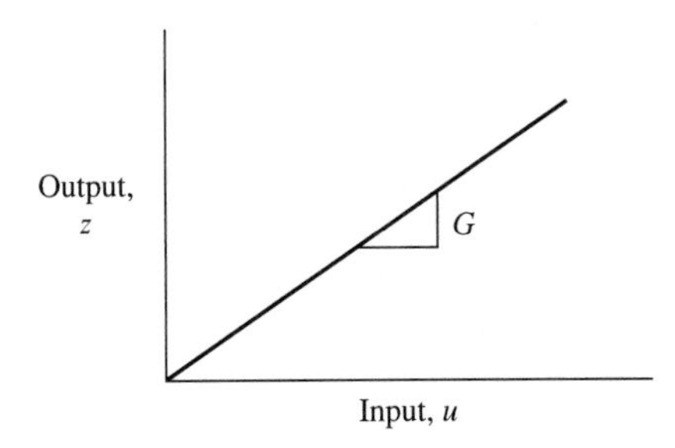


**Figure 2.2** Static gain of a dynamic system.

***Second-Order Differential Equations.*** The general form of a linear, constant-coefficient, second-order differential equation is

$$a_2\frac{d^2x}{dt^2} + a_1\frac{dx}{dt} + a_0x = b_0u \tag{2.16}$$

with initial conditions $x(0)$ and $\dot{x}(0)$.

In $D$-operator notation, this equation can be expressed as

$$[a_2D^2 + a_1D + a_0]x = b_0u \tag{2.17}$$

Again, a more revealing form of the differential equation is obtained by normalizing it with respect to the coefficient of the lowest order dependent variable $a_0$ and redefining the coefficients. This form emphasizes the static and inherent dynamic characteristics of the system and is as follows:

$$\left[\frac{D^2}{\omega_n^2} + 2\zeta\frac{D}{\omega_n} + 1\right]x = Gu \tag{2.18}$$

Here, $\omega_n$ is the **natural frequency** of the system (the frequency of oscillation of the system if there were no damping), and $\zeta$ is the **damping ratio** (a dimensionless ratio of the value of damping in the system to the value of damping which defines the boundary between oscillatory and nonoscillatory behavior.) The **static gain** $G$ has the same meaning and definition for a second- (or higher) order system as it does for a first-order system. (See Appendix E for a further discussion of these dynamic characteristics.)

***Higher Order Differential Equations.*** Higher order linear differential equations can be stated in the following general format, in which the system is $n$th order with $k$th order input derivatives ($k \leq$ n) and has initial conditions $x(0)$, $Dx(0)$, ..., $D^{n-1}x(0)$:

$$[a_nD^n + \ldots + a_1D + a_0]x = [b_kD^k + \ldots + b_1D + b_0]u \quad (2.19)$$

Note that a differential equation might include derivatives of the input, as well as derivatives of the output; however, the physics of real systems dictates that the order of the input derivatives cannot exceed the order of the system.

Here again, it is desirable to normalize the differential equation with respect to the coefficient of the lowest order dependent variable. This will give the static gain $G = b_0/a_0$ of the system, although the dynamic characteristics cannot be determined until the roots of the characteristic equation of the system are found.

**Characteristic Equations and Roots.** It is well known that the solution to the homogeneous form of a linear differential equation (the form with the right-hand side of the equation set to zero) can be written as

$$x_h(t) = Ce^{\lambda t} \quad (2.20)$$

Substituting into the homogeneous equation and simplifying by factoring out $Ce^{\lambda t}$ then gives a polynomial in $\lambda$. The **characteristic equation** of the system is the polynomial function of $\lambda$. This equation establishes the conditions on $\lambda$ for $Ce^{\lambda t}$ to be a solution to the differential equation and is written

$$a_n\lambda^n + a_{n-1}\lambda^{n-1} + \ldots + a_1\lambda + a_0 = 0 \quad (2.21)$$

Notice that the characteristic equation has no input or output variables associated with it and therefore depicts the inherent dynamic characteristics of the system, without regard to the type of system, the discipline being considered, or the particular type of input.

The roots of the characteristic equation reveal the dynamic properties of the system, expressed in terms of time constants, natural frequencies, and damping ratios. The roots are the values of $\lambda$ that satisfy the characteristic equation. (See Appendices D, E and H).

Notice further that roots can be real or complex. First-order systems have real roots, and second-order systems have two real roots or two complex roots (one complex conjugate pair). Since these are the only two possible types of roots for a differential equation of any order, it is important to understand the behavior of first- and second-order systems, because higher order systems may be thought of as combinations of first- and second-order systems. For example, a third-order system could have three real roots, or one real root and one pair of complex roots.

Table E.1 in Appendix E gives equations for roots of second-order differential equations in polynomial, root, and dynamic forms and presents the conversions to and from each form.

**Initial Conditions.** In solving differential equations, a distinction is made in the initial values of the input and response variables at an instant in time just before and just after the time defined as $t = 0$. The notation $t = 0^-$ is used to identify the time just before $t = 0$, and $t = 0^+$ is used to identify the time just after $t = 0$. The initial conditions for the response variables are stated at $t = 0^-$. The reason for this is that impulse types of inputs that occur at $t = 0$ can be considered to affect an instantaneous change in the system response, so it is important to know what the response was prior to any possible impulse inputs.

### 2.2.2 Transfer Function Format for Differential Equations

A convenient way of expressing the dynamics of a system is to convert a linear differential equation into transfer function notation. A **transfer function** is the ratio of the output to input variables, expressed as an algebraic ratio of the polynomials of the $D$-operator (or the variable $\lambda$, or the Laplace operator $s$).

The transfer function of a first-order system can be expressed as

$$\frac{x}{u} = \frac{G}{\tau D + 1} \tag{2.22}$$

The transfer function of a second-order system can be expressed as

$$\frac{x}{u} = \frac{G}{\dfrac{D^2}{\omega_n^2} + 2\zeta\dfrac{D}{\omega_n} + 1} \tag{2.23}$$

A general $n$th-order system with two inputs $u(t)$ and $w(t)$, including derivatives of $u(t)$, can be expressed as

$$x = \frac{\left[\dfrac{b_k}{a_0} D^k + \ldots + \dfrac{b_1}{a_0} D + \dfrac{b_0}{a_0}\right] u + \left[\dfrac{c_0}{a_0}\right] w}{\dfrac{a_n}{a_0} D^n + \ldots + \dfrac{a_1}{a_0} D + 1} \tag{2.24}$$

The **characteristic equation** of the system is the denominator of the transfer function set equal to zero. For the $n$th-order system above, the transfer functions are

$$\frac{x}{u} = \frac{\dfrac{b_k}{a_0} D^k + \ldots + \dfrac{b_1}{a_0} D + \dfrac{b_0}{a_0}}{\dfrac{a_n}{a_0} D^n + \ldots + \dfrac{a_1}{a_0} D + 1} \tag{2.25a}$$

and

$$\frac{x}{w} = \frac{\dfrac{c_0}{a_0}}{\dfrac{a_n}{a_0} D^n + \ldots + \dfrac{a_1}{a_0} D + 1} \tag{2.25b}$$

Nonlinear systems, of course, cannot be represented in transfer function notation.

### 2.2.3 State-Space Format for Differential Equations

The **state-space format** for differential equations consists of a set of simultaneous first-order differential equations. The **state variables** are those dynamic variables of the system which can be arranged to be the variables that appear as first derivatives. There must be at least $n$ state variables to describe an $n$th-order system. For a given system, there is no unique set of state variables; a variety of system variables could be used. Later we illustrate techniques for conveniently selecting state variables that best describe the system.

**Example 2.1 State-Space Format for an Electrical Circuit**

As an example of the state-space format, consider the following modeling equations, which represent an electrical system composed of a voltage source connected to a series resistance-capacitance ($RC$) circuit as shown in Fig. 2.3. (see Chapter 4 for the derivation of these equations.) The $e$'s are voltages and the $i$'s are currents; the input voltage is $e_0(t)$ and the output is $e_1(t)$.

$$e_0 - e_1 = Ri_R \tag{2.26}$$

$$i_C = C\dot{e}_1 \quad \text{with } e_1(0) \tag{2.27}$$

$$i_R = i_C \tag{2.28}$$

To determine a state-space representation for this system (and indeed, for any system), we search all of the system equations for the variables that occur in first-derivative form; these variables are then defined as the state variables. Thus, eliminating the current from the preceeding equations, we develop a single state-space differential equation involving the derivative of the voltage $e_1$:

$$\dot{e}_1 = -\frac{1}{RC} e_1 + \frac{1}{RC} e_0 \tag{2.29}$$

In this first-order model, $e_1$ is the only state variable and $e_0$ is the input.

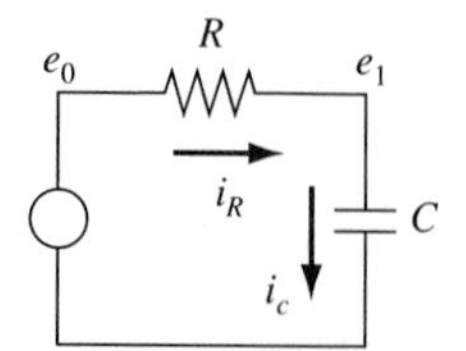


**Figure 2.3** Electrical circuit.

**Example 2.2 State-Space Format for Mechanical System**

Now consider the spring-mass-damper system of Figure 2.4 in which the motion of the mass is described by $x_1$ and the motion of the free end of the spring is described by $x_0$. (See Chapter 3 for the modeling techniques used for this mechanical system.)

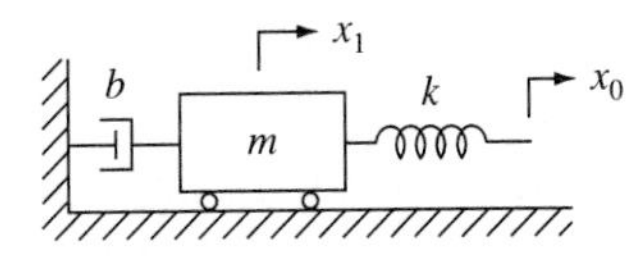


**Figure 2.4** Mechanical system.

The second-order modeling equation is

$$m\ddot{x}_1 + b\dot{x}_1 + kx_1 = kx_0(t) \tag{2.30}$$

By taking note of the relationship between the position $x_1$ and the velocity $v_1 = \dot{x}_1$, we can write

$$v_1 = \dot{x}_1 \text{ and } \dot{v}_1 = \ddot{x}_1 \tag{2.31}$$

Observation of Eqs. (2.30) and (2.31) reveals that there are two variables whose first derivatives occur in the formulation, and thus two state variables can be defined: the position $x_1$ and the velocity $v_1$. Arranging the equations into two first-order differential equations yields

$$\dot{x}_1 = v_1 \tag{2.32}$$

$$\dot{v}_1 = -\frac{k}{m}x_1 - \frac{b}{m}v_1 + \frac{k}{m}x_0(t) \tag{2.33}$$

The two state-variable equations in this example are coupled together because both state variables appear in the equations. The input to the set of equations is $x_0(t)$.

**Example 2.3 State-Space Format for Electrohydraulic Servo System**

We next present the state-space differential equations from an electro-hydraulic servo system, without explaining the origin of the modeling equations (see Chapter 7):

$$\dot{z} = v \tag{2.34}$$

$$\dot{v} = -\frac{G_a G_v A h}{m} z - \frac{RA^2}{m} v + \frac{G_a G_v A}{m} e_a \tag{2.35}$$

$$\dot{e}_a = -\frac{1}{RC} e_a + \frac{1}{RC} e_i(t) \tag{2.36}$$

Here the state variables are the position $z$, velocity $v$, and voltage $e_a$. Note that the three first-order differential equations are coupled and have an input $e_i$.

We see, then, that state-space differential equations are a set of $n$ simultaneous first-order differential equations derived directly from the modeling equations. One must specify the initial conditions of all state variables in order to obtain a solution. Note that, although state variables will allow us to solve for the response of the system, we may be interested in the response of a system variable that is not a state variable. It is possible to write any system variable as a function of the state variables, so a distinction should be made between outputs (response variables of interest) and state variables.

## 2.3 TYPICAL INPUTS OR TEST SIGNALS

Dynamic systems respond to initial conditions and to input signals. The response to initial conditions displays the basic inherent dynamic behavior of the system, and the response to the input illustrates the response character of the system to forced inputs that emulate physically realistic conditions. This section describes typical inputs, or test signals, that we apply to systems and their models. These signals represent physical inputs to dynamic systems. The type of system response can be classified according to the kind of applied input signal. A frequency response is the steady-state response to a harmonic input signal and is discussed in Chapter 8. A time response is the system response to an arbitrary time-varying input and is the subject of Chapter 9.

### 2.3.1 Step Input

In physical systems, it is common to cause a sudden change in the input by throwing a switch (in an electrical system), actuating a valve (in a fluid power system), or releasing a spring preload (in a mechanical system), or even by some other means. In these instances, the input value changes very quickly relative to the time constants or dynamic response time of the system. Plotted against time, the input signal resembles a stair step and hence is termed a step input, as illustrated in Figure 2.5.

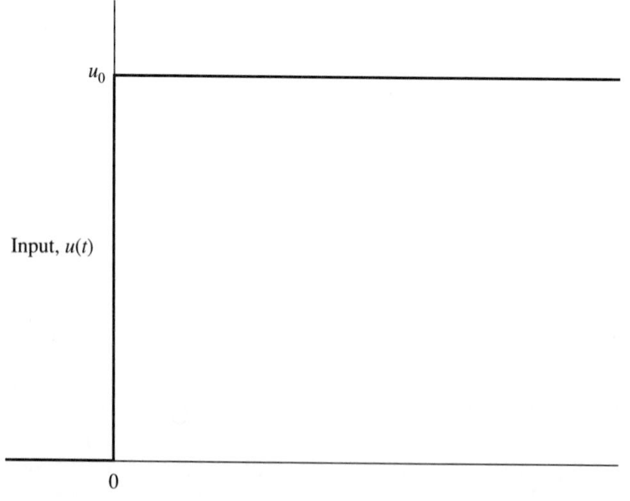


**Figure 2.5** Step input.

Mathematically, the step input can be stated as

$$u(t) = 0 \text{ for } t < 0 \tag{2.37a}$$

$$u(t) = u_0 \text{ for } t \geq 0 \tag{2.37b}$$

Notice that a distinction is made between $t = 0^-$ and $t = 0$. Just before $t = 0$, the input is zero, but the input is nonzero at $t = 0$ and thereafter.

### 2.3.2 Ramp Input

Like step inputs, ramp inputs occur naturally in physical systems and therefore are defined mathematically. An example of a ramp input is a tracking situation in which a target is moving past a pointing system at constant velocity. For example, a TV camera trying to track a float in a parade might require a ramp input of the pointing angle of the camera. In this case, the desired input increases as a linear function of time. If the input started at zero for $t = 0$ and before, then this input, plotted against time, would resemble a ramp, as illustrated in Figure 2.6.

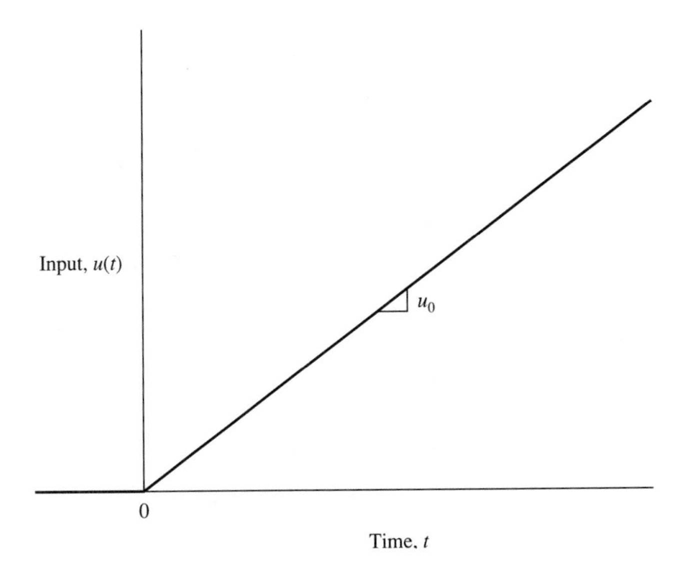


**Figure 2.6** Ramp input.

Mathematically, the ramp input can be stated as

$$u = 0 \qquad \text{for } t < 0 \tag{2.38a}$$

$$u = u_0 t \qquad \text{for t} \geq 0 \tag{2.38b}$$

### 2.3.3 Universal Inputs

If we notice in moving from the step to the ramp that one is the integral of the other, then we are led to a more general progression of input signals that are integrals or derivatives of previous input signals. The ramp is the integral of the step, and inversely, the step is the derivative of the ramp. We can thus define a set of inputs, termed *universal inputs,* that consist of the doublet, impulse, step, ramp, and quadratic inputs. In each of these, the constant $u_0$ has a different meaning and different units, depending upon the type of input. For example, $u_0$ might be a position, in inches, for a step input and might be a velocity, in inches per second, for a ramp input. Therefore, even though we employ the concept of universal inputs as derivatives or integrals of each other, the constant $u_0$ will have a different meaning in each case.

**Impulse Input.** The derivative of the step is termed the **impulse input** and is shown in Figure 2.7. This function is often called a **delta function,** $\delta(t)$. Since the step input is discontinuous at $t = 0$, the derivative of the step at $t = 0$ is infinite. Therefore, the derivative must be written in terms of the integral across the discontinuity.

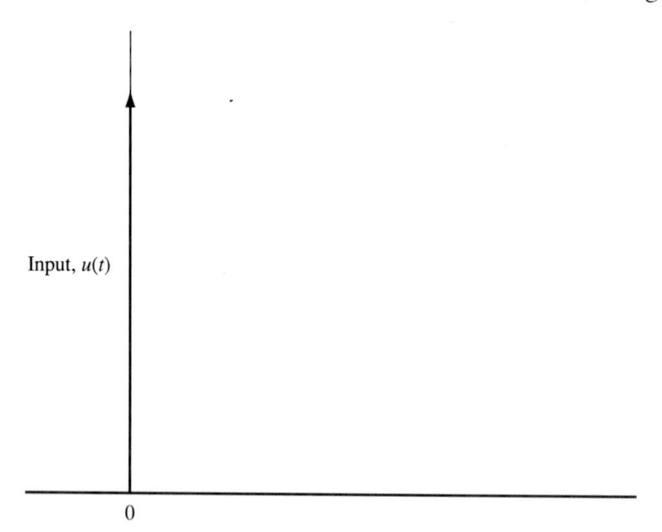


**Figure 2.7** Impulse input.

Mathematically, the impulse input can be stated as

$$u = 0 \quad \text{for} \quad t \neq 0 \tag{2.39a}$$

$$\int_{0^-}^{0^+} u \, dt = u_0 \tag{2.39b}$$

**Doublet and Quadratic Inputs.** Although they are not so useful in testing physical systems, further integrals and derivatives of the foregoing signals can be mathematically defined. The **doublet input** is the derivative of the impulse input and has a positive impulse derivative followed by a negative impulse derivative.

The **quadratic input** is the integral of the ramp input and thus is a quadratic function of time for $t \geq 0$.

### 2.3.4 Pulse Input

It is impossible to give a pure impulse signal to a physical system, since doing so would require an infinite value of the input variable for a zero duration of time. However, it is possible to give a finite input for a finite duration of time. This can be considered an impulse if the duration is very small in comparison with the system response time. Such an input signal, called a **pulse input,** is defined as a signal with a finite input magnitude that occurs over a finite amount of time and is zero before and after. (See Figure 2.8.)

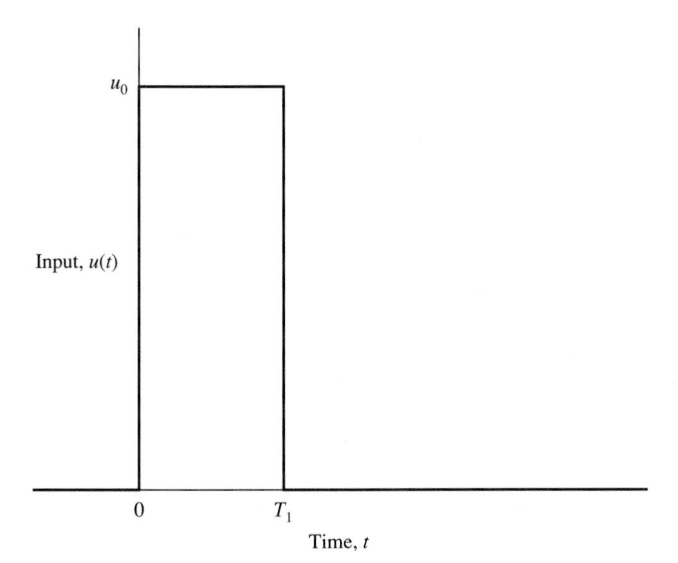


**Figure 2.8** Pulse input.

Mathematically, the pulse input can be stated as

$$u = 0 \qquad \text{for } t < 0 \text{ and } t > T_1 \tag{2.40a}$$

$$u = u_0 \qquad \text{for } 0 \leq t \leq T_1 \tag{2.40b}$$

### 2.3.5 Ramped Step Input

When an input is suddenly changed by an operator or generated by some other dynamic system with a finite response time, the resulting action may not have a zero rise time relative to the system response time. In that case, it might be better to model the signal as a ramp for some duration and a constant thereafter as illustrated in Figure 2.9.

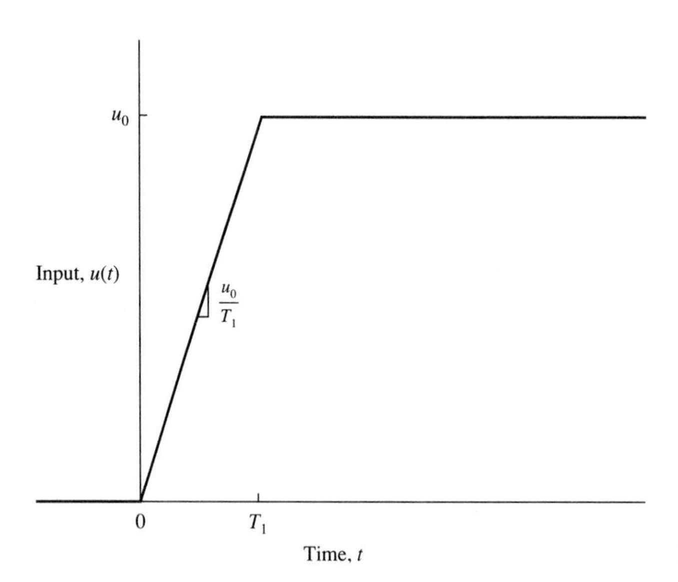


**Figure 2.9** Ramped step input.

Mathematically, the ramped step input can be stated as

$$u = 0 \qquad \text{for } t < 0 \tag{2.41a}$$

$$u = u_0 \frac{t}{T_1} \qquad \text{for } 0 \leq t \leq T_1 \tag{2.41b}$$

$$u = u_0 \qquad \text{for } t > T_1 \tag{2.41c}$$

This model provides a realistic representation of many physical processes.

### 2.3.6 Harmonic Inputs

Harmonic inputs can occur with rotating machinery or with oscillatory electronic circuits. Sine waves are relatively easy to generate and use as test signals for dynamic systems. Sine wave testing leads to the concept of the frequency response, which describes how the system will respond to sine waves of various frequencies. Usually, in sine wave testing, we are interested only in the system's steady-state response to a sine wave input. In other words, we will give a system a sine wave input for a period of time and observe the response of the system after all start-up transients have died away. In this case, we are not interested in the absolute phase of the sine wave with respect to time; rather, we are interested only in the phase *difference* of the output with respect to the input signal and to the amplitude ratio of the output and input sine waves. In such a case, it doesn't matter whether the test signal is a sine or cosine wave.

In other situations, we might be interested in sine wave testing for a transient situation, in which case the input would be zero and the sine wave would start at $t = 0$.

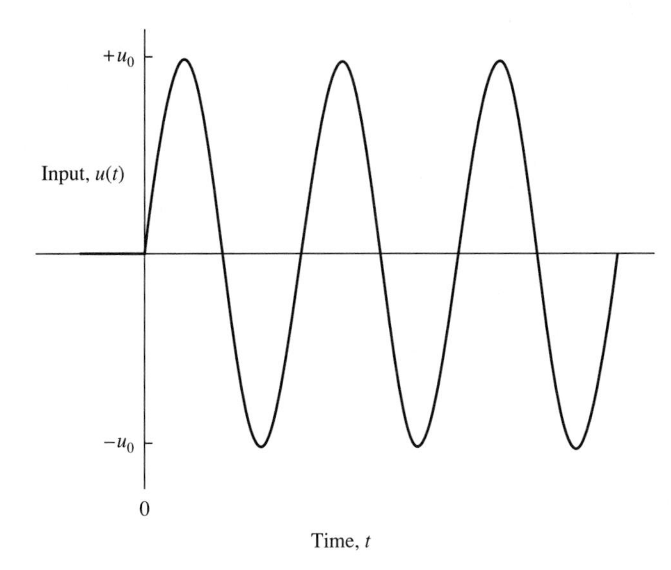


**Figure 2.10** Harmonic input.

Mathematically, the zero-mean harmonic transient test signal of Figure 2.10 is defined as

$$u = 0 \qquad \text{for } t < 0 \tag{2.42a}$$

$$u = u_0 \sin \omega t \text{ for } t \geq 0 \tag{2.42b}$$

Another useful variation of the transient sine wave test signal is a signal that starts at zero and then oscillates between zero and a maximum value. In this case, the mean value is $u_0$ and the maximum value is $2u_0$.

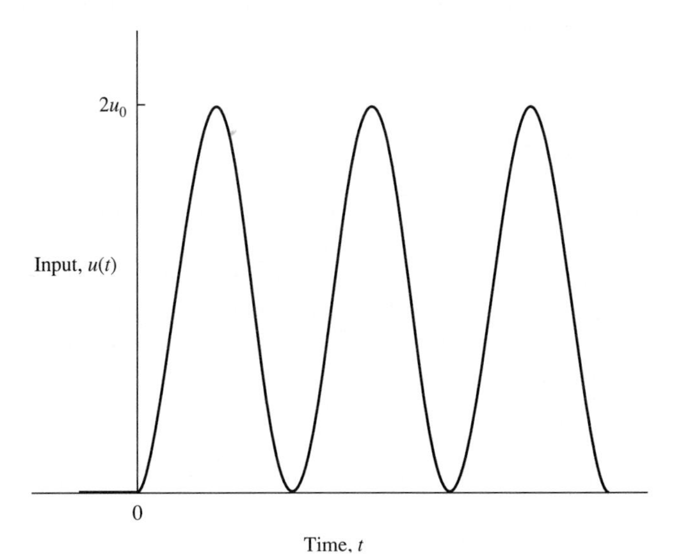


**Figure 2.11** Non-zero-mean harmonic input.

Mathematically, a non-zero-mean harmonic transient test signal is defined as

$$u = 0 \qquad \text{for } t < 0 \tag{2.43a}$$

$$u = u_0(1 - \cos \omega t) \qquad \text{for } t \geq 0 \tag{2.43b}$$

Figure 2.11 illustrates a signal of this type.

### 2.3.7 Common Test Signals

The transient response of a system to initial conditions reveals directly the inherent natural dynamics of the system and is therefore a good response-testing method. For most systems the step input reveals basically the same response character as an initial condition on a response variable and may be easier to generate than an initial condition.

The step input is probably the most common test signal, since it is so easy to generate physically, and, in addition, solving a differential equation with a constant input is not difficult. Ramp inputs are also relatively easy to generate physically and treat analytically. Because impulse signals are generally difficult to produce physically, the "impulsive response" used in many discussions is really more of a mathematical concept than a commonly observed physical response. Sine wave testing is very popular and is not difficult to perform experimentally; however, sine wave testing is usually of interest for evaluating the steady-state frequency response rather than testing the transient response. The solution of differential equations subjected to harmonic input is easy to find.

Often, it is convenient or interesting to test with repetitive input signals that are periodic, but nonharmonic. The repetitive step input of alternating signs is a square wave and is very easy to generate physically. The repetitive ramp input of alternating signs on the slopes is a triangular wave or a sawtooth wave that is also relatively easy to generate.

## 2.4 ENGINEERING SYSTEMS SIMILARITY

One major purpose of this text is to show the techniques and concepts used to model dynamic systems in the major engineering disciplines and to point up the similarity or unity that they have. By **systems similarity,** we mean that modeling of many physically different kinds of dynamic systems result in the same or similar differential equations and have similar response behaviors. Therefore, the dynamic response characteristics studied in one engineering specialty can transfer to other disciplines. Further, dynamic systems in different disciplines contain the same basic elements—resistance, capacitance, and inductance—and have various energy converters and power amplifiers or modulators. The ways in which they handle energy concepts are also similar.

For example, if you understand why your car has an exponential characteristic of gaining speed after you make a constant change in the position of the throttle to pass another car, then you should also understand that there could be a similar dynamic response of an electronic instrument reading some measured variable, even though the time constants for the two systems are completely different. To be specific, if you were passing another car in high gear down the highway and depressed the throttle 1 inch with your foot, it might take your car 20 seconds to accelerate from 50 mph to 65 mph.

Suppose you then pull off the road and stop. You place the car in neutral and make a sudden change in the position of your throttle; then the engine would achieve a high rpm in less than 1 second. However, if you were to watch the tachometer during this acceleration of the engine with essentially no load, you would notice that the reading was actually less than the engine speed. Therefore, if the engine were idling at 1000 rpm, and the throttle change would ultimately result in a steady-state speed of 5000 rpm when the engine was actually passing through 2000 rpm on its way to a higher rpm, the tachometer might be reading only 1500 rpm or so if our eyes were fast enough to catch it. Thus, there is a time delay between the input (the engine speed) and the output (the tachometer reading).

In effect, then, the electronic tachometer has a dynamic response similar to the acceleration of the mechanical system represented by the car. You might even notice that the tachometer could overshoot the maximum speed reading and oscillate slightly before it reached a steady state, even though the engine itself did not overshoot the maximum speed. Thus, the behavior of the car and the behavior of the tachometer are similar in many ways, even though the car and the tachometer are completely different dynamic systems.

The car provides us with additional examples of dynamic system response. When you start your car for the first time of the day, the engine water temperature is low. Over the period of a few minutes, the water temperature would rise from, say, 75° to 190 °F. The temperature rise is a response characteristic similar to the acceleration of a car (in a mechanical system) and the acceleration of a tachometer needle (in an electro-mechanical system), but is much slower (i.e., has a larger time constant).

The power steering system of a car is a hydraulic servo system with a valve and cylinder actuator. It also has a dynamic response similar to those in the preceeding examples; however, it must have a much faster response, to the point where the driver cannot actually perceive any delay. Still, if you were to connect sensitive instrumentation to the steering system input and output, you could definitely measure the lag in the output response in relation to the input command, even though the time base now would be in milliseconds.

The foregoing examples demonstrate that dynamic systems in several disciplines have similar characteristics and a similar dynamic response. As a student of dynamic systems, you will begin to recognize the similarities and analogies that exist among dynamic systems in the various engineering disciplines as we work through this text.

### 2.4.1 Effort and Flow Variables

The basis of systems similarity is that the engineering disciplines can share the concepts of effort and flow quantities in the variables used to describe the system characteristics. In every discipline, the variables that we employ to write the differential equations can be classified as effort or flow variables.

An **effort variable** is the system variable that expresses the effort which can be placed on a component. This could also be regarded as the potential or ability to do work. The effort variable for a mechanical system is considered to be **force** (in units of newtons or pounds force). For an electrical system, it is **voltage** (volts). **Pressure** is the effort variable for a fluid system (newtons/meter$^2$ or pounds force/inch$^2$), and **temperature** (°C or °F) is the effort variable for a thermal system.

A **flow variable** is the system variable that expresses the flow, or rate of change with times, of a system variable. The flow variable usually is a derivative of a variable and thus has units of something per unit time. It represents the rate at which work can be done. The flow variable for a mechanical system is considered to be **velocity** (meter/second or inch/second for linear motion and radian/second or degree/second for angular motion); however, in some cases it may be more desirable to work with **position** instead of the derivative of position. The flow variable for electrical systems is **current** (1 amp = 1 coulomb/second). **Volume flow rate** (meter$^3$/second or inch$^3$/second) is the flow variable for fluid systems while **heat flow** (joule/second or Btu/hr) is the flow variable for thermal systems.

As we write the basic equations for the resistive, capacitive storage, and inductive storage elements in each discipline, we can use similar equations for each in all of the engineering disciplines by employing the effort and flow variables appropriate to the physical situation.

Effort and flow variables provide a basis for the concept of impedance in all of the engineering disciplines. The **impedance** of an element is the ratio of its effort variable to its flow variable. Impedance can be static (resistance) or dynamic (capacitance and inductance). Electrical impedance is well known and understood; however, mechanical impedance may not be as familiar a concept. Fluid impedance and thermal impedance should not be too difficult to grasp, since modeling approaches to them often are similar to electric circuit analysis. Therefore, our basic model for each element in the various engineering disciplines is an impedance relationship such as

$$\text{effort} = \text{impedance} \times \text{flow} \tag{2.44}$$

### 2.4.2 Dissipative Elements

Elements that dissipate energy (or convert it to another form such as heat, light, vibration, etc.) are found in all engineering disciplines. **Dissipative elements** do not store energy and thus are described by algebraic equations rather than differential equations.

In mechanical systems, the dissipative element is the dashpot or damper. In electrical systems, the dissipative element is the resistor. Resistance in fluid systems is presented by a capillary viscous resistor (pressure loss due to viscous shear) or an orifice (Bernoulli velocity head loss due to sudden expansion). In thermal systems, resistance is due to heat transfer through convection, conduction, or radiation. Note that thermal resistors transport heat from one location to another and do not technically dissipate energy; however, they do follow algebraic equations rather than differential equations and therefore are classed as resistors in this discussion.

In all cases the resistance equations are algebraic, and if the system is linear, they can be expressed as

$$\text{effort} = \text{resistance} \times \text{flow} \tag{2.45}$$

Nonlinear resistors also have a static relationship between the effort and flow variables, but it is, of course, not a linear relationship. While the concept of impedance is applicable to a nonlinear system, we cannot use a simple value or variable for the resistance.

The impedance of dissipative elements (resistors) is symbolized by $R$.

### 2.4.3 Effort Storage Elements

**Effort storage** elements store energy by virtue of the effort variable. Effort storage elements are thus capacitive in nature, since they follow a differential equation of the form

$$\text{flow} = \text{capacitance} \times \frac{d}{dt}\text{effort} \tag{2.46}$$

In this form it becomes clear that, since the effort variable is differentiated, it is possible to have an initial condition on the dependent variable (effort), and thus, it is possible to store potential energy in the initial condition of the effort variable by the amount of capacitance $\times$ effort$^2/2$. For this reason, capacitors are termed effort storage elements.

In **mechanical systems,** the component that follows the form of Eq. (2.46) is

$$\text{velocity} = \frac{1}{\text{stiffness}} \times \frac{d}{dt}\text{force} \tag{2.47}$$

Stated in a more familiar way, force $=$ stiffness $\times$ position. (Since position is the integral of velocity, $x = v/D$, just as $Dx = v$.) Thus, the capacitive element in mechanical systems is the spring, which can store potential energy by means of an initial force preload.

In **electrical systems,** the electrical capacitor obeys the differential equation form

$$\text{current} = \text{capacitance} \times \frac{d}{dt}\text{voltage} \tag{2.48}$$

An electrical capacitor can be made to store energy by an initial voltage charge.

In **fluid systems,** the storage differential equation describes an accumulator or a volume of compressible fluid in a tank:

$$\text{flow} = \text{capacitance} \times \frac{d}{dt}\text{pressure} \tag{2.49}$$

This equation is based upon the compressibility of the fluid itself or the ability of the container to expand mechanically with pressure. Thus, an incremental volume of fluid placed in the container will result in an incremental change in pressure according to the capacitance of the system. Since flow is the derivative of volume with respect to time, the derivative of the volume-pressure relationship gives Eq. (2.49). Fluid capacitance, then, is represented by a volume of a compressible fluid or a container of fluid that can expand with pressure. Energy can be stored by an initial pressure in the container.

In **thermal systems,** the storage differential equation represents the property of heat capacity of the material:

$$\text{heat flow} = \text{capacitance} \times \frac{d}{dt}\text{temperature} \tag{2.50}$$

Thermal capacitance is the mass of the element times the specific heat of the material. A thermal capacitance stores energy in proportion to its initial temperature.

The impedance of effort storage components (capacitors) is $1/(CD)$.

### 2.4.4 Flow Storage Elements

**Flow storage** elements store energy by virtue of the flow variable. Flow storage elements are inductive in nature, since they follow the differential equation form

$$\text{effort} = \text{inductance} \times \frac{d}{dt}\text{flow} \tag{2.51}$$

Inductive elements can store kinetic energy as an initial condition of the flow variable. This energy is $L \times (\text{flow}^2)/2$, where $L$ is the inductance.

In **mechanical systems,** the equations that are of the same form as the flow storage differential equation are

$$\text{force} = \text{mass} \times \frac{d}{dt}\text{velocity} \tag{2.52a}$$

and

$$\text{torque} = \text{intertia} \times \frac{d}{dt}\text{angular velocity} \tag{2.52b}$$

Therefore, **mass** or **inertia** is mechanical inductance. A mechanical inductor can store kinetic energy by virtue of its initial condition of velocity. A flywheel is a good example of a mechanical energy storage system.

In **electrical systems,** the inductor is the flow storage element, since the equation that is of the same form as the flow storage differential equation is

$$\text{effort} = \text{inductance} \times \frac{d}{dt}\,\text{current} \tag{2.53}$$

The electrical inductor can store energy with an initial value of current.

In **fluid systems,** the flow of a dense fluid in a long line of small cross-sectional area represents inductance, due to the inertial properties of the fluid attempting to change its flow rate. In this case, the flow storage differential equation is

$$\text{pressure} = \text{inductance} \times \frac{d}{dt}\,\text{flow rate} \tag{2.54}$$

This is the well-known *water hammer effect*, in which a pressure is generated when the flow rate is suddenly changed. Here, kinetic energy is stored in the initial value of the flow rate.

In **thermal systems,** there is no engineering equation that has the form of the flow storage differential equation. Therefore, thermal inductance does not exist. This is a very interesting fact that further emphasizes that thermal systems are slightly different from other engineering systems. It also leads to the observation that passive (noncontrolled) thermal systems cannot have resonance or overshoot (i.e., the characteristic equation of a thermal system cannot have complex roots), as can other systems, even though a thermal system can be described by a second- or higher order differential equation.

The impedance of flow storage components (inductors) is $LD$.

### 2.4.5 Summary of the Elements in Engineering Disciplines

Systems similarity is inherent in the preceding discussions which show that the basic elements in each discipline can be classified as resistive, capacitive, or inductive, depending upon how the discipline treats energy. In each case, the behavior of the linear elements can be expressed as an impedance relationship.

The impedance of the three types of elements is the ratio of the effort to flow variables. In resistive elements, this relationship is algebraic; however, the storage elements (capacitance and inductance) have dynamic impedances that are properly described using the $D$-operator as a part of the definition of impedance. Thus, the capacitive impedance becomes $1/(CD)$, which means that capacitive impedance is the inverse of the product of capacitance and the integral of whatever variable is multiplied by the impedance. Similarly, inductive impedance is $LD$, which means that inductive impedance is inductance times the derivative of whatever variable is being multiplied by the impedance.

As an introduction to the models that are developed in detail in the chapters that follow, the basic elements of the engineering disciplines are listed in Table 2.1, which emphasizes systems similarity.

**TABLE 2.1** BASIC ELEMENTS IN ENGINEERING DISCIPLINES.

| | Dissipative (Resistive) | Effort Storage (Capacitive) | Flow Storage (Inductive) |
|---|---|---|---|
| Mechanical Translation<br>Effort = Force<br>Flow = Velocity | $F = b(v_1 - v_2)$ | $F = \frac{k}{D}(v_1 - v_2)$ | $F = mDv_1$ |
| (Alternative Form)<br>Effort = Force<br>Flow = Position | $F = bD(x_1 - x_2)$ | $F = k(x_1 - x_2)$ | $F = mD^2x_1$ |
| Mechanical Rotation<br>Effort = Torque<br>Flow = Speed | $T = b(\omega_1 - \omega_2)$ | $T = \frac{k}{D}(\omega_1 - \omega_2)$ | $T = JD\omega_1$ |
| (Alternative Form)<br>Effort = Torque<br>Flow = Angle | $T = bD(\theta_1 - \theta_2)$ | $T = k(\theta_1 - \theta_2)$ | $T = JD^2\theta_1$ |
| Electrical<br>Effort = Voltage<br>Flow = Current | $e_1 - e_2 = Ri$ | $e_1 - e_2 = \frac{1}{CD}i$ | $e_1 - e_2 = LDi$ |
| Fluid<br>Effort = Pressure<br>Flow = Volume Flow Rate | $P_1 - P_2 = RQ$ | $P_1 = \frac{1}{CD}Q$ | $P_1 - P_2 = LDQ$ |
| Thermal<br>Effort = Temperature<br>Flow = Heat Flow | $T_1 - T_2 = RQ_h$ | $T_1 = \frac{1}{CD}Q_h$ | Does Not Exist |

## REFERENCES

2.1 Close, Charles M., and Frederick, Dean K. *Modeling and Analysis of Dynamic Systems,* 2d ed. Houghton Mifflin Co., Boston, MA, 1993.

2.2 Dorny, C. Nelson. *Understanding Dynamic Systems: Approaches to Modeling, Analysis, and Design*. Prentice Hall, Englewood Cliffs, NJ, 1993.

2.3 Ogata, Katsuhiko. *System Dynamics,* 2d ed. Prentice Hall, Englewood Cliffs, NJ, 1992.

2.4 Palm, William J., III. *Modeling, Analysis, and Control of Dynamic Systems*. John Wiley & Sons, New York, 1983.

2.5 Shearer, J. Lowen, and Kulakowski, Bohdan T. *Dynamic Modeling and Control of Engineering Systems*. Macmillan Publishing Company, New York, 1990.

## NOMENCLATURE

$a_0, a_1, b_0$ coefficients
$b$ dissipation constant of a viscous damper
$C$ capacitance
$e$ voltage
$g$ gravitational acceleration
$G$ static gain
$h$ specific enthalpy
$i$ current
$J$ moment of inertia
$k$ spring constant
$L$ inductance
$m, \dot{m}$ mass, mass flow rate
$P$ pressure
$Q$ electronic charge; also heat transfer rate and volume flow rate
$R$ resistance
$s$ Laplace transform variable
$t$ time
$u$ specific internal energy; also independent (or input) variable
$u_o$ constant for generalized inputs
$v$ translational velocity
$V$ volume
$W$ work
$x$ dependent ( or response ) variable
$z$ height
$\rho$ density
$\tau$ time constant
$\omega$ rotational speed, or excitation frequency
$\omega_n$ natural frequency
$\zeta$ damping ratio

## PROBLEMS

**2.1** *Identify* the basic conservation laws on which engineering systems design and analysis are based.

**2.2** *Which* are the principal physical properties that influence the dynamic behavior of the following systems?

**a.** Mechanical

**b.** Electrical

**c.** Fluid

**d.** Thermal

**2.3** *Explain* the difference between digital and analog simulation.

**2.4** *Discuss* the advantages and disadvantages of using digital simulation. *Discuss* the advantages and disadvantages of using analog simulation.

**2.5** *Discuss* what is meant by the order of a dynamic system.

**2.6** The differential equations representing the behavior of a dynamic system can take many forms. *What* are the most important forms?

**2.7** *Define* the term time constant as it applies to dynamic systems.

**2.8** *Where* does the characteristic equation of a dynamic system come from?

**2.9** With *what* information do the roots of the characteristic equation provide the engineer?

**2.10** Suppose you hold a ruler or similar object flat against a tabletop, with most of it extending over the edge. You then pull the tip down and let it go so that it will vibrate. *What* are the initial conditions for the dynamic response?

**2.11** *Design* an experiment that performs the motion described in the previous problem such that the vibration continues for about a second or longer. *Do* the experiment. *Give* a description of the recommended ruler or other object, the initial conditions, and the expected response. *What* factors do you think influence the dynamic response?

**2.12** *How* does the state-space differential equation form differ from the classical form?

**2.13** *Derive* the state-space equations given in Example 2.2 for the spring-mass-damper system.

**2.14** *Develop* the state-space equations for the spring-mass-damper system of Example 2.2 if the state variables are

$$z = x_1 - x_0 \quad \text{and} \quad v = \dot{z} = \dot{x} - \dot{x}_0$$

**2.15** Let

$$\dddot{x} + 2\ddot{x} - 3\dot{x} + x = \sin \omega t$$

be a system equation. Set $y_1 = x$, and $y_2 = \dot{x}$, and $y_3 = \ddot{x}$, and *determine* the equivalent state-space representation.

**2.16** The state-space equations for a particular system are

$$\dot{y}_1 = -2y_2 - y_1 + e^{-t}$$

$$\dot{y}_2 = y_1$$

*Find* the equivalent classical second-order differential equation of the form

$$a_2\ddot{y}_2 + a_1\dot{y}_2 + a_0y_2 = f(t)$$

**2.17** Stand on the bumper of your parked automobile and let the suspension settle to its equilibrium position. Step off quickly and observe the response. *Describe* what you see, and characterize the condition of the car's shock absorbers. *What* kind of excitation are you providing to the system?

**2.18** Hold on to one end of a fairly tight cord. (The stretched cord of a wall-mounted telephone handset does pretty well.) Hold the cord tight with one hand, move it up and down rhythmically at different rates with your other hand, and *report* what you observe. Can you cause motions of large amplitude if the rate is just right? Also, *describe* what happens if you move your hand up and down just once with no repetition. *Sketch* the cord displacement response as a function of time.

**2.19** *Describe* three types of system input that are useful in evaluating system performance.

**2.20** *What* is meant by similarity in dynamic systems?

**2.21** *Describe* the meaning of effort and flow variables. *Identify* effort and flow variables for mechanical, electrical, fluid, and thermal systems.

**2.22** *What* mechanical, electrical, fluid, and thermal system components cause energy dissipation? *How* does the thermal model differ from the others?

**2.23** *Describe* the mechanical electrical, fluid, and thermal system components that can store energy.

**2.24** Two different definitions of effort and flow variables are available for mechanical systems. How do they differ? Is either preferable? *Explain.*

**2.25** *Show* mathematically that thermal systems which are composed of first-order systems multiplied together can never have complex roots. Use the characteristic equation

$$(\tau_1 D + 1)(\tau_1 D + 1) = a_2D^2 + a_1D + 1 = 0$$

# Part 2

# Modeling of Engineering Systems

Electro-thermal-mechanical silicon wafer–growing system. (Photo courtesy of Texas Instruments Incorporated.)

"If you can't get to be oncommon through going straight, you'll never get to do it through going crooked. So don't tell no more [lies] on 'em Pip, and live well and die happy."

Joe Gargery to Pip in *Great Expectations* by Charles Dickens, 1861

# CHAPTER 3

# Mechanical Systems

## 3.1 INTRODUCTION

In this chapter, we introduce the fundamental ideas and the basic building blocks commonly used in modeling mechanical systems. While stiffness, energy dissipation, mass, and inertia are basic quantities distributed spatially throughout all components of mechanical systems, the behavior of many dynamic systems can be accurately represented through the intelligent combination of discrete springs, discrete dampers, and point masses or inertias. That is to say, while a spring is certainly not massless, if it supports an object whose mass is very much larger than its own mass, it may well suit the objective of the analysis to ignore the mass of the spring and treat the spring as an element with discrete stiffness only. Sound physical reasoning establishes a foundation from which the engineer develops the judgment necessary to know when quantities can safely be neglected and when they cannot. This evaluative ability lies at the very heart of the engineer's art.

The physical behavior of each mechanical element, together with Newton's laws of motion, provides the fundamental principles governing the development of suitable models for mechanical systems. In what follows, translational systems are discussed first, followed by rotational systems and, last, systems composed of combined translational and mechanical elements.

## 3.2 SYSTEMS OF UNITS

The fundamental quantities used to describe the behavior of systems of discrete particles are mass, distance, time, and force. For mechanical systems with constant mass, the momentum law, Eq. (2.1), may be simplified, since the rate of change of the mass term with respect to time is zero. The expression is restated as Newton's second law of motion, which governs the dynamic behavior of a point mass:

$$\sum \mathbf{F} = m\mathbf{a} \tag{3.1}$$

This equation says that the sum of all forces acting on a particle is equal to the product of the particle's mass and acceleration. For general motion in three-dimensional space, the sum of the forces and the acceleration are taken in the vector sense, and the position vector must be measured with respect to a fixed (nonaccelerating) reference frame. In most applications, we may use a frame attached to the earth. Remember that the displacement of the particle is differentiated with respect to time in order to obtain the velocity, and the derivative of the velocity with respect to time is the acceleration. Thus, acceleration has dimensions of length per unit time, per unit time. It is common to use the second as the unit of time in studying the dynamics of mechanical systems; the meter, the foot, and the inch are all units of length commonly used in describing dynamic systems.

The terms in an equation must combine to form an expression that is consistent in its units if it is to be a valid representation of a physical process. We can select the units of mass, length, and time to suit our needs, but the unit of force must then be defined consistently so that the units of all terms in the equation form a homogeneous system. Alternatively, we can select the units of force, length, and time, but then we are required to select a consistent unit of mass in order for Newton's second law to be homogeneous in its units.

Two homogeneous systems of units are in common use in engineering: the International System, abbreviated SI, and the British system. In the SI convention, the unit of mass is defined, and the unit of force is derived to be consistent with it. This system uses the meter (m) for length, the second (s) for time, and the kilogram (kg) for mass. The SI unit of force is then derived by the second law, Eq. (3.1), and is called the newton (N). One newton is the force required to accelerate 1 kg at a rate of 1 $\text{m/s}^2$. Mathematically,

$$F = ma \tag{3.2}$$

$$1 \text{ N} = (1 \text{ kg})(1 \text{ m/s}^2)$$

**Weight,** $W$, is the force acting on a mass in a gravitational or acceleration field. The acceleration due to gravity on earth is written $g$ and has the value 32.18 $\text{ft/s}^2$ or 9.807 $\text{m/s}^2$ (at sea level at 45° latitude).

In the British system, the unit of weight is defined and the unit of mass is derived. The unit of length is the foot (ft), the unit of time is the second (s), and the unit of force is the pound force (lbf). The unit of mass is then derived from the second law:

$$m = \frac{F}{a} \tag{3.3}$$

$$\text{mass} = \frac{W \text{ lbf}}{32.18 \text{ ft/s}^2}$$

Notice that the weight of the object is divided by the acceleration due to gravity to obtain the mass term. Since the acceleration due to gravity contains a length unit, be sure to use length units that are compatible with it. That is, if you are using inches as the units of length, remember to use $g = 32.18 \times 12 = 386.16$ in/s$^2$ as the acceleration due to gravity.

## 3.3 TRANSLATIONAL SYSTEMS

### 3.3.1 Translational Springs

**Springs** resist applied forces and may be used to store energy. You will find springs in use in a wide variety of common devices and machines. The clip that keeps a pen from falling out of a shirt pocket, the device that keeps the lid of a car trunk open, and the mechanism that returns the key on a computer keyboard to its unpressed position are all examples of springs. A common translational spring is shown in Figure 3.1.

We characterize a spring by its static response to an applied load. You can perform an experimental evaluation of the behavior of a spring using the simple apparatus described in Figure 3.2.

Apply the loads slowly and carefully, so that no significant dynamic motions occur, and record the amount of deflection corresponding to each load. A plot of load vs. deflection indicates whether the spring is a linear, hardening, or softening spring. If the spring is linear, the resulting deflection will be directly proportional to the applied load. If, as the load is increased, a smaller and smaller amount of load is required to achieve a given displacement, the spring is called a softening spring. A hardening spring is defined in a similar manner. Most springs display a high degree of linearity over a range of deflections that is small relative to the length of the spring.

The force in a spring depends upon the relative displacement of the endpoints of the spring. Figure 3.3 shows a spring that is unstretched, and the figure also shows a free-body diagram of the extended spring and the forces acting on the spring in order to cause the extension.

Linear springs are an appropriate representation of the behavior of a vast number of practical system components. If a spring is **linear,** its load is directly

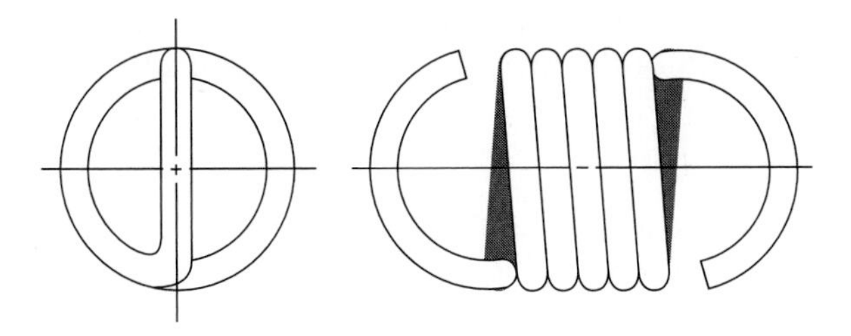

**Figure 3.1** Translational spring.

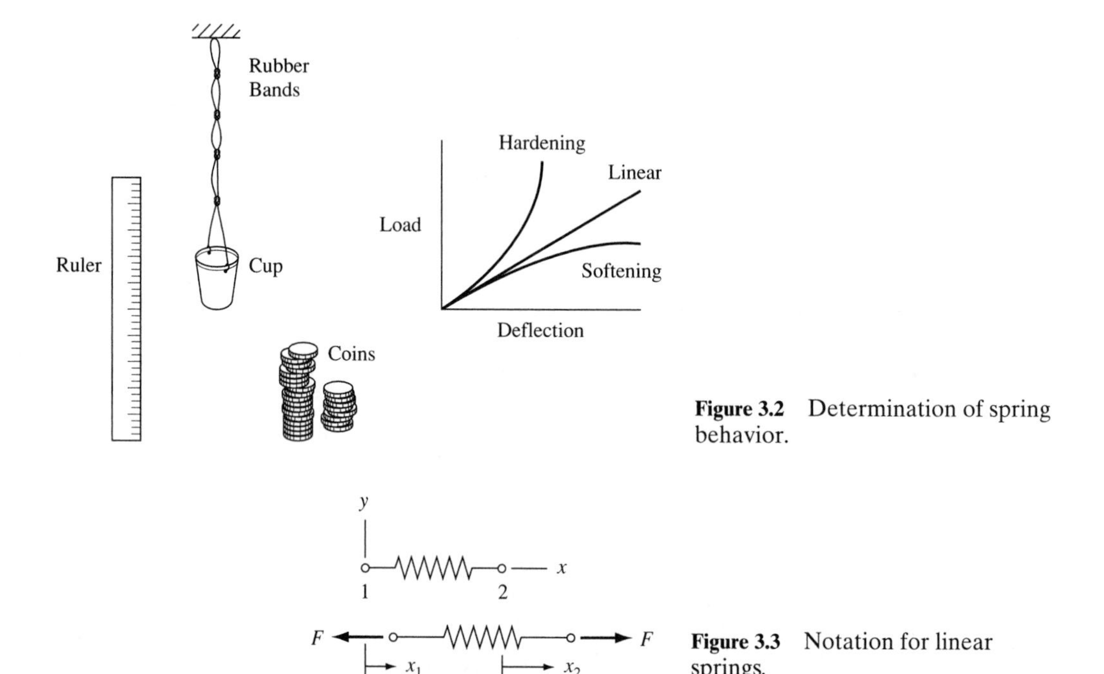


**Figure 3.2** Determination of spring behavior.

**Figure 3.3** Notation for linear springs.

proportional to its deflection, with the spring constant $k$ the proportionality constant. Here, $x_1$ and $x_2$ are the displacements of the ends of the spring, and $F$ is the spring force. The **spring constant** $k$ has dimensions of force/length. Mathematically,

$$F = k(x_2 - x_1) \tag{3.4}$$

Note that if $x_2 - x_1$ is positive, the spring will be in tension, and the force $F$ shown in Figure 3.3 will be positive also. To get the correct modeling equations, it is essential that the sense of the force in the figure and the equation representing it be consistent. It doesn't matter whether we show the spring in tension or in compression, as long as the associated equation is written consistently. Nor does it matter whether we find, in the solution of the problem, that the force is in the opposite sense to what was shown in the sketch. What matters in setting up problems is that we be consistent in the free-body sketches that define the sense of the forces and the equations that describe them. Establish a consistent set of sketches and equations, stick with that arrangement through the solution of the problem, and then interpret your results accordingly.

**Example 3.1 Linear Spring**

A tensile force of 350 N is applied statically to the free end of a linear spring that is fixed at the other end as illustrated in Figure 3.4. The spring constant is 2000 N/m. Find the resulting deflection.

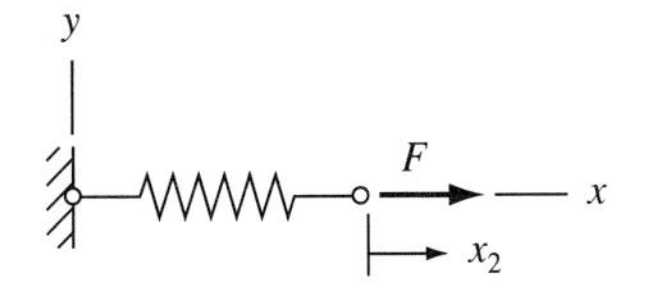


**Figure 3.4** Linear spring.

**Solution**

$$F = k(x_2 - x_1) \tag{3.5}$$

$$x_1 = 0 \tag{3.6}$$

$$x_2 = \frac{F}{k} = \frac{350\ \text{N}}{2000\ \dfrac{\text{N}}{\text{m}}} = 0.175\ \text{m} = 6.89\ \text{inches} \tag{3.7}$$

The load-deflection relationship for any spring, linear or nonlinear, may be written in a general way by specifying that the spring force is a function $h$ of the displacement of the endpoints:

$$F = h(x_1, x_2) \tag{3.8}$$

For a hardening spring, the function could be a simple power law. For a spring whose force is proportional to the third power of the relative displacement, the expression would be

$$F = k(x_2 - x_1)^3 \tag{3.9}$$

where $k$ is a proportionality constant with units of force/length$^3$. Figure 3.2 shows a typical load-deflection curve for a hardening spring.

### 3.3.2 Translational Dampers

A **damper,** or **dashpot,** is a device that generates a force in proportion to the difference in the velocity of the two endpoints of the device. Dampers cannot store energy; they can only dissipate it. Thus, they are not dynamic components. Lumped-parameter translational dampers are used to represent the energy dissipation components of a system. A common energy dissipation device is shown in Figure 3.5.

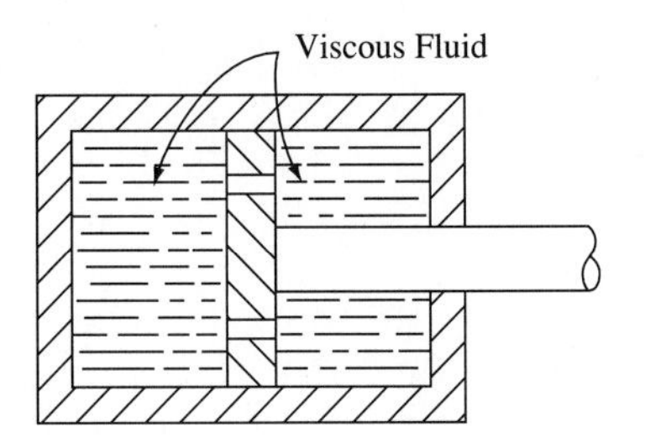


**Figure 3.5** Translational damper.

A schematic representation of this damper is shown in Figure 3.6.

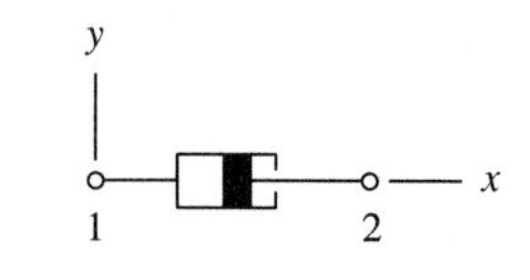


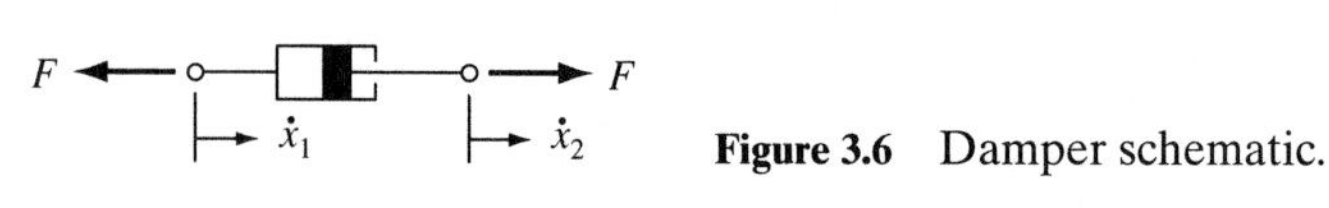


**Figure 3.6** Damper schematic.

For a **linear damper,** the force carried by the element is directly proportional to the difference in velocity of the two endpoints of the device; that is,

$$F = b(\dot{x}_2 - \dot{x}_1) \tag{3.10}$$

where $b$ is the **damping constant,** with units of force/velocity or force/(length/time), and $\dot{x}_2$ and $\dot{x}_1$ are the velocities of the endpoints. This type of device is called a viscous damper. The force generated is due to the pressure drop across a fluid resistor. If the fluid resistor is a capillary or viscous resistor, the pressure drop will be linear with flow (or velocity) because the flow will be laminar. Discrete damping elements are assumed to have no significant stiffness or mass. An example of their use is given next.

**Example 3.2 Linear Damper**

A viscous damper element is initially at rest, with its left end restrained from motion. A constant force of 7.5 lbf is applied to the other end as shown in Figure 3.7. What displacement has occurred at the right end after this force has been applied for 1.5 s if the viscous damping constant is 25 lbf in/s?

**Solution** The left end of the damper is fixed, and the system is initially at rest, so

$$x_1 = \dot{x}_1 = 0 \tag{3.11}$$

$$x_2(t = 0) = 0 \tag{3.12}$$

The load-velocity relationship for the damper element, Eq. (3.10), gives

$$F = b(\dot{x}_2 - \dot{x}_1) = b\dot{x}_2 \tag{3.13}$$

Thus,

$$\dot{x}_2 = \frac{dx_2}{dt} = \frac{F}{b} \tag{3.14}$$

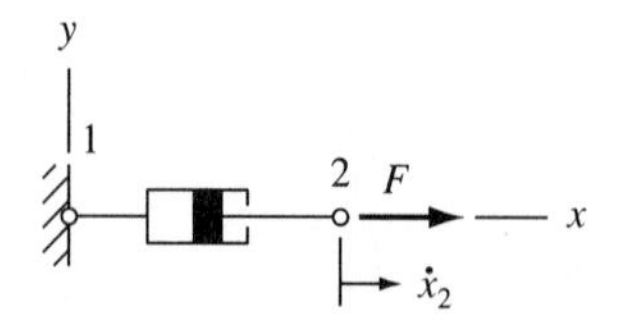


**Figure 3.7** Linear damper.

and integrating produces the desired result:

$$x_2 = \int_0^{1.5} \frac{F}{b}\,dt = \frac{F}{b}(1.5 - 0) = \frac{7.5}{25}(1.5) = 0.45 \text{ inch} = 11.4 \text{ mm} \tag{3.15}$$

If energy dissipation comes about by turbulent flow through an orifice or by fluid dynamic drag, the force on the object is proportional to the square of the velocity. (See Chapter 5.) The load-velocity relationship for a damper constructed using turbulent flow is

$$F = \text{sign}\,(\dot{x}_2 - \dot{x}_1) b (\dot{x}_2 - \dot{x}_1)^2 \tag{3.16}$$

where sign( ) is a function that gives the algebraic sign of the quantity in parentheses to the term that follows in the equation. The sign function is needed because the drag force is proportional to the square of the velocity, which is always positive and must be multiplied by the sign of the velocity if it is to change sign when the velocity changes sign.

A damping element can also be devised to represent the sliding friction between two surfaces. The friction that results depends upon the coefficient of friction and the normal force that presses the two surfaces together. Sliding friction of this type occurs in many mechanical devices, and its behavior is often represented by a frictional process called **Coulomb friction**. In a Coulomb friction idealization, the resulting damping force opposes the motion, but is independent of the relative velocity of the two surfaces.

### 3.3.3 Discrete Mass

**Mass** is the property of matter that causes it to resist acceleration (changes in velocity). You recognize mass when you push a heavy object across the floor; you have to exert a sufficiently large force to get it moving. Part of this applied force is used to overcome the friction between the object and the floor. The other part of the force is used to accelerate the mass from rest (zero velocity relative to the earth) to a velocity necessary to move it where you want it to go. A free-body diagram of this situation is shown in Figure 3.8.

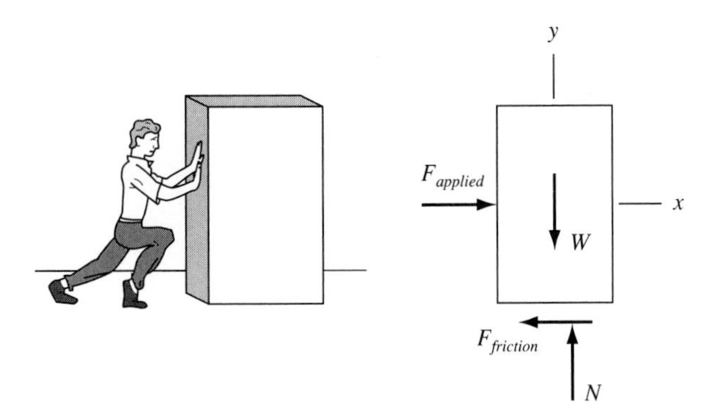


**Figure 3.8** Forces acting on mass.

The relationship between force and acceleration for a **point** or **lumped mass** is given by Newton's second law of motion:

$$\sum \text{forces} = m\ddot{x} \tag{3.17}$$

$$F_{applied} - F_{friction} = m\ddot{x} \tag{3.17a}$$

Note that forces are positive if they are in the direction of the positive reference axis, which is $x$ in this case. For mechanical systems with constant mass properties, mass is the constant of proportionality between force and acceleration and has units of force/acceleration or force/(length/time squared).

**Example 3.3 Acceleration of Mass**

In Figure 3.9, a cargo container is being loaded onto an airplane. The container has a mass of 95 kg. Find its acceleration if a force $F_p$ of 12 N is applied to push it. How far would the container move if the constant pushing force is applied for 5 seconds? It is reasonable to neglect the force of friction acting on the container, since the container is supported by rollers.

**Solution** Sketching the free-body diagram of the container and applying Newton's second law gives the equation of motion:

$$\sum F = F_p = m\ddot{x} \tag{3.18}$$

$$\ddot{x} = \frac{F_p}{m} = \frac{12\text{ N}}{95\text{ kg}} = 0.126\,\frac{\text{m}}{\text{s}^2} = 4.97\,\frac{\text{in}}{\text{s}^2} \tag{3.19}$$

$$\ddot{x} = \frac{d^2x}{dt^2} = \frac{F_p}{m} \tag{3.20}$$

Integrating Eq. (3.20) gives

$$\frac{dx}{dt} = \int \frac{F_p}{m}\,dt + C_1 = \frac{F_p}{m}\int dt + C_1 = \frac{F_p}{m}t + C_1 \tag{3.21}$$

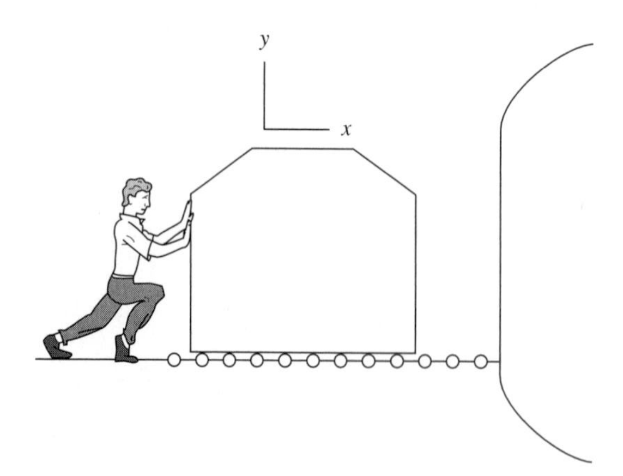


**Figure 3.9** Cargo container.

$$C_1 = 0, \text{ initial velocity} \tag{3.22}$$

$$x = \frac{F_p}{m}\int t\,dt + C_2 \tag{3.23}$$

$$C_2 = 0, \text{ initial displacement} \tag{3.24}$$

The resulting displacement is

$$x = \frac{F_p}{m}\frac{t^2}{2} = \left(0.126\,\frac{\text{m}}{\text{s}^2}\right)\frac{5^2\,\text{s}^2}{2} = 1.575\text{ m} = 5.167\text{ ft} \tag{3.25}$$

### 3.3.4 Modeling Translational Systems

The concepts just discussed can be used to develop mathematical models of systems composed of combinations of spring, dissipation, and discrete mass elements. Methods of constructing mathematical models of translational mechanical systems are illustrated in the next four examples.

**Example 3.4 Spring and Damper**

A component of a photocopy machine is modeled as a viscous damper connected to a linear spring, as shown in Figure 3.10. The mass of the part is judged to be negligible. The spring has a displacement $x_2(t)$ prescribed at its free end. Find the equations governing the displacement of the node at which the spring and damper join together and an expression for the force $p(t)$ that must be applied to cause the motion of the component.

**Solution** The connection nodes are numbered 1 and 2, and a convenient axis system is established. We then draw free-body diagrams of the damper, spring, and connection nodes. Element loads are defined for the spring and the damper in terms of the stiffness and damping constants and the node displacement and velocity variables, using Eqs.

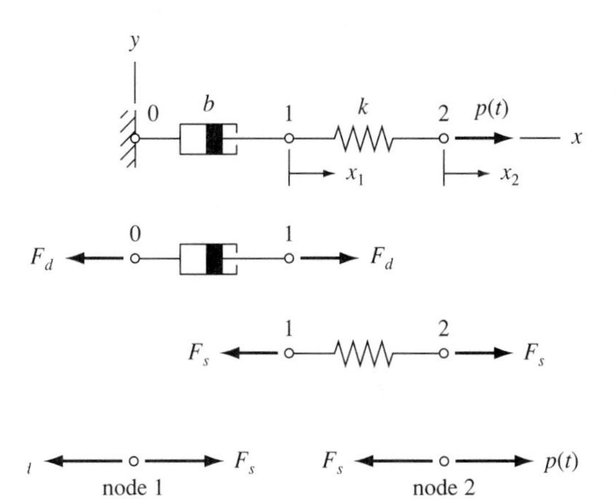


**Figure 3.10** Photocopy component.

(3.4) and (3.10). Since the mass at each node is zero, Newton's second law becomes a statement of the balance of the forces applied to each node.
For node 2:

$$\sum F_x = m\ddot{x} = 0\ddot{x} = 0 \tag{3.26}$$

$$-F_s + p(t) = 0 \tag{3.27}$$

$$F_s = p(t) \tag{3.28}$$

$$F_s = k(x_2 - x_1) = p(t) \tag{3.29}$$

For node 1:

$$\sum F_x = m\ddot{x} = 0\ddot{x} = 0 \tag{3.30}$$

$$-F_d + F_s = 0 \tag{3.31}$$

$$F_d = b\dot{x}_1, \quad F_s = k(x_2 - x_1) \tag{3.32}$$

$$b\dot{x}_1 = k(x_1 - x_2) \tag{3.33}$$

so that

$$b\dot{x}_1 + kx_1 = kx_2 \tag{3.34}$$

Thus,

$$\left[\frac{b}{k}D + 1\right]x_1 = x_2 \tag{3.35}$$

Since $x_2(t)$ is a known quantity, we can solve the last equation for $x_1(t)$. The force $p(t)$ can then be found from the spring stiffness relationship, Eq. (3.29).

As a second example, we consider a common device that may be idealized as a combination of spring and mass elements.

**Example 3.5 Spring Scale**

A schematic of a simple spring scale used to measure items that are sold by weight is shown in Figure 3.11. The weighing pan has a mass of $m$. Develop an equation describing the motion of the pan.

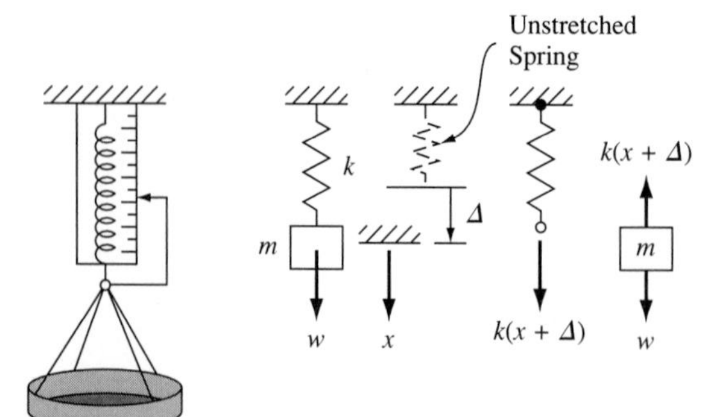


**Figure 3.11** Spring scale.

**Solution** When the spring is unstretched, it exerts no force on the weighing pan. When the pan is installed, its weight stretches the spring an amount $\Delta$ until the spring exerts a force on the pan that is equal to the weight of the pan. That is,

$$k\Delta = W = mg \tag{3.36}$$

The quantity $\Delta$ determines the static equilibrium position of the system. If we use $x$ to measure the displacement of the pan from this *static equilibrium position*, the forces acting on the pan are its weight $W$ and the spring force $F_s = k(x + \Delta)$.

Applying Newton's second law to the mass, we obtain

$$\sum F_x = m\ddot{x} \tag{3.37}$$

or

$$W - F_s = m\ddot{x} \tag{3.38}$$

Thus, we have

$$W - k(x + \Delta) = m\ddot{x} \tag{3.39}$$

which reduces to

$$m\ddot{x} + kx = W - k\Delta = 0 \tag{3.40}$$

Hence, if the motion of a linear mechanical spring-mass-damper system acted on by gravity is measured from its equilibrium position, the equation of motion takes the simple form of Eq. (3.40), in which the weight forces are canceled by the initial spring forces due to the static deflection. Any subsequent motions are calculated with respect to the equilibrium position.

We have neglected energy dissipation in this model, and the equation we derived cannot predict any decay of the motion of the pan once it is disturbed from its equilibrium position. To simulate the decay, the model would have to be modified to include energy dissipation of some kind.

**Example 3.6 Machine Part**

Figure 3.12 shows a machine part that slides along a smooth lubricated surface that itself is attached to a fixed base by a spring. In its operation, the part is subjected to a force that varies harmonically with time at a frequency of $f$ Hz $= \omega$ rad/s. Derive the governing equation of motion for the mass $m_2$.

**Solution** In Figure 3.13, the reference axis $x$ is taken positive to the right, and the masses are labeled $m_1$ and $m_2$. The oil film is modeled as a viscous damping element. Free-body diagrams of the masses, spring, and damper are shown in Figure 3.14, with

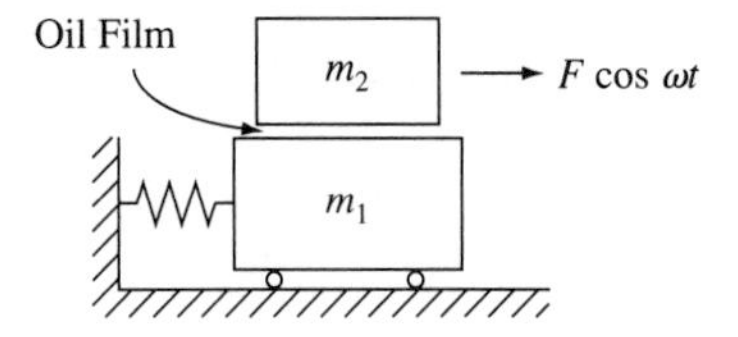


**Figure 3.12** Machine part schematic.

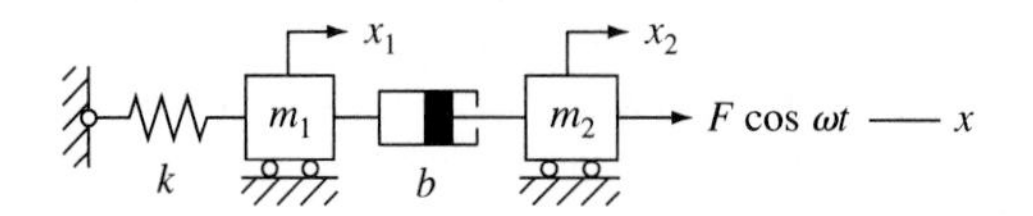


**Figure 3.13** Machine part model.

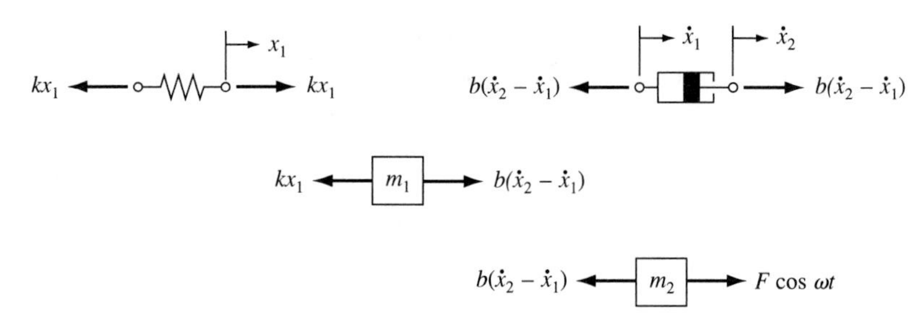


**Figure 3.14** Free-body diagrams.

forces that are consistent with the positive displacements shown in the schematic of the system.

Newton's second law is now applied to each of the masses:

Node 1:

$$\sum F = m_1\ddot{x}_1 \tag{3.41}$$

$$-kx_1 + b(\dot{x}_2 - \dot{x}_1) = m_1\ddot{x}_1 \tag{3.42}$$

Node 2:

$$\sum F = m_2\ddot{x}_2 \tag{3.43}$$

$$-b(\dot{x}_2 - \dot{x}_1) + F\cos\omega t = m_2\ddot{x}_2 \tag{3.44}$$

Thus, the two second-order equations of motion become

$$m_1\ddot{x}_1 + b\dot{x}_1 - b\dot{x}_2 + kx_1 = 0 \tag{3.45}$$

$$m_2\ddot{x}_2 + b\dot{x}_2 - b\dot{x}_1 = F\cos\omega t \tag{3.46}$$

We may also write these equations in the following form:

$$[m_1D^2 + bD + k]x_1 - bDx_2 = 0 \tag{3.47}$$

$$[m_2D^2 + bD]x_2 - bDx_1 = F\cos\omega t \tag{3.48}$$

We solve for $x_1$ from the first equation and substitute into the second equation:

$$x_1 = \frac{bDx_2}{[m_1D^2 + bD + k]} \tag{3.49}$$

$$[m_2D^2 + bD]x_2 - \frac{b^2D^2x_2}{[m_1D^2 + bD + k]} = F\cos\omega t \tag{3.50}$$

After simplification, we obtain the following third-order equation:

$$[m_1m_2D^4 + (m_1 + m_2)bD^3 + m_2kD^2 + bkD]x_2 = [m_1D^2 + bD + k]F\cos\omega t \quad (3.51)$$

We next consider an example in which the input to the system is a displacement instead of a force.

**Example 3.7 Vehicle Suspension System**

A simplified translational model of an automotive suspension system is constructed by considering only the translational motion of one wheel of the vehicle (quarter-car model). The model is shown in Figure 3.15. The stiffness of the tire is modeled by a linear spring, the tire, axle, and moving parts by a mass $m_1$, the suspension system by a spring and viscous damper (shock absorber), and the supported vehicle components by a mass $m_2$.

**Solution** We follow the same basic steps as used in previous examples. A coordinate system for translation is selected, with $y$ taken positive upwards. If the wheels stay in contact with the ground, the lower end of the tire spring follows the surface of the roadway, as described by $y_0(t)$. The springs, damper, and masses are separated from their attached components so that the forces acting on them can be identified in terms of the parameters of the problem. Figures 3.16 and 3.17 illustrate the appropriate free-body diagrams. We next use Newton's second law to write the equations of motion for each mass and imply equilibrium of forces at points where elements join.

We measure $y_1$ and $y_2$ from the at-rest equilibrium position of the springs, acted on by the weight of the vehicle's components. This means that the forces due to the weights of the components are balanced by the preload in the springs, and the two will cancel each other in the equations of motion. Thus, to simplify the process, we write the equations of motion without including the weight terms and understand that $y_1$ and $y_2$ are measured from the equilibrium position. (See Example 3.5.)

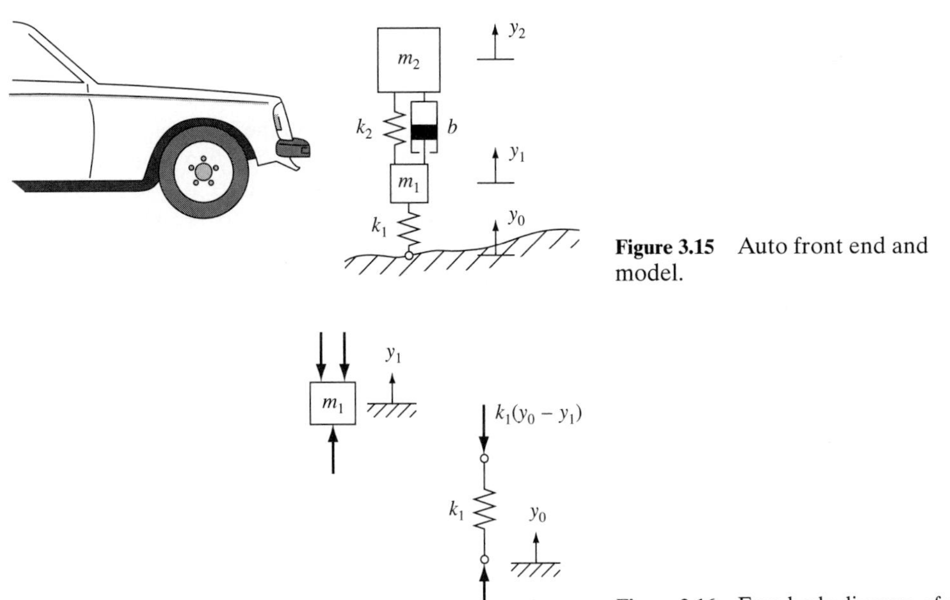


**Figure 3.15** Auto front end and model.

**Figure 3.16** Free-body diagram of $m_1$.

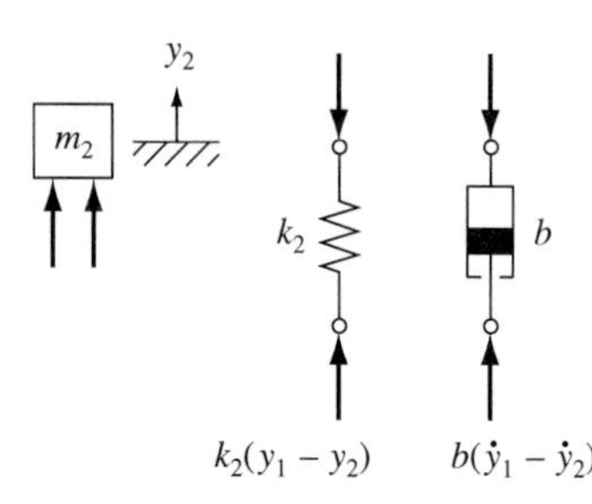


**Figure 3.17** Free-body diagram of $m_2$

Recognizing that the springs will probably be in compression most of the time, we show the spring forces compressive and write the spring element force equations accordingly. In the development of an accurate set of governing equations, it doesn't matter whether elements are shown in tension or compression, as long as the element force equations are consistent with the convention adopted. Thus, for mass $m_1$,

$$\sum F_y = m_1\ddot{y}_1 \tag{3.52}$$

$$k_1(y_0 - y_1) - k_2(y_1 - y_2) - b(\dot{y}_1 - \dot{y}_2) = m_1\ddot{y}_1 \tag{3.53}$$

And mass $m_2$,

$$\sum F_y = m_2\ddot{y}_2 \tag{3.54}$$

$$k_2(y_1 - y_2) + b(\dot{y}_1 - \dot{y}_2) = m_2\ddot{y}_2 \tag{3.55}$$

The resulting equations describing the motion of the two masses are

$$m_1\ddot{y}_1 + b(\dot{y}_1 - \dot{y}_2) + k_2(y_1 - y_2) + k_1y_1 = k_1y_0(t) \tag{3.56}$$

and

$$m_2\ddot{y}_2 - k_2(y_1 - y_2) - b(\dot{y}_1 - \dot{y}_2) = 0 \tag{3.57}$$

These equations represent the mathematical model of the suspension system, a fourth-order system consisting of two second-order equations. The two equations can be combined into one fourth-order equation of the classical form.

Alternatively, we can convert Eqs. (3.56) and (3.57) into a system of four first-order equations by defining a new set of variables such that the first derivative is the highest derivative in the set. This change of variables to produce the **state-space** form is shown next. (See also Appendix G.) We make the following definitions and calculate derivatives:

$$\text{Let}\quad \begin{aligned} x_1 &= y_1 \\ x_2 &= \dot{y}_1 \\ x_3 &= y_2 \\ x_4 &= \dot{y}_2 \end{aligned} \qquad \text{Then}\quad \begin{aligned} \dot{x}_1 &= \dot{y}_1 = x_2 \\ \dot{x}_2 &= \ddot{y}_1 \\ \dot{x}_3 &= \dot{y}_2 = x_4 \\ \dot{x}_4 &= \ddot{y}_2 \end{aligned} \tag{3.58}$$

After substitution, we obtain the following four state-space equations:

$$\dot{x}_1 = x_2 \tag{3.59a}$$

$$\dot{x}_2 = \frac{1}{m_1}[k_1y_0(t) - b(x_2 - x_4) - k_2(x_1 - x_3) - k_1x_1] \tag{3.59b}$$

$$\dot{x}_3 = x_4 \tag{3.59c}$$

$$\dot{x}_4 = \frac{1}{m_2}[b(x_2 - x_4) + k_2(x_1 - x_3)] \tag{3.59d}$$

It may be convenient to write these first-order equations in matrix form:

$$\dot{\mathbf{x}} = \mathbf{Ax} + \mathbf{Bu} \tag{3.60a}$$

$$\begin{bmatrix} \dot{x}_1 \\ \dot{x}_2 \\ \dot{x}_3 \\ \dot{x}_4 \end{bmatrix} = \begin{bmatrix} 0 & 1 & 0 & 0 \\ \dfrac{-k_1 - k_2}{m_1} & \dfrac{-b}{m_1} & \dfrac{k_2}{m_1} & \dfrac{b}{m_1} \\ 0 & 0 & 0 & 1 \\ \dfrac{k_2}{m_2} & \dfrac{b}{m_2} & \dfrac{-k_2}{m_2} & \dfrac{-b}{m_2} \end{bmatrix} \begin{bmatrix} x_1 \\ x_2 \\ x_3 \\ x_4 \end{bmatrix} + \begin{bmatrix} 0 \\ \dfrac{k_1}{m_1} \\ 0 \\ 0 \end{bmatrix} y_0(t) \tag{3.60b}$$

## 3.4 ROTATIONAL SYSTEMS

### 3.4.1 Rotational Springs

Rotational systems are composed of elements that are constrained to rotate about an axis. A pencil sharpener and an automotive drive train are examples. Energy can be stored in rotational systems by springs that experience an angular displacement when subjected to a torque or moment. A shaft can serve as a rotational spring when one end is twisted relative to the other. The torsion bar suspension system in automotive applications and a clock spring are good examples. Figure 3.18 shows some representative rotational system springs.

The load-deflection relationship of a rotational spring can be linear or nonlinear. For a **linear torsional spring,** the torque and deflection angles of twist are linearly related:

$$T = k\theta \tag{3.61}$$

Here $k$ is the **torsional spring constant** and has units of torque/angle = (force × length)/radian. A shaft employed as a torsional spring is shown in Figure 3.19. Sometimes a doubleheaded arrow may be used to represent an angle or a torque. If so, its rotational sense is interpreted according to the right-hand rule.

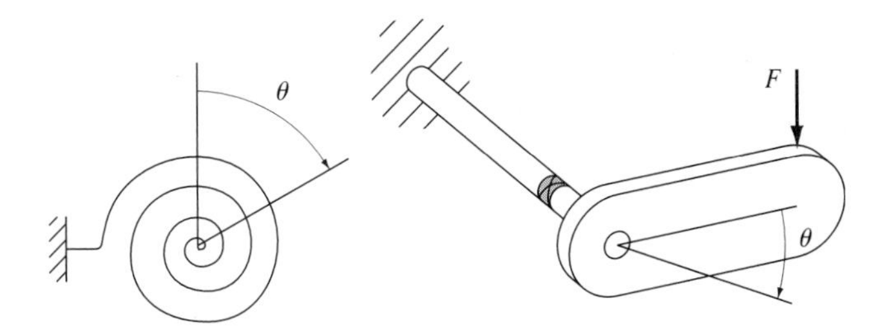


**Figure 3.18** Rotational springs.

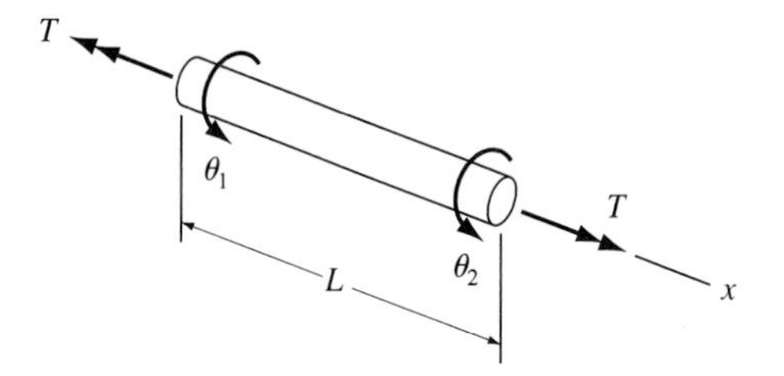


**Figure 3.19** Shaft in torsion.

**Example 3.8 Shaft Stiffness**

Find the torsional spring constant for a steel shaft that is 7.5 inches long and has a circular cross section that is 3/8 inch in diameter. The shear modulus of steel is taken to be $12.5 \times 10^6$ lbf/in$^2$. (See Appendix B.)

**Solution** The torque $T$ and angular twist $\theta$ for a uniform shaft are related (Ref. 3.4) by

$$\theta_2 - \theta_1 = \frac{TL}{GI_{xx}} \tag{3.62}$$

where $G$ is the shear modulus of the material, $I_{xx}$ is the geometric torsional constant for the cross section of the shaft, and $L$ is the length of the shaft. Thus,

$$k = \frac{GI_{xx}}{L} \text{ and } I_{xx} = \frac{\pi d^4}{32} \text{ for a circular cross section} \tag{3.63}$$

Substituting the given values yields

$$k = \left(12.5 \times 10^6 \frac{\text{lbf}}{\text{in}^2}\right)\left(\frac{\pi(3/8)^4}{32}\text{in}^4\right)\frac{1}{7.5 \text{ in}} = 3236 \text{ in lbf/rad} \tag{3.64}$$

(Remember that $I_{xx}$ is equal to the polar moment of inertia of the cross section *only* if the cross section is circular; see Appendix B for information on the other cross-sectional shapes.)

### 3.4.2 Rotational Dampers

Friction caused by a thin film of lubricant between two rotating surfaces can produce a resisting torque that is directly proportional to the relative angular velocity between the surfaces. A schematic of a rotational damper is shown in Figure 3.20.

The torque-velocity relationship for a **linear damper** is

$$T = b(\dot{\theta}_2 - \dot{\theta}_1) \tag{3.65}$$

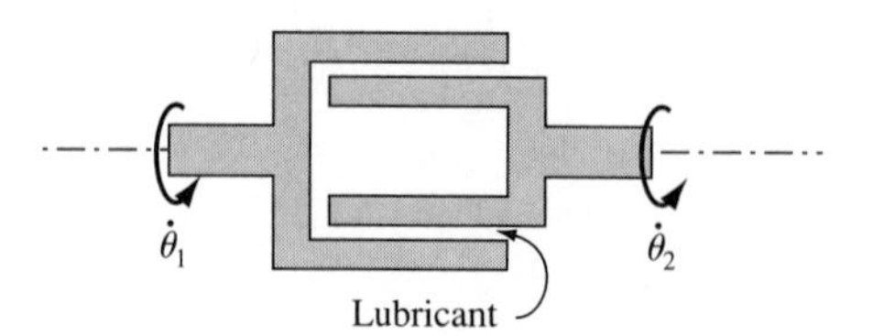


**Figure 3.20** Rotational damper.

where $b$ is the **torsional damping coefficient,** with units of torque/(radian/s), and $\dot{\theta}_1$ and $\dot{\theta}_2$ are the angular velocities of the endpoints of the damper.

### 3.4.3 Discrete Inertias

Rotational inertia is the resistance of an object to angular acceleration and is dependent upon the mass of the object as well as the geometry of the object, i.e., the manner in which the mass is distributed with respect to the axis of rotation. Thus, a wheel with a large rim has a greater inertia than a uniform wheel of the same mass. When applied to all of the particles that make up a rigid body in rotation about a fixed axis, Newton's second law for a particle requires that the sum of the moments about the axis of rotation of the body is equal to the product of the **mass moment of inertia** $J_x$ of the body about that axis and the angular acceleration $\ddot{\theta}_x$ of the body about the axis. Mathematically,

$$\sum M_x = J_x \ddot{\theta}_x \tag{3.66}$$

Hence, discrete inertia has the units of moment/(angular acceleration) = (force $\times$ length)/(radians/second squared).

**Example 3.9 Disk Speed Control**

A uniform brass disk of 250 mm diameter $d$ and 125 mm thickness $h$ is supported on a shaft as shown in Figure 3.21. The disk is spinning at a constant angular rate of 42 rad/s when a constant braking torque of 1.2 N m is applied. What is the resulting angular deceleration? What is the angular speed of the disk if the torque is held constant for 7.5 seconds?

**Solution** A free-body diagram of the disk is shown in Figure 3.21. Positive angular rotations, velocities, and accelerations are taken about the $x$-axis according to the right-hand rule. For brass, $\rho = 8.5 \times 10^3$ kg/m$^3$. Thus,

$$m = \rho \frac{\pi d^2 h}{4} = \left(8.5 \times 10^3 \frac{\text{kg}}{\text{m}^3}\right) \frac{\pi (.25)^2 \text{m}^2}{4} (0.125 \text{ m}) = 52.16 \text{ kg} \tag{3.67}$$

$$J_x = \frac{md^2}{8} = \frac{52.16 \text{ kg } (.25)^2 \text{ m}^2}{8} = 0.4075 \text{ kg m}^2 \tag{3.68}$$

$$\sum M_x = J_x \ddot{\theta} \tag{3.69}$$

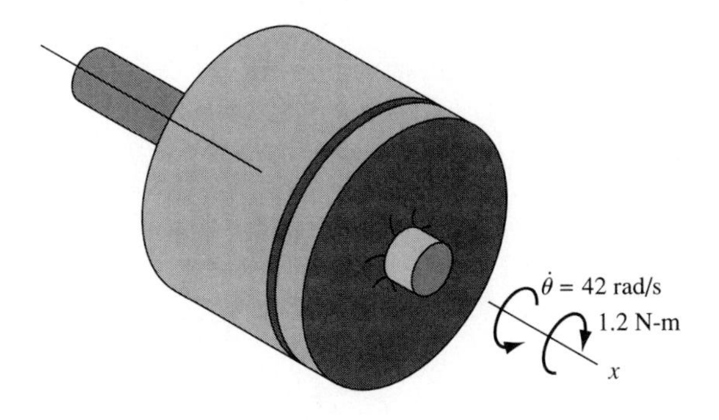


**Figure 3.21** Free-body diagram of disk.

$$\ddot{\theta} = \frac{M_x}{J_x} = \frac{-1.2 \text{ N m}}{0.4075 \text{ kg m}^2} = -2.945 \text{ rad}/s^2 \tag{3.70}$$

(Recall that N = kg m/s².)

Finally,

$$\int \ddot{\theta} dt = \dot{\theta}_2 - \dot{\theta}_1 = \frac{M_x}{J_x} \Delta t = -2.945(7.5) = -22.08 \text{ rad/s} \tag{3.71}$$

$$\dot{\theta}_2 = \dot{\theta}_1 - 22.08 = 42 - 22.08 - 19.92 \text{ rad/s} \tag{3.72}$$

$$19.92 \text{ rad/s} \frac{60 \text{ s/min}}{2\pi \text{rad/rev}} = 190.2 \text{ rpm} \tag{3.73}$$

### 3.4.4 Modeling Rotational Systems

The discrete components just described can be combined to provide the basis for modeling rotational systems. A careful consideration of the sense of the angular displacement and torque parameters is necessary in order to obtain error-free system equations. The procedures follow directly from our discussion of translational systems and are illustrated in the next two examples.

**Example 3.10 Engine and Propeller Model**

A simplified model of a turboprop aircraft engine and propeller is shown in Figure 3.22. The mass moment of inertia of the rotating parts of the engine is represented by $J_e$, the mass moment of inertia of the propeller by $J_p$. The driving torque applied to the engine is $T(t)$. The drive shaft has a small mass moment of inertia in comparison to that of the engine and propeller and is represented by an inertialess discrete torsional spring. The rotation of the propeller is opposed by aerodynamic drag torque, which is proportional to the square of the rotational speed of the propeller. Develop a mathematical model for this system, and write its equations of motion.

**Solution** A reference axis $x$ is established along the drive shaft as shown, and rotations $\theta_1$ and $\theta_2$ of inertias $J_e$ and $J_p$ are taken to be positive along this axis, according to the right-hand rule at the propeller end of the shaft. If it is assumed that $\theta_2 > \theta_1$, then

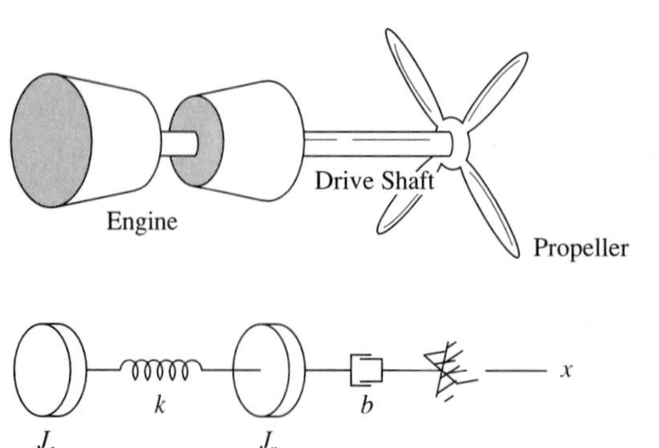


**Figure 3.22** Engine-and-propeller model.

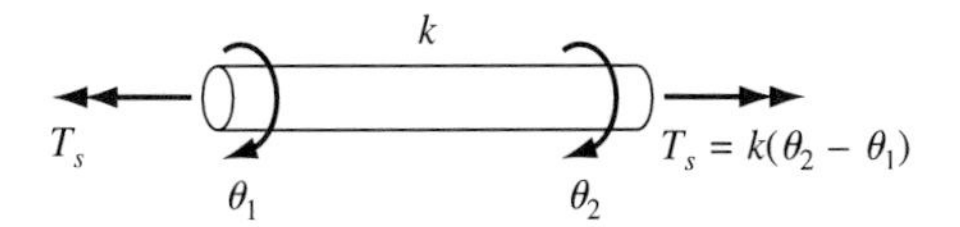


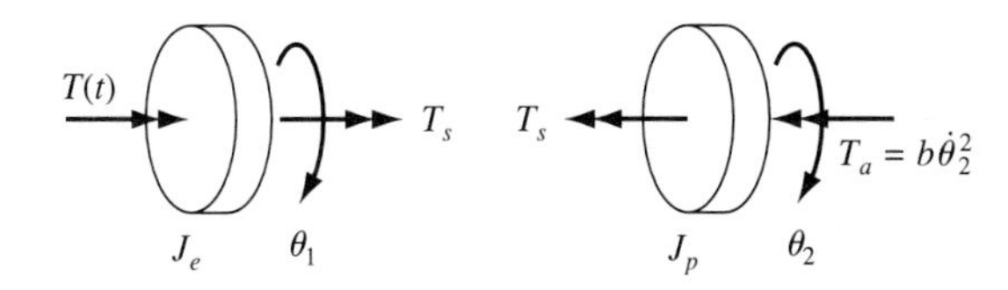


**Figure 3.23** Engine and propeller free-body diagrams.

the drive shaft torques must be in the directions shown in Figure 3.23. Of course, the equations can just as easily and correctly be derived by taking $\theta_2 < \theta_1$ and reversing the torque; just be consistent.

The equations governing the motion of the two inertias become the following:
At the $\theta_1$ node:

$$\sum M_x = J_e\ddot{\theta}_1 \tag{3.74}$$

$$T(t) + k(\theta_2 - \theta_1) = J_e\ddot{\theta}_1 \tag{3.75}$$

At the $\theta_2$ node:

$$\sum M_x = {}_{Jp}\ddot{\theta}_1 \tag{3.76}$$

$$-k(\theta_2 - \theta_1) - b\dot{\theta}_2^2 \text{ sign } (\dot{\theta}_2) = J_p\ddot{\theta}_2 \tag{3.77}$$

Equations (3.75) and (3.77) constitute the fourth-order model of the torsional system.

### Example 3.11 Geared System

The shaft of Example 3.8 is fixed at one end and has the larger gear of a pair of gears at the other end. The pitch radii of the steel gears are $r_1 = 8$ inches (64 teeth) and $r_2 = 4$ inches (32 teeth); the tooth face widths are 0.5 inch. Find the equation of motion of this system if the smaller gear has a torque $T \sin \omega t$ applied to it, where $\omega$ is the frequency, in rad/s, of the excitation torque.

**Solution** The shaft and gears are sketched in Figure 3.24, with angular variables assigned to each gear. The angular coordinates are taken in opposite senses to account for the opposite rotations of two mating gears. The contact force between the mating gear teeth is called $F$.

The equations of motion for the two gears are:
For gear 1:

$$\sum M_x = J_1\ddot{\theta}_1 \tag{3.78}$$

$$-T_s + Fr_1 = J_1\ddot{\theta}_1 \tag{3.79}$$

For gear 2:

$$\sum M_x = J_2\ddot{\theta}_2 \tag{3.80}$$

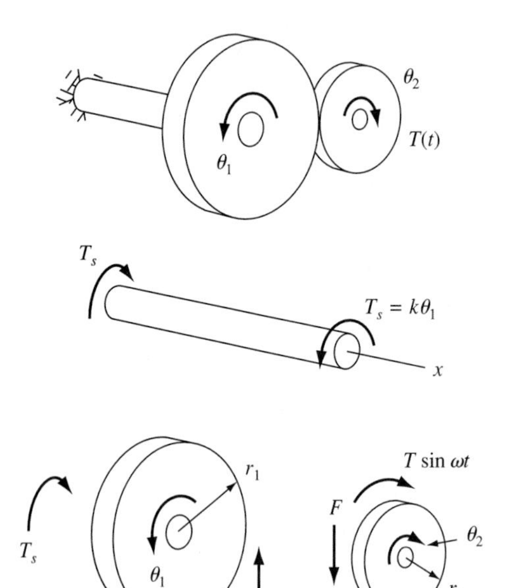


**Figure 3.24** Shaft and gear system.

$$T \sin \omega t - Fr_2 = J_2 \ddot{\theta}_2 \tag{3.81}$$

Solving for $F$ from the last equation and substituting into the equation for gear 1, we obtain

$$F = -\frac{J_2 \ddot{\theta}_2 - T \sin \omega t}{r_2} \tag{3.82}$$

$$J_1 \ddot{\theta}_1 + k\theta_1 = Fr_1 = -\frac{r_1}{r_2}(J_2 \ddot{\theta}_2 - T \sin \omega t) \tag{3.83}$$

$$J_1 \ddot{\theta}_1 + J_2 \ddot{\theta}_2 \frac{r_1}{r_2} + k\theta_1 = \frac{r_1}{r_2}(T \sin \omega t) \tag{3.84}$$

This problem requires only one variable to define the position of the gears. As the two gears turn in contact with each other, the arc lengths they traverse are equal. We have the following constraint equation to consider:

$$r_1 \theta_1 = r_2 \theta_2 \tag{3.85}$$

Thus, there is only one independent coordinate for the problem. We solve for $\theta_2$ and differentiate the result to get $\ddot{\theta}_2$:

$$\theta_2 = \frac{r_1}{r_2}\theta_1 \quad \text{and} \quad \ddot{\theta}_2 = \frac{r_1}{r_2}\ddot{\theta}_1 \tag{3.86}$$

We substitute this result into Eq. (3.84) and obtain an equation in $\theta_1$ only:

$$J_1 \ddot{\theta}_1 + J_2 \ddot{\theta}_1 \left(\frac{r_1}{r_2}\right)^2 + k\theta_1 = \frac{r_1}{r_2}(T \sin \omega t) \tag{3.87}$$

or

$$\left[J_1 + J_2\left(\frac{r_1}{r_2}\right)^2\right]\ddot{\theta}_1 + k\theta_1 = \frac{r_1}{r_2}(T\sin\omega t) \tag{3.88a}$$

If we write this result in terms of the number of teeth on the gears, we get

$$\left[J_1 + J_2\left(\frac{N_1}{N_2}\right)^2\right]\ddot{\theta}_1 + k\theta_1 = \frac{N_1}{N_2}(T\sin\omega t) \tag{3.88b}$$

## 3.5 SYSTEMS OF COMBINED TRANSLATIONAL AND ROTATIONAL ELEMENTS

Mechanical systems are very often composed of combinations of translational and rotational elements rather than elements of just one type. However, the fundamental method of developing correct models for these systems remains the same as that used in previous examples. You should:

**(1)** Establish an inertial coordinate system (one that is not attached to an accelerating object).

**(2)** Identify and isolate the discrete system elements (springs, dampers, masses, rotational inertias).

**(3)** Determine the minimum number of variables needed to uniquely define the configuration of the system. This can be done by subtracting the number of constraints from the number of equations of motion.

**(4)** Establish convenient axis systems and make appropriate free body sketches, showing all variables and all forces and torques acting on the elements.

**(5)** For stiffness and damping elements, write the equations that relate element loadings to element deformation variables.

**(6)** Apply Newton's second law of motion at all nodes of the model.

**(7)** Combine equations as necessary to isolate response variables of interest.

These steps are essential for accurate modeling and simulation. The degree to which you accomplish them correctly, solve the resulting equations, and intelligently interpret the solution will determine your success in modeling and simulation. The next four examples illustrate the ideas we have discussed.

**Example 3.12 Rolling Wheel**

A portion of a mechanical device may be idealized as a uniform, homogeneous wheel rolling without slipping on a horizontal surface, as shown in Figure 3.25. The center of the wheel is fastened to the frame of the device by a linear spring, and a force is applied at the top of the wheel. Find the equation of motion that governs the horizontal position of the center of the wheel.

**Solution** In Figure 3.26, the wheel is shown in a displaced position, and the forces acting on it are indicated in a free-body diagram. Since the wheel rolls without slipping, a

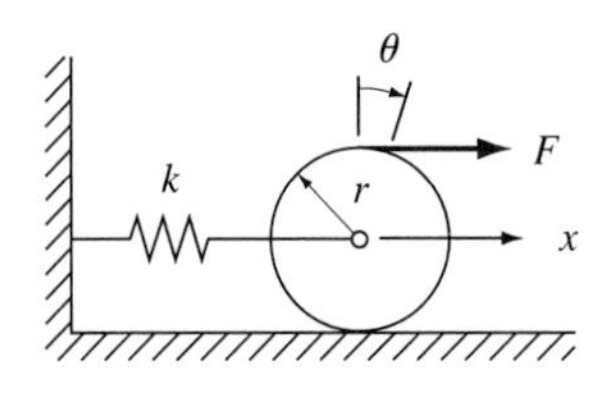


**Figure 3.25** Rolling wheel and spring.

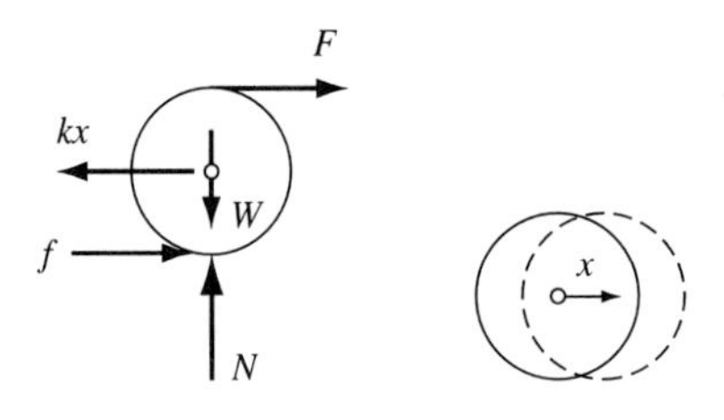


**Figure 3.26** Free-body diagram of wheel.

frictional force $f$ occurs at the point of contact between the wheel and the surface. Only one variable is required to uniquely define the location of the wheel; thus, the wheel displacement $x$ and the wheel angular rotation $\theta$ are not independent. In our solution, we will use $x$. The wheel is not constrained to rotate about a fixed axis, so we use the general form of Newton's second law that includes $x$- and $y$-translations and refer the moment and mass moment of inertia to an axis through the center of gravity of the object:

$$\sum F_x = m\ddot{x}_{cg} \tag{3.89}$$

$$\sum F_y = m\ddot{y}_{cg} \tag{3.90}$$

$$\sum M_{cg} = J_{cg}\ddot{\theta} \tag{3.91}$$

The first and third equations may be used to develop a description of the horizontal motion of the wheel. Because the wheel rolls on a horizontal surface, there is no acceleration in the $y$ direction, and the second equation expresses the equilibrium between the weight of the wheel and the normal force $N$ at the surface. The first and third equations give

$$\sum F_x = m\ddot{x}_{cg} \qquad F + f - kx = m\ddot{x} \tag{3.92}$$

$$\sum M_{cg} = J_{cg}\ddot{\theta} \qquad rF - rf = J_{cg}\ddot{\theta} \tag{3.93}$$

We now solve for $f$ from the moment equation:

$$f = F - J_{cg}\frac{\ddot{\theta}}{r} \tag{3.94}$$

Before substituting the right-hand side of Eq. (3.94) into the translation equation, we find $\theta$ in terms of $x$. Note that as the wheel rolls through an angle $\theta$, the arc length $r\theta$ along the rim of the wheel is equal to the distance $x$ traveled by the center. If it is helpful, think of the wheel on a stationary axis winding up a rope, the analog of the horizontal surface. The amount of rope wound onto the drum is $r\theta$. Thus,

$$x = r\theta \quad \text{and} \quad \ddot{x} = r\ddot{\theta}, \quad \text{so} \quad \ddot{\theta} = \frac{\ddot{x}}{r} \tag{3.95}$$

$$f = F - J_{cg}\frac{\ddot{x}}{r^2} \tag{3.96}$$

Substitution gives

$$F + \left(F - J_{cg}\frac{\ddot{x}}{r^2}\right) - kx = m\ddot{x} \tag{3.97}$$

$$m\ddot{x} + J_{cg}\frac{\ddot{x}}{r^2} + kx = 2F \tag{3.98}$$

$$\left(m + \frac{J_{cg}}{r^2}\right)\ddot{x} + kx = 2F \tag{3.99}$$

For a uniform circular disk, the mass moment of inertia with respect to an axis through its center of gravity is $mr^2/2$. Substituting this for $J_{cg}$ gives

$$\frac{3}{2}m\ddot{x} + kx = 2F \tag{3.100}$$

If a rigid body rotates about a fixed axis $O$, the moment equation in Newton's second law can be written for that axis. Both the sum of the moments and the mass moment of inertia are then written for that axis:

$$\sum M_O = J_O\ddot{\theta} \tag{3.101}$$

**Example 3.13 Trailing Arm Suspension System**

A simplified model of an automotive suspension system is shown in Figure 3.27. The wheel is supported relative to the chassis by a torsion bar spring and a shock absorber. The tire stiffness is to be represented in the model. Develop the equations governing the angular motion of the pivot arm.

**Solution** An $x$-, $y$-, $z$-axis system is established as shown in Figure 3.28. The torsion bar and tire stiffnesses are represented by linear springs, the shock absorber by a viscous damper. The torsion arm is shown in a displaced position. It is assumed that

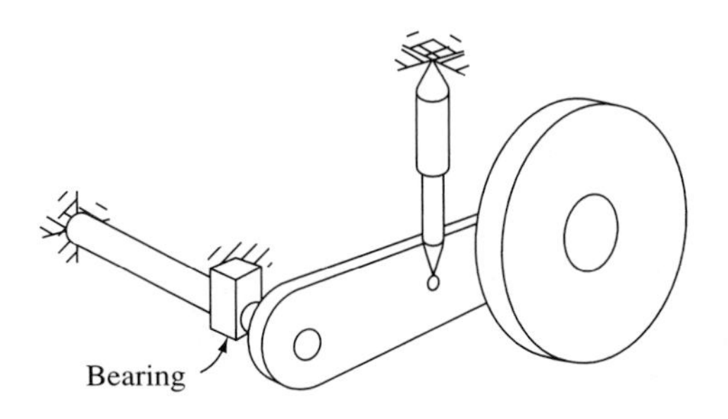


**Figure 3.27** Automotive suspension system.

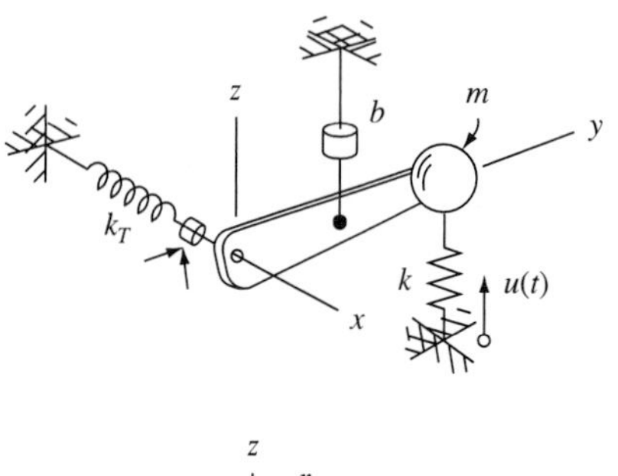


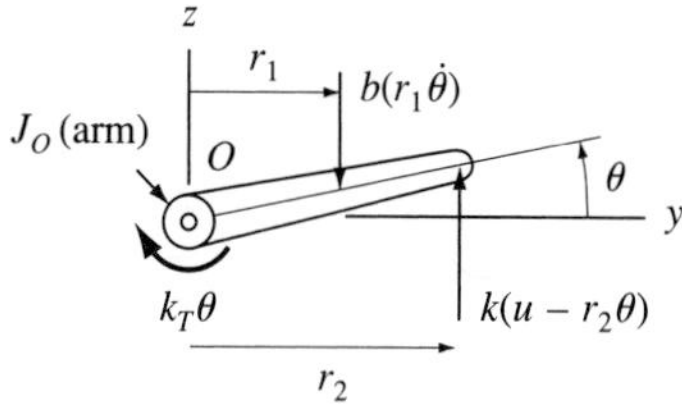


**Figure 3.28** Suspension schematic and free-body diagram.

the angular motion is small enough that forces from the damper and tire spring remain essentially vertical. That assumption will need to be verified when the equations are solved to find the deformation. The pivot arm is assumed to be rigid, and the quantity $u(t)$ represents the variation in the height of the road surface and acts as an input to the system as the car moves along the road.

The tire spring, torsion spring, and shock absorber forces are

$$k(u(t) - r_2\theta), \quad k_T\theta, \quad \text{and} \quad b(r_1\dot{\theta}) \tag{3.102}$$

We use the moment equation form of Newton's second law, written for rotation about a fixed axis $O$:

$$\sum M_O = J_O\ddot{\theta} \tag{3.103}$$

$$-k_T\theta - r_1b(r_1\dot{\theta}) + r_2k(u(t) - r_2\theta) = J_O\ddot{\theta} \tag{3.104}$$

Here the total rotational inertia about the torsion bar axis is the inertia of the arm plus the inertia of the mass. The parallel axis theorem is used to calculate the total inertia. (see Appendix B):

$$J_O = J_{O(arm)} + J_{mass} \tag{3.105}$$

$$J_O = J_{O\,(arm)} + mr_2^2 \tag{3.106}$$

$$J_O\ddot{\theta} + r_1^2b\dot{\theta} + (k_T + r_2^2k)\theta = r_2k\,u(t) \tag{3.107}$$

The quantity $u(t)$ is assumed to be a known function of time as the vehicle moves forward.

As an additional example of modeling mechanical systems we consider a hoisting system that contains a translational mass, a damper, and a spring, as well as a torsional spring and two rotational inertias.

**Example 3.14 Hoisting System**

Find the differential equations describing the motion of the hoisting system shown in Figure 3.29. A torque supplied by the motor at the right end of the shaft raises or lowers the mass $m$. The mass is guided so that it can move only in the vertical direction, and a viscous friction device between the container and its guides is used to damp out possible oscillations.

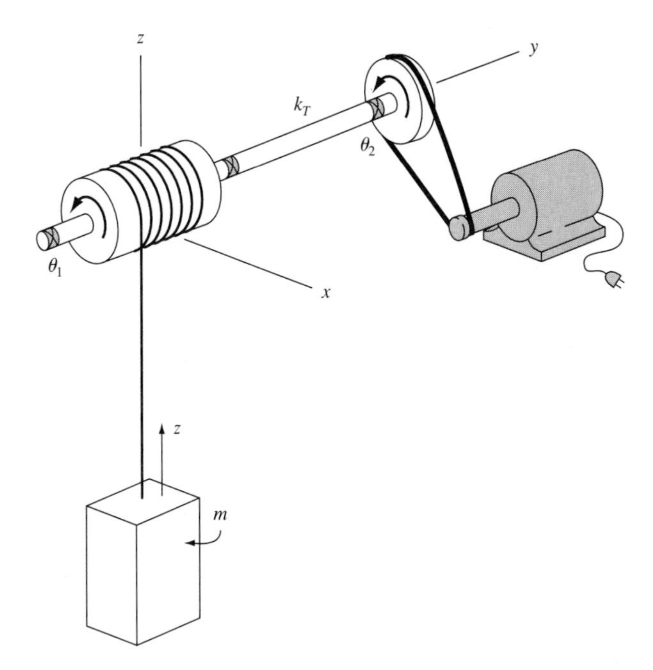


**Figure 3.29** Hoisting system.

**Solution** First, we sketch the two spring elements and the damper element and show the loadings corresponding to positive values of the displacement variables. Next, letting $k$ be the tensile stiffness of the hoisting cable, we write the element load-deformation equations in terms of these variables. Then we show the two inertias and the mass with compatible loadings acting on them. The element loadings are sketched in Figure 3.30. The element loading-deformation relations are

$$T_s = k_T(\theta_1 - \theta_2) \tag{3.108}$$

$$F = k(r_1\theta_1 - z) \qquad F_d = b\dot{z} \tag{3.109}$$

Applying Newton's second law to the two inertias and to the mass, we obtain the following equations:

At $\theta_1$ node:

$$\sum M_y = J_1\ddot{\theta}_1 \tag{3.110}$$

$$-T_s - r_1F = J_1\ddot{\theta}_1 \tag{3.111}$$

$$J_1\ddot{\theta}_1 = -k_T(\theta_1 - \theta_2) - r_1k(r_1\theta_1 - z) \tag{3.112}$$

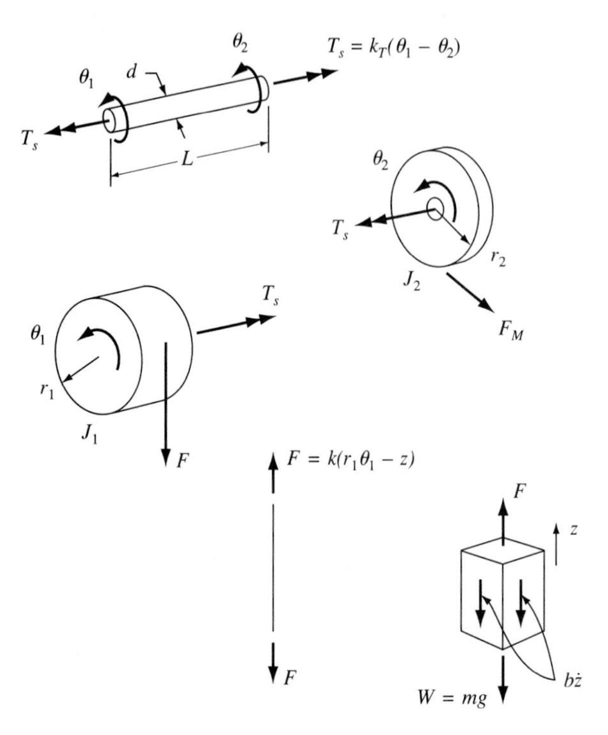


**Figure 3.30** Hoisting system free-body diagrams.

At $\theta_2$ node:

$$\sum M_y = J_2\ddot{\theta}_2 \tag{3.113}$$

$$T_s + r_2 F_M = J_2\ddot{\theta}_2 \tag{3.114}$$

$$J_2\ddot{\theta}_2 = k_T(\theta_1 - \theta_2) + r_2 F_M \tag{3.115}$$

Hoisted mass:

$$\sum F_z = m\ddot{z} \tag{3.116}$$

$$m\ddot{z} = -b\dot{z} + k(r_1\theta_1 - z) - W \tag{3.117}$$

The foregoing three second-order equations of motion can be combined into a single sixth-order equation or written as six first-order equations. Selecting the latter in order to prepare for digital simulation of the system response, we define the following state variables:

$$x_1 = \theta_1, \;\; x_2 = \dot{\theta}_1, \;\; x_3 = \theta_2, \;\; x_4 = \dot{\theta}_2, \;\; x_5 = z, \;\; x_6 = \dot{z} \tag{3.118}$$

With these definitions, the state space representation of the system is given by the following six equations:

$$\dot{x}_1 = x_2 \tag{3.119}$$

$$\dot{x}_2 = [k_T(x_3 - x_1) - r_1 k(r_1 x_1 - x_5)]/J_1 \tag{3.120}$$

$$\dot{x}_3 = x_4 \tag{3.121}$$

$$\dot{x}_4 = [r_2 F_M - k_T(x_3 - x_1)]/J_2 \tag{3.122}$$

$$\dot{x}_5 = x_6 \tag{3.123}$$

$$\dot{x}_6 = [-W - bx_6 + k(r_1 x_1 - x_5)]/m \tag{3.124}$$

**Example 3.15 Planetary Gear System**

The so-called planetary gear system shown in Figure 3.31 is a common means of providing a high gear-reduction ratio in a compact mechanical device. Several input and output combinations are possible, depending on which gear is fixed. Typical applications include helicopter and automotive transmission systems. The system is called a planetary gear system because the small gears move about the central gear in a way that is similar to the motion of planets about the sun in our solar system. In this exercise, we want to develop the equation of motion for one of the planet gears. For our purposes here, we remove the sun gear and just analyze the motion of the planet in mesh with the ring gear. The effect of the sun gear can be incorporated later.

**Solution** The gears can be modeled as wheels that roll on one another without slipping. The notation shown in Figure 3.32 is used to analyze the kinematics of motion of the planet gear as it meshes with the ring gear. Only one variable is required to define the position of the planet. We use its angle of rotation for this purpose.

We can duplicate the motion of the planet gear by means of the following sequence: To arrive at the displaced position, we divide the movement into two steps. First, we rotate the planet gear about the system's central axis $Z$ through an angle $\beta$, but

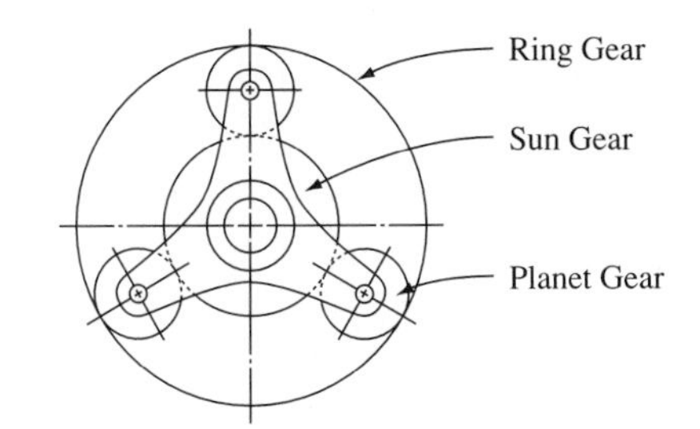


**Figure 3.31** Planetary gear system.

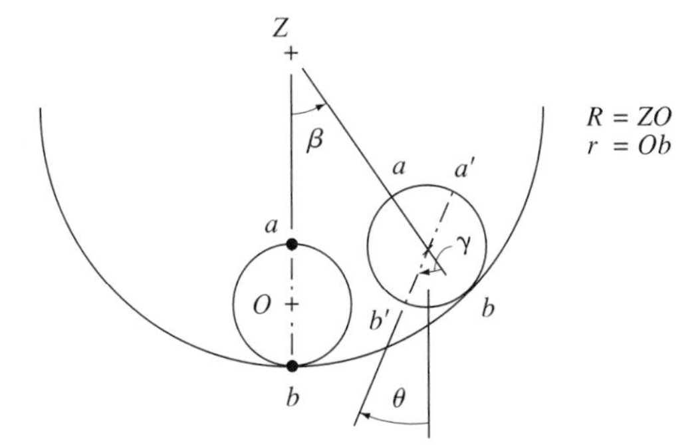


**Figure 3.32** Motion of planet gear.

do not let the planet roll on the ring gear. During this step, the line $a$–$b$ remains a radial line from the central axis. In the second step, we rotate the planet through an angle $\gamma$ about its own axis to bring it into its final displaced position. This brings $a$–$b$ to $a'$–$b'$, and we note that $\gamma$ is in a direction opposite to $\beta$.

Since $a$–$b$ was originally vertical, the net amount through which the planet rotates with respect to its original orientation is $\theta = \gamma - \beta$. This defines the angular motion of the planet gear. Now we need the relationship that must exist between $\theta$ and $\beta$ to satisfy the geometric conditions of rolling without slip. This is found by determining the length of the arc of contact between the planet gear and the ring gear. The arc length can be calculated either by using the angle $\beta$ and the radius $R + r$ or by using the angle $\gamma$ and the planet radius $r$. The two expressions must be the same. We thus have

$$(R + r)\beta = r\gamma \tag{3.125}$$

$$\gamma = \beta\left[1 + \frac{R}{r}\right] \tag{3.126}$$

This result is now used to find the relationship between $\theta$ and $\beta$:

$$\theta = \gamma - \beta = \beta\left[1 + \frac{R}{r}\right] - \beta \tag{3.127}$$

$$\theta = \frac{R}{r}\beta \tag{3.128}$$

This completes the kinematic analysis, which can be the most difficult part of a problem to visualize. (Note that it's not always helpful to think of an object rotating about a point, especially in cases of general motion. Just follow the orientation of a line on the object as it moves from one position to the next. The angular difference in the line between the two positions is the angular motion of the object.) Once the kinematics of the problem have been analyzed, it is easy to get the equations of motion. A free-body diagram of the planet gear is shown in Figure 3.33; here $F$ is the force applied by the carrier arm. The equations of motion are

$$\sum F_x = m\ddot{x}_{cg} \tag{3.129}$$

$$\sum F_y = m\ddot{y}_{cg} \tag{3.130}$$

$$\sum M_{cg} = J_{cg}\ddot{\theta} \tag{3.131}$$

The center of mass of the planet gear is at the geometric center of the gear. Thus, the acceleration of $O$ is the acceleration of its center of mass, and it moves in a

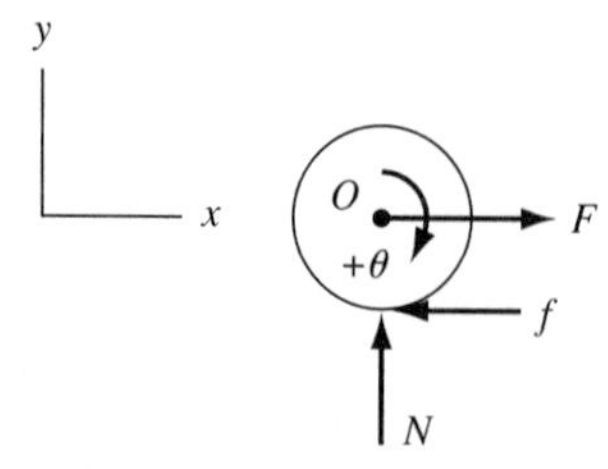


**Figure 3.33** Free-body diagram of planet gear.

circular path with respect to the central axis $Z$, as described by the following equations of motion:

$$\ddot{x}_{cg} = R\ddot{\beta} = R\frac{r}{R}\ddot{\theta} = r\ddot{\theta} \tag{3.132}$$

$$\ddot{y}_{cg} = R\dot{\beta}^2 = R(r\dot{\theta}/R)^2 = \frac{r^2}{R}\dot{\theta}^2 \tag{3.133}$$

The equations of motion become, for the $x$-direction,

$$\sum F_x = m\ddot{x}_{cg} \qquad F - f = mr\ddot{\theta} \tag{3.134}$$

and for the rotation,

$$\sum M_{cg} = J_{cg}\ddot{\theta} \qquad rf = J_{cg}\ddot{\theta} \tag{3.135}$$

Eliminating $f$, we have

$$(J_{cg}/r + mr)\ddot{\theta} = F \tag{3.136}$$

or

$$(J_{cg} + mr^2)\ddot{\theta} = rF \tag{3.137}$$

From this, and knowing the force $F$ applied to planet by the carrier arm we determine the motion of the planet. The normal force between the planet gear and the ring gear can be found from the remaining equation of motion:

$$\sum F_y = m\ddot{y}_{cg} \qquad N = m\frac{r^2}{R}\dot{\theta}^2 \tag{3.138}$$

The use of Newton's second law to develop the equations of motion for mechanical systems is the basic method of doing so, but is not always the most convenient one. In the next two sections, we consider d'Alembert's principle and Lagrange's method.

## 3.6 D'ALEMBERT'S PRINCIPLE

J. d'Alembert (French mathematician and philosopher, 1717–83) suggested that the mass-times-acceleration and inertia-times-angular-acceleration terms be taken to the left side of Newton's second law and considered to be force and moment quantities:

$$\sum F_x = m\ddot{x}_{cg} \qquad \sum F_x - m\ddot{x}_{cg} = 0 \tag{3.139}$$

$$\sum F_y = m\ddot{y}_{cg} \qquad \sum F_y - m\ddot{y}_{cg} = 0 \tag{3.140}$$

$$\sum M_{cg} = J_{cg}\ddot{\theta} \qquad \sum M_{cg} - J_{cg}\ddot{\theta} = 0 \tag{3.141}$$

The resulting equations then take the form of equilibrium statements about the balance of forces and torques. Also, in drawing free-body diagrams, we now include the mass/inertia × acceleration quantities in a direction opposite to the positive directions of acceleration. For example, the free-body diagram of the discrete mass of Figure 3.8 is modified as shown in Figure 3.34. No angular terms are considered, since the object slides without tipping.

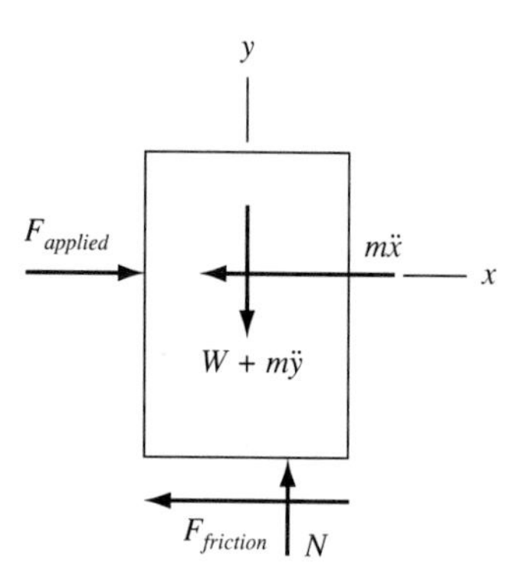


**Figure 3.34** Free-body diagram using d'Alembert's approach.

We now sum the vector components; i.e., we obtain the balance of forces in the $x$- and $y$-directions:

$$x\text{-direction:} \quad F_{applied} - F_{friction} - m\ddot{x} = 0 \tag{3.142}$$

$$y\text{-direction:} \quad N - W - m\ddot{y} = 0 \tag{3.143}$$

When the inertial term $m\ddot{x}$ is placed on the right-hand side of the equation, Eq. (3.142) is the same as Eq. (3.18) found previously in Section 3.3.3.

The value of the d'Alembert approach is more apparent in problems involving combined mechanical components and in cases where rotational motions occur. Consider, for example, the wheel of Example 3.12. The equations previously derived using Newton's second law for translation and rotation are as follows:

$$F + f - kx = m\ddot{x} \tag{3.144}$$

$$rF - rf = J_{cg}\ddot{\theta} \tag{3.145}$$

Now multiply the first equation by $r$, and add the two equations together. This gives

$$2rF - rkx = rm\ddot{x} + J_{cg}\ddot{\theta} \tag{3.146}$$

or

$$2rF - rkx - rm\ddot{x} - J_{cg}\ddot{\theta} = 0 \tag{3.147}$$

Next, apply the d'Alembert method to the problem; the modified free-body diagram of the wheel is sketched in Figure 3.35.

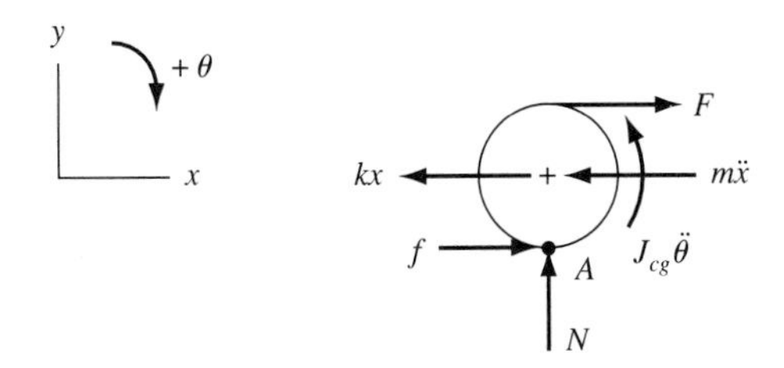


**Figure 3.35** Free-body diagram of wheel for d'Alembert method.

D'Alembert showed that the moment equation for a free body that includes the mass-times-acceleration terms as vector components gives the same result as applying Newton's second law. Furthermore, the advantage of the d'Alembert principle is that *moments can be taken about any point*: We are no longer restricted to the center of mass or a fixed axis of rotation as before. For the free-body diagram of Figure 3.35, we find the moment of all terms about the point of contact with the horizontal surface. We include the moment due to the **inertial force** $m\ddot{x}$ and the moment due to the **inertia torque** $J_{cg}\ddot{\theta}$. The result is

$$\sum M_A = 0 \tag{3.148}$$

$$2rF - rkx - rm\ddot{x} - J_{cg}\ddot{\theta} = 0 \tag{3.149}$$

Plainly, this equation is the same as the one we obtained previously, but now we did not have to combine two equations to get it.

## 3.7 LAGRANGE'S EQUATION

Both Newton's and d'Alembert's methods of developing equations of motion require that a mechanical system consisting of several components be taken apart and all forces acting on the components be identified on the free-body diagram of each element. However, there are cases in which we have no interest in the forces at interconnections, and it would be advantageous to be able to derive the equations of motion by regarding the system from an overall standpoint. J. L. Lagrange (French mathematician, 1736–1813) showed that a consideration of system energies allows such a derivation.

According to Lagrange's approach, we first select the minimum number of independent coordinates necessary to describe the position of the system at any instant. Call these coordinates $q_i$, and let $Q_i$ be the corresponding loading in each coordinate. We select reference positions for the coordinates and express the potential energy $U$ of the system in terms of its coordinates. When the system is displaced from its reference position, the potential energy includes energy stored in springs and the potential energy of weights raised above a reference height. We have

$$U = f_1(q_i) \tag{3.150}$$

where $f_1$ is a function that depends upon the problem under consideration.

Now we express the kinetic energy $T$ in terms of the system masses, mass moments of inertia, linear velocities, and angular velocities:

$$T = f_2(\dot{q}_i^2) \tag{3.151}$$

The loss due to viscous friction devices is written in an energy dissipation function that depends upon the velocities of the system and the damping constants:

$$R = f_3(\dot{q}_i^2) \tag{3.152}$$

Forces can be considered by including them in the force term $Q_i$ associated with each coordinate. Using the d'Alembert principle, we can show that the equations of motion for each coordinate can be found by merely differentiating the energy functions using the equation

$$\frac{d}{dt}\left(\frac{\partial T}{\partial \dot{q}_i}\right) - \frac{\partial T}{\partial q_i} + \frac{\partial R}{\partial \dot{q}_i} + \frac{\partial U}{\partial q_i} = Q_i \tag{3.153}$$

(Since the energies are functions of all the coordinates, partial derivatives must be used.)

Let us consider the spring-mass-damper system as a simple example of the use of Lagrange's equation. We need only one coordinate, $q_1 = x$, to describe the position of the mass shown in Figure 3.36.

The energy functions are

$$U = \frac{1}{2}kx^2 \tag{3.154}$$

$$T = \frac{1}{2}m\dot{x}^2 \tag{3.155}$$

$$R = \frac{1}{2}b\dot{x}^2 \tag{3.156}$$

Taking the indicated differentials gives the following terms:

$$\frac{\partial T}{\partial \dot{x}} = m\dot{x} \qquad \frac{d}{dt}\left(\frac{\partial T}{\partial \dot{x}}\right) = m\ddot{x} \tag{3.157}$$

$$\frac{\partial T}{\partial x} = 0 \tag{3.158}$$

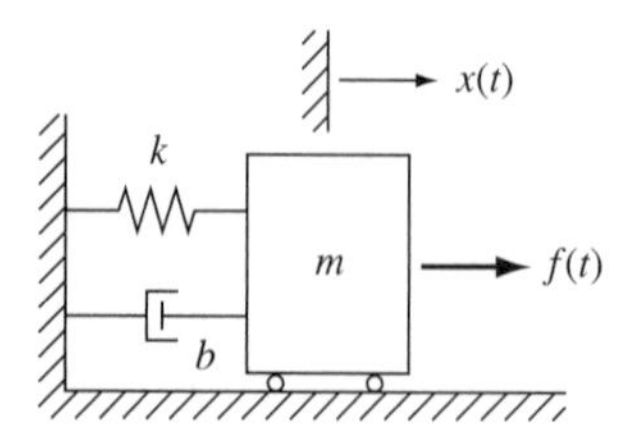


**Figure 3.36** Spring-mass-damper system.

$$\frac{\partial R}{\partial \dot{x}} = b\dot{x} \tag{3.159}$$

$$\frac{\partial U}{\partial x} = kx \tag{3.160}$$

The force in the coordinate $x$ is the excitation force:

$$Q = f(t) \tag{3.161}$$

Summing the terms on the right-hand side of Eqs. (3.157) through (3.160) gives the familiar spring-mass-damper equation of motion:

$$m\ddot{x} + b\dot{x} + kx = f(t) \tag{3.162}$$

This simple problem illustrates the important features of the technique, but Lagrange's method is more useful when employed in higher order systems and systems of interconnected bodies.

**Example 3.16 Vehicle Suspension System**

Consider the system shown in Figure 3.37. This could represent a simplified model of a vehicle and its suspension system. The system requires two degrees of freedom to describe its position: vertical translation and angular pitch. The motion is taken with respect to the equilibrium position of the vehicle, and thus the potential energy of the vehicle's weight is not required in the energy considerations. (See Example 3.5.)

**Solution** The kinetic energy is the sum of its translational and rotational components and is given by the equation

$$T = \frac{1}{2}m\dot{y}^2 + \frac{1}{2}J_{cg}\dot{\theta}^2 \tag{3.163}$$

(Since the energies are scalar quantities, their combined effects are included by simple addition.)

The potential energy stored in the springs is

$$U = \frac{1}{2}k_1\delta_1^2 + \frac{1}{2}k_2\delta_2^2 \tag{3.164}$$

where

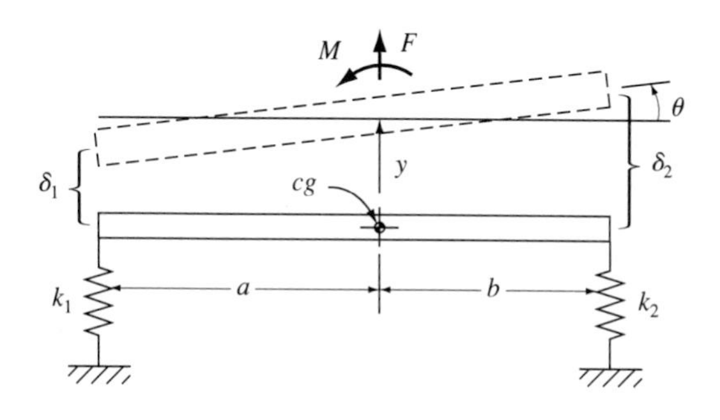


**Figure 3.37** Vehicle suspension system.

$$\delta_1 = y - a\theta \quad \text{and} \quad \delta_2 = y + b\theta \tag{3.165}$$

Taking the appropriate derivatives gives

$$\frac{\partial T}{\partial \dot{y}} = m\dot{y} \quad \text{and} \quad \frac{\partial U}{\partial \dot{y}} = k_1\delta_1\frac{\partial \delta_1}{\partial y} + k_2\delta_2\frac{\partial \delta_2}{\partial y} = k_1(y - a\theta) + k_2(y + b\theta) \tag{3.166}$$

$$\frac{\partial T}{\partial \dot{\theta}} = J_{cg}\dot{\theta} \quad \text{and} \quad \frac{\partial U}{\partial \theta} = k_1(y - a\theta)(-a) + k_2(y + b\theta)(b) \tag{3.167}$$

Combining these results gives the following equations of motion:

$$m\ddot{y} + (k_1 + k_2)y + (bk_2 - ak_1)\theta = F(t) \tag{3.168}$$

$$J_{cg}\ddot{\theta} + (k_1a^2 + k_2b^2)\theta + (bk_2 - ak_1)y = M(t) \tag{3.169}$$

As an example of a situation involving interconnected bodies, we next consider the development of the equations of a robot arm.

**Example 3.17 Robot Arm**

The top view of a selective compliant assembly robot arm (SCARA) is shown in Figure 3.38. The SCARA consists of two arms pinned together at a joint that has a vertical axis. The base joint of link 1 also rotates about a vertical axis. These devices are frequently used in "pick and place" operations, such as inserting integrated circuit chips in printed circuit boards. We wish to derive the equations of motion, considering joint actuation torques and friction effects.

**Solution** In the chip insertion process, the mass of the chip is insignificant in comparison with the mass of the robot arms. For such an application, we can ignore the contribution of the chip to the inertia of the system, and for simplicity, we will do so in this example. First, we find the kinetic energy expressions. The two angles shown are convenient and sufficient to define the motion. The angle at link 2 is measured relative to link 1.

The kinetic energy of link 1 can be written in terms of the link's mass moment of inertia about the fixed axis of rotation $O$:

$$T_1 = \frac{1}{2}J_O\dot{\theta}_1^2 \tag{3.170}$$

Link 2 has general motion, so we must include both translational and rotational terms. Since its total angular velocity is the sum of the rates of the two angles shown, we get

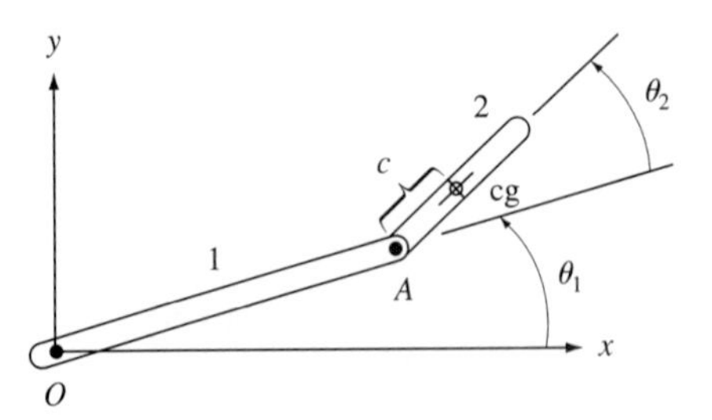


**Figure 3.38** SCARA.

$$T_2 = \frac{1}{2} m_2 v_{cg}^2 + \frac{1}{2} J_{2cg}(\dot{\theta}_1 + \dot{\theta}_2)^2 \tag{3.171}$$

The velocity of the center of mass of link 2 can be found using relative velocity expressions:

$$\vec{v}_{cg} = \vec{v}_A + \vec{v}_{cg/A} \tag{3.172}$$

Resolving the right-hand terms into components in the $x$- and $y$-directions gives

$$v_{Ax} = -L_1\dot{\theta}_1 \sin\theta_1, \qquad v_{Ay} = L_1\dot{\theta}_1 \cos\theta_1 \tag{3.173}$$

and

$$v_{(cg/A)x} = -c(\dot{\theta}_1 + \dot{\theta}_2)\sin(\theta_1 + \theta_2), \qquad v_{(cg/A)y} = c(\dot{\theta}_1 + \dot{\theta}_2)\cos(\theta_1 + \theta_2) \tag{3.174}$$

Now since $v^2 = v_x^2 + v_y^2$, we get

$$T_2 = \frac{1}{2} J_{2cg}(\dot{\theta}_1 + \dot{\theta}_2)^2 + \frac{1}{2} m_2[L_1^2\dot{\theta}_1^2 + 2L_1c\dot{\theta}_1(\dot{\theta}_1 + \dot{\theta}_2)\cos\theta_2 + c^2(\dot{\theta}_1 + \dot{\theta}_2)^2] \tag{3.175}$$

The total kinetic energy is the sum of the two scalar kinetic energy expressions:

$$T = T_1 + T_2 \tag{3.176}$$

If losses occur because of viscous friction at the joints, the energy dissipation function, which provides torques proportional to the relative velocities at the joints, is

$$R = \frac{1}{2} b_1\dot{\theta}_1^2 + \frac{1}{2} b_2\dot{\theta}_2^2 \tag{3.177}$$

In addition, each joint is driven by an actuator, which provides a joint excitation torque. These comprise the loadings in the two defined coordinates.

The application of Lagrange's equation to these energy terms is sufficient to generate the appropriate equations of motion without "disassembling" the device to determine the internal loadings.

The preceding development is brief, and while a number of details were left out, the salient features have been presented. Reference 3.2 gives additional information on the effective use of Lagrange's method for the development of equations of motion for mechanical systems.

## 3.8 THREE-DIMENSIONAL MOTION

All of the translation and rotation problems discussed thus far could be considered from a two-dimensional perspective. A great many practical engineering systems fall in this category, and their design and analysis are simplified thereby.

Many other systems, however, must be treated within a three-dimensional framework. A typical case is the rigid body/flexible body motions of an aircraft maneuvering in the atmosphere. The attendant propulsive, weight, lift, and drag forces cause motions that can become quite complex and must be described by a full three-dimensional analysis. The motion of a three-axis gyroscope gimbal system is another example of a common mechanical system that must be analyzed using a three-dimensional model. The references at the end of the chapter give additional information on the treatment of problems of this type.

## 3.9 SUMMARY

In this chapter, we have considered the significant aspects of modeling translational, rotational, and combined translational/rotational mechanical systems. Newton's second law, d'Alembert's principle, and Lagrange's energy method were introduced, together with examples illustrating their application to typical situations involving two-dimensional motion. In the next chapter, we consider the modeling of electrical system components.

## REFERENCES


3.1 Cochin, Ira, and Plass, Jr. Harold J., *Analysis and Design of Dynamic Systems,* 2d ed. Harper Collins Publishers, New York, 1990.

3.2 Dimarogonas, Andrew D., and Sam Haddad. *Vibration for Engineers.* Prentice Hall, Inc, Englewood Cliffs, N.J., 1992.

3.3 Hibbeler, R. C. *Engineering Mechanics, Dynamics,* 5th ed. Macmillan Publishing Co., Inc., New York, 1989.

3.4 Hibbeler, R. C. *Mechanics of Materials,* 2d ed. Macmillan Publishing Co., Inc., New York, 1994.

3.5 Ogata, Katsuhiko, *System Dynamics,* 2d ed. Prentice Hall, Inc, Englewood Cliffs, N.J., 1992.

3.6 Shearer, J. Lowen, and Kulakowski, Bohdan T. *Dynamic Modeling and Control of Engineering Systems*. Macmillan Publishing Co., New York, 1990.


## NOMENCLATURE

$a$ acceleration

$\mathbf{A}$ matrix

$b$ dissipation constant of a viscous damper

$\mathbf{B}$ matrix

$cg$ center of gravity

$d$ diameter

$f$ friction force
$F$ force
$G$ shear modulus
$h$ function, thickness
$I_{xx}$ torsional stiffness constant
$J$ mass moment of inertia
$L$ length
$k$ spring constant
$k_T$ spring constant
$m$ mass
$M$ moment
$N$ number of gear teeth, normal force
$q_i$ coordinate
$r$ radius
$R$ radius, viscous friction potential
$t$ time
$T$ torque, kinetic energy
$U$ potential energy
$v$ velocity
$W$ weight
$x$ displacement
$y$ displacement
$z$ displacement
$\beta$ angle
$\gamma$ angle
$\delta$ displacement
$\Delta$ change in displacement
$\theta$ angle of twist
$\omega$ excitation frequency

## PROBLEMS

**3.1** *Perform* the spring stiffness experiment described in Section 3.3.1 and shown in Figure 3.2.

**3.2** *Perform* the spring stiffness experiment described in Figure 3.2, but for the spring, use a flexible slat of wood, plastic, or metal clamped or held to a tabletop.

**3.3** *Derive* the equations of motion for the spring-damper system shown in Figure P3.3. The input $\delta(t)$ is a known function of time.

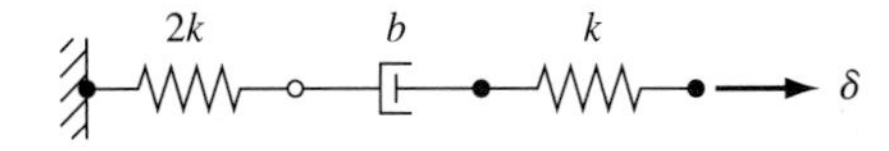


**Figure P3.3** Spring-damper system.

**3.4** A 10,000 lbf boxcar with a velocity of 1 ft/sec approaches an arresting system composed of a linear spring and viscous damper. (See Figure P3.4.) *Derive* the equations of motion, describing what happens after the car contacts the arresting system. Use symbols to represent all variables. *Describe* how to use this model to select values for the spring and damper constants so that the car comes to rest after traveling less than 3 ft.

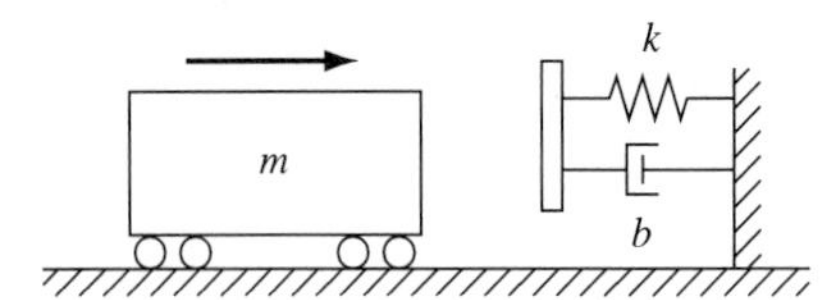


**Figure P3.4** Boxcar.

**3.5** *Complete* the work described in Problem 3.4; that is, *find* spring and damper constants that will meet the requirements outlined.

**3.6** An item in a manufacturing operation is projected onto a flat surface with an initial velocity of $V_1$, as shown in Figure P3.6. The item slides to a point where a robot picks it up and transfers it to another machine for further handling. For successful operation of this system, the package must stop in the target zone within one second after its release, or the robot will not be able to perform its part of the process.

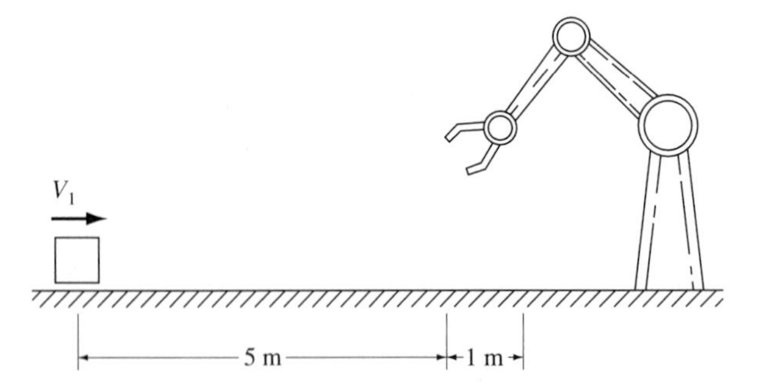


**Figure P3.6** Manufacturing setup.

*Develop* a model describing the motion of the object along the surface. Establish an axis system. Use appropriate symbols to represent the relevant quantities, etc. *Determine* a suitable range for the coefficient of friction between the item and the flat surface if the initial velocity $V_1$ is 6 m/s and the pickup target zone dimensions are as shown in the figure.

**3.7** Consider an arrangement which is the same as that described in Problem 3.6, except that the item starts with a zero initial velocity and is delivered to the flat surface down a chute that is 3 m long. (See Figure P3.7.) *Select* a chute angle and coefficient of friction which will insure that the item has a velocity of about 6 m/s when it reaches the end of the chute.

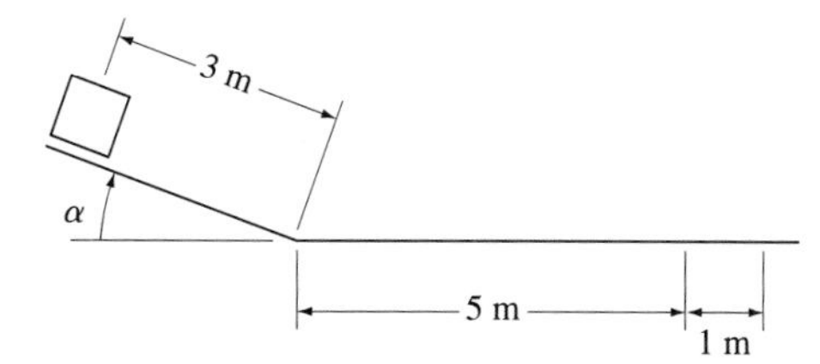


**Figure P3.7** Chute.

**3.8** A 2500 lbf car goes into a skid on slick streets at an initial velocity of 45 mph. A frictional force of 200 lbf is developed between the vehicle's tires and the road. If the car continues to move in a straight line, *develop* an appropriate model and find how far it travels and how long it takes to come to a stop.

**3.9** During the early part of the flight of a rocket (see Figure P3.9), the gravity force and thrust force are constant. The aerodynamic drag varies with the square of the velocity, and because fuel is being burned, the rocket is losing mass at a constant rate. Start with the linear momentum principle of Eq. (2.1), and *devise* a suitable model to describe the motion.

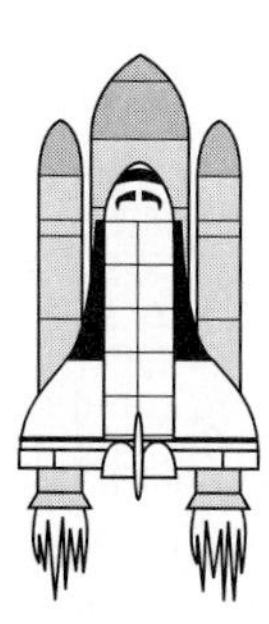

**Figure P3.9** Rocket.

**3.10** *Calculate* the torsional spring stiffness constant for a steel shaft 0.5 inch in diameter and 2.5 ft long. Express your result in in lbf/rad, and in ft lbf/deg.

**3.11** *Repeat* Prob. 3.10, except that the cross section of the shaft is a square 0.5 inch on a side. Consult Appendix B or a reference on the mechanics of deformable bodies for the appropriate torque-deformation relationship for shafts with noncircular cross sections.

**3.12** A beam is to be used as a translational spring. *Design* a steel cantilever beam that will provide a stiffness of 400 lbf/in. To avoid fatigue failure, the beam should not develop a bending stress greater than one-third of its ultimate strength and must carry a maximum load of 75 lbf at its free end.

**3.13** Consider the mechanical system shown in Figure P3.13, where

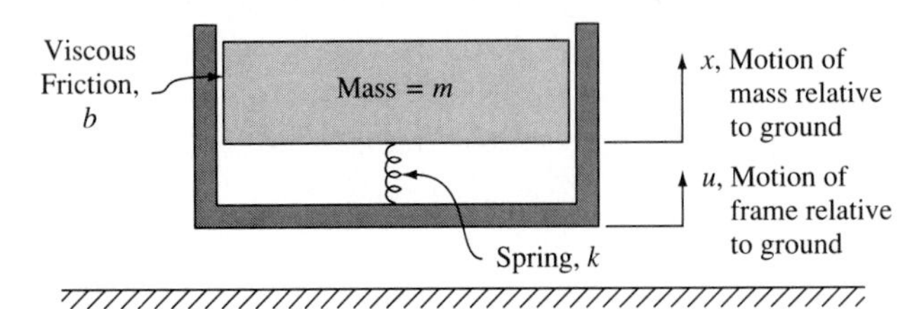


**Figure P3.13** Mechanical system.

$$k = 25\,\frac{\text{N}}{\text{m}} \qquad b = 4\,\frac{\text{N}}{\text{m/s}} \qquad m = 1\text{ kg}$$

(Recall that $\text{N} = \text{kg}\,\frac{\text{m}}{\text{s}^2}$.)

**a.** *Draw* a free-body diagram of the mass.

**b.** *Show* all forces and their direction on the diagram. Neglect gravity.

**c.** *Write* the equation of motion, and *derive* the transfer function of the response $x$ to an input motion $u$.

**d.** *What* is the expression for the static gain, and what is its numerical value?

**3.14** Figure P3.14 shows a mechanical system with two degrees of freedom and two inputs. Considering the effects identified by the coefficients,

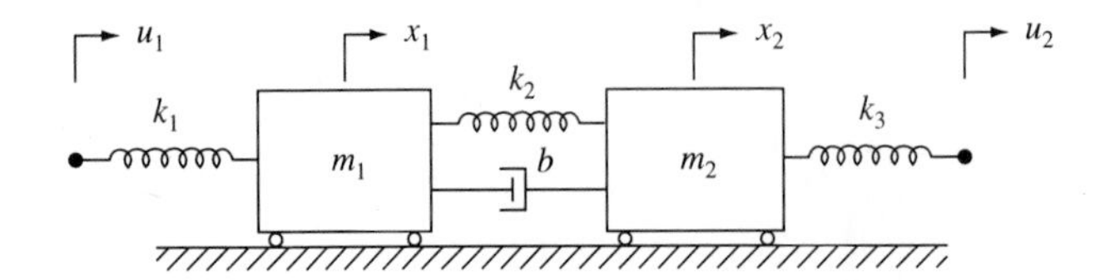


**Figure P3.14** Dual-input system.

**a.** *Write* the basic modeling equations.

**b.** *Derive* a differential equation for $x_2$ as a function of $u_1$ and $u_2$.

**c.** *Normalize* the result and express it as a transfer function. *What* are the static gains?

**3.15** Consider the dynamics of a car accelerating (Figure P3.15). The car has wheels and tires that have a high rotational inertia. Suppose that the aerodynamic drag and rolling resistance forces are linear with velocity.

**a.** *Derive* the differential equation of motion of the car's velocity $v$ as a function of the total driving torque $T_a$ to the rear wheels. Consider the mass $m_c$ of the car itself, the mass $m_w$ of each wheel and tire, the diameter of the tires $d$ and the wind resistance $b$.

**b.** *State* the expression for the time constant of this system.

**3.16** In the previous problem, suppose that the rolling resistance and aerodynamic drag forces are given by $F = b_0 + b_2 v^2$. *Derive* a state-space representation for this nonlinear system.

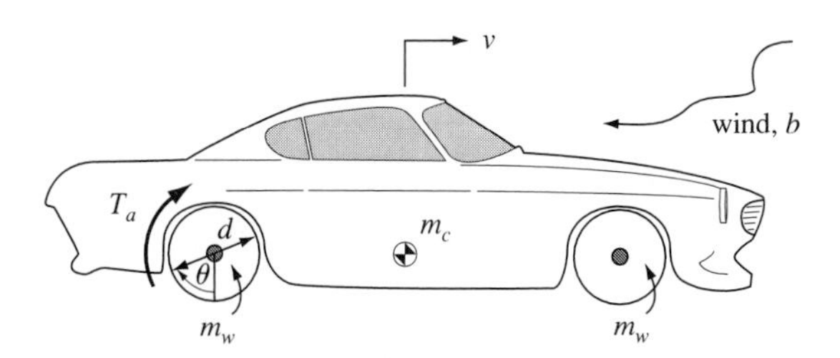


**Figure P3.15** Accelerating car.

**3.17** Consider Figure P3.17, and *derive* the equation of motion for $x_2$ as a function of $F_a$. The indicated damping is viscous.

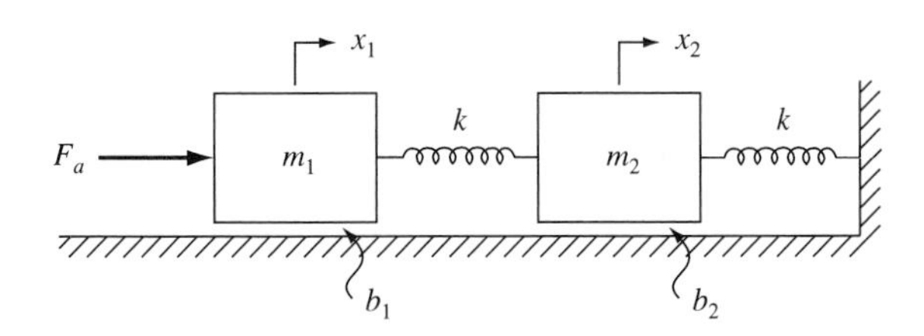


**Figure P3.17** Two-mass system.

**3.18** *Derive* the equations of motion for the system shown in Figure P3.18. *Find* the transfer function between the input $y(t)$ and the output $z(t)$, and *find* the static gain of the system.

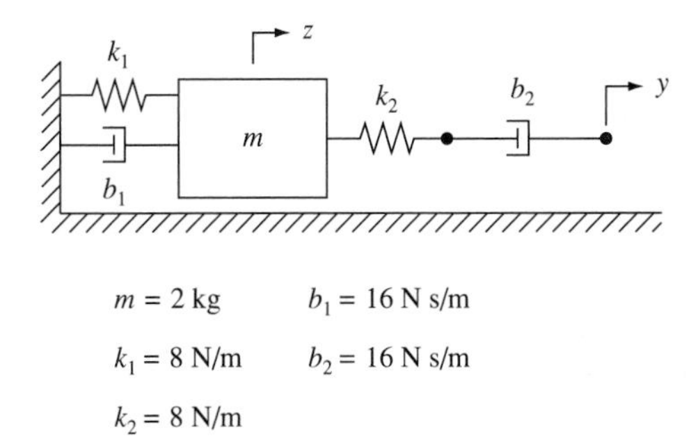


$m = 2$ kg $\quad b_1 = 16$ N s/m

$k_1 = 8$ N/m $\quad b_2 = 16$ N s/m

$k_2 = 8$ N/m

**Figure P3.18** Single-mass system.

**3.19** Suppose that a machine becomes jammed in its operation so that the left end of the shaft shown in Figure P3.19 is fixed and the disk on the right end has a torque applied to it. Assuming that the shaft inertia is very small in comparison to that of the disk, *develop* the equation of motion for the angular position of the disk. Because the machine continues to try to override the malfunction, the torque is repeated and may be idealized as a harmonic function with nonzero mean, as described in Figure 2.11. *Give* your result in the classical equation form, and then *convert* it to the state-space equation form.

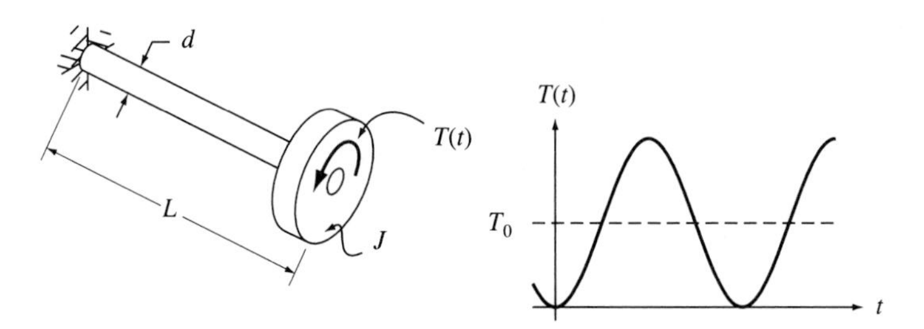


**Figure P3.19** Machine model.

**3.20** An electric motor is attached to a load inertia through a flexible shaft as shown in Figure P3.20. *Develop* a model and associated differential equations describing the motion of the two disks $J_1$ and $J_2$. *Provide* both classical and state-space forms of the resulting equations.

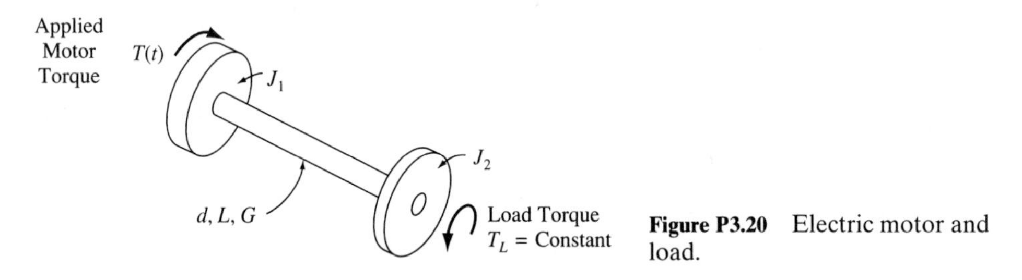


**Figure P3.20** Electric motor and load.

**3.21** Figure P3.21 shows a boat engine driving its propeller through a clutch and flexible shaft. *Find* a suitable model of this mechanical system. *Include* the rotating parts of the engine, the clutch (which we assume is slipping at this moment), the flexible shaft, the propeller, and the water.

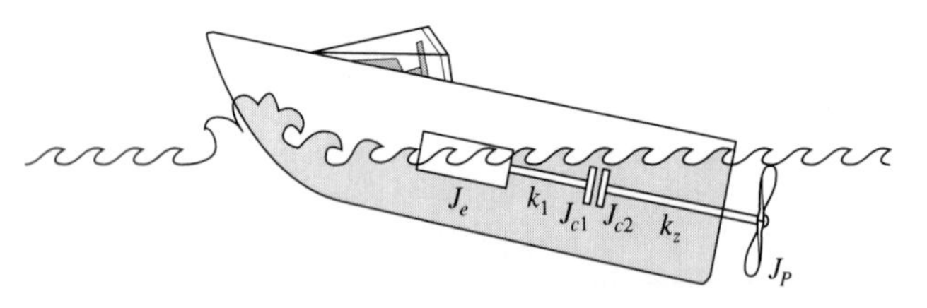


**Figure P3.21** Boat engine and propeller.

**3.22** The crane shown in Figure P3.22 is used to pick up a point mass $m_p$. The uniform boom has a mass $m_b$, and its center of gravity is at its midpoint. The rotational mass moment of inertia of the boom with respect to an axis through its center of gravity is $J_{cg}$. The input is the torque $T$ at the base of the boom. *Draw* a free-body diagram, and *derive* the equation of motion for the angular position of the boom.

**3.23** A device from a copying machine is shown in Figure P3.23. It moves in a horizontal plane. *Develop* the dynamic system model, assuming that the mass of the bar is negligible in comparison to that of the attached mass $m_2$ and that the angular motions are

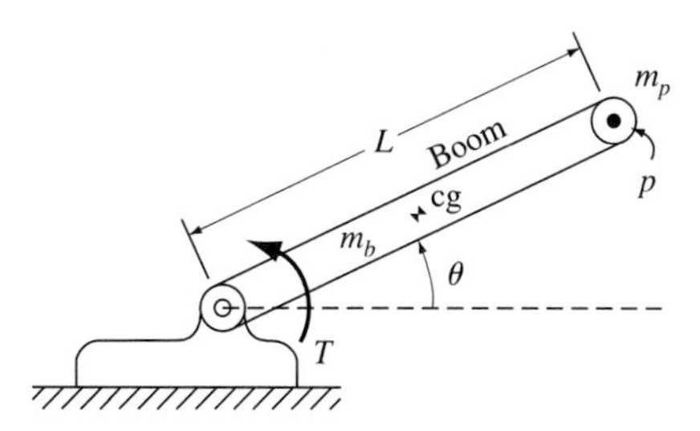


**Figure P3.22** Crane model.

small. The mass is subjected to a step input force $F$. *Find* an expression for the displacement of point $B$ after the transient motions have died out.

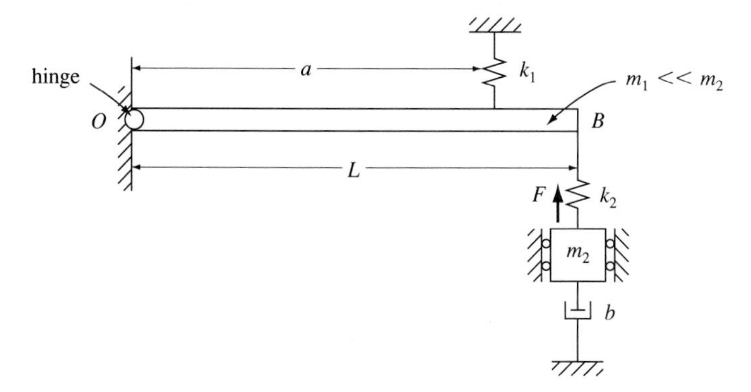


**Figure P3.23** Machine device.

**3.24** The uniform bar shown in Figure P3.24 is attached to its support by two linear springs and is prevented from moving in the $x$-direction. A mass is attached to the bar through a spring as indicated. *Derive* the equations of motion for the system. *Assume* small angular motions of the bar. The motion of the system occurs in a horizontal plane, so you need not consider the effect of gravity.

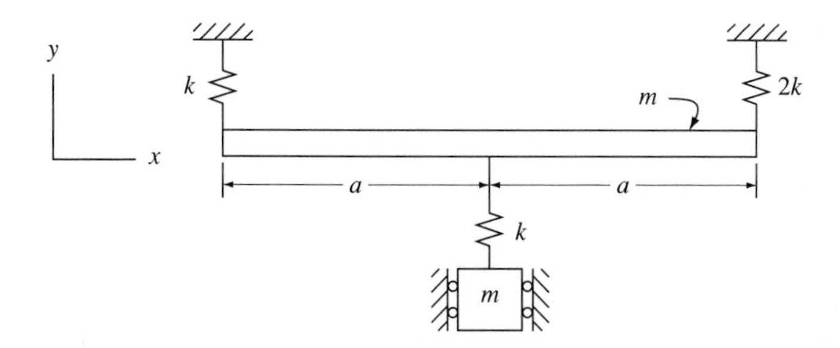


**Figure P3.24** Bar-mass system.

**3.25** *Find* the equations describing the motion of the device shown schematically in Figure P3.25. The angular motions due to the applied force $F$ are small. The damper is a Coulomb friction device; i.e., it develops a constant force that opposes motion. *Express* your answers in both the classical system equation format and the state-space format.

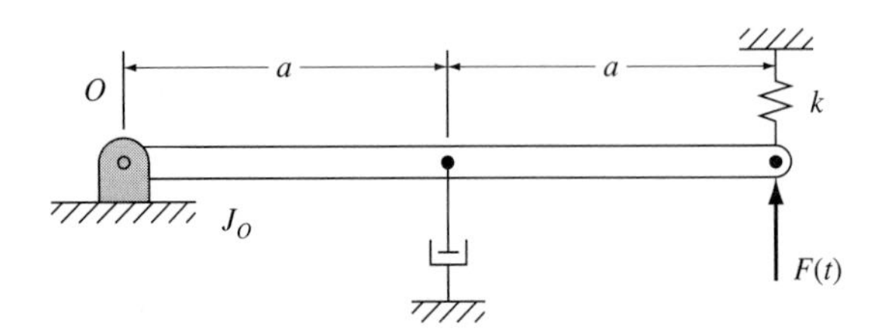


**Figure P3.25** Lever device.

**3.26** A viscous frictional force is developed at the piston-surface interface of the system shown in Figure P3.26. *Develop* the equations of motion, and set them up for solution in state-space form. The angular motions are small.

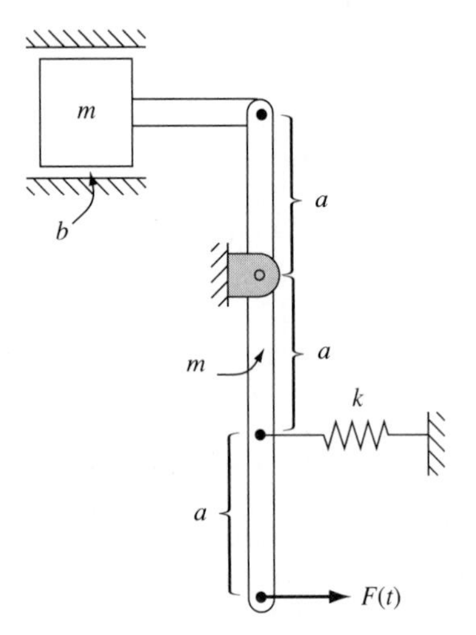


**Figure P3.26** Spring-mass-damper-lever.

**3.27** *Repeat* Problem 3.26, but assume that the piston is acted on by a Coulomb frictional force.

**3.28** Suppose that the apparatus described in Problem 3.25 has a torsional spring $k_T$ added at the bar hinge point $O$. *Develop* the associated dynamic systems model.

**3.29** Shown in Figure P3.29 is a mechanical system with a rotating wheel of mass $m_w$ (with uniform mass distribution). Springs and dampers are connected to the wheel using a flexible cable without slip on the wheel.

**a.** *Write* all of the modeling equations for translational and rotational motion.

**b.** *Derive* the differential equation of motion for the translational motion $x$ as a function of the input motion $u$.

**c.** *State* the expressions for the static gain, natural frequency, and damping ratio.

**3.30** The disk shown in Figure P3.30 rolls without slipping on a horizontal plane. Attached to the disk through a frictionless hinge is a massless pendulum of length $L$ that carries an-

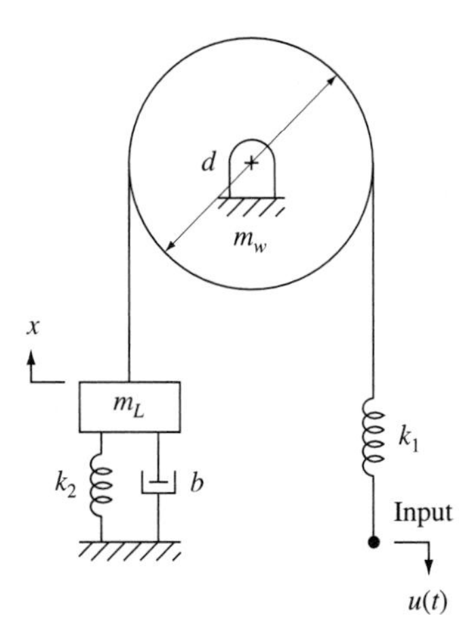


**Figure P3.29** Mass-pulley system.

other disk. The disk at the bottom of the pendulum cannot rotate relative to the pendulum arm. *Draw* free-body diagrams and *derive* the equations of motion for this system.

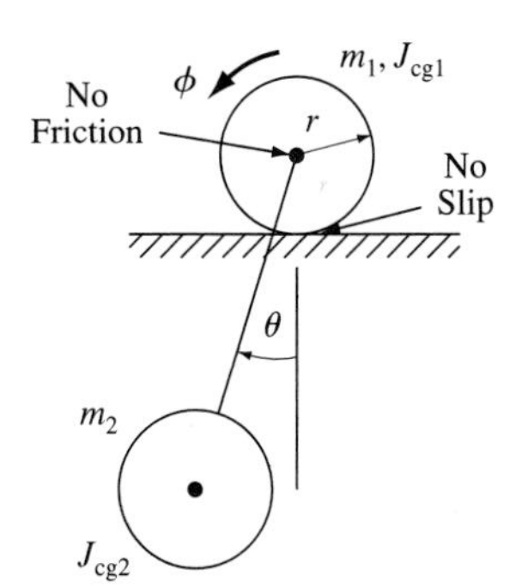


**Figure P3.30** Rolling wheel with disk.

**3.31** A wheel is moved by the mechanical system shown in the Figure P3.31. The mass of the wheel is $m_w$, and the mass of the attachment is $m_c$.

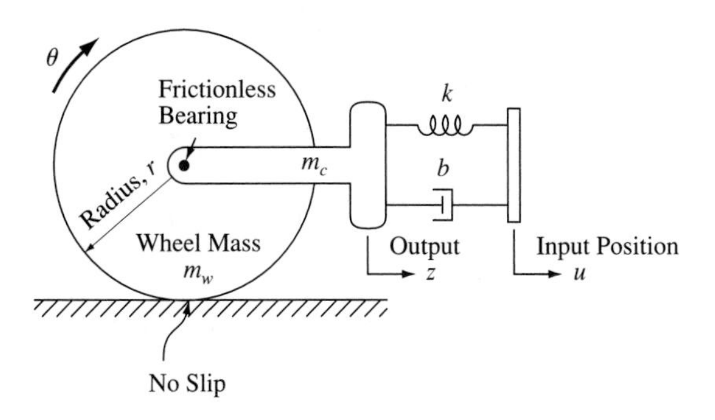


**Figure P3.31** Wheel with position input system.

**a.** *Write* the modeling and geometric equations for the translational and rotational systems.

**b.** *Derive* the transfer function for the output motion $z$ as a function of the input motion $u(t)$.

**c.** *State* the static gain of the system.

**d.** Under what conditions would there be slip between the wheel and the ground? (*State* an equation.)

**3.32** Use Lagrange's equation to *derive* the equations for Problem 3.23.

**3.33** Use Lagrange's equation to *derive* the equations for Problem 3.24.

**3.34** Use Lagrange's equation to *derive* the equations for Problem 3.26.

**3.35** Use Lagrange's equation to *derive* the equations for Problem 3.30.

**3.36** Use Lagrange's equation to *derive* the equations for Problem 3.31.

# CHAPTER 4

# Electrical Systems

## 4.1 INTRODUCTION

**Electrical systems** are composed of resistors, capacitors, inductors, transistors, amplifiers, and power supplies; they use voltage and current to sense, amplify, filter, and compute (various quantities), and to move mechanical loads.

Electrical and electronic circuits range from simple passive circuits to active circuits to complete electronic systems such as stereo amplifiers, servocontrol systems, audio and video communication systems, and digital computers. **Passive circuits** simply respond to an applied voltage or current and thus do not have any amplifying devices or active sources of power. **Active circuits** have transistors and/or amplifiers that require an active source of power to work.

Electronics now pervades our lives and is involved in almost every aspect of our daily activities. The list is long and obvious. The purpose of this chapter is to review the foundations of circuit analysis in order to predict and analyze the response of a circuit to certain input signals. We first cover passive resistance-inductance-capacitance (RLC) circuits and then introduce some of the more simple active circuits using operational amplifiers, or op amps. References [1–7] at the end of the chapter give a more detailed treatment of transistor circuits, logic, communications, and digital computers.

## 4.2 BASIC ELEMENTS

In the analysis of electrical circuits, we must use variables that represent the potential for doing work and the rate at which work can be done. Voltage and current are the two variables used. To better understand these quantities, it is necessary to discuss the character of electrons.

One model of electrical behavior is based upon electrons moving in orbits around the nucleus of atoms. The electrons can break loose and move from atom to atom. The electrons have a very small electrical charge. If a body has an excess of electrons, it has a negative charge; if it has a deficit of electrons, it has a positive charge. The unit of charge is the **coulomb** (in recognition of Charles Augustin Coulomb, French physicist, 1736–1806), which represents $6.24 \times 10^{18}$ electrons.

As the electrons flow from one body to another, a current is established. The **current** $i$ (from the French word *intensité*) is the rate of change of charge $q$ with respect to time and can be positive or negative, depending upon the direction in which the electrons flow. Mathematically,

$$i = \frac{dq}{dt} \tag{4.1}$$

The unit of current is the **ampere,** or simply, amp (in recognition of André Marie Ampère, French physicist and mathematician, 1775–1836), which is defined as a coulomb per second:

$$\text{ampere} = \text{coulomb/second} \tag{4.2}$$

Thus, 1 amp is $6.24 \times 10^{18}$ electrons moving from one body to another in 1 second.

The **voltage** represents the potential energy in the circuit. It is the change in energy $w$ as a charge $q$ is passed through a component; that is,

$$e = \frac{dw}{dq} \tag{4.3}$$

The unit of voltage is the **volt** (in recognition of the Italian physicist Alessandro Volta, 1745–1827), which is defined as a change of 1 joule of energy per coulomb of charge. A **joule** (named in recognition of the English physicist James Joule, 1818–89) is a unit of energy or work and has the units of newton × meter. Thus,

$$\text{volt} = \text{joule/coulomb} \tag{4.4}$$

$$\text{joule} = \text{newton} \times \text{meter} \tag{4.5}$$

To conclude this discussion, it is important to note that the rate of change of energy with respect to time in a component is the power

$$p = \frac{dw}{dt} \tag{4.6}$$

In electrical systems, **power** is the product of voltage and current:

$$p = ei \tag{4.7}$$

The unit of power is the **watt** (named in honor of James Watt, Scottish inventor, 1736–1819), which has units of joules per second; that is,

$$\text{watt} = \text{joule/second} \tag{4.8}$$

Therefore, we can describe electrical components in terms of their treatment of energy and, thus, the relationship between their voltage and current. Components can dissipate or store energy. Resistors dissipate energy, capacitors store energy by virtue of voltage, and inductors store energy by virtue of current.

### 4.2.1 Resistance

If a voltage potential is placed across a device, the ability of the device to conduct electricity depends upon its makeup. An ideal insulator would not conduct any electricity; thus, there would be no exchange of electrons. Insulators isolate voltage potential from other electrical components and protect living things from shock. By contrast, an ideal conductor would freely conduct electricity, with no restriction on the ability of electrons to move. Somewhere in between insulators and conductors are resistors, which show a predictable restriction of electron flow. The current in a resistor is limited and is directly proportional to the applied voltage potential. A resistor dissipates energy by converting it to heat and light and has no ability to store electrical energy.

The fact that the current in a resistive component is proportional to the voltage drop applied to the component was established in 1826 by Georg Simon Ohm (German physicist, 1787–1854). This linear relation between the voltage and current bears the name of **Ohm's law** and is written

$$\delta e = Ri \tag{4.9}$$

$e_1$ $R$ $e_2$

$i$

$$\text{where } \delta e = e_1 - e_2 \tag{4.10}$$

The circuit symbol for a resistor is shown above with its voltages $e_1$ and $e_2$ and its current $i$.

Thus, the **resistance** $R$ of a component is the incremental rate at which voltage changes with current in that component, or

$$R = \frac{de}{di} = \frac{\delta e}{i} \tag{4.11}$$

If the resistance is linear, the slope is constant and the resistance is simply the ratio of the voltage to the current. The unit of measurement of resistance is the **ohm** and is given the symbol Ω.

$$\text{ohm} = \text{volt/amp} \tag{4.12}$$

The amount of power dissipated by a resistor is the product of the voltage drop and the current. This power generally is converted to thermal energy:

$$\text{power dissipated} = \delta e i \tag{4.13}$$

Resistors are manufactured from thin wires or films made of conductive material, carbon, and various other materials. The resistance of such materials is determined by the resistivity $\rho$ of the material and the length, $\ell$ and cross-sectional area $A$ of the conductor. Specifically,

$$R = \frac{\rho \ell}{A}$$

Resistors are categorized by their heat dissipation ability. Thus, 1/8-, 1/4-, and 1/2-watt resistors are common. In typical circuits resistances range from a few hundred ohms to a million ohms. Resistance is stated in ohms, kilohms, and megohms; thus 2,200 ohms would be stated 2.2 kilohms, or simply, 2.2 k.

### 4.2.2 Capacitance

**Capacitance** is the ability of a component to store electrons. Capacitance concepts were established in 1745 when van Mussenbroek of Leyden stored static electricity charges on two metal plates in what came to be known as a Leyden jar.

An electric capacitor can store electrons on two parallel plates that are separated by an insulating dielectric material in an electric field. The total charge on the plates (or the capacity of the component) is basically the number of electrons that can be stored on the plates. Thus, to calculate the charge, we must keep track of the cumulative current into or out of the capacitor. The voltage $\delta e$ across a capacitor is proportional to the integral over time of the current $i$ from $-\infty$ to the present time $t$. Notice that current is in coulomb/second; thus, the integral of the current is in coulomb. Mathematically,

$$\delta e = \frac{1}{C} \int_{-\infty}^{t} i \, dt \tag{4.15}$$

The constant of proportionality is the capacitance $C$. The larger the capacitance, the more charge can be stored at a given voltage.

Because it is not practical for us to keep up with all of the current that has gone into and out of a capacitor since it was built, it is comforting to know that we simply need to know the voltage at some point in time (say, $t = 0$), and then the voltage at some future time $t$ can be calculated. The equation for calculating the voltage expresses the fact that we can perform the integration in two parts; the first part simply gives us an initial condition $e_0$, and the second part gives us the cumulative effect that the current has on the voltage of the capacitor:

$$\delta e = \frac{1}{C} \int_{-\infty}^{0} i \, dt + \frac{1}{C} \int_{0}^{t} i \, dt \tag{4.16}$$

$$\delta e = e_0 + \frac{1}{C} \int_{0}^{t} i \, dt \tag{4.17}$$

Since we usually prefer solving differential equations rather than integral equations, the capacitance equation is normally stated as the following differential equation with the initial condition on the voltage $\delta e(0)$:

$$i = C\frac{d\,\delta e}{dt} \quad \text{with } \delta e(0) \tag{4.18}$$

$C$
$e_1$ $e_2$
$i$

$$\text{where } \delta e = e_1 - e_2 \tag{4.19}$$

The circuit symbol for a capacitor along with its voltages $e_1$ and $e_2$, and its current $i$ are shown above.

As with all differential equations, the quantity differentiated can have an initial condition. It is this property which dictates that energy can be stored in the initial condition. Thus, the electric capacitor can store energy in the initial voltage charge at time $t = 0$ with $\delta e(0)$. Further, the initial condition of the voltage must be known in order to solve for the circuit response.

Capacitors have the ability to store energy by virtue of their voltage. Since power is the rate of change of energy with respect to time, energy is the integral of power over time. Power is the instantaneous product of the voltage drop and the current; therefore, the energy $w$ stored in a capacitor can be expressed as

$$w = \int_{-\infty}^{t} \delta e\; i\; dt = \int_{-\infty}^{t} \delta e(C\,d\,\delta e)\,dt = \int_{0}^{\delta e} C\,\delta e\,d\,\delta e \tag{4.20}$$

Performing this integration reveals that the energy stored in a capacitor is proportional to the capacitance and to the square of the voltage across the capacitance:

$$w = \frac{C\,\delta e^2}{2} \tag{4.21}$$

The unit of measure of capacitance is the **farad** (in recognition of the English chemist and physicist Michael Faraday, 1791–1867) and is represented by the symbol f. A capacitor with a capacitance of 1 farad is a very large capacitor; thus, values of capacitors are stated in picofarads (pf, $10^{-12}$ farad), nanofarads (nf, $10^{-9}$ farad), and microfarads ($\mu$f, $10^{-6}$ farad). Typical values of capacitance in circuits range from nanofarads to hundreds of micro-farads. In terms of its relation to other electrical units of measure, we have

$$\text{farad} = \text{coulomb/volt} = \text{amp second/volt} = \text{second/ohm} \tag{4.22}$$

Modern capacitors are built by using two plates or sheets separated by a dielectric insulator. The sheet can be wrapped in a cylindrical shape to save space. The insulator can be air, but other insulators such as glass or mica provide more efficient capacitance per unit volume. In electrolytic capacitors, the oxidized surface of the metal (such as aluminum or tantalum) sheet itself is used as the insulator. The capacitance $C$ of a capacitor is proportional to the area $A$ of the plates, the spacing or separation distance $d$ of the two plates, and the permittivity $\epsilon$ of the dielectric:

$$C = \frac{\epsilon A}{d} \tag{4.23}$$

### 4.2.3 Inductance

**Inductance** relates the voltage induced in a circuit to the rate of change of a magnetic field with respect to time and hence the rate of change of current with time.

Before discussing inductance, it is necessary to understand some facts about magnetic fields and currents that pass through them. Permanent magnets produce a magnetic field of a given flux, depending upon the configuration and separation distance of the magnets. Current in a conductor or wire produces a magnetic flux that can be intensified by wrapping the wire in a coil.

Michael Faraday established the basic concepts of electromagnetism in 1831. **Faraday's law** states that a voltage $e_i$ is induced in a wire (or conductor) when the magnetic flux $\phi$ changes with time $t$:

$$e_i = \frac{d\phi}{dt} \tag{4.24}$$

In this equation, the induced voltage is measured in volts, the magnetic flux is measured in **webers** (Wb, in recognition of W. E. Weber, German physicist, 1804–91). Therefore, an induced volt is one weber per second:

$$\text{induced volt} = \text{weber/second} \tag{4.25}$$

Induced voltage can be implemented in three ways: (1) when a wire travels through a fixed magnetic field (termed *electromagnetic induction*), (2) when one coil of wire induces a magnetic field in another coil of wire (termed *mutual inductance*), and (3) when a winding in a single coil induces a voltage on an adjacent winding in the same coil (termed *self-inductance*). The first case represents the operational concepts of electric motors, mechanical motion sensors, loudspeakers, and other electromagnetic systems that have motion. The second case represents transformers that reduce or increase voltages. Examples include transformers in electronic systems (radio, TV, etc.), chargers, automotive ignition coils, and power line transformers. The third case represents inductors (often called coils or chokes) used in electronic and tuning circuits. These inductors are used to filter voltages or to provide resonant circuits for tuning frequencies.

Our first modeling concern is with self-inductance. This gives rise to the inductor component, since a coil of wire induces a voltage in opposition to a change in current. In self-inductance, the magnetic field is generated by the current in the wire itself; thus, an opposing voltage drop $\delta e$ is induced by a change in the current in the coil (if $\phi \sim i$, then $d\phi/dt \sim di/dt$, from Faraday's law):

$$\delta e = L\frac{di}{dt} \quad \text{with } i(0) \tag{4.26}$$

$e_1$ $L$ $e_2$

$\longrightarrow i$

$$\text{where } \delta e = e_1 - e_2 \tag{4.27}$$

The circuit symbol for an inductor along with its voltages $e_1$ and $e_2$, and current $i$ are shown above.

The proportionality constant $L$ is termed the inductance. The unit of measure of inductance is the **henry,** named in honor of the American investigator, Joseph Henry (1797–1878), who obtained many of Faraday's results independently. The symbol used for the henry is h, and typical values range from the microhenry ($\mu$h) to the millihenry ($m$h):

$$\text{henry} = \text{volt second/amp} = \text{ohm second} \tag{4.28}$$

As with capacitance, the initial condition $i(0)$ of this differential equation must be known in order to determine a circuit response. Further, energy can be stored in the initial value of the current, since the inductor is a dynamic component.

Energy can be stored in an inductor by virtue of the current passing through it. The amount of energy stored, $w$, is the integral of the power over time:

$$w = \int_{-\infty}^{t} \delta e\, i\, dt = \int_{-\infty}^{t} \left(L\frac{di}{dt}\right) i\, dt = \int_{0}^{i} Li\, di \tag{4.29}$$

Performing this integration reveals that the energy stored in an inductor is proportional to the inductance and to the square of the current in the inductor:

$$w = \frac{Li^2}{2} \tag{4.30}$$

The inductance of a coil depends upon the cross-sectional area $A$ of the coil winding, the length $\ell$ of the coil, the number of turns $n$, and the permeability $\mu_m$ of the magnetic circuit:

$$L = \frac{\mu_m n^2}{\ell} A \tag{4.31}$$

Inductors may be built with either an air core or an iron core. Air core inductors have a low permeability and thus low inductance for a given size, but have greater linearity than iron core inductors because the permeability of ferromagnetic materials changes with flux density.

### 4.2.4 Definition of Impedance

In analyzing circuits, we will work extensively with the concept of impedance of the $RLC$ components. The **impedance** $Z$ is defined as the instantaneous ratio of the voltage difference to the current:

$$Z = \frac{\delta e}{i} \tag{4.32}$$

The impedance of a resistor is quite simple: The **resistive impedance** $Z_r$ is the resistance $R$. Notice that since a resistor is a dissipative device, resistive impedance is not a dynamic function and thus is not a function of time or frequency of excitation. Simply,

$$Z_r = R \tag{4.33}$$

Like the impedance $Z$, the **capacitive impedance** $Z_c$ is defined as the dynamic ratio of the voltage to the current. However, since the equation for a capacitor is an integral equation, capacitive impedance contains an integral term and is not a simple algebraic function. This impedance is expressed with the differential operator $D$, as well as the capacitance $C$:

$$Z_c = \frac{1}{CD} \tag{4.34}$$

The **differential operator** is a shorthand notation for the derivative of a quantity with respect to time. Notice that the inverse of the differential operator represents the integral over time:

$$D = \frac{d}{dt} = \text{differential operator (e.g., } Dx = dx/dt) \tag{4.35}$$

$$\frac{1}{D} = \text{integral operator} \tag{4.36}$$

Thus, capacitive impedance is low during times of high transients and approaches infinity (an open circuit) when the voltage remains constant (static excitation).

The impedance of an inductor is similarly a function of the derivative of the current. Thus, **inductive impedance** is a dynamic term, since the voltage-current relationship is a differential equation:

$$Z_L = LD \tag{4.37}$$

In times of high transients, inductive impedance is large. If the current is constant, then the inductive impedance is zero.

Many textbooks that discuss direct current (DC) and alternating current (AC) circuit analysis state the reactive impedances of the capacitor and the inductor in terms of the frequency of excitation $\omega$ and the complex variable $j = \sqrt{-1}$:

$$Z_c = \frac{1}{j\omega C} \tag{4.38}$$

$$Z_L = j\omega C \tag{4.39}$$

This concept implies steady-state harmonic operation and is very limited and misleading for the more general transient analysis that is presented here. First of all, the foregoing result requires that the circuit be excited by a sinusoidal signal of frequency $\omega$. Secondly, it assumes that all transients of the circuit have died out and that the resulting voltages are fully established and do not change their character with time. (See Appendix E.) Further, the equation expresses a phase relation by the use of the complex variable $j$. It is the purpose of this book to study the transient response of circuits and systems without restricting ourselves to steady-state sinusoidal excitation; therefore, we will not make use of these concepts.

In the circuit-modeling examples that follow, we consider the impedance of the components and treat resistors, capacitors, and inductors the same algebraically. This makes the circuit analysis approach consistent regardless of the components in the circuit. Further, the use of the $D$-operator reduces our mathematical operations to simple algebraic ones that will result in a differential equation representation of the system if it contains capacitors or inductors.

**Series and Parallel Impedance Combinations.** It is very helpful to know the combinations of circuit impedances when the components are connected together in either a series or a parallel configuration. In a series combination, two components are connected in-line such that the current in them is common and the voltage divides between the two. In a parallel combination, the two components are connected side by side such that they share a common voltage excitation and the current divides between them. The general configurations of two impedances $Z_1$ and $Z_2$ in series and in parallel are shown in Figure 4.1. Of interest in series and parallel circuits is how the two impedances combine to form an equivalent impedance $Z_{eq}$ of a presumed single component.

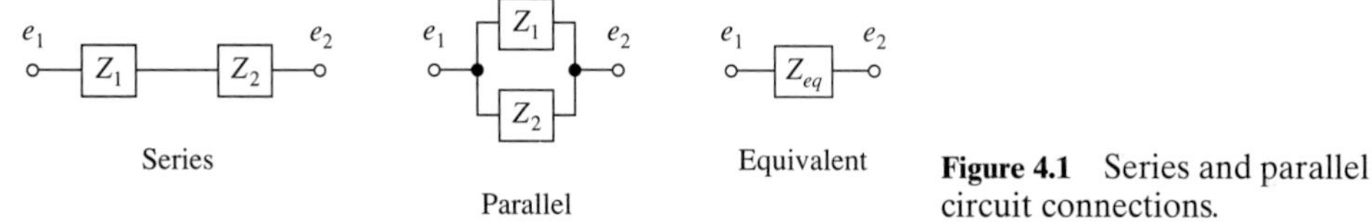


**Figure 4.1** Series and parallel circuit connections.

*Two resistors in series have an equivalent resistance $R_{eq}$ equal to the sum of the two resistances*. Thus, resistors in series add directly:

$$R_{eq} = R_1 + R_2 \tag{4.40}$$

If there are numerous resistors in series, then the equivalent single resistance is the sum of all of the individual resistors:

$$R_{eq} = R_1 + R_2 + R_3 + \cdots + R_n \tag{4.41}$$

*Two resistors in parallel have an equivalent resistance $R_{eq}$ equal to the inverse of the sum of the reciprocals of the two resistances*:

$$\frac{1}{R_{eq}} = \frac{1}{R_1} + \frac{1}{R_2} \tag{4.42}$$

It is easy to transform this equation and remember it as "the equivalent resistance is the product divided by the sum":

$$R_{eq} = \frac{R_1 R_2}{R_1 + R_2} \tag{4.43}$$

When numerous resistors are connected together in series, the reciprocal of the equivalent resistance is the sum of the reciprocals of the individual resistances; or, alternatively, the equivalent resistance is the product of all of the individual resistances divided by their sum. Mathematically,

$$\frac{1}{R_{eq}} = \frac{1}{R_1} + \frac{1}{R_2} + \frac{1}{R_3} + \cdots + \frac{1}{R_n} \tag{4.44}$$

or

$$R_{eq} = \frac{R_1 R_2 R_3 \cdots R_n}{R_1 + R_2 + R_3 + \cdots + R_n} \tag{4.45}$$

*When capacitors are connected in series, the equivalent capacitance $C_{eq}$ is the inverse of the sum of the reciprocals of the individual capacitors*:

$$\frac{1}{C_{eq}} = \frac{1}{C_1} + \frac{1}{C_2} \tag{4.46}$$

or

$$C_{eq} = \frac{C_1 C_2}{C_1 + C_2} \tag{4.47}$$

*When capacitors are connected in parallel, it is the same as increasing the area of the plate, and therefore, the equivalent capacitance $C_{eq}$ is the direct sum of the individual capacitors*:

$$C_{eq} = C_1 + C_2 \tag{4.48}$$

*Inductors in series add directly the same as resistors*:

$$L_{eq} = L_1 + L_2 \tag{4.49}$$

*Inductors in parallel add as reciprocals, the same as resistors do*:

$$L_{eq} = \frac{1}{L_1} + \frac{1}{L_2} \tag{4.50}$$

or

$$L_{eq} = \frac{L_1 L_2}{L_1 + L_2} \tag{4.51}$$

The proof of these results is left as an exercise for the reader.

## 4.3 PASSIVE CIRCUIT ANALYSIS

**Passive circuits** are composed simply of *RLC* components, without amplifiers. The circuits can be excited or driven by voltage or current sources. Normally, we are interested in the general response of the circuit to transient signals, rather than just the sinusoidal response. Thus, we will treat all components as general impedances using the *D*-operator.

Passive circuits are composed of components (*RLC* and voltage or current sources) connected together. The point of connection of two or more components is called a **node**. Physically, this is the point at which the wires or conductors of the components make electrical contact with each other. There is one voltage relative to ground associated with a node, and there are currents to or from each node to the individual components. Thus, we will associate a voltage with each node and currents in or out of the node to each component.

### 4.3.1 Kirchhoff's Laws

Several different approaches are available for analyzing circuits, including node current equations, loop voltage laws, and mesh current techniques. One or the other of these approaches might be more suitable, depending on whether there is a voltage or current input or whether we are interested in a voltage or current output. In this text, we consider only one approach to modeling circuits and rely on algebra to convert the basic modeling equations to any desired form. Our approach to modeling is to use Kirchhoff's current law (Gustav Kirchhoff, German physicist, 1824–87) for each of the significant nodes of the circuit and to write all of the component equations. This will give us the basic equations required, and algebraic methods can be used to rearrange them to an alternative form if desired. Such an approach is best suited for a voltage input and a voltage output, which represents the majority of circuit analysis cases that are of interest to us.

**Kirchhoff's current law** states that the algebraic sum of all currents at a node is zero. This is really a statement of the principle of conservation of charge in physics, but stated in time-differential format. (Recall that the derivative of charge is current.)

The node equations, therefore, are a simple statement that the current into the node must equal the current out of the node. Thus, an assumed direction for current is used to assign polarity for the algebraic sum. This direction is arbitrary for positive current in a component, as long as we are consistent with the polarity in the component equations and the node equations.

It is only necessary to write the node equations of the significant nodes in the circuit; it is not necessary to write node equations for nodes connected to voltage sources. This is true for the positive and negative terminals of the voltage source. The only purpose for writing these equations is to determine the current coming from the voltage source if you are interested in it; however, the equation is not necessary for the solution of the system equations.

Similarly, it is not necessary to write component equations for the components connected to current sources (either the positive or the negative node) to derive the system response equations. The only purpose for writing these equations is to determine the voltage at the current source if you need to know it.

The component equations for the *RLC* components are written in terms of the voltage drop across the component and the current through the component, using an impedance relationship. The impedance is written with *D*-operator notation so that differential equations can be derived directly.

In order to perform this type of circuit analysis, we must first draw the circuit, label the voltages at each node, and assign a variable and a polarity for the current in each component. Next, the component equations are stated in impedance form, and all of the significant node equations are written. The resulting equations are then reduced or manipulated to obtain algebraic or differential equations. This basic procedure is employed throughout the chapter.

### 4.3.2 Resistance Circuits

We seldom have to analyze purely resistive circuits; however, some that commonly occur are worthwhile to mention.

**Voltage Divider.** One of the most basic resistive circuits is the so-called voltage divider circuit, formed by two resistors in series. In the circuit shown in Figure 4.2, a voltage $e_0$ is applied across two resistors, $R_1$ and $R_2$, in series. Of interest is the voltage $e_1$ between the two resistors.

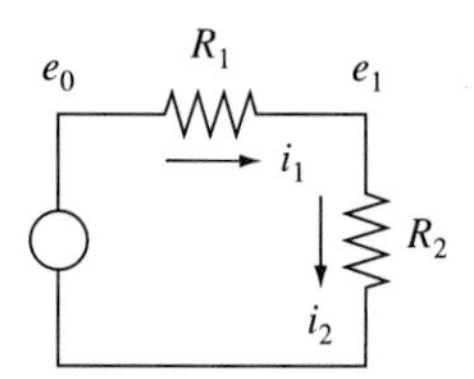


**Figure 4.2** Voltage divider circuit.

The currents in the resistors are given by the basic resistance equations (voltage drop divided by resistance):

$$i_1 = \frac{e_0 - e_1}{R_1} \tag{4.52}$$

$$i_2 = \frac{e_1 - 0}{R_2} \tag{4.53}$$

The node equation at the $e_1$ node is

$$i_1 - i_2 = 0 \tag{4.54}$$

The two component equations for the resistors can be substituted into the node equation:

$$\frac{e_0 - e_1}{R_1} - \frac{e_1}{R_2} = 0 \tag{4.55}$$

Rearrangement of this equation yields the following equation for the voltage divider:

$$e_1 = \frac{R_2}{R_1 + R_2} e_0 \tag{4.56}$$

This equation should be memorized because you will most likely use it or a variation of it many times.

Normalization of the preceding equation reveals that the output voltage $e_1$ is simply a fraction of the supply voltage $e_0$. Thus, the output voltage can vary from zero (when $R_2$ is very small compared to $R_1$) to $e_0$ (when $R_2$ is extremely large compared to $R_1$):

$$e_1 = \frac{1}{1 + \frac{R_1}{R_2}} e_0 \tag{4.57}$$

The voltage divider circuit is common in volume or gain control applications, such as the volume control on a stereo, or the gain or offset in an amplifier circuit, in which case it is common to use a potentiometer, or "pot," to adjust the attenuation or gain. In a pot, a wiper is used to make contact at variable locations along a fixed resistor, as illustrated in Figure 4.3. In this case, $R_1 + R_2$ is a constant, but their ratio can be adjusted.

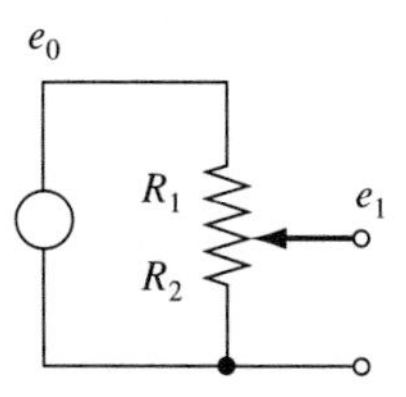


**Figure 4.3** Illustration of a pot being used as a voltage divider.

**Current Divider.** Figure 4.4 illustrates a current divider, in which the current from the current source is split between the load resistance $R_L$ and the shunt resistor $R_s$.

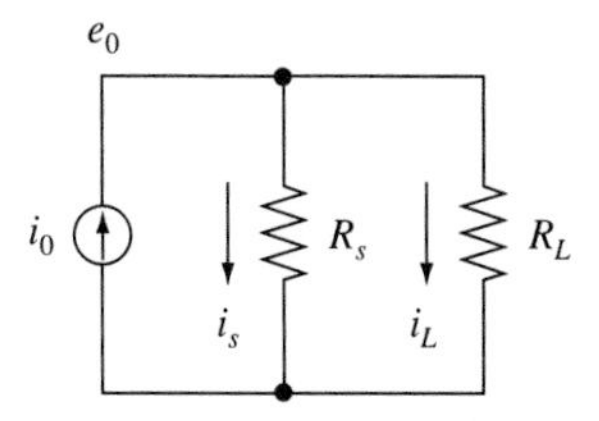


**Figure 4.4** Current divider.

In this case, the node equation reveals that the source current $i_0$ divides between the two resistors:

$$i_0 = i_s + i_L \tag{4.58}$$

The component equations relate the actual voltage at the current source to the component currents:

$$i_s = \frac{e_0}{R_s} \tag{4.59}$$

$$i_L = \frac{e_0}{R_L} \tag{4.60}$$

Substituting these two component equations into the node equation yields

$$i_0 = \frac{R_s R_L}{R_s + R_L} e_0 \tag{4.61}$$

Again using the load current equation in this equation, we obtain the final result:

$$i_L = \frac{R_s}{R_s + R_L} i_0 = \frac{1}{1 + R_L/R_s} i_0 \tag{4.62}$$

Notice that if $R_L/R_s$ is zero, then all of the source current will flow into the load resistor. As $R_L/R_s$ approaches infinity, no current will go to the load resistor. If $R_L$ is equal to the source resistance, half of the current will go to the load.

**Summing Circuit.** Figure 4.5 illustrates a summing circuit, in which the output voltage $e_3$ is the sum of two input voltages, $e_1$ and $e_2$.

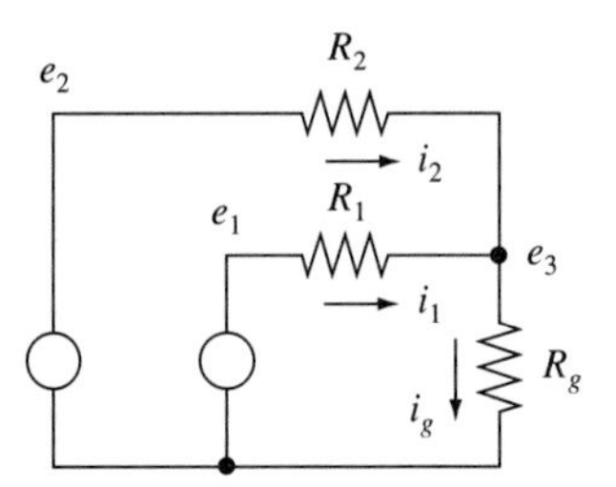


**Figure 4.5** Voltage summing circuit.

The three component equations can be written in terms of the indicated voltages and currents:

$$i_1 = \frac{e_1 - e_3}{R_1} \tag{4.63}$$

$$i_2 = \frac{e_2 - e_3}{R_2} \tag{4.64}$$

$$i_g = \frac{e_3}{R_g} \tag{4.65}$$

The node equation at the interconnection of the three components follows:

$$i_1 + i_2 - i_g = 0 \tag{4.66}$$

Substituting the component equations into the node equation, we obtain

$$\frac{e_1 - e_3}{R_1} + \frac{e_2 - e_3}{R_2} - \frac{e_3}{R_g} = 0 \tag{4.67}$$

Rearrangement yields the following final result, which illustrates that the output voltage is indeed the sum of the two applied voltages:

$$e_3 = \frac{e_1}{\left(1 + \frac{R_1}{R_2} + \frac{R_1}{R_g}\right)} + \frac{e_2}{\left(1 + \frac{R_2}{R_1} + \frac{R_2}{R_g}\right)} \tag{4.68}$$

Notice that each voltage is attenuated from the applied value; for example, if all three resistors were equal, then the output would be one-third of the applied voltage. In practice, $R_g$ should be smaller than $R_1$ and $R_2$ to help isolate the two input voltages, and a greater attenuation will result.

**Bridge Circuit.** Bridge circuits are used in many sensing applications. In a strain gauge circuit, the electrical resistance in one or more of the branches or legs of the bridge varies with the strain of the metal or surface to which the gauge is rigidly attached. This change in resistance causes a change in the voltage differential, which can then be correlated to the strain. Figure 4.6 illustrates a typical bridge circuit.

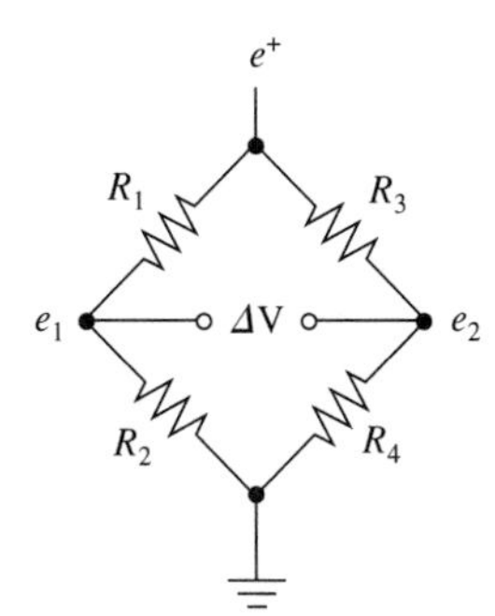


**Figure 4.6** Full bridge circuit.

The bridge is composed of two voltage dividers, so the differential voltage $\Delta e$ can be expressed as the difference in $e_1$ and $e_2$:

$$\Delta e = e_1 - e_2 \tag{4.69}$$

$$\Delta e = \left[\frac{R_2}{R_1 + R_2} - \frac{R_4}{R_3 + R_4}\right]e^+ \tag{4.70}$$

If we observe that the resistance $R_2$ is a base value $R_2^*$, plus a small increment in resistance, $\delta R$, then we can state that

$$R_2 = R_2^* + \delta R \tag{4.71}$$

If all four resistances are equal ($R_1 = R_3 = R_4 = R_2^* = R$), then the bridge equation reduces to

$$\Delta e = \frac{\delta R}{2R} e^+ \tag{4.72}$$

The equivalent resistance from $e^+$ to ground can be calculated by considering two sets of series resistors operated in parallel:

$$R_{eq} = \frac{(R_1 + R_2)(R_3 + R_4)}{(R_1 + R_2) + (R_3 + R_4)} \tag{4.73}$$

If all of the resistances are equal (with value $R$), then the equivalent resistance is simply $R$.

### 4.3.3 Resistance-Capacitance Circuits

Capacitors are generally used in circuits either to filter out high-frequency signals or to store energy. In these cases, the settling time or the dynamic response is of interest.

***RC* Circuit.** Consider the simple $RC$ circuit shown in Figure 4.7 in which a resistor and capacitor are connected in series. Before analyzing this circuit, we should determine what we want from the analysis. For example, we might be interested in the voltage across the capacitor as a function of the supply voltage. If we identify the voltages and currents as shown, we can write the component and node equations as follows.
component equations:

$$i_R = \frac{e_0 - e_1}{R} \tag{4.74}$$

$$i_C = CDe_1 \quad \text{with } e_1(0) \tag{4.75}$$

node equation:

$$i_R = i_C \tag{4.76}$$

Substituting the component equations into the node equation yields

$$\frac{e_0 - e_1}{R} = CDe_1 \tag{4.77}$$

Rearranging results in

$$RCDe_1 + e_1 = e_0 \tag{4.78}$$

This equation is a classical differential equation in $e_1$ as a function of the input $e_0$. It shows that the system has a time constant $\tau = RC$, and a steady-state value $e_1$ equal to the supply voltage $e_0$. To solve the equation, the initial charge $e_1(0)$ on the capacitor must be known.

The differential equation can also be written as a transfer function:

$$e_1 = \frac{e_0}{RCD + 1} \tag{4.79}$$

This result could have been obtained from the voltage divider circuit equation by treating the second resistance in the voltage divider as a capacitive impedance $1/CD$.

**Example 4.1 *RC* Filter**

The circuit shown in Figure 4.7 has a resistance of 8 ohms; find the capacitance necessary to give a settling time of 2.5 milliseconds.

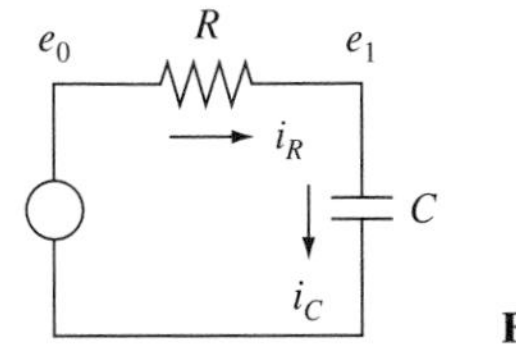


**Figure 4.7** Resistor-capacitor circuit.

From the circuit, the time constant is $\tau = RC$. The settling time is $4\tau$; thus, $\tau = 2.5/4$ ms. Solving for the capacitance then yields

$$C = \frac{\tau}{R} = \frac{2.5/4\ 10^{-3}\,\text{s}}{8\ \text{ohm}} = 78.125\ \mu\text{f}$$

It is important to note that capacitors are not built with very precise tolerance: The value of their capacitance might be $\pm 20\%$ or worse. Therefore, don't go to the distributor and ask for a 78.125 $\mu$f capacitor. You might find a 68 $\mu$f, a 100 $\mu$f, or a 220 $\mu$f capacitor; but you won't find a 78.125 $\mu$f capacitor. If you need the settling time to be precisely 2.5 ms, then you will have to buy the closest capacitor that is smaller than the desired value and add resistance to the circuit to get the settling time exact. Note further that, since the capacitor has a wide tolerance band, you will have to trim the resistance in each circuit to get your precise settling time. However, often the settling time does not have to be exact, and a capacitor close to the desired value will be acceptable.

**Dual *RC* Circuit.** Figure 4.8 shows a circuit made up of two *RC* circuits. In this case, we might be interested in the voltage $e_2$ as a function of the input voltage $e_0$. The component equations are

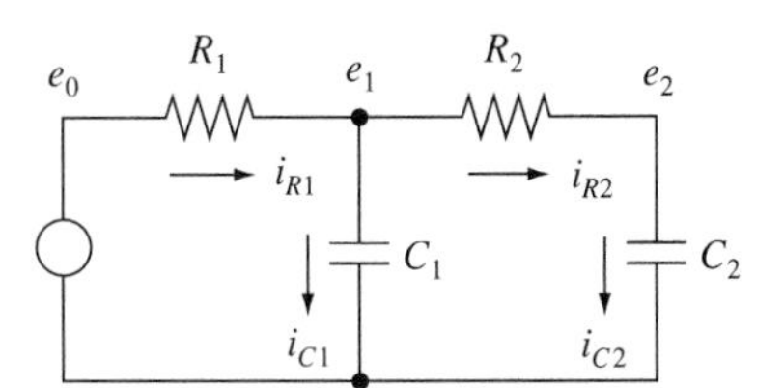


**Figure 4.8** Dual *RC* circuit.

$$i_{R1} = \frac{e_0 - e_1}{R_1} \tag{4.80}$$

$$i_{C1} = C_1De_1 \quad \text{with } e_1(0) \tag{4.81}$$

$$i_{R2} = \frac{e_1 - e_2}{R_2} \tag{4.82}$$

$$i_{C2} = C_2De_2 \quad \text{with } e_2(0) \tag{4.83}$$

The node equations are

$$i_{R1} = i_{C1} + i_{R2} \tag{4.84}$$

$$i_{R2} = i_{C2} \tag{4.85}$$

If we substitute the component equations directly into the node equations and rearrange the result, we will have two equations involving only the voltages of the circuit:

$$[R_1C_1D + (1 + R_1/R_2)]e_1 = (R_1/R_2)e_2 + e_0 \tag{4.86}$$

$$[R_2C_2D + 1]e_2 = e_1 \tag{4.87}$$

Since we are interested in $e_2$ as a function of $e_0$, we can substitute the second equation for $e_1$ into the first and rearrange to obtain

$$[R_1C_1R_2C_2D^2 + (R_1C_1 + R_1C_2 + R_2C_2)D + 1]e_2 = e_0 \tag{4.88}$$

This is a second-order differential equation in $e_2$ as a function of the input $e_0$. As such, the required initial conditions are $e_2(0)$ and $De_2(0)$. Since we only know the initial charges on the capacitors, $e_1(0)$ and $e_2(0)$, we must determine the initial conditions of the differential equation. The component equation for the second capacitor gives the derivative of $e_2$ as a function of the voltages $e_1$ and $e_2$. Since this equation is true for all time, including time $t = 0$, the required initial condition of $De_2$ can be determined:

$$e_2(0) = \text{known and } e_1(0) = \text{known} \tag{4.89}$$

$$De_2(0) = \frac{e_1(0) - e_2(0)}{R_2C_2} \tag{4.90}$$

To avoid the difficulty of determining the initial conditions (which could be very difficult for higher order circuits), the governing equations for this problem could be formulated in state-space format. Notice from the original component equations that the derivative of $e_1$ and the derivative of $e_2$ appear in the equations for the capacitors. If we use $e_1$ and $e_2$ as state variables and use the two component equations for the capacitors, we have

$$De_1 = \frac{i_{C1}}{C_1} \quad \text{with } e_1(0) \tag{4.91}$$

$$De_2 = \frac{i_{C2}}{C_2} \quad \text{with } e_2(0) \tag{4.92}$$

Using the remaining two component equations and the two node equations, we can eliminate $i_{C1}$ and $i_{C2}$ in favor of $e_0$, $e_1$, and $e_2$ to obtain

$$De_1 = -\left(\frac{1}{R_1C_1} + \frac{1}{R_2C_2}\right)e_1 + \frac{1}{R_2C_2}e_2 + \frac{1}{R_1C_1}e_0 \tag{4.93}$$

$$De_2 = -\frac{1}{R_2C_2}e_2 + \frac{1}{R_2C_2}e_1 \tag{4.94}$$

In this case, the required initial conditions are $e_1(0)$ and $e_2(0)$, which are the known initial charges on the capacitors.

### 4.3.4 Resistance-Inductance Circuits

Solenoids, relays, automotive ignition coils, and automotive fuel injectors are all examples of components that exhibit $RL$ circuit behavior. The inductance comes from the coil of wire that is used to make the magnetic circuit, and the resistance is usually due to the resistance of the small wire used in the coil.

Often, the solenoid is driven with a constant voltage source, in which case we would be interested in the current in the inductor and how long it would take to reach a steady state. For example, consider the circuit shown in Figure 4.9, which illustrates a solenoid with an inductance $L$ and a parasitic resistance $R$.

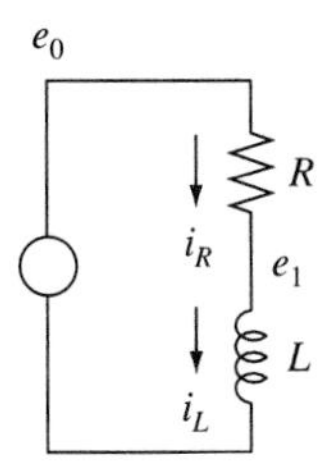


**Figure 4.9** Solenoid circuit.

The equations for the components are

$$i_R = \frac{e_0 - e_1}{R} \tag{4.95}$$

$$i_L = \frac{e_1}{LD} \tag{4.96}$$

The node equation is quite simple:

$$i_R = i_L \tag{4.97}$$

Since we want to derive an equation for $i_L$ as a function of the applied voltage $e_0$, we need to eliminate $i_R$ and $e_1$. Substituting the node equation and the resistance equation into the inductance equation and rearranging yields

$$\left[\frac{L}{R}D + 1\right]i_L = \frac{e_0}{R} \tag{4.98}$$

This equation reveals a system time constant of $\tau = L/R$ and a steady-state current of $e_0/R$.

**Example 4.2 Solenoid Coil Response**

A coil for a solenoid valve has a resistance of 100 ohms and an inductance of 6 mh. Calculate the time constant, settling time, and current required from a 24 volt source.

The time constant is given by the previous equation:

$$\tau = \frac{L}{R} = \frac{0.006 \text{ h}}{100 \text{ ohm}} = 0.06 \text{ ms}$$

The settling time is $4\tau$ or 0.240 ms. The steady-state current required is

$$i = \frac{24 \text{ volt}}{100 \text{ ohm}} = 0.24 \text{ amp}$$

The power required for this solenoid valve is almost 6 watts.

If we drive the $RL$ solenoid of Figure 4.9 with a voltage source that has an associated source resistance $R_0$, the situation is described by the circuit shown in Figure 4.10.

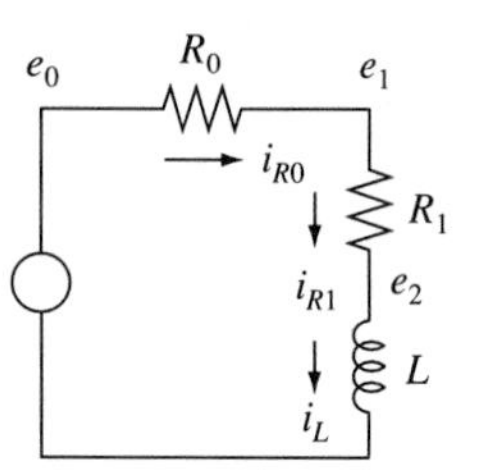


**Figure 4.10** High-voltage drive for $RL$ circuit.

The component equations are

$$i_{R0} = \frac{e_0 - e_1}{R_0} \tag{4.99}$$

$$i_{R1} = \frac{e_1 - e_2}{R_1} \tag{4.100}$$

$$i_L = \frac{e_2}{LD} \tag{4.101}$$

The node equations are

$$i_{R0} = i_{R1} \tag{4.102}$$

$$i_{R1} = i_L \tag{4.103}$$

Starting with the inductor equation, using the two resistor equations and the two node equations, and simplifying yields

$$\left[\frac{L}{R_o + R_1} D + 1\right] i_L = \frac{1}{(R_o + R_1)} e_o \tag{4.104}$$

Notice from this equation that the time constant is now $L/(R_0 + R_1)$ and the steady-state current is $e_0/(R_0 + R_1)$. If we were to increase the voltage and use a relatively high $R_0$, then the time constant could be improved (reduced) and the steady-state current would remain the same, resulting in a faster solenoid actuation time. The purpose of this approach is to drive the solenoid with a constant current source, in which case the electrical part of the actuation time would be very fast.

### 4.3.5 Resistance-Inductance-Capacitance Circuits

Even though simple passive *RC* or *RL* electrical circuits (such as that of Figure 4.8) can be represented by second- or higher order differential equations, such circuits do not exhibit resonance or overshoot, nor do they induce oscillations; in other words, the roots to their characteristic equation are not complex. Complex roots are found for circuits that have a combination of *R, L,* and *C* components. Tuning circuits, signal filters, and similar circuits use inductors and capacitors along with resistors.

**Series *RLC* Circuits.** The classical *RLC* circuit is shown in Figure 4.11.

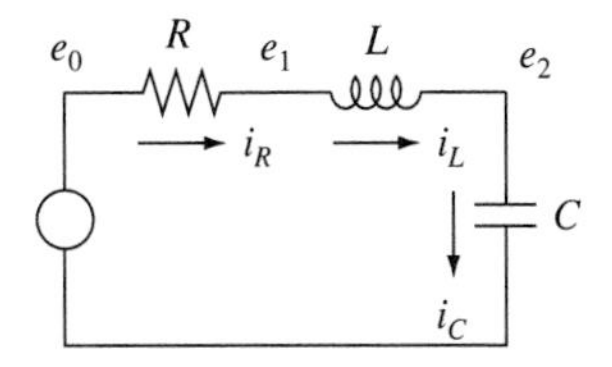


**Figure 4.11** Series *RLC* circuit.

The component equations for this circuit are

$$i_R = \frac{e_0 - e_1}{R} \tag{4.105}$$

$$i_L = \frac{e_1 - e_2}{LD} \quad \text{with } i_L(0) \tag{4.106}$$

$$i_C = CDe_2 \quad \text{with } e_2(0) \tag{4.107}$$

The node equations are

$$i_R = i_L \tag{4.108}$$

$$i_L = i_C \tag{4.109}$$

Substituting the component equations into the node equations and eliminating $e_1$ yields the differential equation for $e_2$ as a function of $e_0$:

$$[LCD^2 + RCD + 1]e_2 = e_0 \tag{4.110}$$

This is a classical second-order differential equation, as discussed in section E.3 of Appendix E. The solution of Eq. (4.110) requires the initial conditions on $e_2$ and $De_2$, which in turn would require some manipulation of the component and node equations, since we know $i_L(0)$ and $e_2(0)$. The capacitor equation gives an expression for $De_2$, and by using the second node equation, we observe the following:

$$e_2(0) = \text{known and } i_L(0) = \text{known} \tag{4.111}$$

$$De_2(0) = \frac{1}{C} i_L(0) \tag{4.112}$$

This system can be put into the state-space form by noting from the component equations that the inductor current $i_L$ and the capacitor voltage $e_2$ are differentiated; therefore, they can be used as state variables. Using the component equations for the inductor and the capacitor, and substituting the resistance equation and the node equations into the $L$ and $C$ component equations, yields the following state-space differential equations:

$$Di_L = -\frac{R}{L} i_L - \frac{1}{L} e_2 + \frac{1}{L} e_0 \tag{4.113}$$

$$De_2 = \frac{1}{C} i_L \tag{4.114}$$

The initial conditions are the natural initial conditions, $i_L(0)$ and $e_2(0)$.

**Example 4.3** ***RLC* Circuit**

A series $RLC$ circuit has an inductance $L = 1$ mh and a capacitance $C = 10\ \mu$f. Calculate the resistance $R$ required to obtain a damping ratio of 0.707.

From the previous differential equation for this circuit, we can find the damping ratio and natural frequency as follows (see Table E–3):

$$2\frac{\zeta}{\omega_n} = RC \quad \text{and} \quad \omega_n = \frac{1}{\sqrt{LC}}$$

Thus, the resistance is

$$R = \frac{\dfrac{2\zeta}{C}}{\sqrt{LC}} = 2\zeta\sqrt{\frac{L}{C}} = 2 \times 0.707\sqrt{\frac{0.001 \text{ s ohm}}{10 \times 10^{-6}\dfrac{\text{s}}{\text{ohm}}}} = 14.14 \text{ ohms}$$

Again, you are not going to buy a resistance of 14.14 ohms; and the variance on the capacitance and inductance are so large that each circuit will require a different value of resistance to achieve a damping ratio of exactly 0.707. Therefore, you will have to settle for a damping ratio around 0.707 or trim the resistance for each circuit. You should opt for the first choice, due to the cost of trimming each circuit and the fact that the circuit responses for damping values around 0.707 are not much different from each other.

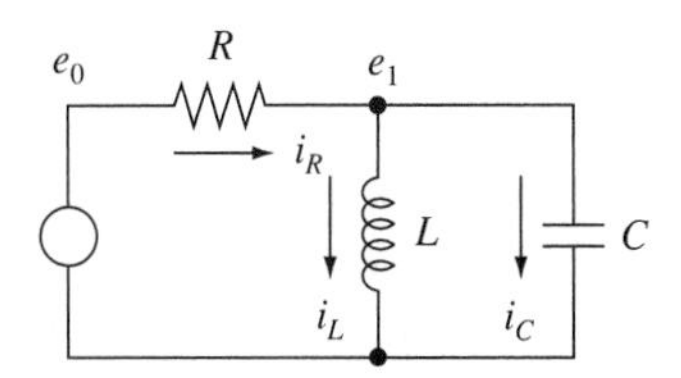


**Figure 4.12** Parallel *RLC* circuit.

**Parallel *RLC* Circuits.** The inductor and capacitor are quite often in parallel in a circuit, as illustrated in Figure 4.12. The component equations for this circuit are

$$i_R = \frac{e_0 - e_1}{R} \tag{4.115}$$

$$i_L = \frac{e_1}{LD} \quad \text{with } i_L(0) \tag{4.116}$$

$$i_C = CDe_1 \quad \text{with } e_1(0) \tag{4.117}$$

The node equation is

$$i_R = i_L + i_C \tag{4.118}$$

Substituting the component equations into the node equation and rearranging yields the classical differential equation for the voltage $e_1$ as a function of $e_0$:

$$\left[LCD^2 + \frac{L}{R}D + 1\right]e_1 = \frac{L}{R}De_0 \tag{4.119}$$

The state-space representation of this system is found by using $i_L$ and $e_1$ as state variables. Substituting the resistor equation and the node equation into the inductor and capacitor equations and rearranging, we arrive at

$$Di_L = \frac{1}{L}e_1 \tag{4.120}$$

$$De_1 = -\frac{1}{C}i_L - \frac{1}{RC}e_1 + \frac{1}{RC}e_0 \tag{4.121}$$

### 4.3.6 Summary of Passive Circuit Analysis Techniques

The **steps for deriving a classical differential equation** from a circuit can be stated in a simple procedure as follows:

1. Draw the schematic of the circuits, and identify each component with a unique symbol (e.g., $R_1$, $R_2$, $C_1$).
2. Assign a variable to represent the voltage at each node in the circuit (e.g., $e_0$, $e_1$, $e_2$).

3. Assign a variable for the current in each component (e.g., $i_{R1}$, $i_{R2}$, $i_{C1}$), and show the direction of current that will be considered positive with an arrow.
4. Write the equation for the current for each component, considering the impedance of the device (e.g., $i_{R1} = (e_1 - e_2)/R_1$, $i_{C1} = CDe_1$).
5. Write the node equations for each significant node (nodes that are not connected to a voltage source).
6. Substitute the component equations into the node equations and reduce the result to a single differential equation, using the variables considered to be inputs and outputs.

The **steps for deriving a set of first-order state-space differential equations** from a circuit can be stated as the following simple procedure:

1. Draw the schematic of the circuit, and identify each component with a unique symbolic value (e.g., $R_1$, $R_2$, $C_1$).
2. Assign a variable to represent the voltage at each node in the circuit (e.g., $e_0$, $e_1$, $e_2$).
3. Assign a variable for the current in each component (e.g., $i_{R1}$, $i_{R2}$, $i_{C1}$), and show the direction of current that will be considered positive with an arrow.
4. Write the equation for the current for each component, considering the impedance of the device (e.g., $i_{R1} = (e_1 - e_2)/R_1$, $i_{C1} = CDe_1$).
5. Write the node equations for each significant node (nodes that are not connected to a voltage source).
6. Using the capacitor voltages and inductor currents as state variables, rearrange the component equations in first-order form (e.g., $De_1 = \ldots$, $Di_3 = \ldots$). Use the remaining component and node equations to reduce the differential equations so that they contain only the state variables and the input voltage or current sources.

The order of the differential equations representing the system will be equal to the number of capacitors and inductors in the circuit that are not connected in a trivial manner. Examples of trivial connections are two capacitors in parallel (or series) with no resistance or inductance between them, and a capacitor connected directly to a voltage source, etc.

## 4.4 ACTIVE CIRCUIT ANALYSIS

**An active circuit** has passive devices, but also has amplifying devices (transistors or amplifiers) that require a power source to operate. Thus, the output voltage may come from the power source rather than an input voltage signal. Typical amplifying devices are op-amps, logic gates, transistors, and other devices that amplify signals and provide isolation from the input and output signals. We discuss the op-amp in this text; transistor operation is beyond its scope.

### 4.4.1 The Operational Amplifier

The **op-amp** is an integrated circuit that amplifies voltages. It has a very high voltage gain (approximately $10^6$ volt/volt), a high input impedance (approximately $10^6$ ohms), and a low output impedance (less than 100 ohms). The high input impedance and low input voltage means that, basically, no current is required by the amplifier; the low output impedance means that the amplifier can drive normal loads as an ideal voltage source. The high gain allows the amplifier to be used as an ideal device for computation.

As shown in Figure 4.13(a), the op-amp has inverting and noninverting inputs $e_a^-$ and $e_a^+$ and a voltage output $e_o$. Normal uses of op-amps require a differential power supply $e^+$ and $e^-$ (e.g., $+10$ volt and $-10$ volt), and the noninverting input is the reference for the differential input signal and is usually grounded. The op-amp is almost always used with an input impedance $Z_i$ and a feedback impedance $Z_f$, as shown in Figure 4.13(b). If the input is grounded and a differential power supply is used, the op-amp circuit symbol is usually simplified to that of Figure 4.13(c).

The derivation of the transfer function for the op-amp can be derived from circuit analysis by considering the op-amp as a component with the characteristics just discussed. From the notation shown in Figure 4.13(b), the following component equations can be written:

$$i_i = \frac{e_i - e_a^-}{Z_i} \tag{4.122}$$

$$i_f = \frac{e_a^- - e_o}{Z_f} \tag{4.123}$$

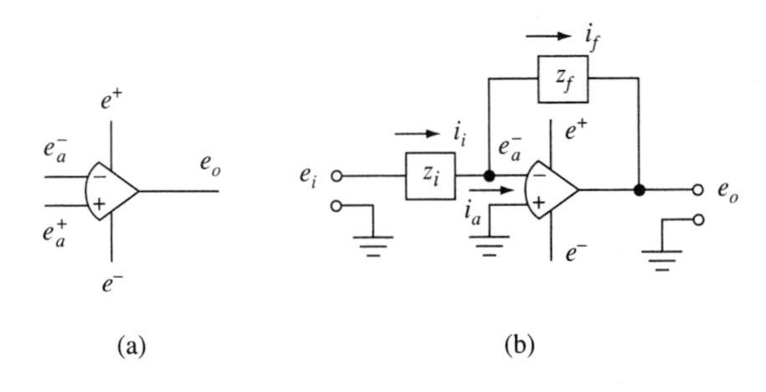


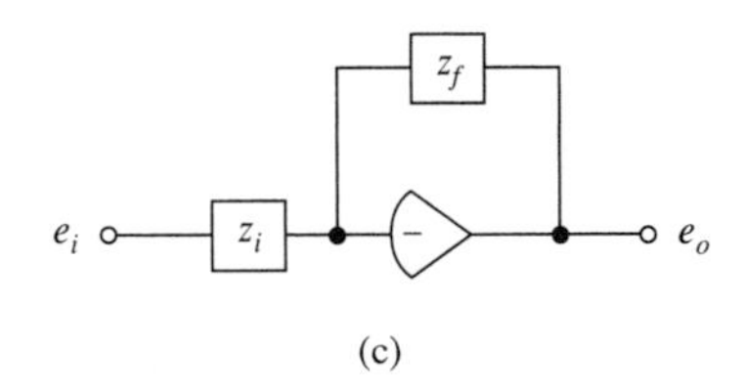


**Figure 4.13** Op-amp symbology. (a) Basic op-amp. (b) Typical application. (c) Simplified notation.

$$e_o = -Ge_a^- \quad \text{(amplifier equation)} \tag{4.124}$$

The node equation at the input port $e_a^-$ of the amplifier is

$$i_i - i_f = i_a \approx 0 \tag{4.125}$$

Substituting the component equations into the node equation and rearranging yields

$$e_o = \frac{-\dfrac{Z_f}{Z_i}e_i}{1 + \dfrac{1 + \dfrac{Z_f}{Z_i}}{G}} \tag{4.126}$$

If the impedance ratio $Z_f/Z_i$ is small (compared to $G$, which is about $10^6$), then the op-amp equation reduces to the classic result,

$$e_o = -\frac{Z_f}{Z_i}e_i \tag{4.127}$$

The reader should memorize this equation, since it is used often.

### 4.4.2 Typical Circuits

Table 4.1 illustrates a variety of uses for the op-amp with different impedances. Notice that resistive input and feedback impedances result in a voltage amplifier, a resistive input and a capacitive feedback impedance results in an integrator, and a capacitive input and a resistive feedback impedance results in a differentiator. Other combinations of series and parallel impedances in the input and feedback impedances result in a variety of interesting transfer functions.

The op-amp circuits described so far in this section can be used in more complex circuits. The transfer functions for the op-amp circuits can be inserted in the circuit analysis, since the op-amp acts as a voltage source that provides an isolation in voltages. For example, consider an op-amp driving a solenoid, as shown in Figure 4.14. The circuit analysis for this system can be divided into two parts, since the op-amp acts as a voltage driver for the solenoid $RL$ circuit.

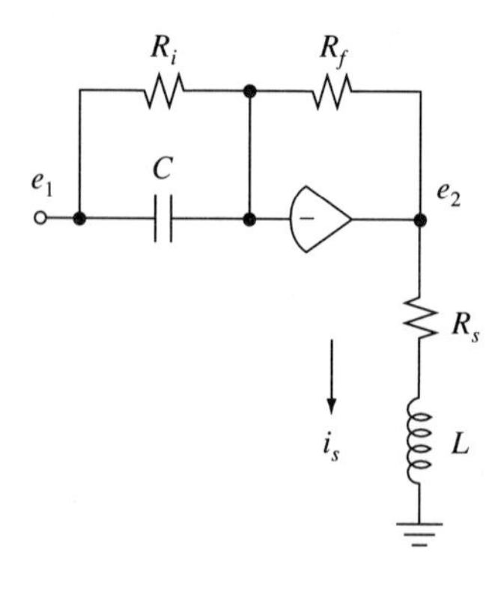


**Figure 4.14** Op-amp solenoid driver.

**TABLE 4.1** Op-Amp Circuits.

| Description | Transfer Function | Circuit |
|---|---|---|
| Sign Changer | $e_o = -e_i$ | 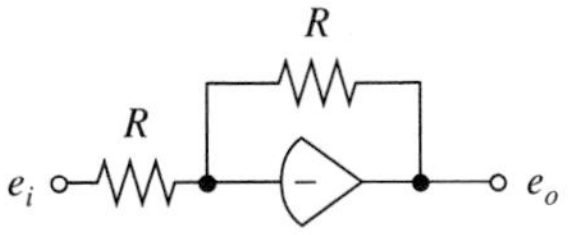 |
| Amplifier | $e_o = -\frac{R_f}{R_i} e_i$ | 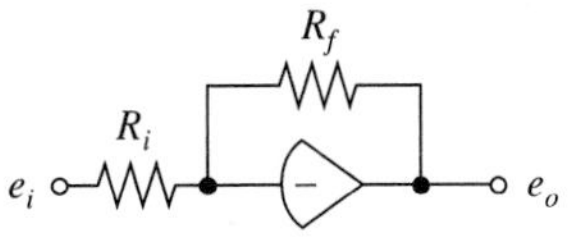 |
| Integrator | $e_o = \frac{-e_i}{\tau D}$<br>$\tau = RC$ | 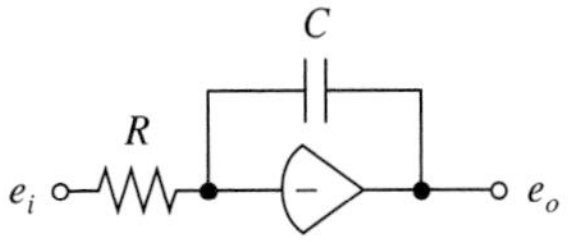 |
| Differentiator | $e_o = -\tau D e_i$<br>$\tau = RC$ | 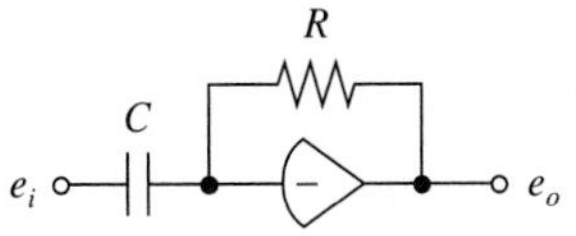 |
| Lag | $e_o = \frac{-\frac{R_f}{R_i} e_i}{(\tau D + 1)}$<br>$\tau = R_f C$ | 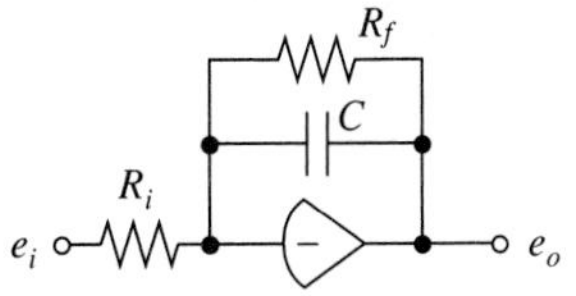 |
| Lead | $e_o = -\frac{R_f}{R_i}(\tau D + 1)$<br>$\tau = R_i C$ | 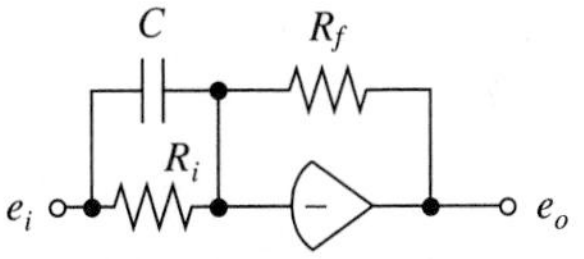 |
| Lead-Lag<br>or<br>Lag-Lead | $e_o = -\frac{R_f}{R_i}\frac{(\tau_i D + 1)e_i}{(\tau_f D + 1)}$<br>$\tau_i = R_i C_i$<br>$\tau_f = R_f C_f$ | 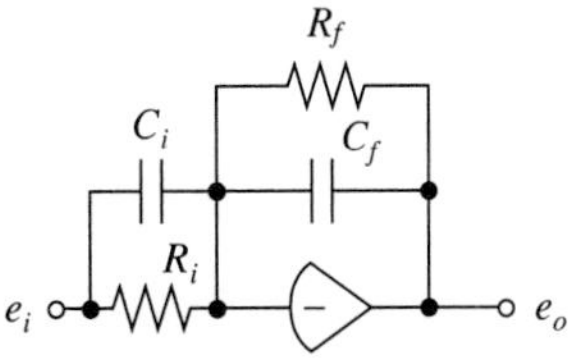 |
| Bandwidth-Limited Integrator | $e_o = \frac{-(\tau_f D + 1)\,e_i}{\tau_i D}$<br>$\tau_f = R_f C$<br>$\tau_i = R_i C$ | 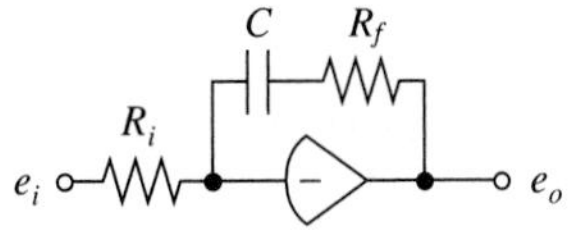 |
| Bandwidth-Limited Differentiator | $e_o = \frac{-\tau_f D e_i}{(\tau_i D + 1)}$<br>$\tau_f = R_f C$<br>$\tau_i = R_i C$ | 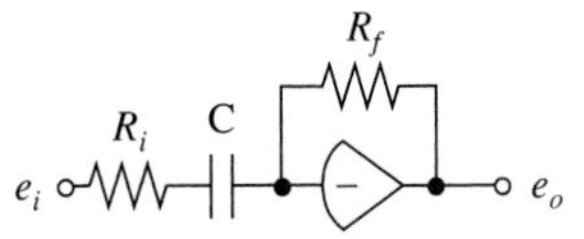 |

We have

$$e_2 = -\frac{R_f}{R_i}(R_iCD + 1)e_1 \tag{4.128}$$

The solenoid circuit can be treated as a series *RL* circuit, yielding

$$i_s = \frac{e_2}{(R_s + LD)} \tag{4.129}$$

Thus, the solenoid current has the following transfer function:

$$i_s = \frac{-\frac{R_f}{R_i}(R_iCD + 1)}{\left(\frac{L}{R_s}D + 1\right)}\frac{e_1}{R_s} \tag{4.130}$$

Now consider the circuit shown in Figure 4.15, in which a sign-changing op-amp is used to drive a filtering op-amp circuit. The transfer functions for these two circuits can be multiplied together, since the first amplifier acts as an isolated voltage source driving the second op-amp circuit.

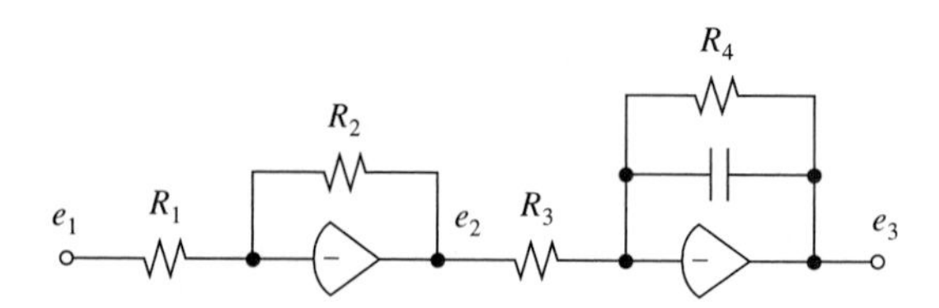


**Figure 4.15** Dual op-amp circuit.

The equations for the two op-amps can be written independently as

$$e_2 = -\frac{R_2}{R_1}e_1 \tag{4.131}$$

$$e_3 = \frac{-\frac{R_4}{R_3}e_2}{(R_4CD + 1)} \tag{4.132}$$

Thus, the overall transfer function from $e_1$ to $e_3$ is

$$e_3 = \frac{\frac{R_2}{R_1}\frac{R_4}{R_3}}{(R_4CD + 1)}e_1 \tag{4.133}$$

Notice that the sign changes cancel each other.

**Example 4.4 Op-Amp Low-Pass Filter**

A low-pass filter is a circuit or system that will pass low frequencies, but will attenuate high frequencies. Generally speaking, the frequency at which the circuit switches from passing to attenuating the signal is termed the break frequency, or the bandwidth. For a first-order system, the break frequency is $1/\tau$ (rad/s).

Design an op-amp low-pass filter that will have a break frequency of 1000 Hz.

The low-pass filter is the lag circuit shown in Table 4.1. A break frequency of 1000 cycles/s corresponds to 6283 rad/s. The break frequency (in units of rad/s) is the inverse of the time constant, from Table 4.1:

$$\omega_b = \frac{1}{R_f C} = 6283 \frac{\text{rad}}{\text{s}}$$

The resistance values used in op-amp circuits are large compared to the output impedance, but are small compared to the input impedance of the op-amp itself. Thus, values ranging from 1k to 100k are common. If we use a nominal value of 10k for the resistors, then the value of capacitance that is required is

$$C = \frac{1}{6283 \frac{\text{rad}}{\text{s}} 10{,}100 \text{ ohm}} = 0.016\ \mu\text{f}$$

Since a 0.016 $\mu$f capacitor is not common, we could use a capacitance value of 0.022 $\mu$f and a 10 kohm potentiometer trimmed down to 7.23 kohm to achieve the desired break frequency.

## 4.5 SUMMARY

In this chapter, we have identified the basic components of passive circuits—resistance, capacitance, and inductance—and have illustrated their fundamental nature. Resistors dissipate power, while capacitors and inductors store energy. Techniques for modeling passive circuits were developed, and methods were presented for the derivation of differential equations, transfer functions, and state-space differential equations.

Operational amplifiers were examined, with particular attention to their circuit characteristics and capabilities. The modeling of op-amp circuits to obtain transfer functions, filters, and amplifier circuits was discussed, with the engineering design of typical circuits presented in examples.

## REFERENCES

4.1 Bell, David A. *Fundamentals of Electric Circuits,* 4th ed. Prentice Hall, Englewood Cliffs, NJ, 1988.

4.2 Boylestad, Robert L. *Introduction to Circuit Analysis,* 6th ed. Macmillan Publishing Co., New York, 1990.

4.3 Jackson, Herbert W. *Introduction to Electric Circuits,* 6th ed. Prentice Hall, Englewood Cliffs, NJ, 1986.

4.4 Neudorfer, Paul O., and Hassui, Michael. *Introduction to Circuit Analysis*. Allyn and Bacon, Boston, 1990.

4.5 Nilsson, James W. *Electric Circuits,* 4th ed. Addison-Wesley Publishing Co., Reading, MA, 1993.

4.6 Scott, Ronald E. *Linear Circuits*. Addison-Wesley, Reading, MA, 1960.

4.7 Suprynowicz, V. A. *Electrical and Electronics Fundamentals*. West Publishing, St. Paul, 1987.

## NOMENCLATURE

$A$ cross-sectional area
$C$ capacitance
$d$ distance
$D$ differential operator $= d/dt$
$e$ voltage
$G$ op-amp gain
$i$ current
$j$ complex variable; $j = \sqrt{-1}$
$\ell$ length
$L$ inductance
$n$ number of turns
$P$ power
$Q$ charge
$R$ resistance
$t$ time
$W$ energy
$Z$ impedance
$\epsilon$ permittivity of dielectric
$\phi$ magnetic flux
$\mu$ permeability
$\rho$ resistivity of a material
$\tau$ time constant
$\omega$ frequency

## PROBLEMS

**4.1** A resistor of 1 kΩ has a voltage of 10 v placed across it. *Calculate* the current. *What* is the power dissipated?

**4.2** A high-voltage resistor of 1 MΩ is connected to a 20-kv voltage source. *Calculate* the current and power in the resistor.

**4.3** A stereo system is advertised as having 50 watts of music power. *Calculate* how much voltage the amplifier must be putting out to have that much power go into an 8-Ω speaker.

**4.4** A 100 $\mu$f capacitor is charged to 10 volts. *Calculate* the amount of energy (in joules) stored in the capacitor. To get a better feel for how much energy this is, calculate the equivalent mechanical potential energy of a weight elevated above a reference height. In other words, *what* mass (in gm) must be lifted 1 m to be the equivalent amount of energy?

**4.5** A large capacitor is used to store energy. The capacitance is 1000 $\mu$f and the voltage is 100 volts. *Calculate* the amount of energy (in joules) stored in the capacitor. To get a better feel for how much energy this is, calculate the equivalent mechanical potential energy of a weight elevated above a reference height. In other words, *what* mass (in kg) must be lifted 1 m to be the equivalent amount of energy?

**4.6** In an electric vehicle, the kinetic energy of the vehicle in motion is converted to electrical potential energy by regenerative braking. The vehicle has a mass of 1000 kg and is traveling at a speed of 50 km/hr. If the combined electric motor and controller is 90% efficient in converting the mechanical energy into electrical energy, *what* size of capacitor must be used to store the electrical energy at 120 volts?

**4.7** An inductor is used in a filter circuit to help suppress voltage transients. During the start-up of the circuit, the 100 mh inductor experiences a current transient of 3 amps. *Calculate* the energy stored in the inductor during this transient. To get a better feel for how much energy this is, calculate the equivalent mechanical potential energy of a weight elevated above a reference height. In other words, *what* mass (in gm) must be lifted 1 m to be the equivalent amount of energy?

**4.8** *Derive* an expression for the equivalent resistance for the circuit shown in Figure P4.8.

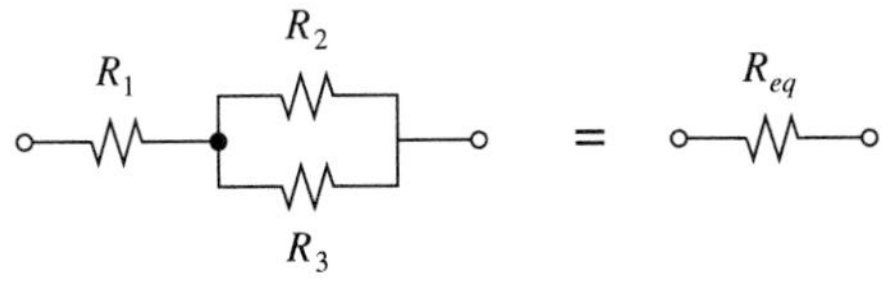

**Figure P4.8** Resistor circuit.

**4.9** *Derive* an expression for the equivalent resistance for the circuit shown in Figure P4.9.

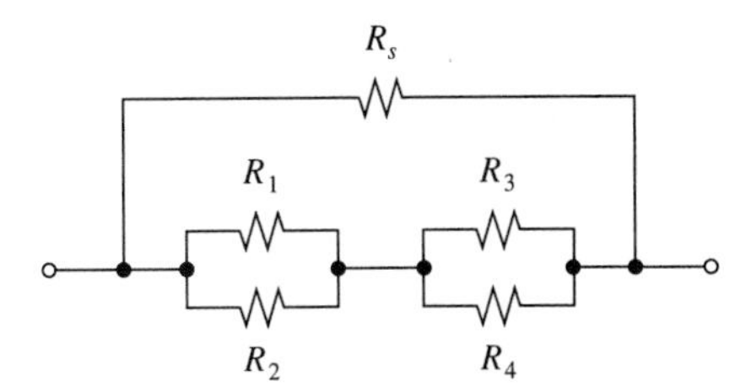


**Figure P4.9** Resistor circuit.

**4.10** *Calculate* the current from the voltage source in the circuit shown in Figure P4.10 with $R_1 = 8\Omega, R_2 = 8\Omega, R_3 = 4\Omega, R_4 = 4\Omega$.

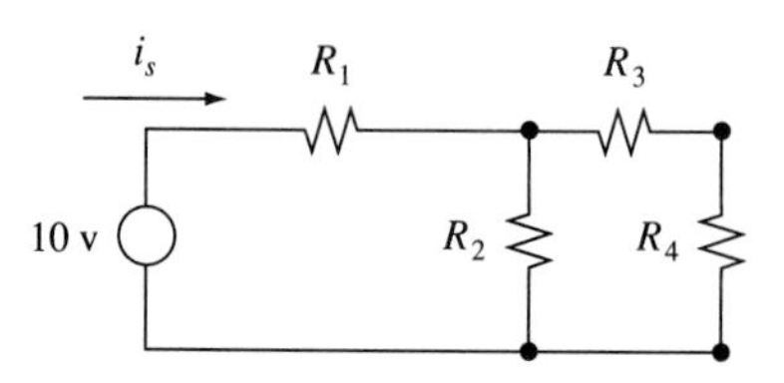


**Figure P4.10** Resistor circuit.

**4.11** *Calculate* the amount of power dissipated in resistor $R_3$ in the circuit shown in Figure P4.11.

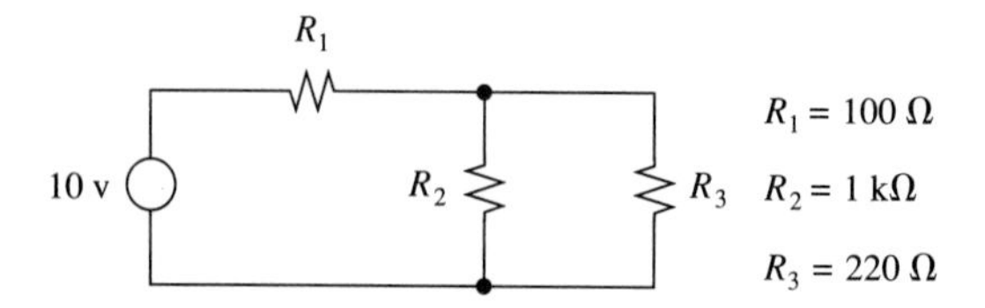


**Figure P4.11** Resistor circuit.

**4.12** For the *RC* network shown in Figure P4.12(a),

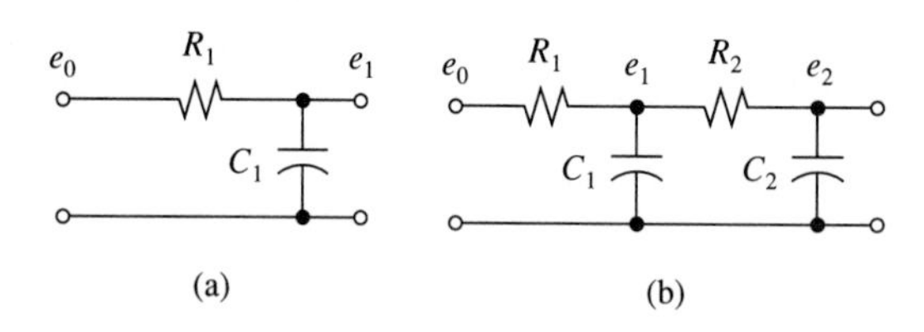


**Figure P4.12** (a) *RC* circuit. (b) Dual *RC* circuit.

**a.** *Write* the modeling equations.

**b.** *Derive* the differential equation of the form $[\tau D + 1]e_1 = Ge_0$.

**c.** *What* are the mathematical expressions for the time constant $\tau$ and the gain $G$?

Suppose two *RC* networks are lumped together as shown in Figure P4.12(b).

**a.** *Write* the modeling equations.

**b.** *Derive* the differential equation for the circuit.

**c.** You might be tempted to think that the differential equation for this circuit is simply the product of two *RC*'s. *Prove* that it is not by asking whether

$$(\tau_1 D + 1)(\tau_2 D + 1)e_2 = e_0?$$

**4.13** Shown in Figure P4.13 is an electric circuit.

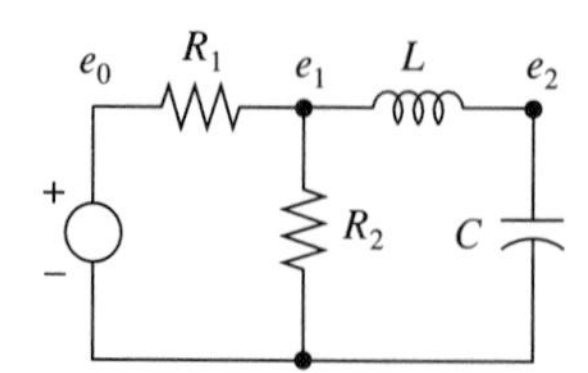


**Figure P4.13** *RLC* circuit.

**a.** *Write* the modeling equations.

**b.** *Derive* the differential equation for $e_2$ as a function of $e_0$.

**c.** *State* the initial conditions necessary for the differential equation.

**d.** *State* the natural frequency, the damping ratio, and the static gain of the circuit.

**4.14** For the circuit shown in Figure P4.14, *write* all of the modeling equations and *derive* the transfer function $e_2/e_0$. All initial conditions are zero.

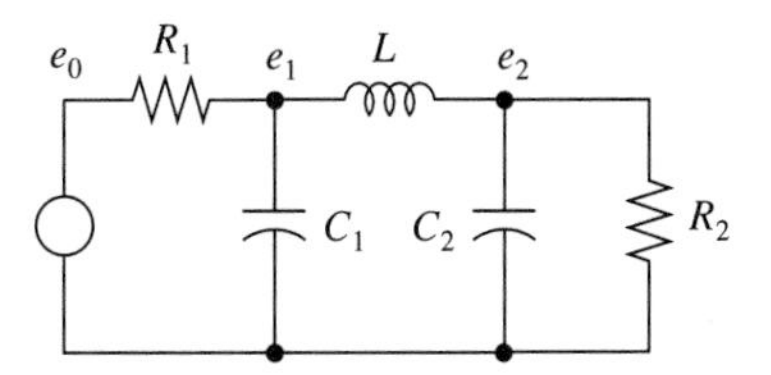


**Figure P4.14** *RLC* circuit.

**4.15** For the circuit shown in Figure P4.14, *write* all of the modeling equations, *select* state variables, and *derive* the state-space representation for the system, using the natural dynamic variables.

**4.16** For the circuit shown in Figure P4.16, *write* the modeling equations and *derive* the transfer function for $e_3$ as a function of $e_0$. *What* is the static gain?

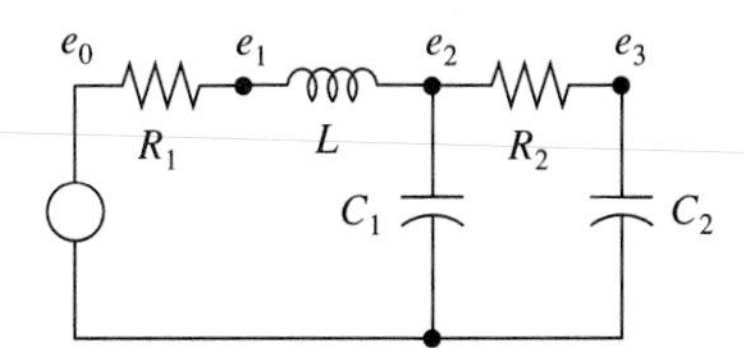


**Figure P4.16** *RLC* circuit.

**4.17** For the circuit shown in Figure P4.16, *write* all of the modeling equations and *select* state variables based on the natural dynamic variables. *Derive* a state-space representation of the system. *State* the initial conditions of the state variables.

**4.18** Shown in Figure P4.18 is an *RLC* circuit with a parallel bypass resistor. *Write* the modeling equations and *derive* a differential equation for $e_1$ as a function of $e_0$. *Express* the required initial conditions of this second-order differential equation in terms of the known initial conditions $e_1(0)$ and $i_L(0)$.

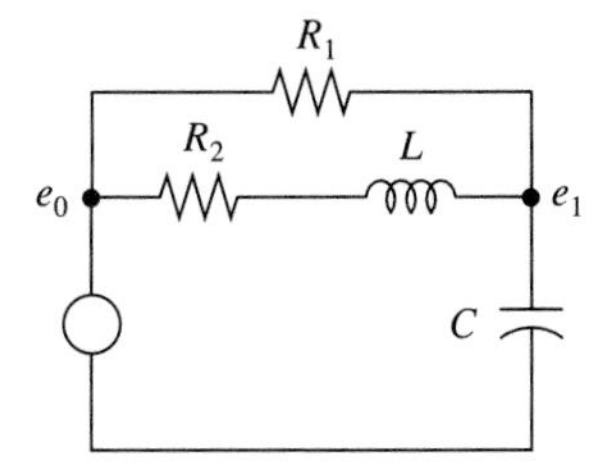


**Figure P4.18** *RLC* circuit.

**4.19** *Derive* a state-space representation of the circuit shown in Figure P4.18, using the natural dynamic variables $e_1$ and $i_L$. *State* the initial conditions of the state variables.

**4.20** Shown in Figure P4.20 is an *RLC* circuit with two voltage inputs. *Write* all of the modeling equations and *derive* a transfer function equation relating the output voltage $e_4$ as a function of the inputs $e_1$ and $e_2$.

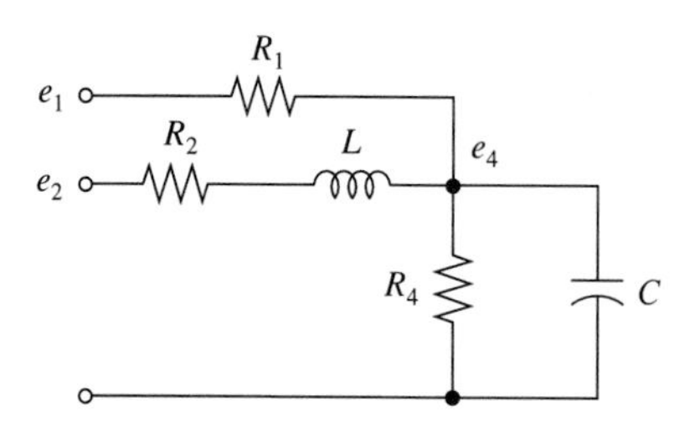


**Figure P4.20** Dual-input *RLC* circuit.

**4.21** For the *RLC* circuit of Figure P4.20, *write* the necessary modeling equations and *select* state variables based upon the natural dynamic variables of the system. *Derive* a set of state-space differential equations that has two state variables and two inputs. *What* are the initial conditions of the state variables?

**4.22** Shown in Figure P4.22 is an op-amp circuit with impedances on the plus and minus inputs.

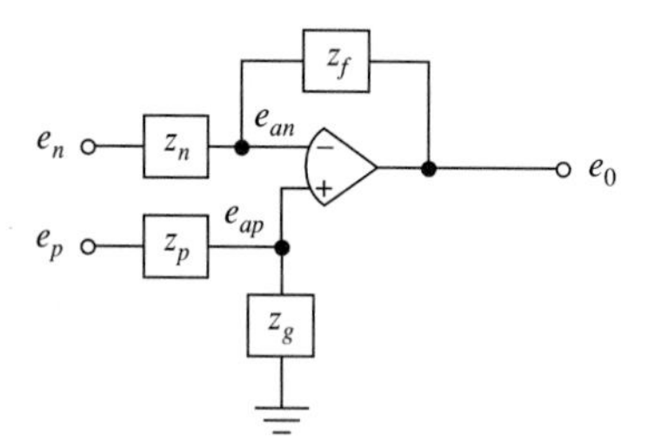


**Figure P4.22** Basic op-amp circuit.

**a.** *Derive* the output equation $e_0$ as a function of $e_n$ and $e_p$. The amplifier has the characteristic $e_0 = G(e_{ap} - e_{an})$, where G >> 1.

**b.** *Show* that if all impedances are resistive and equal to the value R, then $e_0 = e_p - e_n$.

**4.23** Shown in Figure P4.23 is an op-amp filter circuit used to drive an electromagnetic coil on a servo valve. *Write* the modeling equations and *derive* the transfer equation for $i_v$ as a function of the input voltage, $e_i$.

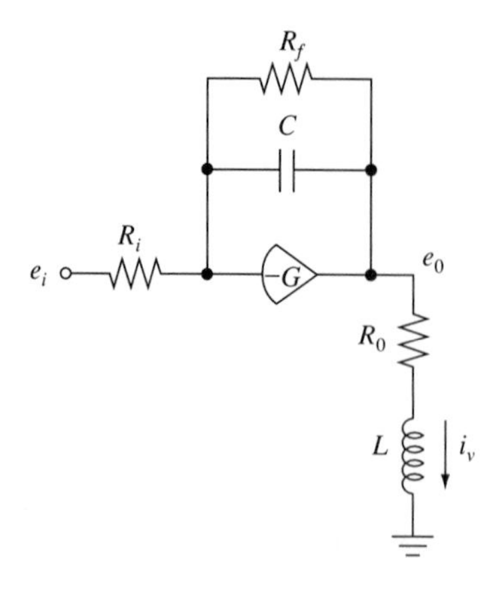


**Figure P4.23** Op-amp filter driving solenoid valve.

**4.24** *Write* the circuit equations and *derive* a state-space representation of the filter circuit shown in Figure P4.23.

**4.25** Shown in Figure P4.25 is an electronic circuit with an op-amp buffer. *Derive* the differential equation for $e_0$ as a function of the input $e_i$. *Calculate* the static gain, natural frequency, and damping ratio if $R_f = 10\ k\Omega$, $R_i = 10\ k\Omega$, $C_f = 1\ \mu f$, $R_L = 500\ \Omega$, and $C_L = 10\ \mu f$.

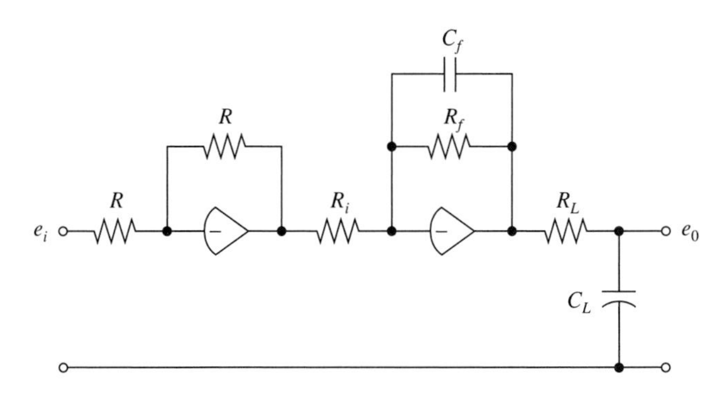


**Figure P4.25** Op-amp filter circuit.

**4.26** *Derive* a state-space representation for the circuit of Figure P4.25.

**4.27** Shown in Figure P4.27 is an op-amp circuit with two capacitors. *Write* the modeling equations and *derive* the transfer function for $e_0$ as a function of $e_i$.

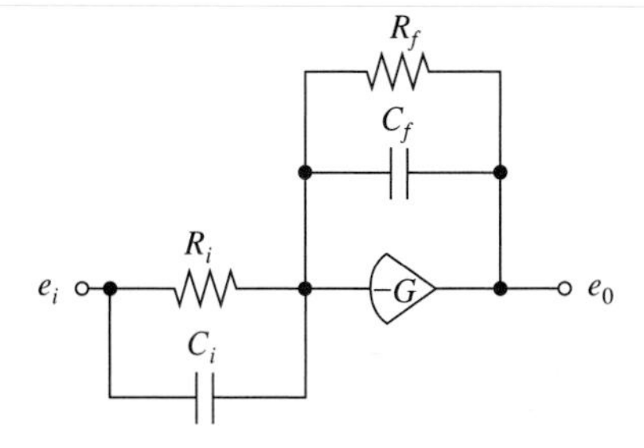


**Figure P4.27** Op-amp filter circuit.

**4.28** A full-bridge strain gauge circuit is used to measure the force applied to a bar by measuring the strain in the small steel bar. The strain gauge resistance elements are 350 Ω. The bridge is driven with a 5 volt power source. As the strain is applied, one of the bridge resistors changes resistance according to the relationship $R_2 = 350 + cF$, where $F$ is the applied force and the coefficient $c$ works out to be 0.7 $\Omega/N$ (taking into consideration the size and stiffness of the bar and the sensitivity of the strain gauge element). From this basic bridge circuit, a differential voltage $\Delta e_b$, can be expressed as a function of the force to be measured. Since this voltage differential is very small, it is desired to use an op-amp circuit to amplify the signal into the 10 volt range. In addition, because there are some unwanted vibrations in the bar, it is required to filter the measured force signal to eliminate the high-frequency vibrations, which are expected to be a few hundred hertz. A first-order system will suffice.

The exact requirements are that a ±20 N force should produce a ±10 volt output signal from the op-amp circuit. The overall circuit should have a dynamic time constant of 1.5 milliseconds. *Draw* a complete circuit diagram of this system, including the strain gauge bridge circuit with amplifier. *Derive* a complete mathematical model of the system from force input to voltage output. *Select* values for all resistors and capacitors.

**4.29** A flashing light is to be placed on the output of an audio amplifier to show the intensity of the sound coming from the speaker. You must select an impedance for this lightbulb that will not degrade the voltage going to the speaker. Intuitively, if the impedance is very large relative to the output impedance of the amplifier, then there will be no degradation; however, if the light resistance starts approaching the impedance of the speaker, then a considerable amount of power will be going into the light, and the speaker and the sound will be degraded. This is clearly undesirable.

*Derive* an expression for the voltage at the speaker in the undisturbed circuit (Figure P4.29(a)) and the circuit loaded with the light (Figure P4.29(b)). The output impedance $R_o$ of the amplifier is 8 Ω, and the impedance of the speaker, $R_{speaker}$, is also 8 Ω. At *what* value of light impedance $R_{light}$ is there a 1% degradation in the voltage going to the speaker (i.e., at what value of light impedance will the gain be 0.99 of the undisturbed gain)?

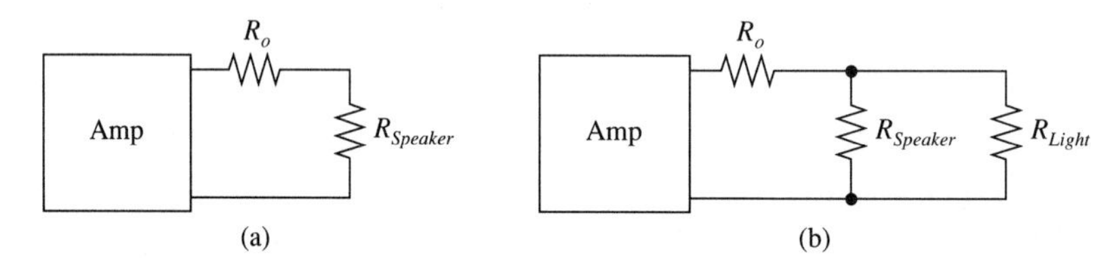


**Figure P4.29** Audio amplifier circuit with light bulb. (a) Normal audio circuit of amplifier and speaker. (b) Modified circuit with lightbulb.

Assuming that the maximum voltage to the speaker is 12 volts, *can* you find a lightbulb with the correct impedance and voltage range in a catalog? If so, *what* is the wattage of the bulb?

**4.30** If an analog voltmeter having an input impedance $R_{meter}$ of 100 kΩ is connected to the circuit shown in Figure P4.30(a) (which would result in the circuit of Figure P4.30(b)), *would* the voltage $e_1$ that you measured be accurate? In other words, would the voltage in the circuit actually go down because of the loading of the voltmeter? If a digital voltmeter having an input impedance of 1 MΩ were used, *would* the voltage in the circuit be degraded?

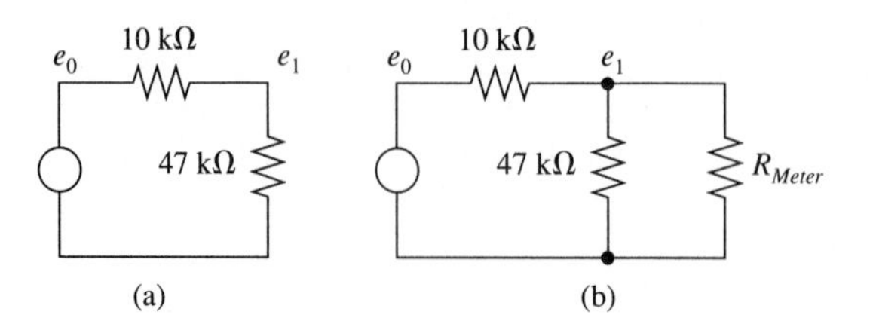


**Figure P4.30** Measurement of voltage in a circuit. (a) Undisturbed circuit. (b) Circuit with voltmeter attached.

# CHAPTER 5

# Fluid Systems

## 5.1 INTRODUCTION

**Fluid systems** range from simple systems involving fluid flow with valves and lines, such as garden hoses with nozzles, gas lines and burners, and gasoline pumps and nozzles, to hydraulic and pneumatic control systems involving pumps, pressure regulators, control valves, actuators, and servocontrols. Industrial applications include automation, logic and sequence control, holding fixtures, high-power motion control, etc. Automotive applications include power steering, power brakes, hydraulic brakes, ventilation controls, and more. Aerospace applications include flight controls systems, steering control systems, air conditioning, brake control systems, and numerous other applications [See Refs. 1–5].

The main reason that hydraulic and pneumatic systems are so popular compared to their electro-mechanical counterparts is the power density capability of the pump and actuators. Electromagnetic motors, generators, and actuators are limited by magnetic field saturation and can produce up to about 200 pounds per square inch of actuator. In hydraulic systems, 3000 to 8000 pounds per square inch of actuator is common in aircraft applications, and 1000 pounds per square inch is common in industrial applications. Therefore, the hydraulic systems required to reproduce a given force output are much smaller. This advantage is true in the generation, as well as the actuation, of the hydraulic power. The hydraulic pumps are smaller than the counterpart electric generator; therefore, the overall system benefits doubly from the larger ratio of force per unit volume. Electromagnetic systems require a ferrous metal to generate a force or torque, and quite often this metal has to move with the motor or actuator; therefore, electromagnetic motors or actuators have a large inertia associated with their motion, so they cannot accelerate quickly. Hence, hydraulic and pneumatic systems are more responsive and have a greater bandwidth of operation at the same power output levels.

A second advantage of fluid control systems is that the fluid circulating to and from an actuator removes the heat generated by the actuator that is doing work. This heat follows the fluid back to a reservoir to be dissipated in a better location than inside the actuator. Electro-mechanical actuators and motors have limited ability to dissipate the heat generated inside the device and rely on free or forced convection to the surrounding environment. Heat is the predominate damaging mechanism in electric and electronic systems. Therefore, the reliability of electromagnetic devices is limited compared to that of hydraulic and pneumatic systems.

Hydraulic and pneumatic systems generally have more significant nonlinearities than do electric or mechanical systems, so we meet a new challenge in modeling and simulating them. In this chapter, we discuss the fluid properties and the various components common to fluid control systems. Hydraulic and pneumatic systems are treated concurrently in the text.

## 5.2 PROPERTIES OF FLUIDS

A **fluid** is a substance that is characterized by its inability to support a static shear and includes both liquids and gases. Liquids are nearly incompressible, whereas gases are highly compressible. Other than orders of magnitude difference in their density, absolute viscosity, and bulk modulus, liquids are distinguished from gases by their **surface tension** effects. In the presentation that follows, incompressible and compressible fluids are discussed in a comparative manner.

### 5.2.1 Density

The **density** $\rho$ of a fluid is defined as the mass $m$ per unit volume $V$ under specified conditions of pressure and temperature:

$$\rho = \left. \frac{m}{V} \right|_{P_0,\, T_0} \tag{5.1}$$

### 5.2.2 Equation of State: Liquids

The density of any fluid is a function of pressure and temperature. An **equation of state** is used to relate the density, pressure, and temperature of a fluid.

An equation of state applicable to all liquids over a large range of pressures and temperatures does not exist in an exact form. However, a relation derived from a Taylor series expansion is quite valid over limited ranges of pressure and temperature. The first-order Taylor series of density $\rho$ as a function of pressure $P$

and temperature $T$ can be stated as follows ($\rho_0$, $P_0$, and $T_0$ are the reference conditions):

$$\rho = \rho_0 + \left.\frac{\partial \rho}{\partial P}\right|_{P_0,\,T_0}(P - P_0) + \left.\frac{\partial \rho}{\partial T}\right|_{P_0,\,T_0}(T - T_0) \tag{5.2}$$

The partial derivatives in this equation have been clearly identified as related to the bulk modulus, $\beta$, and the thermal expansion coefficient, $\alpha$ of the liquid. Mathematically,

$$\rho = \rho_0\left[1 + \frac{1}{\beta}(P - P_0) - \alpha(T - T_0)\right] \tag{5.3}$$

where

$$\beta = \rho_0\left.\frac{\partial P}{\partial \rho}\right|_{P_0,\,T_0} = \left.\frac{\partial P}{\partial \rho/\rho_0}\right|_{P_0,\,T_0} \tag{5.4}$$

$$\alpha = -\left.\frac{1}{\rho_0}\frac{\partial \rho}{\partial T}\right|_{P_0,\,T_0} \tag{5.5}$$

The **bulk modulus** is the inverse of the compressibility of the fluid and is related to Young's modulus (Thomas Young, English philosoper on nature, 1773–1829) for solid materials. The bulk modulus can be measured by noting the change in pressure with a fractional change in volume of a fixed mass of fluid:

$$\beta = -\frac{\partial P}{\partial V/V_0} \tag{5.6}$$

Here, $\beta$ is the **isothermal bulk modulus** (or merely the bulk modulus) and can be used when the pressure changes occur at slow enough rates during heat transfer to maintain constant temperature. The **adiabatic bulk modulus** $\beta_a$ is related to the isothermal bulk modulus by the equation

$$\beta_a = \frac{C_p}{C_v}\beta \tag{5.7}$$

and can be used when the rate of pressure change is rapid enough to prevent significant heat transfer. The ratio of specific heats, $C_p/C_v$, is only slightly greater than 1.0 for liquids. The isothermal bulk modulus for typical liquids is given in Figure 5.1.

The **thermal expansion coefficient** relates the incremental change in volume with changes in temperature and can be expressed in terms of volumes for a fixed mass of fluid as:

$$\alpha = \left.\frac{\partial V/V_0}{\partial T}\right|_{P_0,\,T_0} \tag{5.8}$$

The thermal expansion coefficient has an approximate value of $\alpha = 0.5 \times 10^{-3}/°\text{F}$ for most liquids, as can be observed from Figure 5.2.

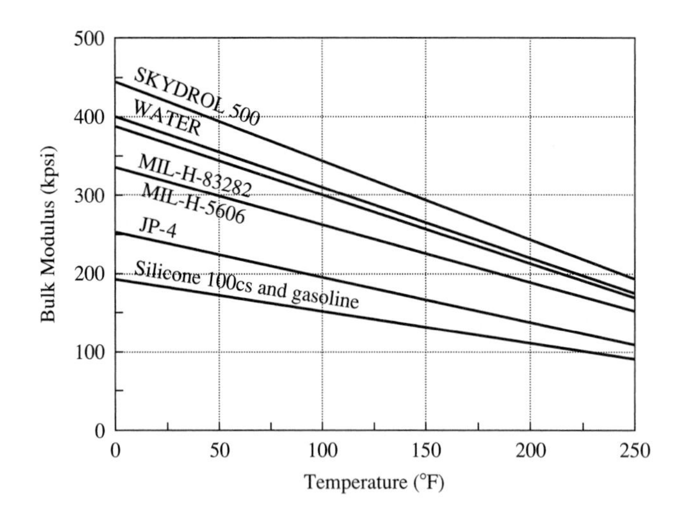


**Figure 5.1** Bulk modulus of typical liquids (from the Lee Company).

### 5.2.3 Equation of State: Gases

The **equation of state** for an ideal gas is

$$\rho = \frac{P}{RT} \tag{5.9}$$

where $P$ and $T$ are in absolute terms and $R$ is the gas constant. Most gases follow this ideal behavior for considerable ranges of pressure and temperatures. Figure 5.3 gives the densities of several typical gases as a function of temperature at one atmosphere. The density at any other pressure can be calculated using pressure ratios.

A gas undergoing a polytropic process follows the relationship

$$\frac{P}{\rho^n} = C = \text{ constant} \tag{5.10}$$

where

$n = 1.0$ for an isothermal process
$n = k$ for an adiabatic process ($k$ = ratio of specific heats)
$n = 0.0$ for an isobaric process
$n = \infty$ for an isovolumetric process

Using the preceding polytropic process law, we can interpret the definition of bulk modulus for a gas:

$$\beta = \rho_0 \left.\frac{\partial P}{\partial \rho}\right|_{P_0,\, T_0} \tag{5.11}$$

Taking the partial derivative in this equation and noting that $P = C\rho^n$ yields

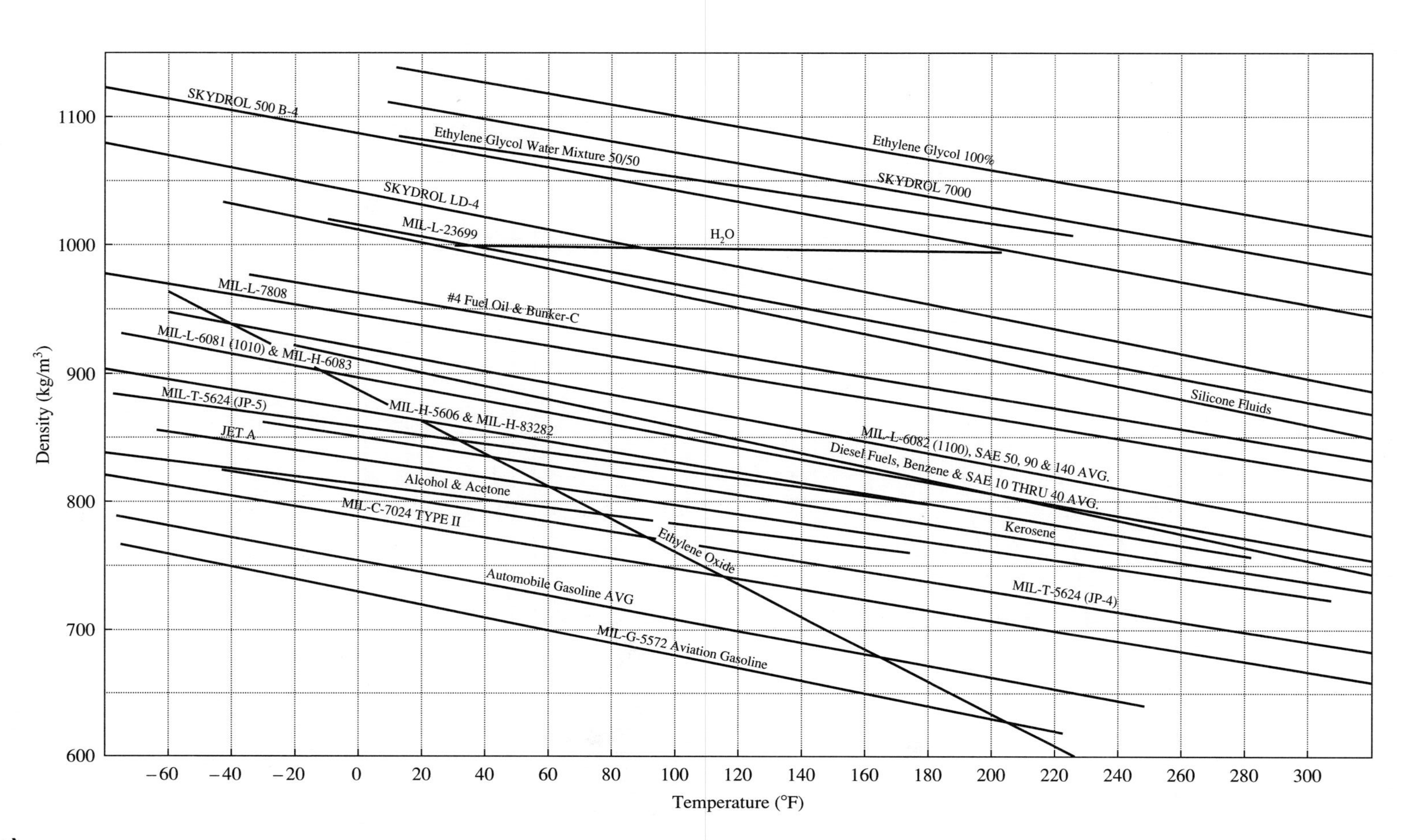


**Figure 5.2** Density of typical liquids (reproduced with permission from the Lee Company [Ref. 6]).

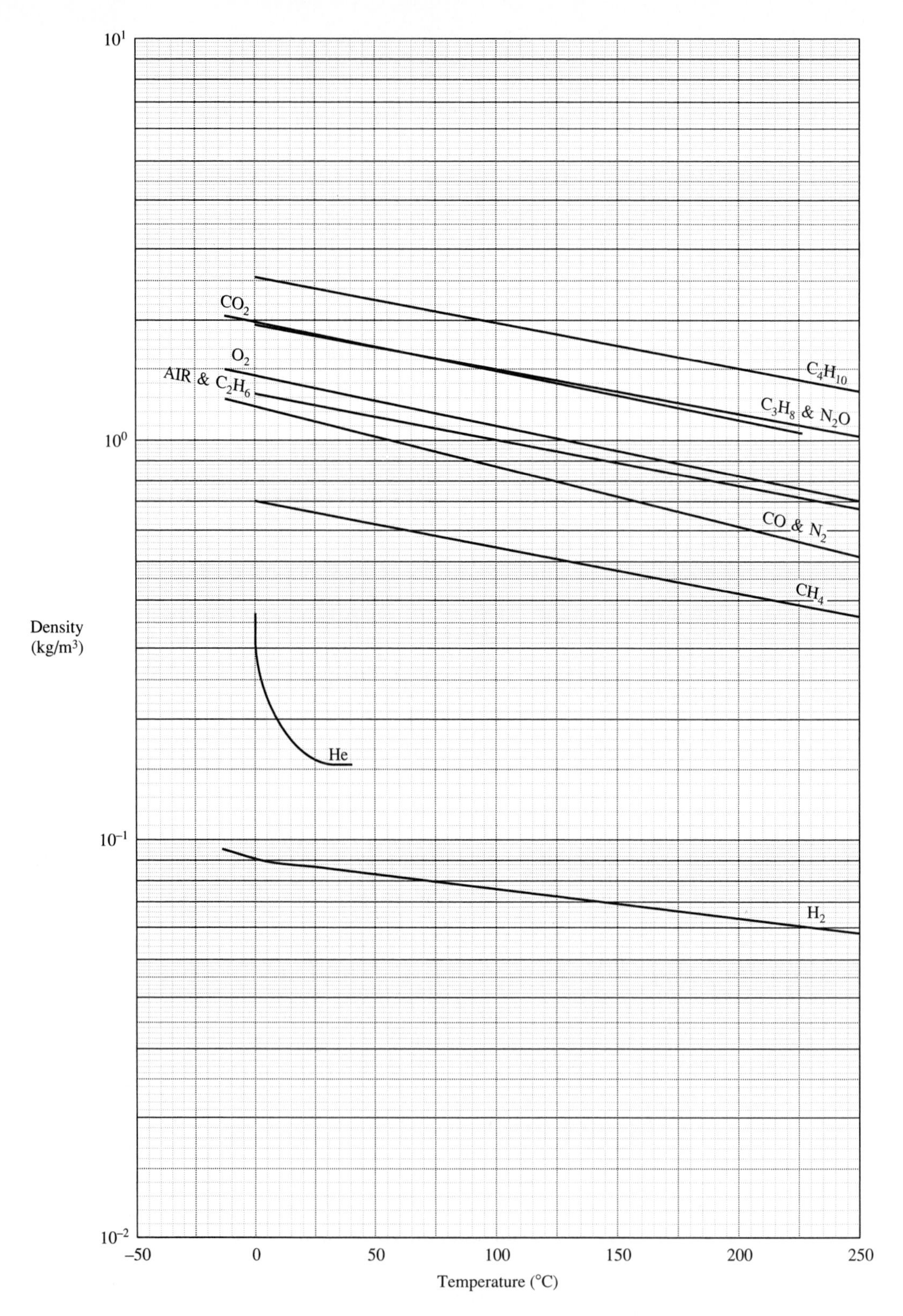


**Figure 5.3** Densities of typical gases [Ref. 7].

$$\beta = \rho_0[nC\rho^{n-1}]\Big|_{P_0,\,T_0} = \rho_0 n \frac{C\rho^n}{\rho}\Big|_{P_0,\,T_0} \tag{5.12}$$

$$\beta = nP_0 \tag{5.13}$$

The bulk modulus of a gas is thus related to the absolute pressure of the gas. Notice that the bulk modulus of a liquid is on the order of 5000 to 15,000 bar, whereas $\beta$ for a gas is usually on the order of 1 to 10 bar.

### 5.2.4 Viscosity

The **absolute viscosity** (or dynamic viscosity) $\mu$ of a fluid represents the ability of the fluid to support a shear stress $\tau$ between a relative velocity $u$ of a fluid and a solid boundary, as shown in Figure 5.4. ($y$ is the coordinate direction normal to the direction of flow.) Mathematically,

$$\mu = \frac{\text{shear stress}}{\text{shear rate}} = \frac{\tau}{\partial u/\partial y} \tag{5.14}$$

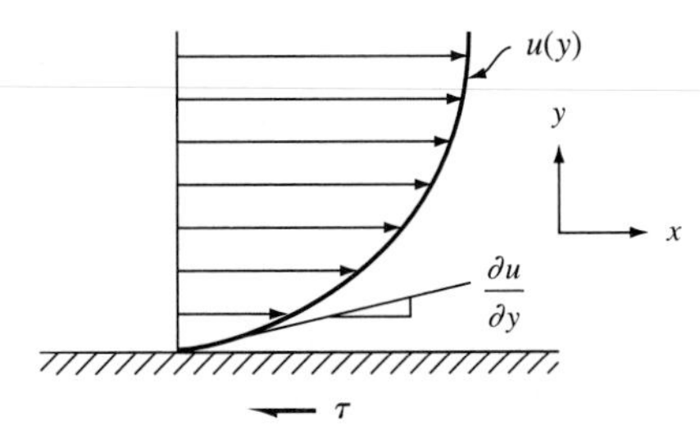


**Figure 5.4** Viscous shear velocity profile.

A **Newtonian fluid** is a fluid in which the absolute viscosity is independent of the shear rate. A non-Newtonian fluid has a variable viscosity, depending upon the shear rate.

The **kinematic viscosity** $\nu$ is the ratio of absolute viscosity to density:

$$\nu = \frac{\mu}{\rho} \tag{5.15}$$

This ratio often arises naturally as equations are developed.

**Liquids.** The absolute viscosity of a liquid decreases markedly with temperature, as shown in Figure 5.5. This decrease can best be approximated as an exponential decay, as described by the following equation ($\mu_0$ and $T_0$ are the values at the reference conditions, and $\lambda_L$ is a constant that depends upon the liquid):

$$\mu = \mu_0 e^{-\lambda_L(T-T_0)} \tag{5.16}$$

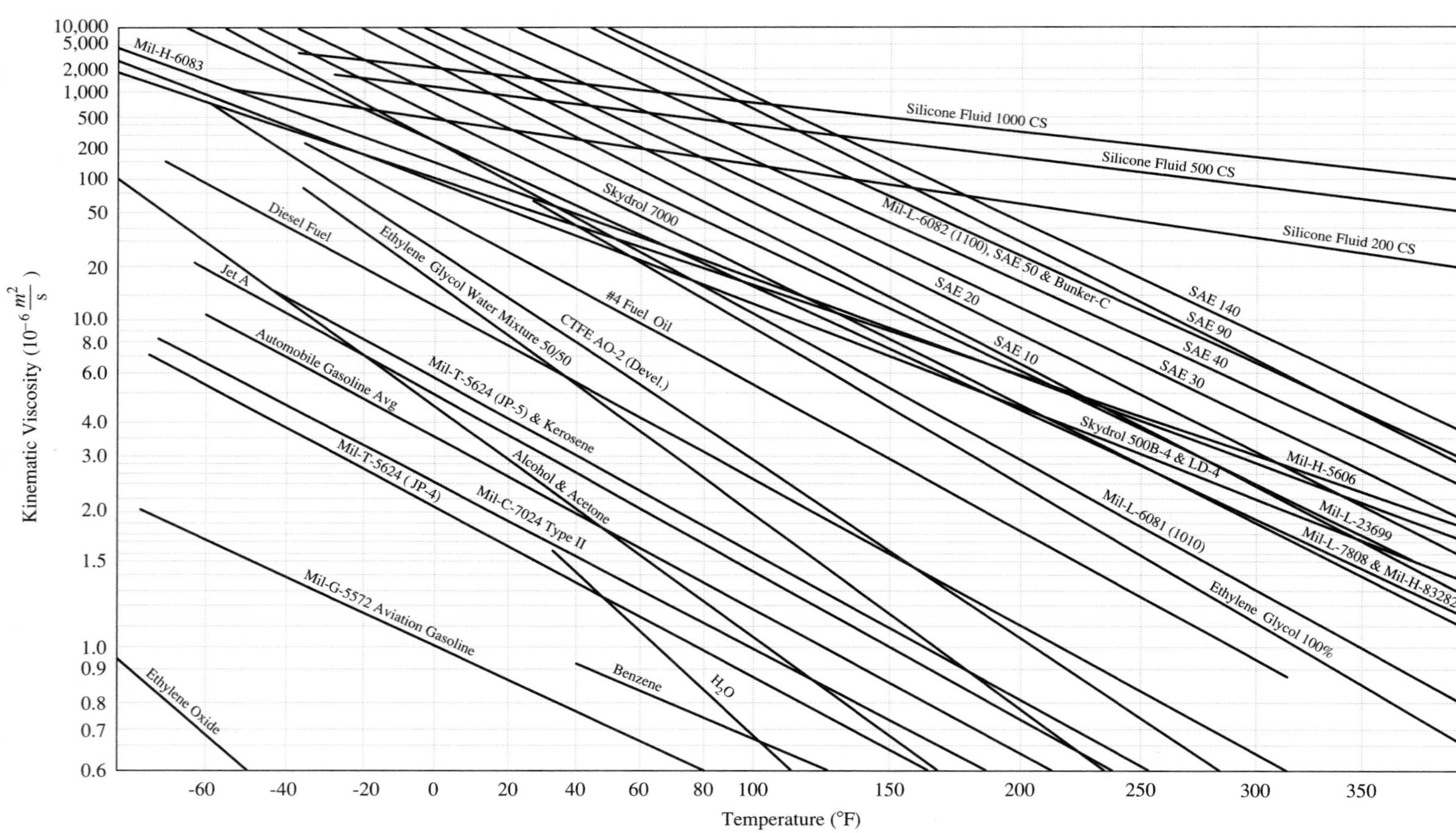


**Figure 5.5** Kinematic viscosity of typical liquids. (Reproduced with permission from the Lee Company [Ref. 6].)

**Gases.** Whereas the absolute viscosity of a liquid decreases with temperatures, that of a gas increases with temperature. This function can best be approximated as a straight line ($\mu_0$ and $T_0$ are the values at the reference conditions, and $\lambda_G$ is a constant that depends upon the gas):

$$\mu = \mu_0 + \lambda_G(T - T_0) \tag{5.17}$$

The viscosities of typical gases are given in Figure 5.6. The absolute viscosity of most gases is almost independent of pressure (from 1 to 30 bar).

### 5.2.5 Propagation Speed

The **speed of propagation** (or speed of sound), $c_0$, of a pressure signal in a fluid depends upon the bulk modulus and the density:

$$c_0 = \sqrt{\frac{\beta}{\rho}} \tag{5.18}$$

The speed of sound in a typical hydraulic fluid (e.g., MIL-H-5606, red oil) is 1370 m/s at 25°C.

The bulk modulus of a gas being perturbed at a high speed is the ratio of specific heats, $k$ (see Section 5.2.6), times the absolute pressure of the gas, $P$. Thus, $\beta = kP$, and the ratio $\beta/\rho$ reduces to the familiar expression

$$c_0 = \sqrt{kRT} \tag{5.19}$$

The speed of sound in air at 25°C is 347 m/s.

### 5.2.6 Thermal Properties

The **specific heat** of a fluid is the amount of heat required to raise the temperature of a unit mass of the fluid by 1 degree. The specific heat at constant pressure is $C_p$ and the specific heat at constant volume is $C_v$. The ratio of specific heats, or the **specific heat ratio** $k$ is

$$k = \frac{C_p}{C_v} \tag{5.20}$$

For some liquids, the specific heat ratio is approximatley 1.04. For gases, $k$ is larger (e.g., $k = 1.4$ for air).

## 5.3 REYNOLDS NUMBER EFFECTS

If a small amount of colored dye is introduced into a stream of oil or water that is flowing slowly through a transparent tube, the filament of color formed by the dye will flow smoothly with the fluid. However, if the flow rate is increased, a rate would be reached at which the filament colored by the dye would break up and be lost in swirls. The initial, smooth, streamlined flow is called **laminar**. The swirling flow is called **turbulent**.

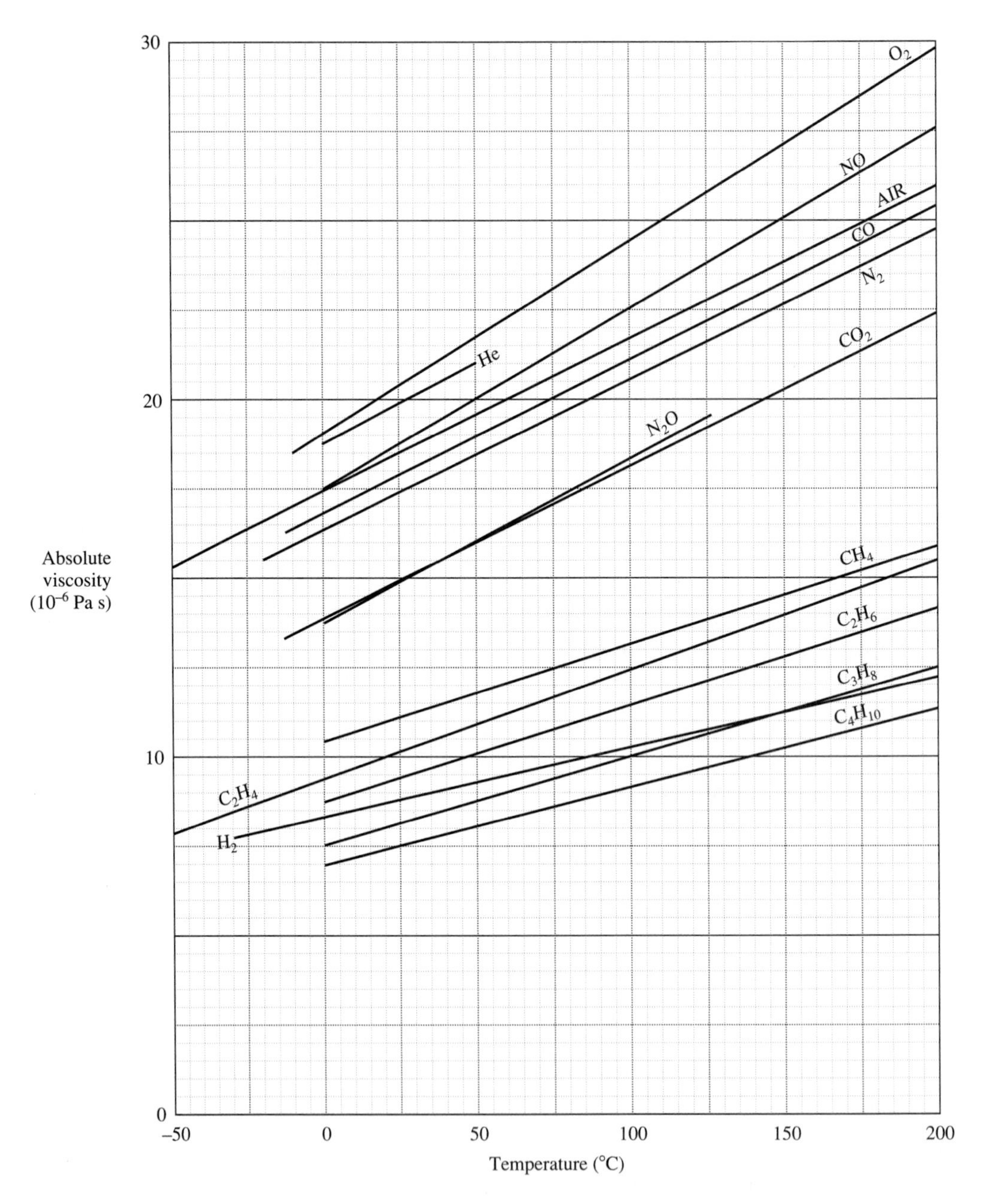


**Figure 5.6** Absolute viscosities of typical gases [Ref. 7].

In laminar flow, the viscous flow forces dominate over the inertial forces. In turbulent flow, the inertial forces are larger than the viscous forces.

Flow conditions are often defined by means of a dimensionless ratio called the **Reynolds number** $N_r$ (Osborne Reynolds, English engineer, 1842–1912). This number relates the inertial flow forces ( $\propto \rho A v^2$) to the viscous flow forces ( $\propto \mu d v$), where $v$ is the velocity of the fluid and $d$ is a characteristic dimension associated with the physical situation. Mathematically,

$$N_r = \frac{(\rho d^2 v^2)}{(\mu d v)} = \frac{v d}{\nu} \tag{5.21}$$

At small values of the Reynolds number, the viscous forces predominate and the flow is laminar. As the Reynolds number is increased, the effect of the inertial forces becomes sufficiently great to break up streamlined flow. In general, flow is laminar when the Reynolds number is less than 1400 and turbulent when it is greater than 3000. Flow conditions in the $1400 < N_r < 3000$ range depend on local conditions and the previous flow history, and the range is termed the transition range from laminar to turbulent flow. For $N_r > 3000$, the flow is turbulent and has energy loss due to fluid collisions and mixing. The two flow conditions are very dissimilar. In laminar flow, the pressure loss due to friction has a first-order relationship with the flow; it is analogous to electrical resistance in which the voltage is linear with current. When the flow is turbulent, the pressure loss becomes proportional to the square of the flow, due to energy losses associated with the Bernoulli velocity head, $\rho v^2/2$. Obviously, systems with laminar flow are far simpler than those with turbulent flow; however, laminar flow is generally impractical for most systems because of the low pressures and small dimensions it requires.

## 5.4 DERIVATION OF PASSIVE COMPONENTS

The **passive fluid components** are resistance, capacitance, and inductance. Each is derived in this section and is discussed in the context relative to both hydraulic and pneumatic systems.

### 5.4.1 Capacitance

**General Fluid Capacitance.** **Fluid capacitance** relates how fluid energy can be stored by virtue of pressure. For the control volume (c.v.) shown in Figure 5.7, the law of conservation of mass (also called the continuity equation) can be stated as

$$\dot{m}_{net} = \frac{d}{dt}(M_{cv}) = \frac{d}{dt}(\rho_{cv} V_{cv}) \tag{5.22a}$$

$$\dot{m}_{net} = \rho Q_{net} = \rho_{cv} \dot{V}_{cv} + V_{cv} \dot{\rho}_{cv} \tag{5.22b}$$

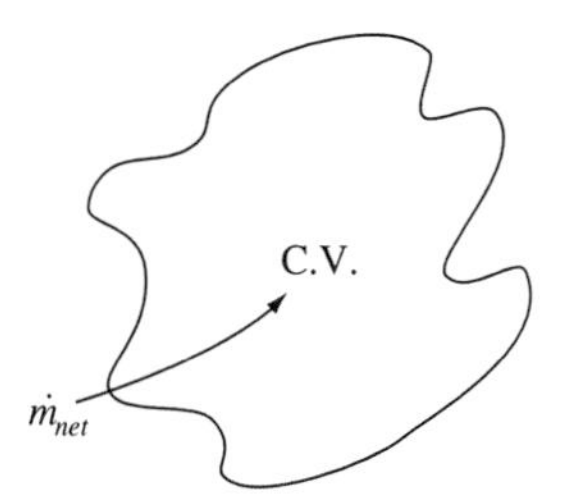


**Figure 5.7** Control volume for derivation of capacitance.

The net mass flow $\dot{m}_{net}$, is the sum of the flow entering (+) the system and the flow leaving (−) the system. If all of the densities of the system (the inlet flow, outlet flow, and control volume densities) are constant and equal to $\rho$, the following equation results:

$$Q_{net} = \dot{V} + \frac{V}{\rho}\dot{\rho} \tag{5.23}$$

The assumption that all of the densities of the system are constant and equal to $\rho$ is justified for incompressible fluids and is quite accurate for compressible fluids if pressure variations are not too large and the temperature of flow into the control volume is almost equal to the temperature of flow out of the control volume.

The $\dot{\rho}$ term in Eq. (5.23) needs to be expressed in terms of pressure. This is accomplished through an equation of state relating $\rho$ to pressure and temperature and by recalling the definition of the bulk modulus $\beta = \rho_0\, \partial P/\partial \rho|_{P_0,\,T_0}$. The equation of state is

$$\rho = \rho(P,T) \tag{5.24}$$

Differentiating this function yields

$$\dot{\rho} = \frac{\partial \rho}{\partial P}\frac{\partial P}{\partial t} \tag{5.25}$$

or

$$\dot{\rho} = \frac{\rho_0}{\beta}\dot{P} \tag{5.26}$$

Using this result, we can express the flow equation as

$$Q = \dot{V} + \frac{V}{\beta}\dot{P}_{cv} \tag{5.27}$$

This equation is called the **continuity equation** and can be used in systems analysis without repeating all of the derivation required to obtain it.

If the container is rigid, then $\dot{V} = 0$, and the continuity equation reduces to the familiar capacitance form

$$Q = \frac{V}{\beta}\dot{P}_{cv} \tag{5.28}$$

(Recall that $i = C(de/dt)$ for electrical systems.)

In Eq. (5.28), $V/\beta$ is the **fluid capacitance**. Therefore, any large volume of a compressible fluid becomes a capacitance; that is,

$$C_f = \frac{V}{\beta} \tag{5.29}$$

In pneumatics, an air tank used with a compressor is a capacitance.

The continuity equation gives us the differential equation for the pressure inside the control volume. The following equation should be used in systems analysis (for systems with an inlet flow with the same density as the control volume), without repeating its derivation:

$$\dot{P}_{cv} = \frac{\beta}{V}(Q - \dot{V}) \tag{5.30}$$

The $\dot{V}$ term can take several forms, depending upon the exact configuration of the system (control volume). An unloaded piston actuator or area $A$ results in $A\dot{x}$ (where $x$ is the position of the piston) for the $\dot{V}$ term.

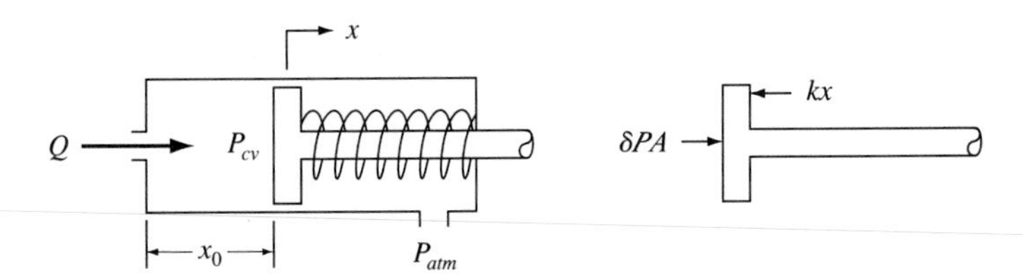


**Figure 5.8** Spring-loaded piston capacitor.

A spring-loaded piston of stiffness $k$ and area $A$ as shown in Figure 5.8 can be used as a capacitor. By applying a force balance equation, neglecting inertial and frictional effects, we obtain

$$A\delta P - kx = 0 \tag{5.31}$$

where $\delta P = P_{cv} - P_{atm}$ is the gauge pressure inside the actuator. Solving for $x$ yields

$$\dot{x} = \frac{A}{k}\delta\dot{P} \tag{5.32}$$

The volume of the cylinder is

$$V = A(x + x_0) \tag{5.33}$$

$$\dot{V} = A\dot{x} \tag{5.34}$$

where $x_0$ is the minimum stroke representing the dead volume space in the actuator and fittings.

Substitution yields

$$\dot{V} = \frac{A^2}{k}\delta\dot{P} \tag{5.35}$$

The continuity equation can now be rewritten in final form:

$$Q = \left(\frac{V}{\beta} + \frac{A^2}{k}\right)\delta \dot{P} \tag{5.36}$$

The total capacitance is therefore $(V/\beta) + (A^2/k)$, which is a combination of the fluid compressibility effects and the mechanical compliance of the container. (The $V/\beta$ term represents the fluid compressibility and the $A^2/k$ term is the capacitance of the compliant container.)

Thus, fluid capacitance can be either a compliant container or a volume of fluid. Notice that increasing $A$ or decreasing $k$ increases the mechanical capacitance term, and increasing $V$ or decreasing $\beta$ increases the fluid compressibility capacitance. Note that $\beta$ for a gas is on the order of 1 to 10 bar, and $\beta$ for a liquid is on the order of 10,000 bar. That is,

$$\beta = nP_{abs} \quad \text{for a gas} \tag{5.37a}$$

where

$n = 1.0$ for an isothermal or very slow process
$n = k$ for an adiabatic or very fast process

$$\beta \approx 10{,}000 \text{ bar for a liquid} \tag{5.37b}$$

Capacitors for liquids are called **accumulators**. Typical accumulators are spring-loaded pistons, bellows, and gas-filled bladders for hydraulic systems. Since $\beta$ is large for incompressible fluids, mechanical types of capacitors are used. To obtain significant compressibility capacitance with a liquid, $V$ would have to be very large. Capacitance for gaseous systems can be that of mechanical capacitors or volume-type capacitance, since $\beta$ is low for compressible fluids.

The effect of fluid capacitance must be considered relative to the rest of the system. For example, the resistance connected to the capacitor and the bandwidth of interest determine how significant the capacitive effects are.

**Capacitance of Gases.** Whenever the density of the fluid flowing into the control volume is different from the density of the fluid inside the control volume (due to high inlet pressure or to significantly different temperatures of the inlet gas and the control volume), additional considerations must be made regarding the continuity equation and hence the differential equation for the pressure inside the control volume.

In this case, the energy equation must be considered in addition to the equation for the conservation of mass flow (Eq. (5.22)). The heat transfer that might occur between the surroundings and the control volume is $q_{net}$. The energy equation for this system, neglecting kinetic and potential energy terms, results in

$$q_{net} + C_p(\dot{m}_{in}T_{in} - \dot{m}_{out}T_{cv}) - \dot{W} = \frac{d}{dt}U_{cv} \tag{5.38}$$

Note that there could be several input flows from different temperatures; therefore, the $\dot{m}_{in}T_{in}$ term should be written as the sum of *all* input flows, but is simplified in this presentation.

The internal energy of the control volume can be expressed as

$$U = m_{cv}C_vT_{cv} = \rho_{cv}V_{cv}C_vT_{cv} \tag{5.39}$$

Therefore, using the ideal gas law for the density inside the control volume, we can express the derivative of the internal energy of the control volume as

$$\dot{U} = \frac{C_v}{R}\frac{d}{dt}[P_{cv}V_{cv}] \tag{5.40}$$

By noting that $C_v \approx R/(k-1)$, and expanding the derivative of $P_{cv}V_{cv}$, we obtain

$$\dot{U} = \frac{1}{k-1}[P_{cv}\dot{V}_{cv} + V_{cv}\dot{P}_{cv}] \tag{5.41}$$

The work done by the control volume is $P_{cv}\dot{V}_{cv}$; therefore, by noting that $C_P \approx kR/(k-1)$ and substituting Eq. (5.41) and the expression for the internal energy (Eq. (5.39)) back into the energy equation, we obtain

$$q_{net} + \frac{kR}{(k-1)}\left(\sum \dot{m}_{in}T_{in} - \dot{m}_{out}T_{cv}\right) = \tag{5.42}$$

$$\frac{1}{(k-1)}[P_{cv}\dot{V}_{cv} + V_{cv}\dot{P}_{cv}] + P_{cv}\dot{V}_{cv}$$

Rearranging yields a differential equation for the pressure inside the control volume:

$$\frac{V_{cv}}{(k-1)}\dot{P}_{cv} = q_{net} + \frac{kR}{(k-1)}\left(\sum \dot{m}_{in}T_{in} - \dot{m}_{out}T_{cv}\right) - \frac{k}{(k-1)}P_{cv}\dot{V}_{cv} \tag{5.43}$$

Since calculation of the actual heat flow $q_{net}$ is often difficult, it is convenient to represent the heat flow by a percentage or ratio of $q_{net}$ to the maximum heat flow that could occur $q_{max}$. This maximum heat flow is the amount of heat transfer required to have the temperature of the inlet flow heated or cooled to equal the temperature of the control volume. If we use the variable $r$ to represent the ratio of the actual heat flow to the maximum possible heat flow, we can express this ratio as

$$r = \frac{q_{net}}{q_{max}} \tag{5.44}$$

If there is no heat transfer, then $r = 0$; if there is time for full heat transfer, then $r = 1$.

The maximum amount of heat transfer could be expressed as [Ref. 10]

$$q_{max} = P_{cv}\dot{V}_{cv} - R\left[\frac{k}{(k-1)}\dot{m}_{in}\left(T_{in} - \frac{T_{cv}}{k}\right) - \dot{m}_{out}T_{cv}\right] \tag{5.45}$$

Using this result with the ratio $r$, together with the ideal gas law and some algebraic manipulation, results in the differential equation

$$\dot{P}_{cv} = \frac{nP_{cv}}{V_{cv}}\left[\sum \frac{\dot{m}_{in}}{\rho_{cv}}\left\{\frac{k(1-r)\dfrac{T_{in}}{T_{cv}} + r}{k(1-r) + r}\right\} - \frac{\dot{m}_{out}}{\rho_{cv}} - \dot{V}_{cv}\right] \tag{5.46}$$

where

$$n = k(1-r) + r \tag{5.47}$$

and

$$\rho_{cv} = \frac{P_{cv}}{RT_{cv}} \tag{5.48}$$

This is the final result that should be used for a gas in which there is a significant temperature or pressure difference between the inlet gas and the control volume. The variable $r$ can be used to estimate how close you think the process is to an isothermal or adiabatic process. Notice from the equation that if the temperature of the inlet flow is equal to the temperature of the control volume, the equation reduces to Eq. (5.30).

### 5.4.2 Inductance

**Fluid inductance** is often called fluid "inertance," since its effect is due to the inertia of a moving fluid. Consider the one-dimensional flow of an incompressible fluid in a fluid line of constant area, as shown in Figure 5.9. A control volume can be taken, and the momentum (force balance) equation and continuity equations can be written. The control volume can be fixed in space or can be moving along with the fluid and hence be a control volume of fixed identity. Either approach will yield the same final result.

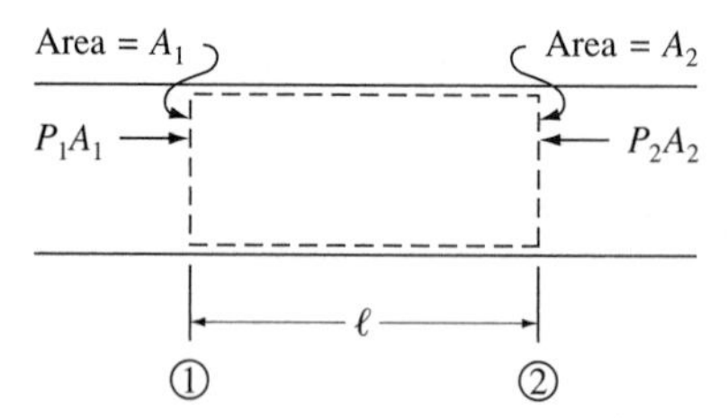


**Figure 5.9** Transient flow in a long tube, fluid inductance.

For the derivation that follows, we will consider a control volume of fixed identity (constant mass) moving down the tube at a velocity $v$. The equation for the conservation of momentum is

$$\sum F_{ext} = \frac{d}{dt}(mv)_{cv} = m\dot{v} + \dot{m}v \tag{5.49}$$

Since the mass of the control volume is fixed, $\dot{m} = 0$. The external forces acting on the control volume are due to the pressures; thus,

$$\sum F_{ext} = P_1A_1 - P_2A_2 = m\dot{v} \tag{5.50}$$

For a flow tube of constant area, $A_1 = A_2 = A$, and the mass of the control volume is $\rho A\ell$. The velocity $v$ of the control volume is related to the flow rate by

$$Q = Av \tag{5.51}$$

Thus, the momentum equation reduces to

$$(P_1 - P_2)A = \rho\ell A\frac{d}{dt}(v)_{cv} \tag{5.52}$$

or

$$\delta P = \frac{\rho\ell}{A}\dot{Q} \tag{5.53}$$

where $\delta P = P_1 - P_2$.

Equation (5.53) is an inductive-type equation (recall that $e = L(di/dt)$ for an electrical system) in which energy is stored by virtue of the flow rate. The **fluid inductance** is $(\rho\ell)/A$; therefore, a long tube with a small cross-sectional area with a dense fluid represents fluid inductance, and

$$L = \frac{\rho\ell}{A} \tag{5.54}$$

This equation is valid for liquids and gases; however, if a gas is used, the density must be evaluated at the upstream conditions.

The significance of fluid inductance must also be evaluated relative to the rest of the system and the bandwidth or frequency of interest.

### 5.4.3 Resistance

A **fluid resistor** dissipates power and can have a large variety of forms: laminar flow (viscous-dominated flow) resistance, orifice-type or head loss resistance (inertia-dominated flow), and compressible flow resistance.

**Laminar Flow Resistance.** A long **capillary** flow tube such as illustrated in Figure 5.10 with low enough flow rates (or a low enough pressure drop) gives rise to viscous flow.

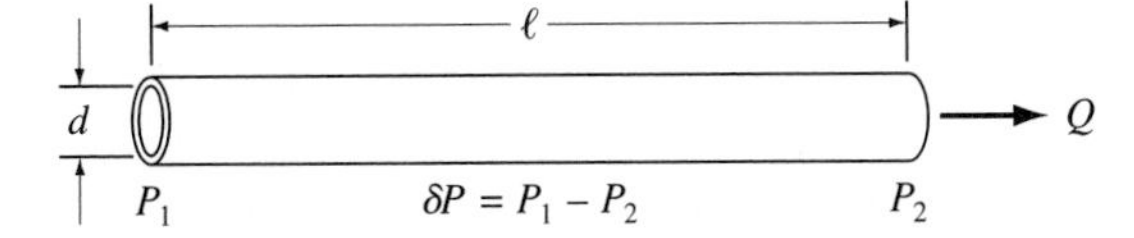


**Figure 5.10** Laminar flow, viscous fluid resistance.

Laminar resistors are characterized by a linear relationship between the pressure drop $\delta P$ and the flow $Q$ (recall that $e = Ri$ for an electrical resistor):

$$\delta P = RQ \tag{5.55}$$

The fluid resistance $R$ can be derived from basic principles with no difficulty, but is only summarized here. The resistance of a general passage having hydraulic diameter $d_h$, area $A$, and length, $\ell$ ($\mu$ is the absolute viscosity of the fluid) can be expressed as

$$R = \frac{32\mu\ell}{Ad_h^2} \quad \text{(general resistance)} \tag{5.56}$$

where the **hydraulic diameter** is

$$d_h = \frac{4 \text{ area}}{\text{perimeter}} \tag{5.57}$$

The resistance of a passage with circular cross section of diameter $d$ reduces to

$$R = \frac{128\mu\ell}{\pi d^4} \quad \text{(circular section)} \tag{5.58}$$

A square flow passage of width $w$ has resistance

$$R = \frac{32\mu\ell}{w^4} \quad \text{(square section)} \tag{5.59}$$

The resistance of a rectangular cross section of width $w$ and height $h$ is

$$R = \frac{8\mu\ell}{\dfrac{wh^3}{(1 + h/w)^2}} \quad \text{(rectangular section)} \tag{5.60}$$

A popular approximation for small values of $h/w$ is

$$R = \frac{12\mu\ell}{wh^3} \quad \text{(rectangular section approximation)} \tag{5.61}$$

An annular flow passage formed by concentric tubes with the outer tube having an inner diameter of $d$ and a symmetrical clearance of $b$ on either side of the internal tube has resistance

$$R = \frac{8\mu\ell}{\pi d b^3 \left(1 - \dfrac{b}{d}\right)} \quad \text{(annular section)} \tag{5.62}$$

A popular approximation of this function for small values of $b/d$ is

$$R = \frac{12\mu\ell}{\pi d b^3} \quad \text{(annular section approximation)} \tag{5.63}$$

If the tubes are displaced such that they have an eccentricity of $e$, then the resistance decreases slightly and is expressed as

$$R = \frac{12\mu\ell}{\pi d b^3\left[1 + 1.5\left(\frac{e}{b}\right)^2\right]} \quad \text{(eccentric annular section approximation)} \qquad (5.64)$$

In laminar flow resistance, the viscous terms are dominant, so the Reynolds number $N_r$ is low. Thus, for laminar flow, and hence linear resistance, the Reynolds number should be

$$N_r < 1000$$

The establishment of a fully developed laminar flow is a function of the length of the capillary. A rule of thumb is for $\ell/d_h$ to be greater than 50; more specifically, as a function of the Reynolds number,

$$\frac{\ell}{d_h} > 0.0575 N_R \qquad (5.65)$$

**Example 5.1 Flow in a Capillary Resistor**

A hydraulic line is 1.0 m long with an internal diameter of 2.0 mm and has hydraulic fluid flowing in it. The fluid has the following properties:

$$\rho = 840\ \text{kg/m}^3, \quad \nu = 20 \times 10^{-6}\ \text{m}^2/\text{s}, \quad \mu = 16.8 \times 10^{-6}\ \text{kPa s}$$

The resistance of a circular tube is

$$R = \frac{128\mu\ell}{\pi d_4} = \frac{128(16.8 \times 10^{-6}\ \text{kPa s})1\text{m}}{\pi 2^4\ \text{mm}^4} = \frac{128\ 16.8 \times 10^{-6}\ \text{kN s 1m}}{\pi 2^4\ 10^{-12}\ \text{m}^4\ \text{m}^2}$$

$$= \frac{42.78\ 10^6\ \text{kN s}}{\text{m}^5}$$

It is desired to calculate the maximum pressure drop that could be used to maintain laminar flow (note that flow can be related to pressure drop through the resistance equation $\delta P = RQ$):

$$N_r = \frac{vd}{\nu} = \frac{Qd}{A\nu} = \frac{\delta P d}{RA\nu}$$

To be laminar, the flow, must have a Reynolds number below 1000; thus, the maximum pressure for laminar flow with area 3.1416 mm$^2$ can be calculated as

$$\delta P < \frac{N_r RA\nu}{d} = \frac{1000 \times 42.8 \times 10^6\ \text{kN s } 3.1416\ \text{mm}^2\ 20 \times 10^{-6}\text{m}^2}{2\ \text{mm m}^5\ \text{s}} = 1345\ \text{kPa}$$

The maximum allowable pressure drop is therefore 1345 kPa (195 psi), which is a relatively low pressure in hydraulic systems which normally have pressures in the thousands of pounds per square inch.

**Orifice Flow Resistance.** An **orifice** with a very short length in the direction of flow gives rise to a head loss flow characteristic when operated with turbulent flow.

For the one-dimensional, steady, incompressible, frictionless flow through an orifice shown in Figure 5.11, we see that the lossless Bernoulli equation can be applied for most sections of the flow. The equation is valid between sections 1 and 2, and between sections 2 and 3, since streamlines can be readily identified; however, a velocity head loss is experienced between sections 3 and 4; thus, the equation does not apply there without considering energy losses.

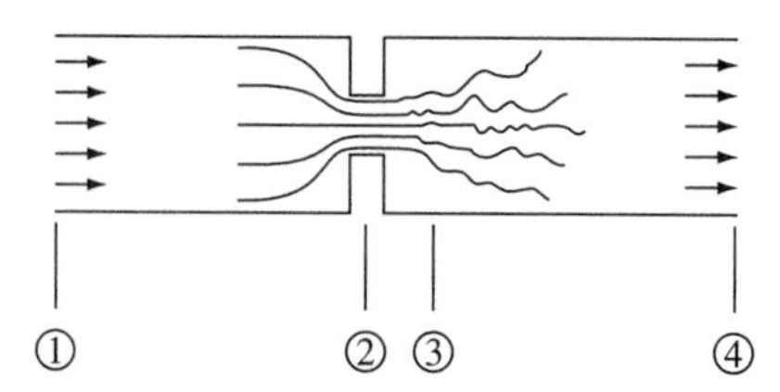


**Figure 5.11** Orifice flow sections.

Accelerating flow is experienced between sections 1 and 2. The flow is further "pinched down" to the *vena contracta* (minimum flow area) at section 3. The expanding flow between sections 3 and 4 causes a velocity head loss.

Neglecting differences in elevation, we can express Bernoulli's equation for sections 1, 2, and 3 as

$$\frac{P_1}{\rho} + \frac{v_1^2}{2} = \frac{P_2}{\rho} + \frac{v_2^2}{2} = \frac{P_3}{\rho} + \frac{v_3^2}{2} \tag{5.66}$$

By equating the flows at all sections, the following relations for the average fluid velocities can be observed:

$$Q = A_1 v_1 = A_2 v_2 = A_3 v_3 = A_4 v_4 \tag{5.67}$$

The ideal Bernoulli equation applies between sections 1 and 3. The geometry of the exit from section 2 to section 4 determines the relationship between the pressures at sections 3 and 4. If there is a smooth, gradual expansion of the area of the orifice (a diffuser), then the static pressure at section 3 will be lower than the static pressure at section 4, and more advanced methods must be used to calculate the flow based upon $P_4$. If the orifice has a sudden expansion between sections 2 and 4, then there is a total velocity head loss, and the pressure at section 3 will be essentially equal to the pressure at section 4. This makes our calculations easier, since we can measure $P_4$ and we cannot measure $P_3$. Thus,

$$P_3 = P_4 \tag{5.68}$$

Assuming that $A_1 > A_2$ and $A_1 > A_3$, it can be seen that $v_1 < v_3$ and $v_4 < v_3$. Hence, the velocity heads at sections 1 and 4 are negligible. Consequently, combining the Bernoulli equation, the continuity equation, and the preceding assumptions, we obtain the equation

$$P_1 - P_4 = \frac{\rho}{2} v_3^2 \tag{5.69}$$

Due to the *vena contracta,* $A_3 < A_2$. Since $A_2$ is easily measured, $A_3$ can be expressed in terms of $A_2$ using a contraction coefficient. Further, the acceleration of the flow from section 1 to section 3 does have some losses for real fluids. Both of these effects can be combined into a "discharge coefficient" $C_d$, so that the orifice flow equation becomes

$$\delta P = \frac{\rho}{2C_d^2 A_2^2} Q^2 \quad \text{5.70)}$$

where $\delta P = P_1 - P_4$.

In considering the sign of the pressure drop and hence the direction of flow, we can set forth a more appropriate form of the orifice flow equation. We get

$$\delta P = KQ^2 \operatorname{sign}(Q) \tag{5.71a}$$

or

$$Q = \sqrt{\frac{|\delta P|}{K}} \operatorname{sign}(\delta P) \tag{5.71b}$$

where

$$K = \frac{\rho}{2C_d^2 A^2}$$

$$\operatorname{sign}(x) = +1 \quad \text{if } x > 0$$

$$\operatorname{sign}(x) = -1 \quad \text{if } x < 0 \tag{5.72}$$

The discharge coefficient $C_d$ has a value that depends upon the geometry, roughness, and configuration of the orifice and the properties and Reynolds number of the fluid. The vlaue of $C_d$ ranges from 0.5 to 1.0, depending upon the flow geometry in front of and behind the orifice. A sharp-edged orifice has $C_d = 0.6$, while a smooth, convergent nozzle could have $C_d = 1.0$.

For laminar flow, an orifice behaves as a first-order or viscous-dominated resistance and is thus linear. Most orifices are not operated in the laminar region and hence follow Eq. (5.70) for $N_r > 2000$.

**Example 5.2 Orifice Flow Calculation**

A sharp-edged orifice has a diameter of 0.030 inch with a sudden expansion after the hole. (Thus, we expect the discharge coefficient to be 0.6.) The same hydraulic fluid from Example 5.1 is used at a 1000 psi pressure drop. It is necessary to confirm that the flow is turbulent and to calculate the flow rate.

The flow rate can be calculated from the orifice equation as follows (recall that lbf = lbm 386 in/s²):

$$Q = C_d A \sqrt{\frac{2\ \delta P}{\rho}} = 0.6\frac{\pi d^2}{4}\sqrt{\frac{2 \cdot 1000\ \text{lbf/in}^2}{0.0303\dfrac{\text{lbf/in}^3}{386\ \text{in/s}^2}}} = 2.14\frac{\text{in}^3}{\text{s}} = 35.1\frac{\text{cm}^3}{\text{s}}$$

For the flow to be turbulent, the Reynolds number must be greater than about 2000:

$$N_r = \frac{Q}{A}\frac{d}{\nu} = \frac{2.14\dfrac{\text{in}^3}{\text{s}}0.030\ \text{in}}{706.9 \times 10^{-6}\ \text{in}^2\ 0.031\dfrac{\text{in}^2}{\text{s}}} = 2930$$

Since the Reynolds number is just greater than the critical value of about 2000, the flow is (just slightly) turbulent.

**Compressible Fluid Flow.** The **compressible fluid flow equation** is similar to the orifice flow equation; however, it does consider the variation in density of the gas. As such, the compressible flow equation has a higher degree of nonlinearity and is quite involved. In aerospace applications, both of these equations are normally written as functions of the **Mach number** $N_m$, which is the ratio of the fluid velocity $v$ to the speed of sound, $c_0$:

$$N_m = \frac{v}{c_0} \tag{5.73}$$

Shown in Figure 5.12 is a converging-diverging nozzle having a throat defined as the minimum flow area. In the incompressible flow equation, the pressure drop $P_u - P_d$ determines the flow; in the compressible fluid flow equation, the pressure ratio $P_t/P_u$ is the important property. Thus, we can write the equations in terms of the pressure ratio instead of the Mach number.

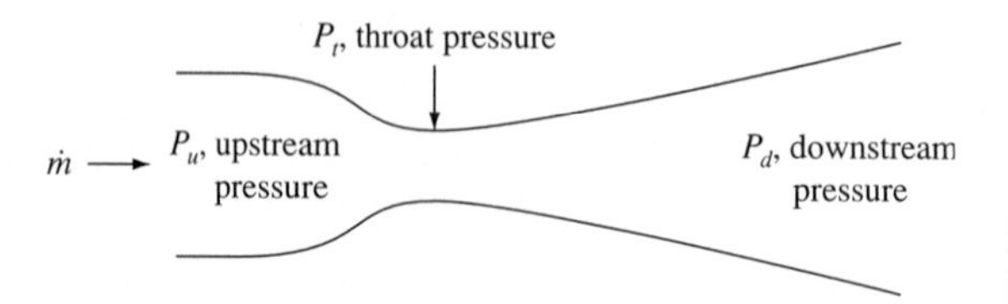


**Figure 5.12** Compressible flow nozzle.

The interesting property of compressible fluid flow is that *the Mach number cannot exceed 1.0 at the throat* (minimum flow area) of the restriction. For air, this occurs at a value of $P_t/P_u = 0.528$, and the flow is then **choked**. The flow-choking condition continues to exist even if the downstream pressure decreases. Once the flow becomes choked, the Mach number at the throat is always 1.0, and consequently, the volumetric flow rate is a constant ($Q = A_t c_0$). However, since the fluid is compressible, the density can vary, and hence, the mass flow rate ($\dot{m} = \rho_t Q$) can vary if the upstream conditions change. The mass flow rate of a choked orifice is determined totally by the upstream pressure (and temperature), and further decreases in the downstream (or exit) pressure have no effect upon the volume flow, the mass flow, or the pressure at the throat.

In the analysis that follows, the equations are written from the upstream pressure $P_u$ to the pressure at the throat or minimum flow area, $P_t$. The exit pressure, or the pressure downstream at a large flow area, can be higher or lower than the throat pressure, depending upon the geometry from the throat to the larger exit flow area. For example, a slowly expanding diffuser can obtain nearly 100% recovery of the total pressure, while a sudden expansion from the throat has very little pressure recovery and thus loses most of the velocity head energy.

The complete compressible flow equation for an orifice of area $A$ with discharge coefficient $C_d$, subjected to a pressure ratio $P_t/P_u$, is given next [Refs. 1, 5, 8, 10].

If the pressure ratio is less than the critical pressure $P_{cr}$, the flow is unchoked, and the mass flow per unit area is

$$\frac{\dot{m}}{C_d A} = \sqrt{\frac{2}{RT^*}}\sqrt{\frac{T^*}{T_u}}\sqrt{\frac{k}{k-1}}P_u\left[\left(\frac{P_t}{P_u}\right)^{\frac{2}{k}} - \left(\frac{P_t}{P_u}\right)^{\frac{k+1}{k}}\right]^{\frac{1}{2}} \quad \left(\text{unchoked flow, } \frac{P_t}{P_u} > \mathrm{P}_{cr}\right) \tag{5.74}$$

where $R$ is the gas constant, $k$ is the ratio of specific heats, and $T^*$ is a reference temperature for the purposes of evaluating the coefficients.

If the downstream pressure is low enough, the throat pressure will be a minimum (as determined by the critical pressure ratio), and the flow will be choked. The throat pressure cannot be any lower than this critical value, even though the downstream or exit pressure could be lower. In this case, the flow is

$$\frac{\dot{m}}{C_d A} = \sqrt{\frac{2}{RT^*}}\sqrt{\frac{T^*}{T_u}}\sqrt{\frac{k}{k-1}}P_u\left[(P_{cr})^{\frac{2}{k}} - (P_{cr})^{\frac{k+1}{k}}\right]^{\frac{1}{2}} \quad \left(\text{choked flow, } \frac{P_t}{P_u} = P_{cr}\right) \tag{5.75}$$

Notice that this equation is identical to that of the unchoked, except that the critical pressure ratio is substituted for the actual pressure ratio.

The **critical pressure ratio** that determines whether the flow is choked is

$$P_{cr} = \left(\frac{2}{k+1}\right)^{\frac{k}{k-1}} \tag{5.76}$$

Notice that the terms in the foregoing equations for compressible flow have been grouped for computational convenience. For example, the equations could have been written in terms of $RT_u$; however, a reference temperature $T^*$ is introduced so that the square root term can be calculated once and then corrected by the square root of a temperature ratio (if the actual temperature is different from the reference temperature), rather than be calculated for each problem.

When the flow is choked, the pressure ratio is a function of the ratio of specific heats. It is interesting to know the maximum flow available. The choked flow equation can be expressed as follows when Eq. (5.76) for the critical pressure ratio is substituted:

$$\frac{\dot{m}}{C_d A} = \sqrt{\frac{2}{RT^*}}\sqrt{\frac{T^*}{T_u}} f_{kc} P_u \quad \left(\text{choked flow, } \frac{P_t}{P_u} = P_{cr}\right) \tag{5.77}$$

Here $f_{kc}$ is the compressible flow equation constant, which is a dimensionless function of the ratio of specific heats, $k$:

$$f_{kc} = \sqrt{\frac{k}{(k-1)}\left[\left(\frac{2}{k+1}\right)^{\frac{2}{k-1}} - \left(\frac{2}{k+1}\right)^{\frac{k+1}{k-1}}\right]} \tag{5.78}$$

The value of this constant is 0.4842 for air ($k = 1.40$); Table 5.1 gives $f_{kc}$ for other values of $k$.

**TABLE 5.1** COMPRESSIBLE FLOW EQUATION CONSTANT, $f_{kc}$.

| $k$ ratio of specific heats | $f_{kc}$ compressible flow equation constant |
|---|---|
| 1.05 | 0.4368 |
| 1.10 | 0.4443 |
| 1.15 | 0.4516 |
| 1.20 | 0.4586 |
| 1.25 | 0.4640 |
| 1.30 | 0.4705 |
| 1.35 | 0.4781 |
| 1.40 | 0.4842 |
| 1.45 | 0.4901 |
| 1.50 | 0.4957 |

Equation (5.74) or equation (5.76) should be used for compressible flow calculations. Recall that these equations are written from the upstream pressure to the throat pressure. Normally, we do not know the throat pressure; we know the exit or downstream pressure $P_d$. Therefore, our challenge is to relate the downstream pressure to the throat pressure. In an orifice that exits into a large chamber, the throat pressure is almost equal to the downstream pressure and thus is more easily handled. In a converging-diverging nozzle that has the proper expansion angle and ratio, the flow can actually be choked when the pressure ratio $P_d/P_u$ is 0.9 or even higher. In the case of a converging-diverging nozzle, more advanced analysis techniques must be used [Ref. 12]. In the case of an orifice exiting into a large area, we can assume that $P_t = P_d$.

An interesting approximation to the *compressible* flow equation is possible by rearranging the *incompressible* flow equation used for an orifice, Eq. (5.70). If we evaluate the density at the *downstream conditions* and substitute the ideal gas equation into the incompressible flow equation expressed in mass flow terms, then the following equation results:

$$\frac{\dot{m}}{C_d A} = \frac{\rho_t Q}{C_d A} = \rho_t \sqrt{\frac{(P_u - P_t)}{\rho_t}} = \sqrt{\frac{2}{RT_t}} \sqrt{P_t P_u - P_t^2} \tag{5.79}$$

The temperature at the throat is related to the upstream temperature by an exponential function, depending upon how much heat transfer there is. However, as a further approximation, if we use the upstream temperature instead of the throat temperature in the incompressible equation, the resulting equation is slightly simpler and gives good results when compared to the compressible equation.

For pressure ratios greater than the critical (unchoked) pressure, the following equation should be used as the incompressible approximation:

$$\frac{\dot{m}}{C_d A} = \sqrt{\frac{2}{RT^*}} \sqrt{\frac{T^*}{T_u}} P_u \sqrt{\frac{P_t}{P_u} - \left(\frac{P_t}{P_u}\right)^2} \quad \text{(unchoked incompressible approx.)} \tag{5.80}$$

When the downstream pressure is low enough to cause choked flow, the pressure ratio in this equation should be replaced by the critical pressure ratio, resulting in

$$\frac{\dot{m}}{C_d A} = \sqrt{\frac{2}{RT^*}} \sqrt{\frac{T^*}{T_u}} P_u \sqrt{P_{cr} - P_{cr}^2} \quad \text{(choked incompressible approx.)} \tag{5.81}$$

Since the critical pressure ratio is a function of $k$, this equation can be rearranged in terms of a constant $f_{ki}$ for the incompressible approximation. The constant is similar to the function $f_{kc}$ for the compressible flow function. Thus, if the orifice is choked, the mass flow per unit area is

$$\frac{\dot{m}}{C_d A} = \sqrt{\frac{2}{RT^*}} \sqrt{\frac{T^*}{T_u}} f_{ki} P_u \quad \text{(choked incompressible approx.)} \tag{5.82}$$

where

$$f_{ki} = \sqrt{\left(\frac{2}{k+1}\right)^{\frac{k}{k-1}} - \left(\frac{2}{k+1}\right)^{\frac{2k}{k-1}}} \tag{5.83}$$

is the incompressible flow equation constant, which is a function only of $k$, as given in Table 5.2.

A comparison of the incompressible equations with the compressible equations reveals that they are very similar. When the flow is choked, the $f_{kc}$ constant for the compressible equation is 0.4842 for air, while the $f_{ki}$ constant for the incompressible equation is 0.4992, a difference of 3%. The incompressible approximation equation is often easier to apply and gives quite accurate results, as illustrated by Figure 5.13.

One of the advantages of the incompressible approximation is the ability to solve for the pressures in terms of the flow when this relationship is needed. Notice

**TABLE 5.2** INCOMPRESSIBLE FLOW EQUATION CONSTANT, $f_{ki}$.

| $k$<br>ratio of specific heats | $f_{ki}$<br>incompressible flow equation constant |
|---|---|
| 1.05 | 0.4908 |
| 1.10 | 0.4928 |
| 1.15 | 0.4944 |
| 1.20 | 0.4958 |
| 1.25 | 0.4970 |
| 1.30 | 0.4979 |
| 1.35 | 0.4986 |
| 1.40 | 0.4992 |
| 1.45 | 0.4996 |
| 1.50 | 0.4999 |

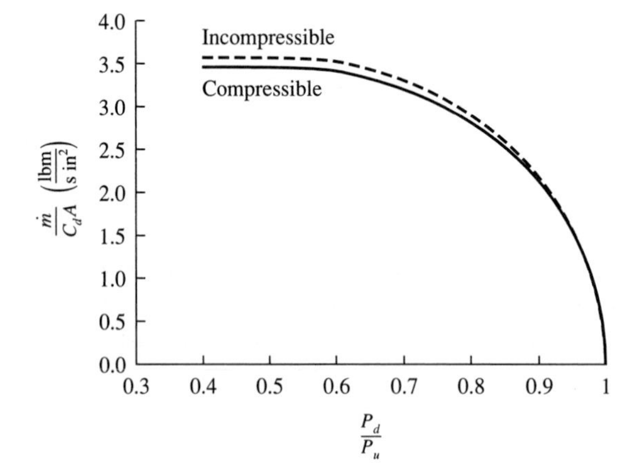


**Figure 5.13** Comparison of compressible and incompressible flow equations.

that because of the nonlinear nature of the compressible flow equation, one can only calculate the flow based upon pressures; the converse is not possible. However, the incompressible approximation can be solved backwards, as is shown next.

If the flow is less than the critical value required to choke, as given by Eq. 5.80, then the upstream and throat pressures can be related to the flow. If the throat pressure is known, the upstream pressure is

$$P_u = P_t + \frac{\left[\dfrac{\dot{m}}{C_d A \sqrt{\dfrac{2}{RT_u}}}\right]^2}{P_t} \quad \text{(unchoked approximation)} \tag{5.84}$$

And if the upstream pressure is known, the throat pressure is

$$P_t = \frac{P_u}{2}\left[1 + \sqrt{1 - \left(\frac{2\dot{m}}{C_d A \sqrt{\frac{2}{RT_u}} P_u}\right)^2}\right] \quad \text{(unchoked approx.)} \tag{5.85}$$

If the flow is choked and the flow rate is known, the pressures are

$$P_u = \frac{\dot{m}}{C_d A \sqrt{\frac{2}{RT_u}}} \frac{1}{\sqrt{P_{cr} - P_{cr}^2}} \quad \text{(choked approx.)} \tag{5.86}$$

and

$$P_t = \frac{\dot{m}}{C_d A \sqrt{\frac{2}{RT_u}}} \frac{P_{cr}}{\sqrt{P_{cr} - P_{cr}^2}} \quad \text{(choked approx.)} \tag{5.87}$$

## 5.5 SUMMARY

This chapter has presented the fundamental characteristics and mathematical modeling of fluid systems. Since fluid properties determine the performance of a dynamic system, graphs of the fluid properties (density, bulk modulus, and viscosity) were presented as a function of temperature. The basic passive components are resistance (composed of laminar flow viscous resistance, head loss or orifice flow resistance, and compressible fluid flow), capacitance (composed of the compressibility of the fluid in a volume or a container that has compliance with respect to an internal pressure), and inductance (the "water hammer effect," composed of the inertia of fluid flowing in a pipe). These basic components appear in fluid systems such as water distribution systems and a water tower and also set the stage for modeling control valves, actuators, and other components used in fluid power controls systems (hydraulic and pneumatic systems).

## REFERENCES

5.1 Blackburn, J. R., G. Reethof, and J. L. Shearer. *Fluid Power Control.* The MIT Press, Cambridge, Massachusetts, 1960.

5.2 McCloy, D., and Martin H., *The Control of Fluid Power*. John Wiley, New York, 1980.

5.3 Merritt, Herbert E. *Hydraulic Control Systems*. John Wiley and Sons, New York, 1967.

5.4 Watton, John. *Fluid Power Systems*. Prentice Hall, New York, 1989.

5.5 Andersen, Blaine W. *The Analysis and Design of Pneumatic Systems*. John Wiley and Sons, New York, 1967.
5.6 The Lee Company, Product Literature, 2 Pettipaug Rd., Westbrook CT.
5.7 Bolz, Ray E., and Tuve, George L. *Handbook of Tables for Applied Engineering Science,* 2nd ed. CRC Press, Boca Raton, Florida, 1979.
5.8 Ogata, K. *System Dynamics,* 2d ed. Prentice Hall, Englewood Cliffs, New Jersey, 1992.
5.9 Kirshner, J. M., and Katz, S., *Design Theory of Fluidic Components*. Academic Press, New York, 1975.
5.10 Hullender, David A., and Woods, Robert L. "Modeling of Fluid Circuit Components." Proceedings of the First Conference on Fluid Control and Measurement, FLUCOME '85. Permagon Press, Tokyo, London, 1985.
5.11 Keller, G. R. *Hydraulic System Analysis*. Industrial Publishing Co., Cleveland, OH, 1974.
5.12 Shapiro, Ascher H. *The Dynamics and Thermodynamics of Compressible Fluid Flow*. Ronald Press, New York, 1954.

## NOMENCLATURE

$A$ area
$b$ clearance of concentric tubes
$c_0$ speed of propagation of sound
$C_d$ discharge coefficient of orifice flow
$C_f$ fluid capacitance
$C_p$ constant-pressure specific heat
$C_v$ constant-volume specific heat
$d$ diameter of a circular cross section
$d_h$ hydraulic diameter
$e$ eccentricity
$F$ force
$h$ height
$k$ spring stiffness; ratio of specific heats
$\ell$ length
$L$ fluid inductance
$m$ mass
$\dot{m}$ mass flow rate
$n$ polytropic coefficient
$N_m$ Mach number
$N_r$ Reynolds number

$P$ pressure
$P_u$ upstream pressure
$P_t$ throat pressure
$P_d$ downstream pressure
$P_{cr}$ critical pressure ratio
$q$ heat transfer rate
$Q$ volume flow rate
$r$ ratio of the actual heat flow to the maximum heat flow possible
$R$ fluid resistance; gas constant
$T$ temperature
$U$ internal energy
$v$ velocity
$V$ volume
$w$ width
$W$ work
$\alpha$ thermal expansion coefficient
$\beta$ bulk modulus
$\beta_a$ adiabatic bulk modulus
$\mu$ absolute ( or dynamic ) viscosity
$\rho$ density
$\tau$ shear stress
$\nu$ kinematic viscosity

## PROBLEMS

**5.1** *Mention* applications of fluid controls in the following areas:

**a.** automotive
**b.** aerospace
**c.** industrial
**d.** consumer
**e.** commercial

**5.2** *Show* that the expression for the viscosity (Sutherland's equation) of a gas as a function of the absolute temperature,

$$\frac{\mu}{\mu_o} = \left(\frac{T}{T_o}\right)^{1.5}\left(\frac{T_o + c_1}{T + c_1}\right)$$

can be linearized (using a Taylor series expansion) to the form $\mu/\mu_o = 1 + c_2(T - T_o)$. For air, $c_1 = 110°K$.

*Calculate* the value of $c_2$. Use a spreadsheet to *illustrate* the range of validity and the accuracy of the linearization in a graph.

**5.3** Consider a fluid of fixed identity (i.e., constant mass) undergoing a compression. Use the definitions of bulk modulus and fluid density to *show* that

$$\beta = -\frac{\partial P}{\partial(V/V_o)}\bigg|_{P_o, T_o}$$

**5.4** *Perform* all of the algebraic steps in the derivation of Eq. 5.46.

**5.5** As shown in Figure P5.5, a volume of a gas, $V_g$, is trapped in a rigid container that has a volume of liquid, $V_L$. (The total volume $V_T = V_L + V_g$.) The bulk modulus of the liquid is $\beta_L = 1.4 \times 10^6$ kPa.

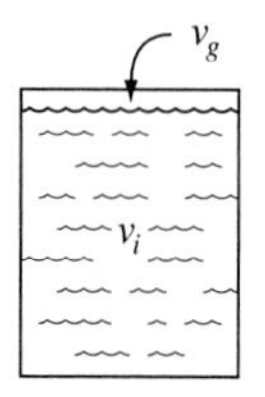


**Figure P5.5** Air trapped in liquid container.

**a.** Use the basic definition of bulk modulus to *derive* an expression for the equivalent bulk modulus $\beta_e$. Assume that the container is rigid.

**b.** Use the ratio $x = V_g/V_T$ and *normalize* your result for the ratio of $\beta_e/\beta_L$ as a function of $x$ and $\beta_g/\beta_L$.

**c.** *What* volume ratio $x$ is required to reduce the equivalent bulk modulus to one-half the liquid bulk modulus? Assume that the gas is air at high frequency and the mean pressure is one bar (100 kPa). *What* is the ratio $x$ if the mean pressure is $20 \times 10^3$ kPa?

**5.6** The radius expansion, $R - R_0$, of a balloon filled with a gas is directly proportional to the internal pressure of the gas. Let us write this proportionality as $\delta P_i = k(R - R_0)$. *Derive* an expression for the total capacitance of the ballon that considers the change in volume of the balloon and the effect of compressibility of the gas. (Volume $= (4/3)\pi R^3$).

**5.7** A circular tube of length $\ell$ is used to hold fluid pressure. If the tube has an internal diameter $d_i$, a wall thickness $t$, and a Young's modulus $E$, *derive* (a) The capacitance of the tube, using an incompressible fluid, and (b) The total capacitance $C_T$, which includes the volume capacitance of the fluid, $C_F$ (with a fluid of bulk modulus $\beta$), and the mechanical capacitance, $C_M$. Write your result in normalized form, $C_T/C_F$, using normalized variables.

Now *write* an expression for the normalized equivalent bulk modulus $\beta_{eq}/\beta_F$ resulting from the tube capacitance.

**5.8** A 10-gallon air tank is used with an air compressor to smooth pulses from the compressor and to store energy so that the instantaneous flow rate for short bursts can be larger than the continuous flow rate provided by the compressor. *Calculate* the energy stored in the air tank at 125 psig, using the equation for $(C/2)\delta P^2$, where the capacitance is $V/\beta$. *Express* your energy in ft-lbf. In order to get a feel for how much energy this is, *calculate* how much weight lifted to 1 foot above a reference height this amount of energy represents. *Calculate* the speed at which a 1 lb mass would be moving if all of the energy could be converted to kinetic energy. *Express* your speed in miles per hour.

**5.9** Consider steady, incompressible, viscous, one-dimensional, fully developed laminar flow between flat parallel plates, as depicted in Figure P5.9. The Navier-Stokes (Claude Navier, French engineer, 1785–1836, George Gabriel Stokes, English physicist, 1819–1903) equation for this case reduces to $\partial P/\partial x = -\mu\ \partial^2 u/\partial y^2$. The nomenclature for the situation is as follows: $u$ = fluid velocity in the $x$-direction, $P$ = pressure of the fluid at the given location, $\mu$ = absolute viscosity of the fluid. Since the flow is fully developed and is one dimensional, there is no fluid velocity component in the $y$- and $z$-directions; however, the velocity in the $x$-direction varies with the distance from the centerline $y$ of the flow.

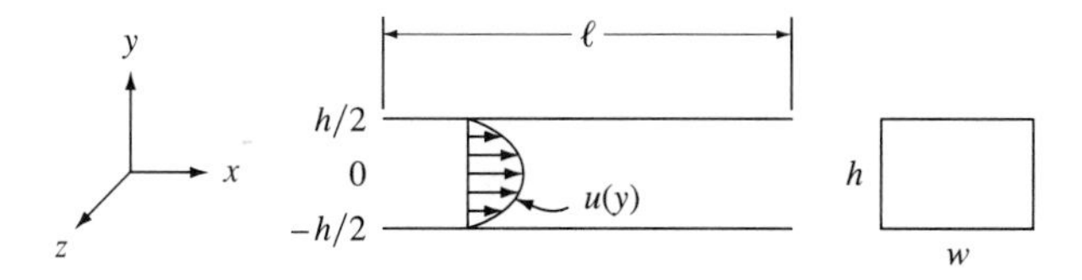


**Figure P5.9** Viscous flow in parallel plates.

The Navier-Stokes equation is a partial differential equation. You must first solve for the velocity profile at a given location $x$. This is done by solving the equation for $\partial^2 u/\partial y^2$ (treating $x$ and $\partial P/\partial x$ as constant) and integrating twice. Next, you can solve for $\partial P/\partial x$ and then integrate the result with respect to $x$ (notice that the velocity profile does not vary with $x$ and is therefore treated as a constant) to solve for the variation in the pressure as a function of $x$.

**a.** *Integrate* the Navier-Stokes equation at a given location $x$ to derive the velocity profile $u(y)$ of the fluid at that point. Use the boundary conditions $u = u_{max}$ and $\partial u/\partial y = 0$ at $y = 0$. Note further that the velocity at both walls is zero; that is $u(\pm h/2) = 0$. After you perform the integration and apply all of the known boundary conditions, you should have an expression for the velocity $u$ as a parabolic function of $y$ (with constants $h/2$ and $u_{max}$). *Calculate* the average velocity $\bar{u}$ over the profile by integrating $u(y)$ over $y$.

**b.** In the process of deriving part a, you should have an expression for $\partial P/\partial x$ as a function of $\mu$, $h/2$, and $u_{max}$. Separate the derivative variables and *integrate* the resulting equation over the length $\ell$, using the pressures at both ends. (Notice that $\mu$, $h/2$, and $u_{max}$ are all constants with respect to $x$ and $P$.)

**c.** In parts a and b, the flow was considered over an infinitely wide parallel plate. Assuming that the profile still holds for a finite plate of width $w$, *calculate* the average fluid flow $Q$ over the cross-sectional area, using the average flow velocity $\bar{u}$. Based upon the pressure drop over the length and the average flow, *state* an expression for the fluid resistance of this rectangular passageway.

**5.10** Using the definition of general resistance given by (Eq. 5.56) and the definition of hydraulic diameter given by (Eq. 5.57), *derive* the expression for the resistance of circular, square, rectangular, annular, and eccentric annular cross sections.

**5.11** A capillary tube is used as a laminar flow resistance, as illustrated in Figure P5.11. Its length is 100 mm and its diameter is 0.5 mm. Use a Reynolds number (based upon average velocity and line diameter) of $(500 \times 10^3)/(\ell/d)$ as the critical Reynolds number for length.

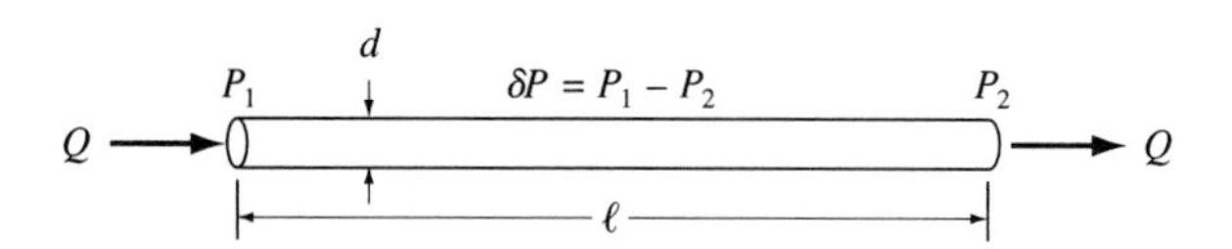


**Figure P5.11** Laminar flow resistance.

**a.** *Calculate* the resistance of the line (in units of $10^6$ kN s/m$^5$) and the maximum pressure drop allowable (in units of kPa), using air ($\rho = 1.2$ kg/m$^3$, $\mu = 0.018 \times 10^{-6}$ kPa s.

**b.** Repeat part a for oil ($\rho = 850$ kg/m$^3$, $\mu = 17 \times 10^{-6}$ kPa s).

**5.12** *Confirm* all of the conversions and calculations in Example 5.2.

**5.13** Water flows through an orifice that has a pressure drop of 1000 psi. The orifice has a sharp edge and very short longitudinal length; therefore, we expect the discharge coefficient to be 0.6. The diameter of the orifice is 0.050". *Calculate* the flow rate through the orifice. *Express* your result in gallons per minute (GPM). Recall that 231 in$^3$ is a gallon.

**5.14** An air blowgun has an orifice of 0.075" diameter. The back side of the orifice has been drilled with a taper, so the fluid experiences a gradual taper in front of the orifice. We suspect that this will result in a discharge coefficient of about 0.9. If the air pressure at the gun is 75 psig, *calculate* the mass flow (lbm/hour) of air through the orifice, using the compressible flow equation or its approximation. *Convert* this mass flow to volume flow at standard conditions, and *express* your answer in SCFM (standard cubic feet per minute).

**5.15** Consider compressible flow through a nozzle that is choked. The maximum mass flow for a given upstream pressure $P_u$ occurs when the throat pressure ratio $P_t/P_u$ is equal to the critical pressure ratio $P_{cr}$. It is desired to calculate the expression for the critical pressure ratio.

In order to derive the expression for the critical pressure ratio, find the maximum of the function by *taking* the derivative of the square root term of the compressible flow equation with respect to the pressure ratio and *setting* the result equal to zero. Then *solve* for the pressure ratio that satisfies this maximum condition, which is the critical pressure ratio. The mathematics here is tricky, be sure you have a positive exponent. Did you obtain the result given by Eq. 5.76?

**5.16** *Derive* Eq. 5.79.

**5.17** A rigid tank of compressed air is discharged through an orifice to atmospheric pressure. Using state-space notation and digital simulation, *obtain* the transient response of the pressure inside the tank. *Plot* your results for the following cases:

**a.** Using the model for the incompressible flow equation

$$\delta P = KQ^2 \text{ sign } (Q) \quad \text{where } K = \frac{\rho}{2(C_d A_0)^2}$$

**b.** Using the model for the compressible flow equation.

In both cases, *compare* your results with the experimental data shown in Figure P5.17. *Try* different values for $n$ (the polytropic constant) and $\beta$ (constant $\beta$ or variable $\beta$), and determine which is the best. The initial pressure is 100 kPa gauge.

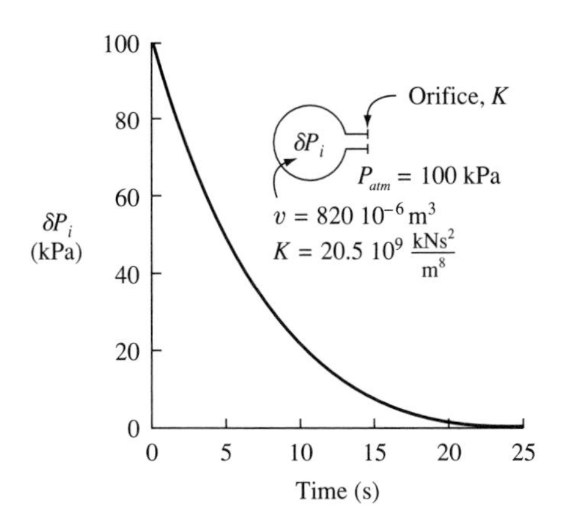


**Figure P5.17** Volume of gas discharging through an orifice.

**5.18** The system shown in Figure P5.18(a) is a single-acting rolling diaphragm actuator with a spring return. There is no preload on the spring. The actuator is driven from a pressure regulator through a fast-acting solenoid valve. In series with the solenoid valve is a small orifice. The system has been built and instrumented in the lab. The components have been measured, and their values are as follows:

$m$ = mass of actuator = 0.1 kg $\qquad$ $k_s$ = stiffness of spring = 1.33 N/mm

$A$ = mass of actuator = 1774 mm$^2$ $\qquad$ $K_0$ = orifice coefficient = $23.8 \times 10^9$ kN s$^2$/m$^8$

$$V_0 = 15{,}000 \text{ mm}^3$$

**a.** First, *model* the system, neglecting the effects of the mass and considering the incompressible flow equation. (See problem 5.17.) *Derive* the system dynamics equations for the actuator. *Express* the equation in the state-space format, and *state* equations that can be used for the internal pressure and position of the actuator, based upon the state variables. Next, *model* the system, considering the mass. *Derive* a state-space representation of the system, and *state* the equations for pressure and position as a function of the state variables.

b. *Obtain* the simulated response with a step input of pressure 20 kPa for both cases in part a. *Plot* $\delta P_1$ and $z$. *Compare* your results with the experimental response given in Figure P5.18(b). *What* can you suggest that has not been modeled or taken into consideration that would account for the differences in the simulated and experimental responses?

c. From part b, *discuss* which model should be used (i.e., can the mass be neglected in this case?) and *how* could you determine from the equations whether mass would be important.

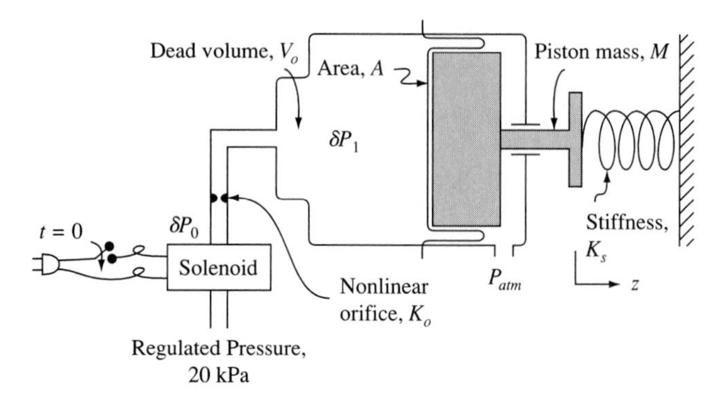


(a) Configuration.

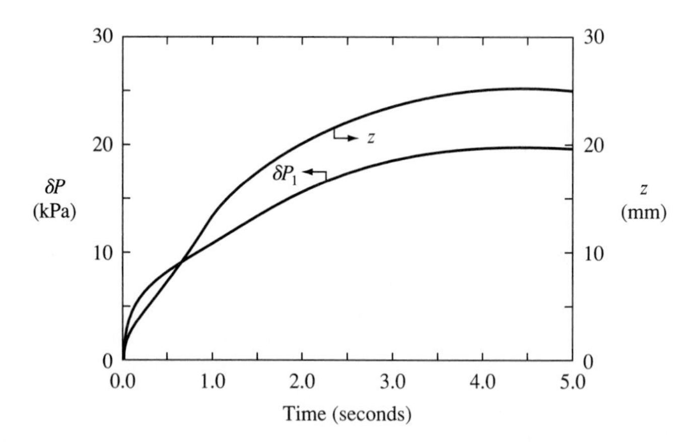


(b) Transient response.

**Figure P5.18** Spring-loaded diaphragm actuator.

# CHAPTER 6

# Thermal Systems

## 6.1 INTRODUCTION

**Thermal systems** transfer or store thermal energy by virtue of temperature and heat flow rate. The **thermal effects** are conduction, convection, radiation, and heat storage capacity.

Thermal systems have a static and dynamic behavior similar to mechanical, electrical, and fluid systems, but in some ways they are quite different from these other systems. Thermal systems are similar in that they exhibit resistance and capacitance effects, can be analyzed by circuit analysis, and have dynamic responses that can be characterized by system time constants, etc. However, they often require nonlinear, variable-coefficient, or distributed-parameter models. Also, although thermal systems exhibit resistance and capacitance, there is no thermal inductance.

The analysis of thermal systems often requires the combination of three technologies: thermodynamics, heat transfer, and fluid mechanics. In our presentation in this chapter, we will idealize thermal systems for applications that result in linear systems, in order to emphasize the similarity to systems discussed in previous chapters. When these simple models are not sufficient, the reader should consult comprehensive treatments of the various aspects of heat transfer [Refs. 1–5].

## 6.2 BASIC EFFECTS

The basic effects exhibited by thermal systems are thermal conduction, thermal convection, thermal radiation, and thermal storage. In our study of thermal systems, we will consider temperature to be the effort variable, or the potential variable, as discussed in Chapter 2. The temperature $T$ can be expressed in degrees Celsius or Kelvin, or in degrees Fahrenheit or Rankine (Anders Celsius, Swedish astronomer, 1701–44; Lord Kelvin is William Thomson, British physicist, 1824–1907; Gabriel Daniel Fahrenheit, German physicist, 1686–1736; William John M. Rankine, English engineer, 1820–72). The heat transfer will be the variable that relates to the flow variable or the rate variable. The heat flow rate $Q_h$ has the units of power, e.g., joules per second or Btu (British thermal units) per hour.

Thermal conduction, convection, and radiation all display an algebraic relation between the temperature and the heat flow; therefore, these effects represent thermal resistance and have no dynamic effects.

### 6.2.1 Thermal Conduction

**Thermal conduction** is the ability of solid or continuous media to conduct heat. The heat transfer $Q_h$ is related to the temperature gradient in the direction of heat flow $dT/dx$ as shown by Figure 6.1 (The figure illustrates one-dimensional heat transfer; a more general case would be two or three-dimensional heat transfer.)

The steady-state relationship between the temperature and heat flow in a material is given by **Fourier's law of heat conduction** (named after the French mathematician and physicist Joseph Fourier, 1768–1830). In this one-dimensional relationship, the heat transfer per unit area is related to the temperature gradient in the direction of heat flow by the equation

$$\frac{Q_h}{A} = -k_t \frac{dT}{dx} \tag{6.1}$$

Note that the negative sign indicates that there is a temperature drop in the direction of flow.

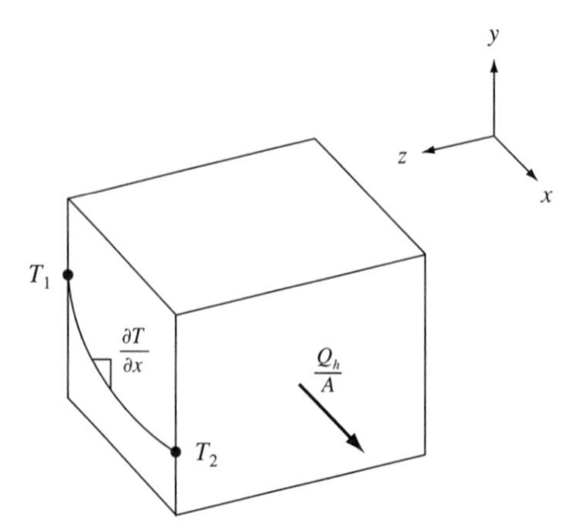


**Figure 6.1** Conduction heat transfer in a continuous material.

The proportionality constant is termed the **thermal conductivity** $k_t$ of the material. Thermal conductivity is a material property and is listed in Table 6.1 for a variety of metals that are good conductors. Table 6.2 lists the thermal properties of a number of insulators.

Figure 6.2 indicates that thermal conductivity is a function of temperature for some metals. Figure 6.3 shows how the thermal conductivity of gases and liquids varies with temperature and how the conductivity of gases is about one order of magnitude lower than that of liquids. This means that the coefficient in Fourier's equation will be variable if the temperature gradient is very high (greater than 50°F); the result will be nonlinear equations.

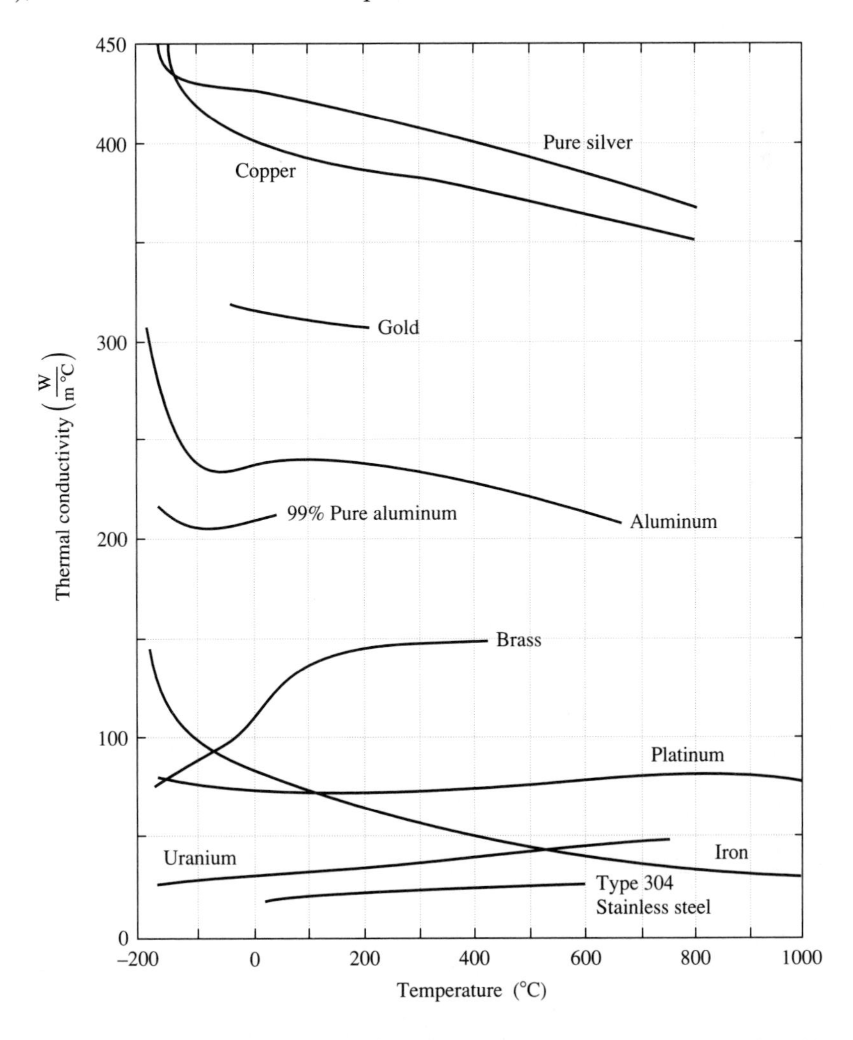


**Figure 6.2** Thermal conductivity of various metals. (This figure was taken from *A Heat Transfer Textbook*, by John H. Lienhard, and published by Prentice Hall Publishing Company, ©1987.)

**TABLE 6.1** THERMAL PROPERTIES OF METALS

| Element | Thermal conductivity $\left(\frac{W}{m\ °K}\right)$ 200°K −73°C | 273°K 0°C 32°F | 400°K 127°C 261°F | 600°K 327°C 621°F | 800°K 527°C 981°F | 1,000°K 727°C 1,341°F | Properties at 293°K or 20°C or 68°F: $\rho$ kg/m$^3$ | $C_p$ J/(kg °K) | $k$ W/(m °K) |
|---|---|---|---|---|---|---|---|---|---|
| | × 0.5777 = Btu/(hr ft °F) | | | | | | × 6.243 × 10$^{-2}$ = (lbm/ft$^3$) | × 2.388 × 10$^{-4}$ = Btu/(lbm °F) | × 0.5777 = Btu/(hr ft °F) |
| Aluminum | 237 | 236 | 240 | 232 | 220 | | 2,702 | 896 | 236 |
| Antimony | 30.2 | 25.5 | 21.2 | 18.2 | 16.8 | | 6,684 | 208 | 24.6 |
| Beryllium | 301 | 218 | 161 | 126 | 107 | 89 | 1,850 | 1750 | 205 |
| Bismuth | 9.7 | 8.2 | | | | | 9,780 | 124 | 7.9 |
| Boron | 52.5 | 31.7 | 18.7 | 11.3 | 8.1 | 6.3 | 2,500 | 1047 | 28.6 |
| Cadmium | 99.3 | 97.5 | 94.7 | | | | 8,650 | 231 | 97 |
| Cesium | 36.8 | 36.1 | | | | | 1,873 | 230 | 36 |
| Chronmium | 111 | 94.8 | 87.3 | 80.5 | 71.3 | 65.3 | 7,160 | 440 | 91.4 |
| Cobalt | 122 | 104 | 84.8 | | | | 8,862 | 389 | 100 |
| Copper | 413 | 401 | 392 | 383 | 371 | 357 | 8,933 | 383 | 399 |
| Germanium | 96.8 | 66.7 | 43.2 | 27.3 | 19.8 | 17.4 | 5,360 | | 61.6 |
| Gold | 327 | 318 | 312 | 304 | 292 | 278 | 19,300 | 129 | 316 |
| Hafnium | 24.4 | 23.3 | 22.3 | 21.3 | 20.8 | 20.7 | 13,280 | | 23.1 |
| Indium | 89.7 | 83.7 | 74.5 | | | | 7,300 | | 82.2 |
| Iridium | 153 | 148 | 144 | 138 | 132 | 126 | 22,500 | 134 | 147 |
| Iron | 94 | 83.5 | 69.4 | 54.7 | 43.3 | 32.6 | 7,870 | 452 | 81.1 |
| Lead | 36.6 | 35.5 | 33.8 | 31.2 | | | 11,340 | 129 | 35.3 |
| Lithium | 88.1 | 79.2 | 72.1 | | | | 534 | 3391 | 77.4 |
| Magnesium | 159 | 157 | 153 | 149 | 146 | | 1,740 | 1017 | 156 |
| Manganese | 7.17 | 7.68 | | | | | 7,290 | 486 | 7.78 |
| Mercury | 28.9 | | | | | | 13,546 | | |
| Molybdenum | 143 | 139 | 134 | 126 | 118 | 112 | 10,240 | 251 | 138 |
| Nickel | 106 | 94 | 80.1 | 65.5 | 67.4 | 71.8 | 8,900 | 446 | 91 |
| Niobium | 52.6 | 53.3 | 55.2 | 58.2 | 61.3 | 64.4 | 8,570 | 270 | 53.6 |
| Palladium | 75.5 | 75.5 | 75.5 | 75.5 | 75.5 | 75.5 | 12,020 | 247 | 75.5 |

*Continued*

**TABLE 6.1** (CONTINUED)*

| Element | Thermal conductivity $\left(\frac{W}{m\ °K}\right)$ | | | | | | Properties at 293°K or 20°C or 68°F | | |
|---|---|---|---|---|---|---|---|---|---|
| | 200°K<br>−73°C | 273°K<br>0°C<br>32°F | 400°K<br>127°C<br>261°F | 600°K<br>327°C<br>621°F | 800°K<br>527°C<br>981°F | 1,000°K<br>727°C<br>1,341°F | $\rho$<br>kg/m$^3$ | $C_p$<br>J/(kg °K) | $k$<br>W/(m °K) |
| | × 0.5777 = Btu/(hr ft °F) | | | | | | × 6.243 × $10^{-2}$ = (lbm/ft$^3$) | × 2.388 × $10^{-4}$ = Btu/(lbm °F) | × 0.5777 = Btu/(hr ft °F) |
| Platinum | 72.4 | 71.5 | 71.6 | 73.0 | 75.5 | 78.6 | 21,450 | 133 | 71.4 |
| Potassium | 104 | 104 | 52 | | | | 860 | 741 | 103 |
| Rhenium | 51 | 48.6 | 46.1 | 44.2 | 44.1 | 44.6 | 21,100 | 137 | 48.1 |
| Rhodium | 154 | 151 | 146 | 136 | 127 | 121 | 12,450 | 248 | 150 |
| Rubidium | 58.9 | 58.3 | | | | | 1,530 | 348 | 58.2 |
| Silicon | 264 | 168 | 98.9 | 61.9 | 42.2 | 31.2 | 2,330 | 703 | 153 |
| Silver | 403 | 428 | 420 | 405 | 389 | 374 | 10,500 | 234 | 427 |
| Sodium | 138 | 135 | | | | | 971 | 1206 | 133 |
| Tantalum | 57.5 | 57.4 | 57.8 | 58.6 | 59.4 | 60.2 | 16,600 | 138 | 57.5 |
| Tin | 73.3 | 68.2 | 62.2 | | | | 5,750 | 227 | 67.0 |
| Titanium | 24.5 | 22.4 | 20.4 | 19.4 | 19.7 | 20.7 | 4,500 | 611 | 22.0 |
| Tungsten | 197 | 182 | 162 | 139 | 128 | 121 | 19,300 | 134 | 179 |
| Uranium | 25.1 | 27 | 29.6 | 34 | 38.8 | 43.9 | 19,070 | 113 | 27.4 |
| Vanadium | 31.5 | 31.3 | 32.1 | 34.2 | 36.3 | 38.6 | 6,100 | 502 | 31.4 |
| Zinc | 123 | 122 | 116 | 105 | | | 7,140 | 385 | 121 |
| Zirconium | 25.2 | 23.2 | 21.6 | 20.7 | 21.6 | 23.7 | 6,570 | 272 | 22.8 |

* This table was taken from *Principles of Heat Transfer,* Fifth Edition, Revised Printing by Frank Kreith and Mark S. Bohn. Copyright © 1997 by PWS Publishing Company, a division of International Thomson Publishing Inc.

## TABLE 6.2 THERMAL PROPERTIES OF INSULATORS*

Properties at 293°K or 20°C or 68°F

| Material | $\rho$ kg/m$^3$ $\times 6.243 \times 10^{-2}$ = lbm/ft$^3$ | $C_p$ J/(kg °K) $\times 2.388 \times 10^{-4}$ = Btu/(lbm°F) | $k$ W/(m °K) $\times 0.5777$ = Btu/(hr ft°F) | $\alpha \times 10^5$ (m$^2$/s) $\times 3.874 \times 10^4$ = ft$^2$/hr |
|---|---|---|---|---|
| Asbestos | 383 | 816 | 0.113 | 0.036 |
| Asphalt | 2,120 | | 0.698 | |
| Bakelite | 1,270 | | 0.233 | |
| Brick | | | | |
| Common | 1,800 | 840 | 0.38–0.52 | 0.028–0.034 |
| Carborundum (50% SiC) | 2,200 | | 5.82 | |
| Magnesite (50% MgO) | 2,000 | | 2.68 | |
| Masonry | 1,700 | 837 | 0.658 | 0.046 |
| Silica (95% $SiO_2$) | 1,900 | | 1.07 | |
| Zircon (62% $ZrO_2$) | 3,600 | | 2.44 | |
| Cardboard | | | 0.14–0.35 | |
| Cement, hard | | | 1.047 | |
| Clay (48.7% moisture) | 1,545 | 880 | 1.26 | 0.101 |
| Coal, anthracite | 1,370 | 1,260 | 0.238 | 0.013–0.015 |
| Concrete, dry | 500 | 837 | 0.128 | 0.049 |
| Cork, boards | 150 | 1,880 | 0.042 | 0.015–0.044 |
| Cork, expanded | 120 | | 0.036 | |
| Diatomaceous earth | 466 | 879 | 0.126 | 0.031 |
| Glass fiber | 220 | | 0.035 | |
| Glass, window | 2,800 | 800 | 0.81 | 0.034 |
| Glass, wool | 50 | | 0.037 | |
| | 100 | | 0.036 | |
| | 200 | 670 | 0.040 | 0.028 |
| Granite | 2,750 | | 3.0 | |
| Ice (0°C) | 913 | 1,830 | 2.22 | 0.124 |
| Kapok | 25 | | 0.035 | |
| Linoleum | 535 | | 0.081 | |
| Mica | 2,900 | | 0.523 | |
| Pine bark | 342 | | 0.080 | |
| Plaster | 1,800 | | 0.814 | |
| Plexiglas® | 1,180 | | 0.195 | |
| Plywood | 590 | | 0.109 | |
| Polystyrene | 1,050 | | 0.157 | |
| Rubber, Buna | 1,250 | | 0.465 | |
| Hard (ebonite) | 1,150 | 2,009 | 0.163 | 0.0062 |
| Spongy | 224 | | 0.055 | |
| Sand, dry | | | 0.582 | |
| Sand, moist | 1,640 | | 1.13 | |
| Sawdust | 215 | | 0.071 | |
| Soil | | | | |
| Dry | 1,500 | 1,842 | ~ 0.35 | ~ 0.0138 |
| Wet | 1,500 | | ~ 2.60 | 0.0414 |
| Wood | | | | |
| Oak | 609–801 | 2,390 | 0.17–0.21 | 0.0111–0.0121 |
| Pine, fir, spruce | 416–421 | 2,720 | 0.15 | 0.0124 |
| Wood fiber sheets (celotex) | 200 | | 0.047 | |
| | 400 | | 0.055 | |
| Wool | 200 | | 0.038 | |

*This table was taken from *Principles of Heat Transfer*, Fifth Edition, Revised Printing by Frank Kreith and Mark S. Bohn. Copyright © 1997 by PWS Publishing Company, a division of International Thomson Publishing Inc.

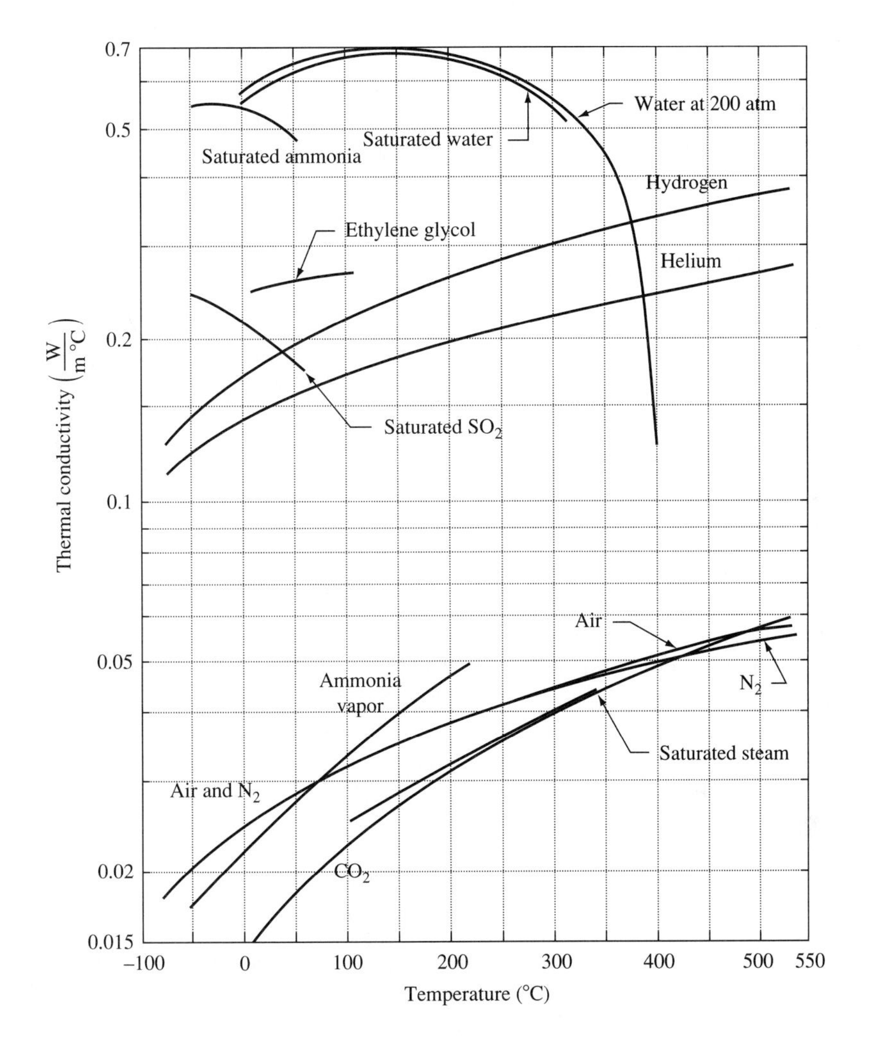


**Figure 6.3** Thermal conductivity of various fluids. (This figure was taken from *A Heat Transfer Textbook*, by John H. Lienhard, and published by Prentice Hall Publishing Company, ©1987.)

Fourier's law, Eq. (6.1), can take a variety of forms, depending upon the geometric shape of the system studied. Common configurations are transverse heat flow in a flat plate, axial heat flow in a rod, and radial heat flow in a cylinder.

**Conduction through a Flat Plate.** For a flat plate of thickness $L$ and cross-sectional area $A$ shown in Figure 6.4, if $k_t$ is constant over the temperatures considered, Fourier's law reduces to

$$Q_h = -k_t A \frac{\partial T}{\partial x} \approx -k_t A \frac{\Delta T}{\Delta x} = -\frac{k_t A}{L} \Delta T \tag{6.2}$$

in which $\Delta T = T_2 - T_1$

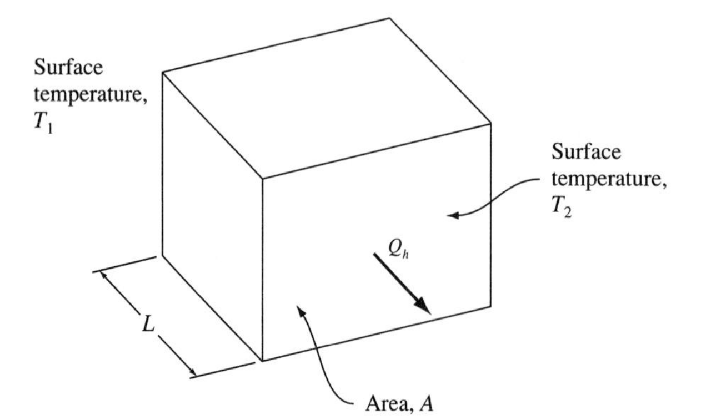


**Figure 6.4** One-dimensional conduction through a flat plate.

Since it is more natural to think of the temperature difference as we do in circuit analysis, Eq. (6.2) can be rewritten to eliminate the negative sign as follows:

$$Q_h = \frac{k_t A}{L} \delta T \tag{6.3}$$

The temperature difference is $\delta T = T_1 - T_2$.

If we consider the heat flow $Q_h$ as the flow variable and the temperature difference $\delta T$ as the effort variable, we can write Eq. (6.3) in the form of an impedance relation:

$$Q_h = \frac{\delta T}{R} \tag{6.4}$$

In this case, we take the thermal resistance $R$ due to conduction to be

$$R = \frac{L}{k_t A} \tag{6.5}$$

This leads to a simple concept for heat transfer in which we consider thermal circuit analysis of interacting thermal components just as we do for electrical circuits. For example, if we had a metal plate coated on one side as shown in Figure 6.5, the total heat transfer would be related to the total temperature difference and the combined thermal resistance, $R_c + R_m$ (see Chapter 4), by

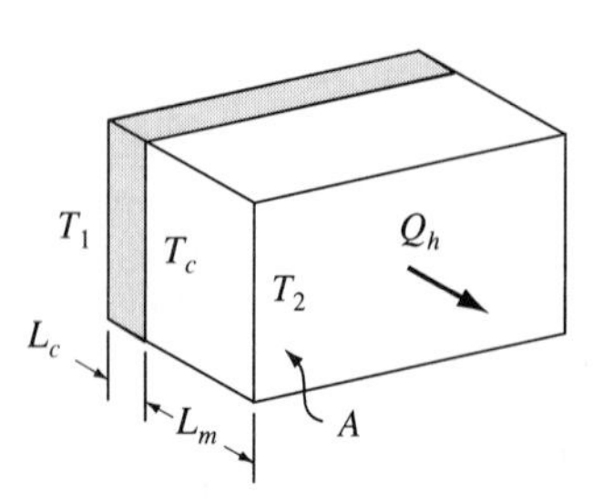


**Figure 6.5** Thermal resistance of two plates.

$$Q_h = \frac{T_1 - T_2}{R_c + R_m} \tag{6.6}$$

where $R_c = L_c/(k_cA)$ and $R_m = L_m/(k_mA)$.

**Axial Conduction in a Rod.** Figure 6.6 shows a rod of length $L$ and cross-sectional area $A$ that has an insulated suface, so there is no radial heat transfer and the heat transfer is only axial. The Fourier law relating to this configuration gives the following result.

$$Q_h = \frac{k_tA}{L}(T_1 - T_2) \tag{6.7}$$

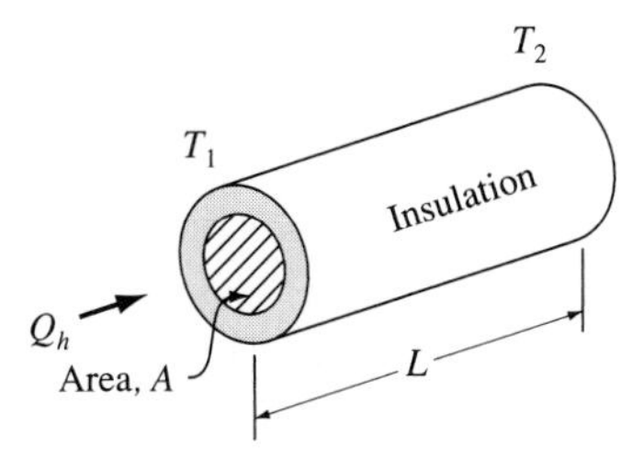


**Figure 6.6** Axial heat transfer in a rod.

**Radial Conduction in a Cylinder.** A common configuration is a pipe with a constant temperature source on its inside and outside surfaces, as shown in Figure 6.7. In this case the incremental area for heat flow is $2\pi rL$, and the temperature gradient is $dT/dr$. Thus, Fourier's law reduces to

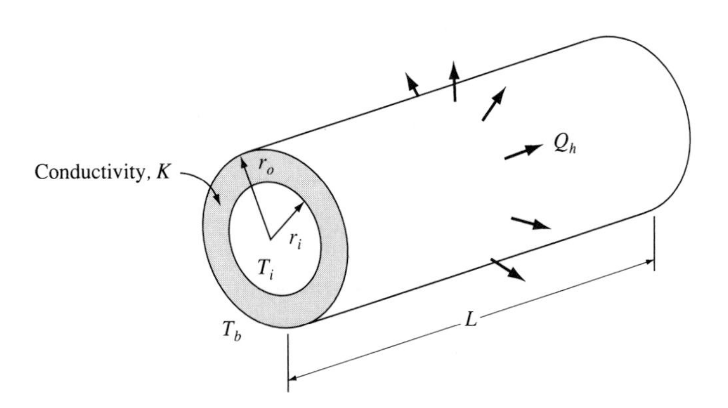


**Figure 6.7** Heat conduction in a cylinder.

$$Q_h = -k_t 2\pi r L \frac{dT}{dr} = -k_t 2\pi L \frac{dT}{\left(\frac{dr}{r}\right)} \tag{6.8}$$

Assuming a constant heat flow, separation of variables and integration from the inside to the outside surface of the cylinder yields

$$Q_h = \frac{2\pi k_t L}{\ln(r_o/r_i)}(T_i - T_o) \tag{6.9}$$

From this result, it is clear that the thermal resistance is

$$R = \frac{\ln(r_o/r_i)}{2\pi k_t L} \tag{6.10}$$

If the surface of the pipe is coated with a material different from that of the rest of the pipe and is of thickness $r_c - r_0$, as shown in Figure 6.8, the configuration will be equivalent to a series combination of two radial resistances. Thus, the total heat transfer will be

$$Q_h = \frac{(T_i - T_o)}{\frac{\ln(r_c/r_i)}{2\pi L k_p} + \frac{\ln(r_o/r_c)}{2\pi L k_c}} \tag{6.11}$$

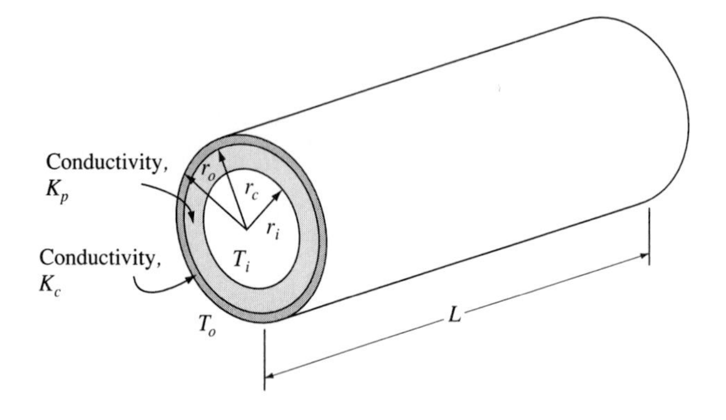


**Figure 6.8** Pipe with a coated surface.

### 6.2.2 Thermal Convection

**Thermal convection** is the process of heat transfer between a surface of a solid material and a fluid that is exposed to the solid surface. The actual process is one of heat transfer by conduction between the solid surface and the fluid at the immediate vicinity of the surface. However, once the fluid has been subjected to the heat transfer, the fluid in the vicinity of the surface can move to another location and be replaced with fresh fluid, and the process can repeat itself. The temperature of the surface of the solid is $T_s$; the temperature of the free-stream fluid at some distance away from the surface $T_\infty$ is constant. Thus, the overall effect is that heat can transfer from the surface temperature $T_s$ to some fluid temperature $T_\infty$ at some distance away from the surface. The heat flow $Q_h$ per unit surface area $A$ is given by **Newton's law of cooling,**

$$\frac{Q_h}{A} = h(T_s - T_\infty) \tag{6.12}$$

and is illustrated in Figure 6.9.

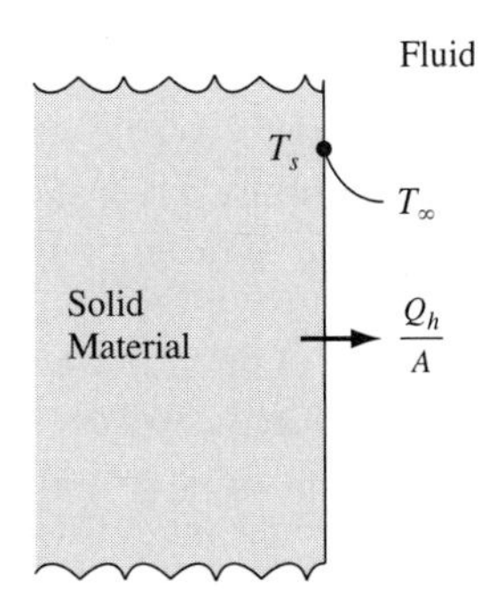


**Figure 6.9** Convection heat transfer.

In Eq (6.12), the **convection coefficient** $h$ represents the overall effects of the mechanism of heat conduction from the surface to the fluid and the motion of the fluid to relocate the heat. The convection coefficient is a function of the velocity of the fluid passing across the surface and the velocity profiles near the surface. The velocity profiles depend upon the viscosity of the fluid, the dimensions of the geometric configuration, and the velocity of the fluid. Therefore, it is expected that the convection coefficient will vary with the properties of the fluid (thermal conductivity, viscosity, density, and specific heat), the geometric configuration of the solid, and the details of the free-stream fluid flow patterns.

In **free convection,** the fluid next to the surface does not have an externally driven flow pattern. Instead, the temperature of the surface will change the density of the fluid immediately adjacent to the surface, and the differential density of the surface fluid and the surrounding fluid will cause a local motion in the presence of a gravity field. The motion of the fluid near the surface can cooperate to cause a net fluid motion and, hence, a relocation of the heat. In **forced convection,** the fluid is forced to flow across the surface and thereby carry away any heat transferred between itself and the surface.

From your own practical experience, it should be intuitively clear that forced convection is much more effective than free convection in heat transfer. If we want to cool a hot object, we blow on it or shake it about in the air; for example, we may install blowers or fans on electronic equipment that generates heat.

If we consider the microscopic behavior of the heat transfer at the surface, we can see that the convection coefficient for cooling the surface might be different from that for heating the surface.

The heat transfer due to convection is proportional to the convection coefficient $h$, the surface area $A$, and the temperature difference $\delta T$. Thus, we may derive the thermal resistance due to convection $R$ as follows:

$$Q_h = hA\delta T \tag{6.13}$$

$$Q_h = \frac{\delta T}{R} \tag{6.14}$$

so

$$R = \frac{1}{hA} \tag{6.15}$$

While the concept of convection heat transfer is very simple, the calculation of the convection coefficient is often quite difficult and is dependent upon the geometric configuration of the solid and flow conditions. Further, the convection coefficient is normally variable in the direction of flow, so we would have variable or dependent coefficients in the convection equation; however, it is often possible to use averaged convection coefficients that will relate the overall or average heat transfer from a surface. Table 6.3 illustrates the typical range of values of the convection coefficient for a variety of conditions and configurations. In free convection, the convection coefficient increases with temperature difference (which increases the buoyancy forces in the fluid due to density difference). In forced convection, the convection coefficient increases with the free-stream velocity of the fluid. The reader is referred to textbooks on heat transfer [Refs. 1–5] for a discussion of determination of convection coefficients (expressed as a dimensionless grouping called the Nusselt number) as a function of buoyancy forces for free convection (expressed as a dimensionless Rayleigh number) and free-stream velocity for forced convection (expressed as a dimensionless Reynolds number).

**TABLE 6.3** APPROXIMATE CONVECTION COEFFICIENTS, *h*

| Fluid and condition | $\frac{\text{Btu}}{\text{hr ft}^2\,°\text{F}} \times 5.698 =$ | $\frac{\text{watt}}{\text{m}^2\,°\text{C}}$ |
|---|---|---|
| Air, free convection | 1–5 | 6–30 |
| Air, forced convection | 5–100 | 30–600 |
| Water, free convection | 10–50 | 60–300 |
| Water, forced convection | 50–1000 | 300–6000 |
| Water, boiling | 500–10,000 | 3000–60,000 |
| Superheated steam, free convection | 1–5 | 6–30 |
| Superheated steam, forced convection | 5–50 | 30–300 |
| Steam, condensing | 1000–20,000 | 6,000–120,000 |
| Oil, free convection | 5–25 | 30–150 |
| Oil, forced convection | 25–500 | 150–3000 |
| Oil, boiling | 250–5000 | 1500–30,000 |

### 6.2.3 Thermal Radiation

**Thermal radiation** is the process of heat transfer in which the energy is high enough to transfer heat without the presence of a surrounding medium, such as a fluid or a solid. Unlike conduction or convection, thermal radiation can transfer heat from one body to another without any physical contact or any intervening fluids between the bodies. Thus, thermal radiation can take place in a vacuum.

A body of matter radiates electromagnetic energy at a rate proportional to the fourth power of the absolute temperature of the body. The heat transfer $Q_h$ from an ideal blackbody of surface area $A$ is given by

$$\frac{Q_h}{A} = \sigma T^4 \tag{6.16}$$

Figure 6.10 illustrates the concept of radiation heat transfer.

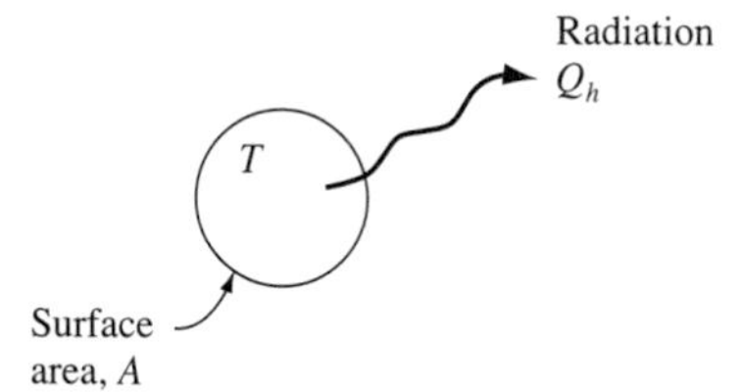


**Figure 6.10** Radiation heat transfer.

The constant in Eq. (6.16) is the **Stefan-Boltzmann** constant, which has the value

$$\sigma = 1.714 \times 10^{-9}\,\frac{\text{Btu/hr}}{\text{ft}^2\,{}^\circ\text{R}^4} = 56.68 \times 10^{-9}\,\frac{\text{watt}}{\text{m}^2\,{}^\circ\text{K}^4} \tag{6.17}$$

A **blackbody** is a perfect emitter or receiver of thermal radiation. Other surfaces or **colors** of surfaces have less **emissivity** than a blackbody. Thus, the heat transfer to or from a surface that has an emissivity less than that of a blackbody will be less by a factor $F_e$. Unless the body is isolated in space, it will exchange radiation with neighboring objects. Therefore, the net heat transfer will be related to the difference in the fourth power of the temperatures of each body. Knowing that not all of the radiation that leaves one body may impinge upon another body forces us to use a **view factor** $F_v$ to account for the lost radiation between the two bodies. Thus, the net radiant heat transfer between two bodies at different temperatures is

$$Q_h = F_e F_v \sigma A (T_H^4 - T_L^4) \tag{6.18}$$

where $T_H$ is the high temperature and $T_L$ the low temperature.

It is interesting to express the radiation heat transfer in terms of an equivalent convection equation. In convection, the driving temperature difference is between the surface temperature and the free-stream temperature of the fluid some distance away from the surface. In radiation heat transfer between one body and another, the driving temperature difference is between the surface temperature of the high-temperature body $T_H$ and the temperature of the low-temperature body $T_L$. Therefore, there is enough similarity between the convection equation and the concept of radiation to warrant the derivation of an equivalent convection coefficient. If we consider the heat flow $Q_h$ from a surface area $A_h$ of the first body, we can derive the equivalent convection coefficient $h_{eq}$ as follows (assuming $T_H > T_L$, and $\Delta T = T_H - T_L$):

$$Q_h = F_e F_v h_{eq} A_h \Delta T \tag{6.19}$$

We can factor the expression $(T_H^4 - T_L^4)$ and in Eq. 6-18 extract a term $(T_H - T_L)$ to be the $\Delta T$ term in the equivalent convection equation. The remaining terms can be then rearranged in terms of $T_H$ and $\Delta T$. Equating the result of these operations to the form of the equivalent convection equation yields the following expression for the equivalent convection coefficient:

$$h_{eq} = 4\sigma T_H^3 \left[1 - \frac{3}{2}\frac{\Delta T}{T_H} + \left(\frac{\Delta T}{T_H}\right)^2 - \frac{1}{4}\left(\frac{\Delta T}{T_H}\right)^3\right] \tag{6.20}$$

This expression can be used in two ways: to express radiation heat transfer as a linear type of equation with a variable coefficient or to determine whether the effects of radiation can be neglected.

The $(\Delta T/T_H)^3$ term in Eq. (6.20) is negligible for almost all systems. The $(\Delta T/T_1)^2$ term is a relatively small term for most systems. Figure 6.11 illustrates the

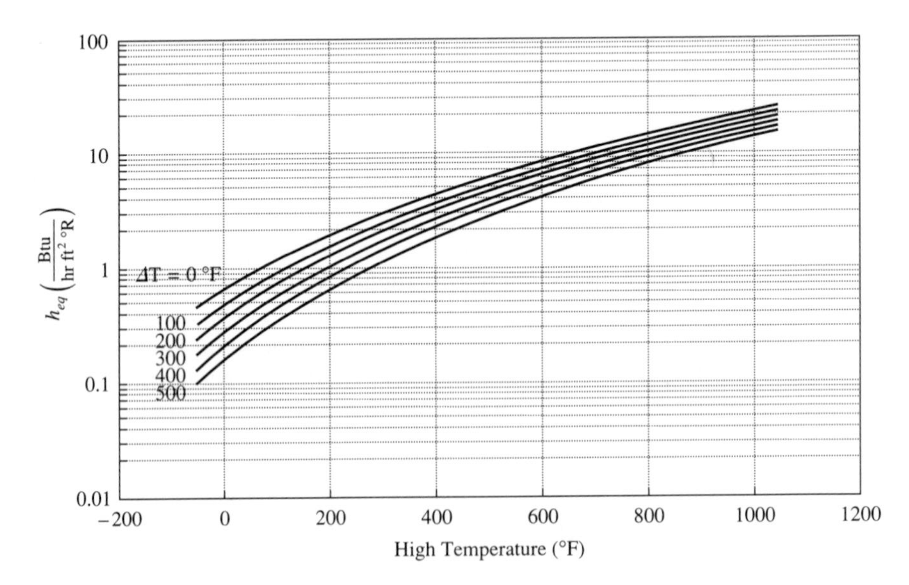


**Figure 6.11** Equivalent convection coefficient for radiation.

equivalent convection coefficient as a function of the temperature of the first body with parametric values of temperature difference between the two bodies. If the equivalent convection coefficient from radiation is much smaller than the actual convection coefficient, the radiation can be neglected. For example, radiation can be neglected if the high temperature is less than a few hundred degrees Fahrenheit with low differential temperatures, and it can be neglected up to several hundred degrees Fahrenheit with higher differential temperatures.

### 6.2.4 Thermal Capacitance

The ability of a substance to hold or store heat is the **heat capacity** of the material and behaves like a **thermal capacitance**. Since the specific heat is stated on a per-mass basis, the total heat storage ability is related to the specific heat $C_p$ times the mass $M$ of the object. When the stored energy is released, the temperature changes as a function of time such that the derivative of the temperature $T$ with respect to time is proportional to the heat flow $Q_h$. That is,

$$Q_h = C_p M \frac{dT}{dt} \tag{6.21}$$

In stating this simple relationship, we must assume that the internal temperature of the entire mass can be accurately represented by one single temperature, and thus, there is no temperature gradient in the mass. This means that the thermal conduction inside the mass is very high relative to the transfer of heat by convection at the surface of the mass.

Equation (6.21) is a dynamic equation that is a differential equation. Since it has the form of a flow variable that is proportional to a rate of change of an effort variable $T$ with time, the heat capacity represents a capacitance.

The specific heat and specific heat ratio is presented for various substances in Table 6.4.

## 6.3 STATIC THERMAL SYSTEMS

In thermal systems, we are interested either in how much heat might be lost by the components or in the resulting temperature of a system with internal heat generation. Examples of thermal systems that do not involve thermal storage or capacitance are illustrated in Sections 6.3.1 and 6.3.2.

### 6.3.1 Thermal Conduction Circuits

#### Example 6.1 Conduction in a Rod

Consider the system shown in Figure 6.12 in which a 0.25 inch diameter steel rod serves as a structural member to locate a steam boiler and a water tank. The one foot long rod

**TABLE 6-4** SPECIFIC HEAT OF VARIOUS SUBSTANCES

| | Specific heat ratio, $k = C_p/C_v$ | Specific heat, $C_p$ kJ/(kg°C) | Specific heat, $C_p$ Btu/(lbm °F) |
|---|---|---|---|
| Liquids | | | |
| ethylene glycol | | 2.36 | 0.565 |
| octane | | 2.15 | 0.514 |
| propane | | 2.41 | 0.576 |
| salt water | | 3.93 | 0.940 |
| water | | 4.18 | 0.998 |
| Gases | | | |
| air | 1.40 | 1.00 | 0.240 |
| carbon dioxide | 1.30 | 0.86 | 0.205 |
| carbon monoxide | 1.40 | 1.05 | 0.250 |
| methane | 1.31 | 2.26 | 0.540 |
| propane | 1.20 | 1.63 | 0.390 |
| Solids | | | |
| brick | — | 0.92 | 0.220 |
| concrete | — | 0.96 | 0.230 |
| earth | — | 1.26 | 0.300 |
| glass | — | 0.84 | 0.200 |
| paper | — | 1.38 | 0.330 |
| plastic | — | 1.67 | 0.400 |
| wood, white pine | — | 2.51 | 0.600 |
| Metals | | | |
| aluminum | — | 0.90 | 0.215 |
| copper | — | 0.38 | 0.092 |
| iron | — | 0.45 | 0.108 |
| lead | — | 0.13 | 0.031 |
| silver | — | 0.24 | 0.057 |
| zinc | — | 0.39 | 0.057 |

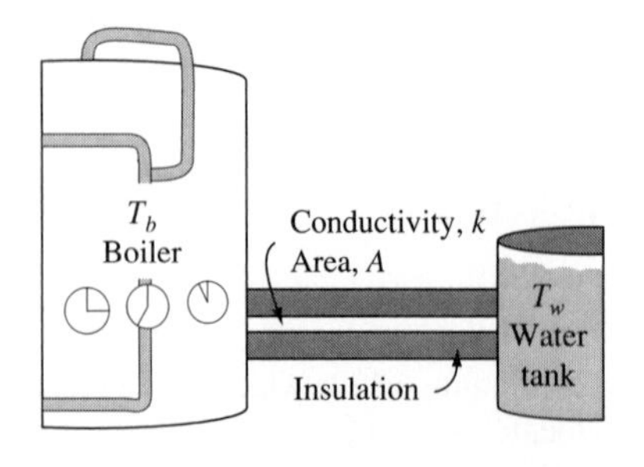


**Figure 6.12** Conductivity in a structural rod.

is insulated along its length to prevent convection from the surface; however, there can be conduction along the axis of the rod. The boiler temperature is 220°F, and the water tank is 100°F. The question is: How much heat will be transferred from the boiler to the water supply tank?

**Solution** For this problem, we can use the conduction equation, Eq. (6.7), and the properties of the steel rod to obtain

$$Q_h = \frac{k_t A}{L}(T_b - T_w) \tag{6.22}$$

From Table 6.2, we find that $k_t = 23$ Btu/(hr ft °F). Therefore, the heat loss is

$$Q_h = \frac{23 \dfrac{\text{Btu}}{\text{hr ft °F}} \pi 0.25^2 \text{ in}^2 (220 - 100)\text{°F}}{4\,(1.0 \text{ ft})\, 144 \dfrac{\text{in}^2}{\text{ft}^2}} \tag{6.23}$$

or

$$Q_h = 0.941 \frac{\text{Btu}}{\text{hr}} = 0.277 \text{ watt} \tag{6.24}$$

This loss is quite small.

### 6.3.2 Thermal Conduction and Convection Circuits

**Example 6.2 Plexiglas® plate**

Consider the system shown in Figure 6.13, in which a plate of Plexiglas® that is a wall of a container is exposed to an internal temperature $T_i = 50$°C on one side and is subjected to free convection to room temperature, 25°C, on the other side. We want to know how much heat is lost and what will be the outer surface temperature $T_s$. The plate is 100 mm by 100 mm and is 6 mm thick. The thermal conductivity of Plexiglas® is 0.195 W/(m °K).

**Solution** Since there is some very slight air motion on the outer surface, we will assume that the convection coefficient is 20 W/(m$^2$ °K).

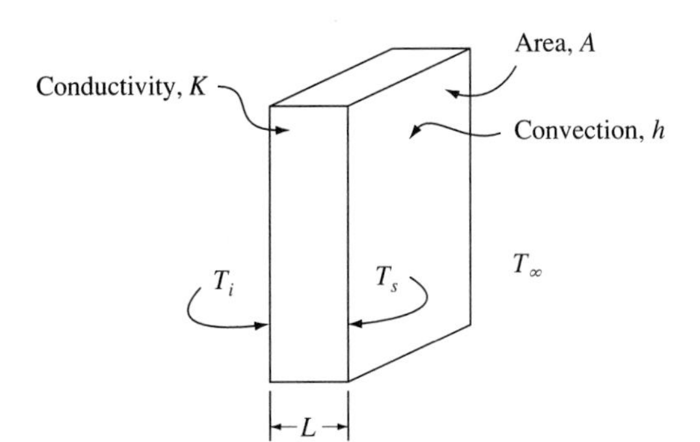


**Figure 6.13** Conduction and convection from a plate.

The resistances for the conduction and convection can be calculated from Eq. (6.5) and Eq. (6.15) as follows:

$$R_{plex} = \frac{L}{kA} = \frac{0.006\text{ m}}{0.195\dfrac{W}{\text{m °K}}0.10\text{ m }0.10\text{ m}} = 3.08\frac{\text{°K}}{\text{W}} \tag{6.25}$$

$$R_{conv} = \frac{1}{hA} = \frac{1}{20\dfrac{\text{W}}{\text{m}^2\text{ °K}}0.10\text{ m }0.10\text{ m}} = 5.00\frac{\text{°K}}{\text{W}} \tag{6.26}$$

Therefore, the heat flow can be calculated as follows (note that $\Delta T\text{ °C} = \Delta T\text{ °K}$):

$$Q_h = \frac{\Delta T}{(R_{plex} + R_{conv})} = \frac{(50 - 25)\text{ °K}}{(3.08 + 5.00)\dfrac{\text{°K}}{\text{W}}} \tag{6.27}$$

$$Q_h = 3.1\text{ watts} = 1.05\text{ Btu/hr} \tag{6.28}$$

The outer surface temperature can be calculated by first equating the heat flows from the Plexiglas® and the convection and then solving for $T_s$:

$$T_s = \frac{R_{conv}T_i + R_{plex}T_\infty}{R_{plex} + R_{conv}} = \frac{5.00[50°C] + 3.08[25°C]}{3.08 + 5.00} \tag{6.29}$$

$$T_s = 30.9°C + 9.5°C = 40.4°C \tag{6.30}$$

Therefore, the outer surface temperature, is closer to the internal temperature, since the resistance of the Plexiglas® is lower.

**Example 6.3 Steam Pipe**

Consider the system shown in Figure 6.14 in which steam at 220°F is flowing in a 6.0 foot steel pipe (2.0 inch ID, 2.25 inch OD) with free convection to room temperature, 75°F. We are considering whether we should insulate the pipe.

**Solution** The uninsulated pipe has three resistances to heat flow from the inside to the outside: the convection on the inside of the pipe $h_i$; the conduction through the steel $k_s$; and the convection on the outside of the pipe $h_o$. These resistances can be calculated from Eq. (6.15) and Eq. (6.10) as follows:

$$R_i = \frac{1}{h_iA_i} = \frac{12\dfrac{\text{in}}{\text{ft}}}{25\dfrac{\text{Btu}}{\text{hr ft}^2\text{ °F}}\pi 2.0\text{ in }6\text{ ft}} = 0.0127\frac{\text{°F}}{\text{Btu/hr}} \tag{6.31}$$

$$R_s = \frac{\ln(d_o/d_i)}{2\pi kL} = \frac{0.1178}{2\pi 30\dfrac{\text{Btu}}{\text{hr ft °F}}6\text{ ft}} = 0.000104\frac{\text{°F}}{\text{Btu/hr}} \tag{6.32}$$

$$R_o = \frac{1}{h_oA_o} = \frac{12\dfrac{\text{in}}{\text{ft}}}{5\dfrac{\text{Btu}}{\text{hr ft}^2\text{ °F}}\pi 2.25\text{ in }6\text{ ft}} = 0.0566\frac{\text{°F}}{\text{Btu/hr}} \tag{6.33}$$

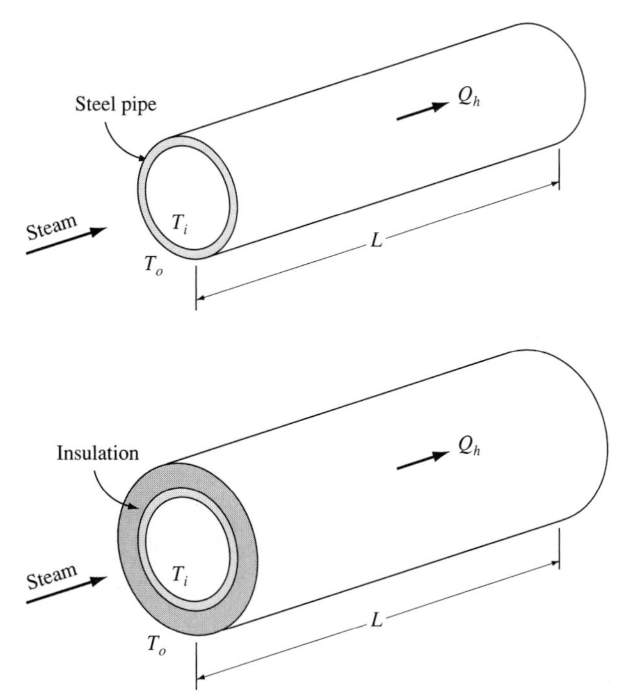


**Figure 6.14** Steam pipe convection/conduction example.

Since the total thermal resistance is the sum of the individual resistances, the total heat flow can be calculated as follows:

$$Q_h = \frac{(220 - 75)°\text{F}}{(0.0127 + .000104 + .0566)} \frac{\text{Btu/hr}}{°\text{F}} \tag{6.34}$$

$$Q_h = 2089 \text{ Btu/hr} = 614 \text{ watts} \tag{6.35}$$

Now consider insulation that could be added to the outside of the pipe. If we use a wrap that has 0.50 inch thickness (3.25 OD, 2.25 ID) and a thermal conductivity of 0.06 Btu (hr ft °F), the resistance is

$$R_c = \frac{\ln(d_c/d_o)}{2\pi kL} = \frac{0.3677}{2\pi 0.06 \dfrac{\text{Btu}}{\text{hr ft °F}} 6 \text{ ft}} = 0.1626 \frac{°\text{F}}{\text{Btu/hr}} \tag{6.36}$$

Since the resistances are in series, they are additive, and the resulting heat transfer can be calculated as follows:

$$Q_h = \frac{(220 - 75)°\text{F Btu/hr}}{(0.0127 + 0.000104 + 0.0566 + 0.1626)°\text{F}} \tag{6.37}$$

$$Q_h = 625 \text{ Btu/hr} = 184 \text{ watts} \tag{6.38}$$

We note that the insulation provides a 70% reduction in heat loss.

## 6.4 DYNAMIC THERMAL SYSTEMS

**Dynamic thermal systems** are thermal systems in which the heat capacity is large enough or the time under consideration is short enough for there to be a rate of change of temperature with time during a process. The actual modeling equations for such a process come from the **diffusion equation,** which is a second-order partial differential equation in space and time. Solving this equation for a variety of different configurations is beyond the scope of this work; the reader is referred to textbooks on heat transfer [Refs. 1–5].

Under certain conditions, we can take a lumped-parameter approach to solving the system dynamics. A **lumped-parameter model** supposes that all of the properties of thermal resistance and capacitance are lumped at selected points in space and produces a set of ordinary differential equations in time. Since we are trying to approximate a partial differential equation in space with ordinary differential equations at certain points in the system, we must break the system into a certain number of lumps. The requirement to break the system into lumps arises only from its dynamic operation, since the steady-state solution to this diffusion equation results in Fourier's law. Therefore, we might have a single-lumped capacitance model or a multiple-lumped capacitance model, depending upon the convection and conduction resistances.

The number of lumps required for accuracy depends upon the relative values of the conduction and convection resistances; the ratio of these resistances gives rise to the definition of the Biot number $N_b$ (Jean B. Biot, French physicist, 1774–1862). The **Biot number** is a dimensionless ratio of the conduction resistance to the convection resistance:

$$N_b = \frac{R_{cond}}{R_{conv}} = \frac{hL_c}{k} \tag{6.39}$$

Here,

$$L_c = \text{a characteristic length of the solid material} = \frac{\text{volume}}{\text{surface area}} \tag{6.40}$$

$$L_c = \text{thickness for a plate}$$

$$L_c = \text{thickness}/2 \text{ for a fin}$$

$$L_c = \text{diameter}/4 \text{ for a long cylinder}$$

$$L_c = \text{diameter}/6 \text{ for a sphere}$$

If the Biot number is small ($N_b < 0.1$), then the resistance in the solid material is very low compared to the convection, and the entire solid material will be at almost the same temperature. Thus, there will be a very small temperature difference in the material from one side to the other, and the dominant temperature difference will be from the surface to the free-stream fluid temperature. In this case, we can consider a single lump of capacitance and only consider the heat capacity of the solid material and the heat transfer due to convection; we do not

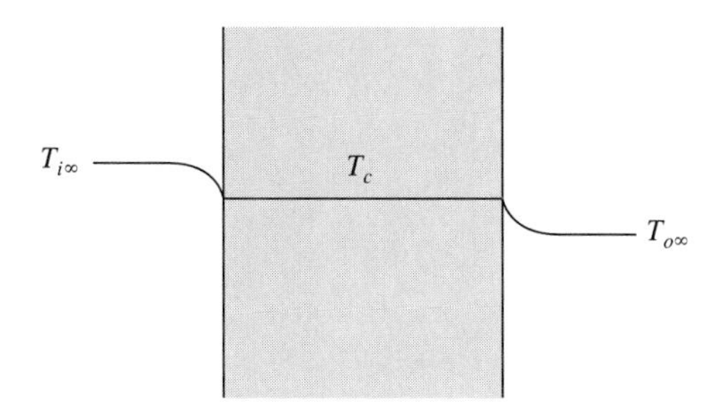


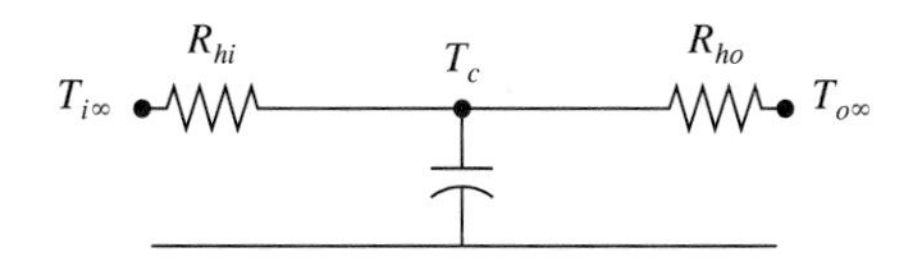


**Figure 6.15** Temperature variations in a flat plate with small Biot number.

have to consider the thermal conduction resistance in the solid. Figure 6.15 illustrates the temperature variations in a flat plate when the Biot number is small.

If the Biot number is large, ($N_b > 0.1$), then the resistance in the solid material is significant compared to the convection, and there will be a temperature difference inside the solid material (in addition to a temperature difference between the surface and the free-stream fluid). In this case, we must consider several lumps of capacitance and determine the temperatures at several points inside the solid as a function of time. The requirement to approximate the partial differential equation arises only from the dynamic characteristics of the system, since the static characteristics do not necessitate internal nodes.

### 6.4.1 Single-Lumped Capacitance Modeling

Many applications in engineering in which the heat capacity is obviously significant can be treated by the single-lumped capacitance analysis technique. In these cases, the Biot number is small.

**Example 6.4 Watermelon Warming**

Suppose that we are interested in predicting how long a watermelon, such as the one depicted in Figure 6.16, will maintain its temperature at a picnic. A 5 kg watermelon was initially cooled to 5°C, but is exposed to 30°C with free convection at the picnic. How long will it take to get to 63% of the temperature rise from 5°C to 30°C, i.e., to 20.75°C?

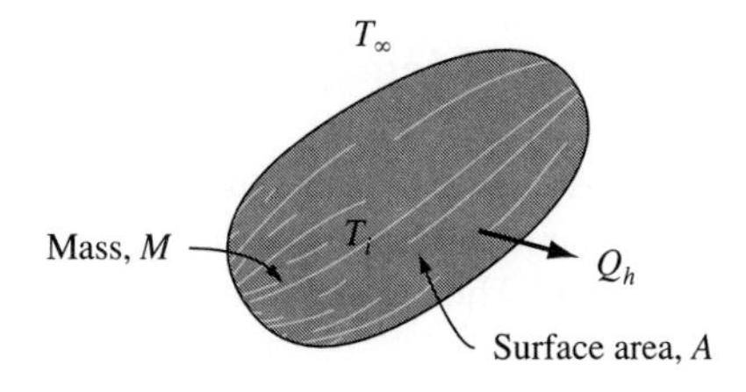


**Figure 6.16** Watermelon-warming problem.

**Solution** The inside of a watermelon is basically water and so should have a specific heat about equal to that of water. Therefore, we will assume that the specific heat $C_p$ is 4200 J/(kg°K) and the density is 1000 kg/m$^3$. With free convection, $h$ will be about 15 W/(m$^2$°K.) Based upon the mass and density of the watermelon, the diameter of the watermelon is 0.21 m, and the surface area of the watermelon is approximately 0.15 m$^2$.

The first thing to do is calculate the Biot number. The thermal conductivity of water is 600 W/(m°K), so

$$N_b = \frac{hL_c}{k} = \frac{15\frac{W}{m^2\,°K}\frac{0.21\,m}{6}}{600\frac{W}{m\,°K}} = 0.00088 \tag{6.41}$$

Since this Biot number is very much below 0.1, we can safely assume that the temperature at every point inside the watermelon is the same at any instant in time. Thus, a single-lumped capacitance model can be used.

The mass of the melon acts as a thermal capacitor. From Eq. (6.21), the heat transfer from the capacitor is

$$Q_h = C_p M \frac{dT_i}{dt} \tag{6.42}$$

To solve this problem, we must equate the heat transfer released by the thermal capacitance to the convection heat transfer:

$$Q_h = MC_p DT_i = hA(T_i - T_\infty) \tag{6.43}$$

Equation (6.43) can be rearranged to produce a differential equation in $T_i$, namely,

$$(\tau D + 1)T_i = T_\infty \tag{6.44}$$

where

$$\tau = \frac{MC_p}{hA} \tag{6.45}$$

This is a first-order differential equation with a time constant of $\tau$. The system will reach 63% of its response in one time constant. Using the earlier-given numerical values, we can calculate the time constant as follows:

$$\tau = \frac{MC_p}{hA} = \frac{5\text{ kg } 4200\frac{\text{J}}{\text{kg °K}}}{15\frac{\text{W}}{\text{m}^2\text{ °K}}0.15\text{ m}^2} = 9333\text{ s} = 2.6\text{ hr} \tag{6.46}$$

Therefore, it will take 2.6 hours for the watermelon to warm up to 20.75°C.

### 6.4.2 Multiple-Lumped Capacitance Modeling

We may model the process of one-dimensional conduction by replacing the continuum under consideration with an **equivalent thermal circuit** with lumped capacitance and resistance elements. The thermal capacitance is concentrated at discrete nodes at which the temperature will be determined. Thermal resistance elements are placed between nodes to represent the resistance to heat flow. Thus, the two properties of energy storage and heat conduction may be separated into distinct lumps for purposes of modeling. This is analogous in mechanical systems to dividing a rod into a collection of discrete masses connected by massless springs. The accuracy of the results, of course, depends upon how many lumps are used.

If the Biot number is not small, then we must divide the solid material into several nodes. (Actually, a node is not a point in space; it is a surface, but we will call it a node for convenience.) At each node, the heat capacitance is represented by the mass on either side of the node halfway to the next node, and the resistance is that associated with the distance between the nodes. For the purposes of multiple-lumped capacitance modeling, we will suppose that the thermal properties are homogeneous.

We will consider one-dimensional heat transfer in a flat plate of mass $M$ and of thickness $L$ with the length and width large compared to the thickness. From Section 6.2.1, the thermal conduction elements are divided into $n$ elements as defined by

$$Q_h = \frac{1}{R_n}\,\delta T \tag{6.47}$$

The resistance is the conduction resistance associated with the distance between nodes:

$$R_n = \frac{L/n}{kA} \tag{6.48}$$

The capacitive elements have a mass of $M/n$, and the thermal capacitance $C_n$ is described by the expression (see Section 6.2.4)

$$Q_h = C_n \frac{dT}{dt} \tag{6.49}$$

where

$$C_n = C_p \frac{M}{n} \tag{6.50}$$

**Constant Surface Temperature Sources.** If the convection coefficients on each side of the wall are very good, the convection resistance will be small, and the surface temperatures will equal the free-stream fluid temperatures and therefore act as ideal temperature sources.

Figure 6.17 illustrates the general configuration of a flat plate driven by two temperature sources and divided into $n$ lumps. If there are $n$ nodes, then there will be $n - 1$ internal nodes at which we must calculate the temperature. There will also be two surface nodes. The resistance between each node is $R_n$. The capacitance associated with each internal node is $C_n$; the capacitance associated with each surface node is one-half of $C_n$.

However, if the surface temperatures act like a temperature source, then any capacitance connected to a temperature source can be neglected (just as in electron-

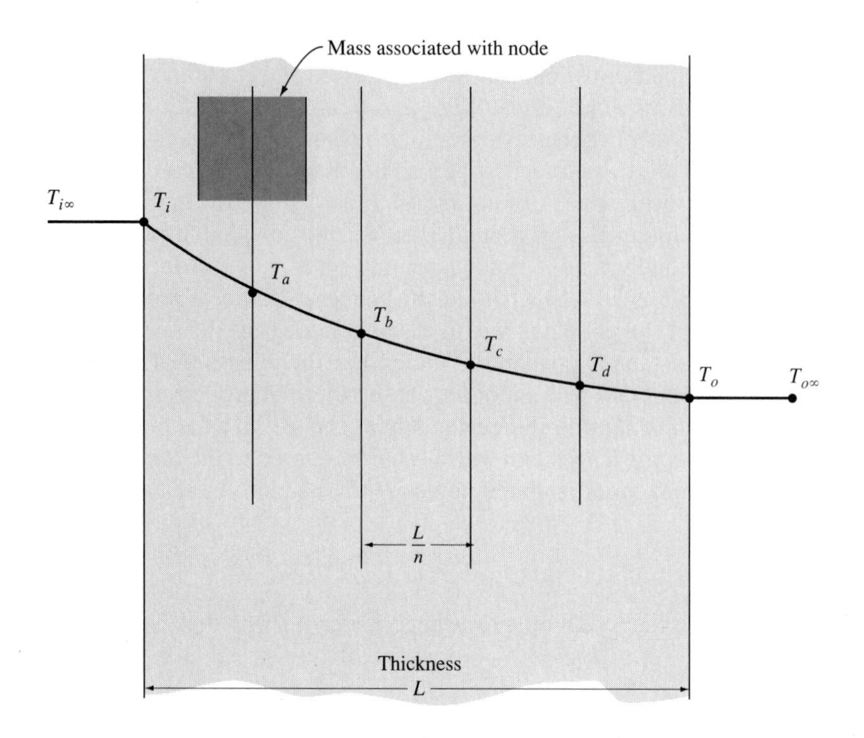


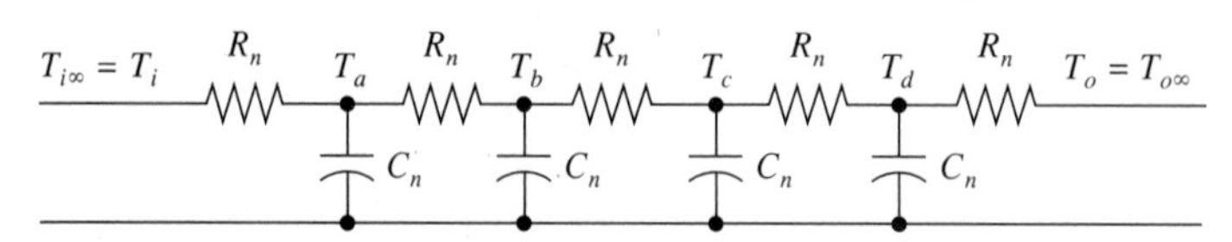


**Figure 6.17** Temperatures in a flat plate with constant surface temperatures.

ics). Therefore, if the surface temperatures are prescribed (due to excellent convection), then we will generate a differential equation for each interior node as follows:

Resistance between $T_i$ and $T_a$:
$$Q_{ia} = \frac{1}{R_n}(T_i - T_a) \tag{6.51}$$

Capacitance at $T_a$:
$$Q_{ia} - Q_{ab} = C_p \frac{M}{n} \dot{T}_a \tag{6.52}$$

Resistance between $T_a$ and $T_b$:
$$Q_{ab} = \frac{1}{R_n}(T_a - T_b) \tag{6.53}$$

Capacitance at $T_b$:
$$Q_{ab} - Q_{bc} = C_p \frac{M}{n} \dot{T}_b \tag{6.54}$$

⋮

Resistance between $T_{n-1}$ and $T_o$:
$$Q_{n-1,o} = \frac{1}{R_n}(T_{n-1} - T_o) \tag{6.55}$$

These equations can be summarized into differential equations for each interior node:

$$\dot{T}_a = \frac{n}{C_p M R_n}[-2T_a + T_i + T_b] \tag{6.56}$$

$$\dot{T}_b = \frac{n}{C_p M R_n}[-2T_b + T_a + T_c] \tag{6.57}$$

$$\vdots$$

$$\dot{T}_{n-1} = \frac{n}{C_p M R_n}[-2T_{n-1} + T_{n-2} + T_o] \tag{6.58}$$

If we want to calculate the heat transfer through the flat plate, we can use any of the preceding nodal heat transfer equations to find the instantaneous heat transfer at that node. For example, the heat going into the inside wall is

$$Q_i = \frac{1}{R_n}(T_i - T_a) \tag{6.59}$$

In this equation, the temperature at the first node, $T_a$, is the temperature found from the differential equation for $T_a$ and is coupled to the temperatures of the surfaces, as well as all of the other internal nodes.

**Convection Surface Temperature Sources.** If the convection coefficients on each side of the wall are not very good, the convection resistance will be large, and there will be a convection resistance between the free-stream fluid temperature and the temperature of the surface. In this case, there will be a significant temperature difference between the free-stream temperatures and the surface temperatures. Figure 6.18 illustrates the multiple-lumped-parameter flat plate with convection resistances at the surface. Here, the capacitance associated with each surface node will be significant and will be one-half of the capacitance associated with each internal node.

The resulting circuit equations are as follows:

Resistance between $T_{i\infty}$ and $T_i$:

$$Q_i = \frac{1}{R_{hi}}(T_{i\infty} - T_i) \tag{6.60}$$

where

$$R_{hi} = \frac{1}{h_i A} \tag{6.61}$$

is the convection resistance.

Capacitance at $T_i$:

$$Q_i - Q_{ia} = \frac{C_p}{2}\frac{M}{n}\dot{T}_i \tag{6.62}$$

Resistance between $T_i$ and $T_a$:

$$Q_{ia} = \frac{1}{R_n}(T_i - T_a) \tag{6.63}$$

Capacitance at $T_a$:

$$Q_{ia} - Q_{ab} = C_p \frac{M}{n}\dot{T}_a \tag{6.64}$$

$$\vdots$$

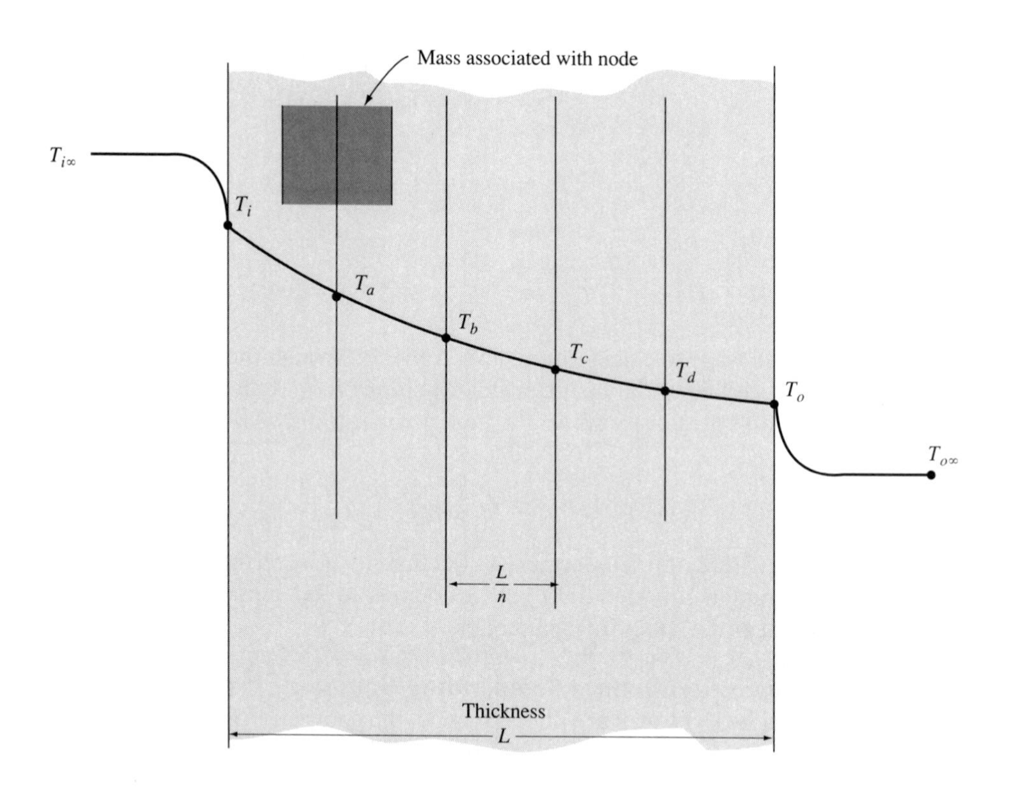


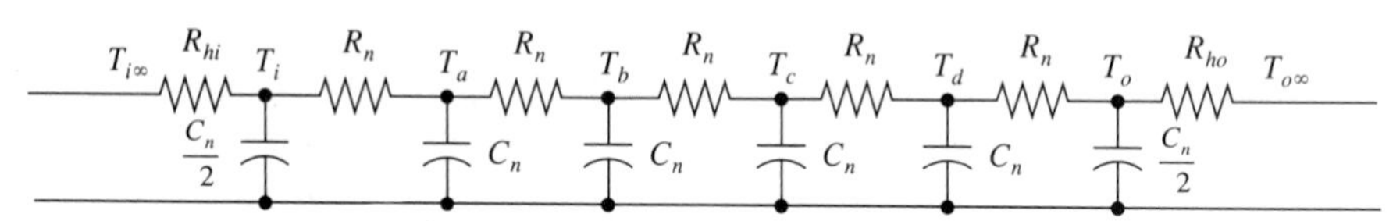


**Figure 6.18** Temperatures in a flat plate with surface convection.

Resistance between $T_{n-1}$ and $T_o$:

$$Q_{n-1,o} = \frac{1}{R_n}(T_{n-1} - T_o) \tag{6.65}$$

Capacitance at $T_o$:

$$Q_{n-1,o} - Q_o = \frac{C_p}{2}\frac{M}{n}\dot{T}_o \tag{6.66}$$

Resistance between $T_o$ and $T_{o\infty}$:

$$Q_o = \frac{1}{R_{ho}}(T_o - T_{o\infty}) \tag{6.67}$$

where

$$R_{ho} = \frac{1}{h_o A} \tag{6.68}$$

is the convection resistance.

These equations can be summarized into differential equations for each node (including both surface and interior nodes) as follows:

$$\dot{T}_i = \frac{2n}{C_p M}\left[-\left(\frac{1}{R_{hi}} + \frac{1}{R_n}\right)T_i + \frac{T_{i\infty}}{R_{hi}} + \frac{T_a}{R_n}\right] \tag{6.69}$$

$$\dot{T}_a = \frac{n}{C_p M R_n}[-2T_a + T_i + T_b] \tag{6.70}$$

$$\vdots$$

$$\dot{T}_{n-1} = \frac{n}{C_p M R_n}[-2T_{n-1} + T_{n-2} + T_o] \tag{6.71}$$

$$\dot{T}_o = \frac{2n}{C_p M}\left[-\left(\frac{1}{R_n} + \frac{1}{R_{ho}}\right)T_o + \frac{T_{n-1}}{R_n} + \frac{T_{o\infty}}{R_{ho}}\right] \tag{6.72}$$

**Example 6.5 Temperature Response of a Steel Plate**

Use thermal circuit modeling to determine equations governing the temperature response of a 200 mm square steel plate with a thickness of 40 mm if the inside plate surface is exposed to force convection of water and the outside surface is exposed to free convection of air. Assume the following physical characteristics:

$$k = 75\,\frac{\text{W}}{\text{m °K}} \qquad h_i = 1000\,\frac{\text{W}}{\text{m}^2\text{ °K}}$$

$$C_p = 450\,\frac{\text{J}}{\text{kg °K}} \qquad h_o = 25\,\frac{\text{W}}{\text{m}^2\text{ °K}}$$

$$L = 0.040\text{ m} \qquad A = 0.040\text{ m}^2$$

$$\rho = 7850\,\frac{\text{kg}}{\text{m}^3} \qquad M = 12.56\text{ kg}$$

**Solution** We first calculate the Biot number for the system. The Biot number for the inside and outside wall convection effects will be different. We will use the larger Biot number:

$$N_{bi} = \frac{h_i L}{k} = \frac{1000\,\frac{\text{W}}{\text{m}^2\text{ °K}}\,0.040\text{ m}}{75\,\frac{\text{W}}{\text{m °K}}} = 0.533 \tag{6.73}$$

$$N_{bo} = \frac{h_o L}{k} = \frac{25\,\frac{\text{W}}{\text{m}^2\text{ °K}}\,0.040\text{ m}}{75\,\frac{\text{W}}{\text{m °K}}} = 0.013 \tag{6.74}$$

Since the Biot number for the inside surface is the larger, we will base our analysis on that value. A Biot number of 0.533 does require multiple-lumped capacitance modeling. Using five nodes will reduce our Biot number down to approximately 0.1. Since the convection is so good on the inside surface, we will suppose that the inside wall temperature is constant and is equal to $T_{i\infty}$. Thus, the differential equations for the system are:

$$\dot{T}_a = \frac{10}{C_p M R_n}\left[-T_a + \frac{T_{i\infty} + T_b}{2}\right] \quad (6.75)$$

$$\dot{T}_b = \frac{10}{C_p M R_n}\left[-T_b + \frac{T_a + T_c}{2}\right] \quad (6.76)$$

$$\dot{T}_c = \frac{10}{C_p M R_n}\left[-T_c + \frac{T_b + T_d}{2}\right] \quad (6.77)$$

$$\dot{T}_d = \frac{10}{C_p M R_n}\left[-T_d + \frac{T_c + T_o}{2}\right] \quad (6.78)$$

$$\dot{T}_o = \frac{10}{C_p M R_n}\left[-\left(1 + \frac{R_n}{R_{ho}}\right) T_o + T_d + \frac{R_n}{R_{ho}} T_{o\infty}\right] \quad (6.79)$$

Now we calculate the conductivity resistances. These are the resistances between each node; the total resistance is 5 times this value. We obtain

$$R_n = \frac{L/n}{kA} = \frac{\frac{0.040\ \text{m}}{5}}{75\,\frac{\text{W}}{\text{m}\ ^\circ\text{K}}\,0.040\ \text{m}^2} = 0.00267\,\frac{^\circ\text{C}}{\text{W}} \quad (6.80)$$

The convection resistances are

$$R_{hi} = \frac{1}{h_i A} = \frac{1}{1000\,\frac{\text{W}}{\text{m}^2\ ^\circ\text{K}}\,0.040\ \text{m}^2} = 0.025\,\frac{^\circ\text{C}}{\text{W}} \quad (6.81)$$

and

$$R_{ho} = \frac{1}{h_o A} = \frac{1}{25\,\frac{\text{W}}{\text{m}^2\ ^\circ\text{K}}\,0.040\ \text{m}^2} = 1.00\,\frac{^\circ\text{C}}{\text{W}} \quad (6.82)$$

The coefficients for the differential equations for the internal nodes and for the outside surface node can be calculated from the time constant:

$$\tau = \frac{C_p M R_n}{10} = \frac{450\,\frac{\text{J}}{\text{kg}\ ^\circ\text{K}}\,12.56\ \text{kg}\ 0.00267\,\frac{^\circ\text{C}}{\text{J/s}}}{10} = 1.51\ \text{s} \quad (6.83)$$

With the foregoing differential equations and the values for the time constant and the resistances, we can solve for the five temperatures as a function of time. This system is a fifth-order state-space dynamic system with two inputs: the free-stream temperatures on the inside and outside surfaces of the flat plate.

This section has discussed circuit methods for modeling one-dimensional thermal problems. When working in two or three dimensions, we use finite difference or finite element methods. (For further information, see the references at the end of the chapter.)

## 6.5 SUMMARY

This chapter has presented the basic thermal system modeling concepts, together with examples of their use in typical situations. Conduction, convection, and radiation heat transfer were examined, along with the dynamic effects of heat capacity. Lumped-capacitance methods of dynamic modeling were presented, and the similarities and differences between thermal systems and other dynamic systems noted. These concepts are considered further in later sections of the book.

## REFERENCES

6.1 Haberman, William L., and John, James E. A. *Engineering Thermodynamics with Heat Transfer*. Prentice Hall, Englewood Cliffs, NJ, 1989.

6.2 Holman, Jack P. *Heat Transfer,* 7th ed. McGraw-Hill, New York, 1990.

6.3 Incropera, Frank P. and DeWitt, David P. *Fundamentals of Heat and Mass Transfer,* 3rd ed. John Wiley & Sons, New York, 1990.

6.4 Kreith, Frank, and Bohn, Mark S. *Principles of Heat Transfer,* 5th ed. West Publishing Co., New York, 1993.

6.5 Lienhard, John H. *A Heat Transfer Textbook,* 2nd ed. Prentice Hall, Englewood Cliffs, NJ, 1987.

## NOMENCLATURE

$A$ cross-sectional area
$C_p$ constant-pressure specific heat
$F_e$ emissivity factor
$F_v$ view factor
$h$ convection coefficient
$h_{eq}$ equivalent convection coefficient
$k$ thermal conductivity
$L$ length
$L_c$ characteristic length
$M$ mass
$N_b$ Biot number
$Q_h$ heat flow rate
$r$ radius
$R$ thermal resistance
$T$ temperature
$x$ position coordinate

$\alpha$ thermal diffusivity
$\sigma$ Stefan-Boltzmann constant
$\rho$ density
$\tau$ time constant

## PROBLEMS

**6.1** Several metals are being considered for use in an application that conducts heat from one location to another through a solid rod of metal at room temperature. *Prepare* a table that lists the thermal conductivity of the following metals: aluminum, steel (use iron), copper, and silver. *Which* metal has the best thermal conductivity?

If the best conductor were used with a given rod diameter $d^*$, this would represent a reference thermal resistance per unit length. Now *calculate* what diameter $d$ of rod would be required in the other metals to have the same thermal resistance per unit length as the best conductor. Based upon the density of each metal, *what* will be the weight per unit length for each? *Which* metal has the lowest weight for the given thermal resistance?

*Obtain* a rough estimate for the cost per pound of each of the preceding metals from industrial suppliers, and *calculate* the relative cost of each metal that is required to have the same resistance as discussed in the previous paragraph. *Which* metal has the lowest cost for the given thermal resistance?

*Which* metal would you use and why?

**6.2** *Perform* an analysis similar to that in Problem 6.1 for the following insulators used as thin sheets: asbestos, glass fibers, plywood, and wool. Now, however, we are looking for the worst conductor (i.e., the best insulator), the lowest weight for a given insulation, and the lowest cost for a given insulation.

**6.3** *Derive* Eq. 6.9 to confirm Eq. 6.10.

**6.4** The actual convection coefficient can be calculated for a variety of situations using the dimensionless parameters of the Nusselt number, Prandtl number, Reynolds number, and Grashof number (Ernst Nusselt, German engineer, 1882–1951; Ludwig Prandtl, German engineer, 1875–1953; Osborne Reynolds, English engineer, 1842–1912, Franz Grashof, German engineer, 1826–1893). *Refer* to a textbook on heat transfer, and *state* the definitions for these terms and the parameters they use.

**6.5** The Nusselt number is a dimensionless ratio for the convection coefficient. Continue Problem 6.4, and *state* the relationship between the Nusselt number and the Prandtl and Grashof numbers for the following convection situations: free convection from a vertical flat plate, horizontal flat plate, vertical cylinder, and horizontal cylinder.

**6.6** *Calculate* the amount of heat lost (in watts) through an uncovered window in a home. The temperature difference between the air inside and outside of the house is 20°C, with moderate convection on each side of the glass. The size of the glass is 0.75 m by 1.1 m, and the thickness is 3 mm.

**6.7** It is desired to replace a 1/8 inch glass window with a sheet of Plexiglas®. *What* thickness of Plexiglas® will be required to have the same insulation as the window has?

**6.8** A dual-pane window glass in a home is used to reduce the heat loss due to thermal conduction. You are asked to compare three situations: (1) the heat loss through a single pane of glass, (2) the heat loss through two panes of glass touching each other, and

(3) two panes of glass separated with an air space. The glass is 3 mm thick, and the air gap in part 3 is 10 mm. *Perform* your calculations on a per-unit area basis. *How* much reduction in heat transfer from part 1 is observed in parts 2 and 3. *Is* the air gap worthwhile compared to two panes?

**6.9** The radiator hose on an automobile is made of something similar to Buna™ rubber. The internal diameter of the hose is 50 mm, its length is 200 mm, and its thickness is 6 mm. The water on the inside is maintained at 100°C, and the outside air temperature is 50°C. *Calculate* how much heat is conducted through the hose.

**6.10** A 3/4 inch iron pipe (ID = 0.824 inch, OD = 1.050 inches) is used to transmit hot water. The internal water temperature is 120°F, and the ambient air temperature is 75°F. Consider a pipe that is 5 feet in length.

**a.** *Calculate* how much heat is transferred from the water to the air.

**b.** If a spongy rubber tape that is 0.100 inch thick is placed on the outside of the pipe, *calculate* how much heat is transferred.

**c.** *What* is the reduction in heat loss with the use of the rubber insulation?

**6.11** A steel plate is heated with a torch. When we are interested in cooling the plate, we can allow normal free convection, we can blow on the plate with our breath, or we can squirt a mist of water on the surface of the plate from a spray bottle. Looking at Table 6.3, *comment* on how fast the plate can be cooled with these three approaches.

**6.12** *Derive* equation 6.20.

**6.13** A copper ball 1 inch in diameter is heated to 500°F with a torch. *Calculate* how much heat will be transferred at this temperature by radiation. Assume that the ball is placed in the center of a room with walls at 75°F. *Calculate* how much heat would be transferred by convection if we were to use the upper end of free convection from Table 6.3. *What* convection coefficient did you select? *What* is the equivalent convection coefficient due to radiation from Fig. 6.11 (or Eq. 6.20)? *How* do these two coefficients compare with each other and *which* mode of heat transfer dominates in this case?

**6.14** Shown in Figure P6.14 is a swimming pool operating in cool weather. The pool has a heater that provides a heat input of $q_h$ watts. The water depth varies from 1.2 m to 3 m, and the pool is 5 m wide by 10 m long.

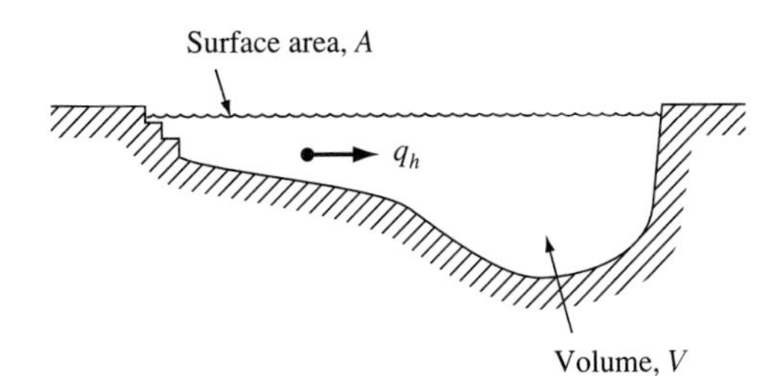


**Figure P6.14** Swimming pool dynamics.

**a.** *Calculate* the Biot number, and *justify* the assumption that the pool can be treated as a single-lumped capacitance model.

**b.** Assuming that the significant heat transfer is the convection at the water surface and that no heat transfers out the sides and bottoms, *write* the modeling equations for the system using a single-lumped capacitor model, and *derive* a differential equation for the temperature of the water as a function of the ambient temperature and the heat input.

**c.** *Calculate* the value of the time constant of the system. If a period of one time constant would be enough time to "take the chill off," *would* you recommend leaving the heater on all the time or turning it off overnight?

**d.** *State* the expression for the steady-state temperature of the water. If the ambient temperature were 20°C and the desired water temperature were 25°C, *what* size heater would be required (in watts)?

**e.** Consider the heat transfer to the ground in the system, and *derive* the differential equation for the temperature of the pool as a function of the ambient temperature, the ground temperature, and the heat input. *State* the expressions for the time constant and the steady-state temperature of the water. If the pool is constructed with concrete 100 mm thick, *is* the heat transfer to the ground significant? *Which* model should we use?

**6.15** Shown in Figure P6.15 is a boiler used in a thermal energy plant. *Write* the modeling equations for the system, and *derive* a differential equation for the temperature inside the boiler as a function of the ambient temperature and the heat input. *State* the expressions for the time constant and the steady-state temperature.

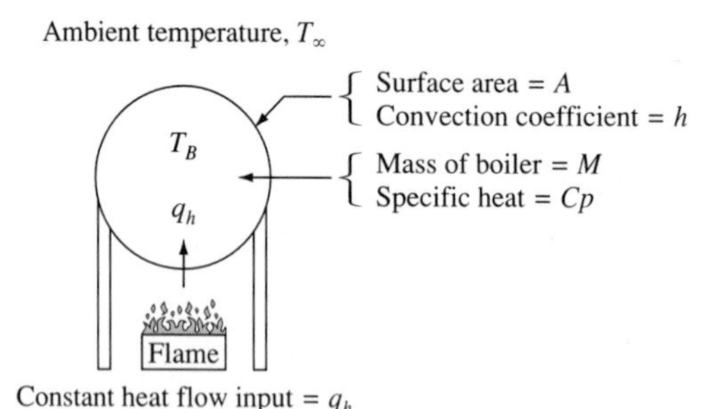


**Figure P6.15** Thermal plant boiler.

**6.16** Shown in Figure P6.16 is a mass suspended at the end of a rod. The mass is exposed to ambient temperature and experiences heat transfer from the rod. The rod is insulated, but is exposed to a constant temperature source $T_0$ at the other end. *Write* the modeling equations for this system, and *derive* the transfer function for the temperature of the mass as a function of the ambient temperature and the temperature of the source. *State* the expressions for the time constant and steady-state temperature.

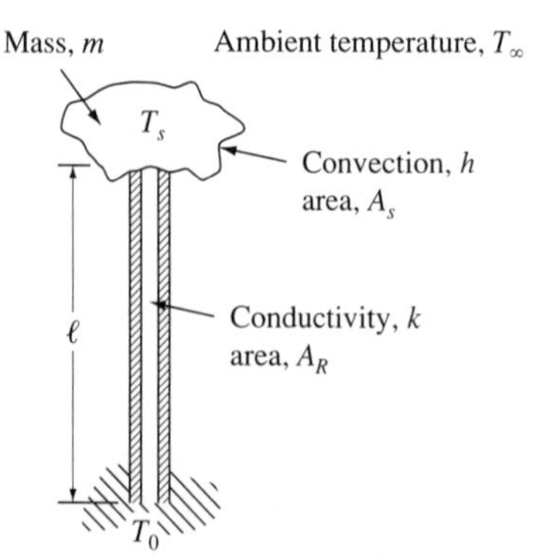


**Figure P6.16** Suspended-mass thermal system.

**6.17** A tank 75% (by volume) full of liquid propane is taken from a warehouse and placed in the direct sunlight on a hot summer day. We are concerned about the internal temperature (and therefore the internal pressure) of the tank as it sits in the sun. We want to know how long it takes to heat up and what is its final steady-state temperature.

Model the tank as a sphere 1 foot in diameter. The walls of the tank are steel 0.100 of an inch thick. The temperature inside the warehouse is 90°F, and the outside air temperature is 100°F. (The tank is in Texas!) The solar insolation for this particular day is 900 watts/m$^2$ = 285 (Btu/hr)/ft$^2$. You should consider the sun as a constant heat flow source of energy over the exposed surface area of the sphere. The density of liquid propane is 30.8 lbm/ft$^3$. The density of steel is 490 lbm/ft$^3$. The specific heat of liquid propane is 0.58 Btu/(lbm °F). The specific heat of steel is 0.11 Btu/(lbm °F). The wind velocity is such that you should use a convection coefficient for high free convection or low forced convection.

*Model* this system, and *derive* a differential equation for the internal temperature of the propane as a function of the solar heat input and the ambient temperature (with a given initial temperature). *Calculate* the time constant for the system, and *state* how long it will take for the tank to be at its steady-state temperature. *What* will the steady-state temperature be for the given conditions?

**6.18** *Derive* the differential equation for the temperature at the center of a can of soft drink that is sitting upright. *What* is the Biot number for the wall of the aluminum can, and *do* you have to model the thermal conductivity of the wall of the can? *What* is the Biot number for the fluid in the can, and can you use a single-lump capacitance model?

Suppose that the can is placed upright in a frost-free refrigerator. *Measure* the dimensions of the can yourself, and use the properties of water to approximate the soft drink. If the air circulation fan inside the refrigerator is running, *what* do you estimate the convection coefficient to be? Using an analytic solution to the differential equation, *plot* the response of the internal temperature over time if the can is taken from room temperature (75°F) and placed in the refrigerator (40°F). *What* is the time constant of the system? *What* is the settling time. *How* long will it take the drink to reach an internal temperature of 50°F? If the can were placed in the freezer section of the refrigerator (10°F), *how* long should it remain there to cool to 50°F?

**6.19** *Work* problem 6.18 using digital simulation (see Chapter 9).

**6.20** *Perform* an experiment to determine the temperature response of water placed in a soft drink can as described in problem 6.18. Use any suitable thermometer to measure the temperature at the center of the volume of water. Before measuring the water temperature, *measure* the temperature of the air inside the refrigerator so that you will know what the steady-state temperature of the water will be. *Record* the observed temperature every 5 to 10 minutes for the first hour and every 15 minutes thereafter. *Plot* your response using a spreadsheet, and compare it to your original guess of the convection coefficient. After seeing the data, *can* you use better values for the convection coefficient, etc., to match the data more closely? *Plot* your revised theoretical response, and *compare* all three responses.

*Perform* the same experiment as in the previous paragraph, except remove the can that has been in the refrigerator for a long time (several hours) and let it heat to room temperature. (*Measure* the temperature of the room beforehand.) Record the temperature every 3 to 6 minutes for the first hour and every 10 minutes thereafter. *Do* you see any difference in the initial response (the first 30 minutes) of heating compared to cooling? Based upon the data you have recorded, is the convection coefficient constant?

Look closely at the surface of the can during the initial heating, and see whether you can *justify* any differences in the time constant over the initial response.

**6.21** It is claimed that if you place a soft drink can on its side in the refrigerator, it will cool faster than if it were placed upright. *What* theoretical justification can you offer for this contention?

**6.22** A fish tank is fabricated using 4 mm Plexiglas® walls. The water on the inside of the tank is maintained at a constant temperature of 26°C. The ambient room temperature is 22°C. We are interested in modeling the transient heat transfer through the walls of the tank if there is a sudden change in room temperature. The inside of the tank has water that is circulated with a pump and therefore should have a convection coefficient on the lower end of forced convection with water. The outside of the tank has air that should have a convection coefficient on the upper end of free convection with air.

**a.** *Calculate* the Biot number for the inside and outside surfaces. Based upon these numbers, *can* lumped parameter modeling be used to model the temperature distribution in the Plexiglas®? *Can* you assume that the convection heat transfer is so good in the water, that the water temperature on the inside surface is constant?

**b.** *Based* upon the Biot number, how many lumps should be considered in the lumped-parameter analysis?

**6.23** A window glass has 25°C air temperature on the inside and 20°C air temperature on the outside. The convection coefficient on both sides is 30 $W/(m^2\ °C)$. The thickness of the glass is 7.5 mm, and the thermal conductivity of the glass is 0.75 $W/(m\ °C)$. The windowpane is 0.8 m by 1.1 m. It is desired to calculate the transient temperature distribution in the glass with a sudden change in outside temperature in order to determine the internal stresses caused by a temperature gradient.

**a.** *Calculate* the Biot number to determine whether a single lumped-parameter analysis is possible. If a multiple lumped-parameter analysis is necessary, *how* many lumps should be used?

**b.** *Set up* the differential equations to simulate the transient response of all temperatures of interest.

**c.** If the outside temperature were suddenly changed from 20°C to 15°C, use digital simulation to *calculate* the transient response of all temperatures. *Plot* your results.

**6.24** Using numerical integration, *solve* the state-space differential equations described in Example 6.5, and *plot* your results as a function of time. Use $T_{i\infty} = 50°C$, $T_{o\infty} = 25°C$, with all initial conditions = 25°C. *What* is the observed settling time of this system? *Compare* the observed settling time to the time constant for an individual internal node.

# CHAPTER 7

# Mixed Discipline Systems

## 7.1 INTRODUCTION

Previous chapters have emphasized single-discipline systems in order to fully develop modeling techniques for each discipline; however, most practical systems actually represent a combination of disciplines. This chapter emphasizes systems of mixed disciplines.

## 7.2 ELECTROMECHANICAL SYSTEMS

Numerous applications of speed and motion control in industry use electric motors. There are a variety of motors, including DC (direct current) with permanent-magnet or electromagnet operation, and AC (alternating current) with synchronous or induction operation. The permanent-magnet DC motor has the fewest nonlinearities of these motors and is the easiest to control to variable speeds. An ideal permanent-magnet DC motor model is used in the next three examples. The mathematical characteristics are discussed in the following.

A DC electric motor has a natural relationship between its torque $T_m$ and its current $I_m$, as well as between its voltage $E_m$ and its angular speed $\omega_m$ [Ref. 1]. For an ideal motor, the current required is directly proportional to the applied torque, and the speed is directly proportional to the applied voltage. These relationships are normally expressed as follows:

$$T_m = K_t I_m \tag{7.1}$$

$$E_m = K_v \omega_m \tag{7.2}$$

The constants of proportionality, $K_t$ and $K_v$, are actually the same in both equations. On the surface, it would appear that they are not the same, either numerically or by units; however, a close examination of the units reveals that if speed is expressed in rad/s, torque in N m, and electrical excitation in volts and amps, then the constants in the two equations do indeed have the same units. Further, assuming 100% efficiency, it becomes clear that the two constants must be the same numerically.

### 7.2.1 DC Motor Speed Control

**Without Feedback (Open-Loop Control)** In many applications, it is desirable to have a variable-speed electric motor that can provide an adjustable speed to a process, which is then maintained constant. This can be accomplished with a DC electric motor with a voltage controller, as depicted in Figure 7.1.

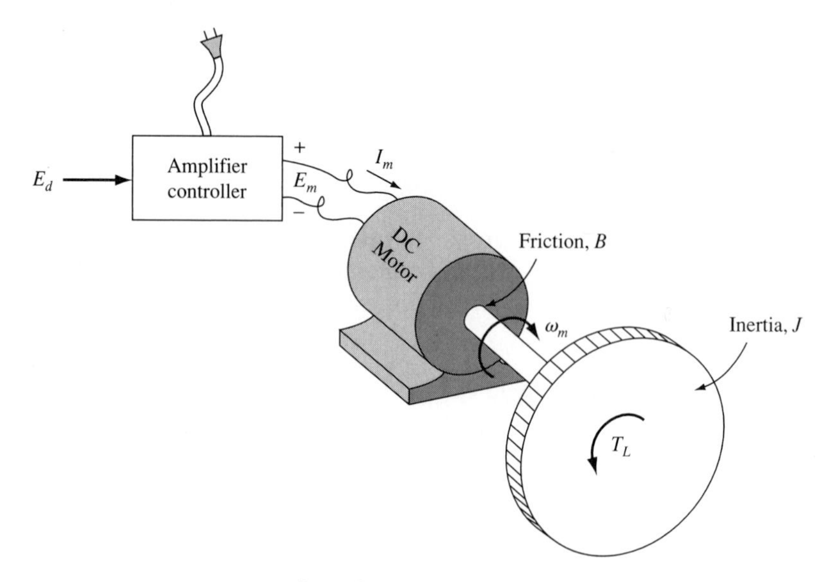


**Figure 7.1** Open-loop DC motor speed control.

The model for the motor is based on the ideal permanent-magnet DC motor (we simply let $K_t = K_v = K$):

$$T_m = KI_m \tag{7.3}$$

$$E_m = K\omega_m \tag{7.4}$$

The motor must be driven by an electronic power source. In this case, we will consider an electronic amplifier or controller. An ideal voltage source would be able to deliver any amount of current at constant voltage; however, real amplifiers lose voltage as the output current increases, due to an output resistance $R_o$. This is equivalent to an ideal voltage source driving the motor through $R_o$. Assuming an overall amplifier gain $G$, we can state the equation for the amplifier as

$$E_m = GE_d - R_o I_m \tag{7.5}$$

The mechanical load connected to the motor is an inertia with linear friction and, possibly, some unknown load disturbance torque $T_L$. The connecting shaft has negligibly small flexibility. Since the torque driving the load is the motor torque $T_m$, and since the inertia, friction, and the disturbance torque oppose motion, the torque balance can be stated as

$$T_m - J\dot{\omega}_m - B\omega_m - T_L = 0 \tag{7.6}$$

If we solve this equation for speed and use the other equations to solve for speed as a function of the input $E_d$ and the torque disturbance $T_L$, we can proceed as follows:

$$J\dot{\omega}_m + B\omega_m = T_m - T_L = K\left[\frac{GE_d - (K\omega_m)}{R_o}\right] - T_L \tag{7.7}$$

$$\left[JD + B + \frac{K^2}{R_o}\right]\omega_m = \frac{GK}{R_o}E_d - T_L \tag{7.8}$$

Normalizing this equation with respect to the lowest order term of $\omega_m$ and stating it in transfer function format, we can see that the speed is a first-order function of the input and the disturbance:

$$\omega_m = \frac{\dfrac{G/K}{\left(1 + \dfrac{B}{K^2/R_o}\right)}E_d - \dfrac{R_o/K^2}{\left(1 + \dfrac{B}{K^2/R_o}\right)}T_L}{\dfrac{\dfrac{J}{K^2/R_o}}{\left(1 + \dfrac{B}{K^2/R_o}\right)}D + 1} \tag{7.9}$$

From this transfer function, it is plain that the static gain, disturbance sensitivity, and time constant can be stated as follows (See Appendix E, Section E.2.1). Two of these perfomance factors are defined in terms of the static (nondymanic) behavior of the system, (i.e., with all derivatives equal to zero).

$$\text{Static gain} = \left.\frac{\partial\,\omega_m}{\partial\,E_d}\right|_{D=0} = \frac{G/K}{\left(1 + \dfrac{B}{K^2/R_o}\right)} \tag{7.10}$$

$$\text{Disturbance sensitivity} = \left.\frac{\partial\,\omega_m}{\partial\,T_L}\right|_{D=0} = -\frac{R_o/K^2}{\left(1 + \dfrac{B}{K^2/R_o}\right)} \tag{7.11}$$

$$\text{Time constant} = \tau = \frac{\dfrac{J}{K^2/R_o}}{\left(1 + \dfrac{B}{K^2/R_o}\right)} \tag{7.12}$$

These static and dynamic factors are the **basic performance factors** for the system.

**With Speed Feedback (Closed-Loop Control)** The foregoing system has a sensitivity to load disturbances that might be objectionable in terms of holding the motor speed constant in the presence of torque disturbances. This sensitivity can be reduced by feedback of the actual motor speed as depicted in Figure 7.2.

The purpose of *feedback control* in a system is to allow the output to track the input and to compensate for any error from the command input (the desired value of the output) and the actual output. In this manner, the system will take action to correct the output when it is not at its desired value. Feedback is normally accomplished by a sensor that measures the actual output variable and sends a signal back to be compared to the input. The comparison is normally done by what is called a *summing junction,* which has the ability to subtract the feedback signal from the input signal to obtain an error signal. The error signal is usually amplified to drive an actuator, which in turn causes the output response.

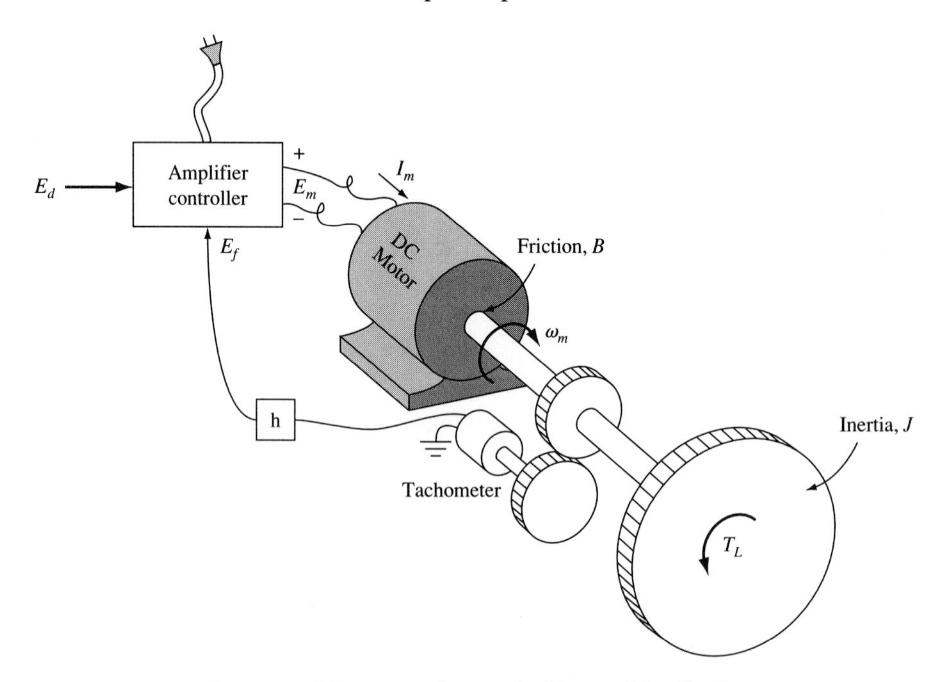


**Figure 7.2** Motor speed control with speed feedback.

The feedback in the preceding system can be accomplished by the use of a small tachometer that spins in proportion to the motor speed and produces a voltage that is linearly proportional to the speed. Notice that this could be a generator with the same behavior as the DC motor, or it could be some other type of speed sensor that has an electronic output. The gain of the tachometer is $h$, and the voltage is

$$E_f = h\omega_m \tag{7.13}$$

As before, the motor equations are

$$T_m = KI_m \tag{7.14}$$

and

$$E_m = K\omega_m \tag{7.15}$$

In this case, the electronic amplifier or controller has provisions for the inputs from the set-point and from the feedback voltage $E_f$. In this servoamplifier, the voltage representing the desired speed $E_d$ is compared to the voltage representing the actual speed $E_f$ by subtracting the actual from the desired. The result, the error voltage $e$, is amplified and is used to drive the motor to make the error zero or, at least, as small as possible. The equation for the amplifier can be stated as

$$E_m = Ge - R_o I_m \tag{7.16}$$

where

$$e = E_d - E_f \tag{7.17}$$

Again assuming an inertial mechanical load with friction and a disturbance torque $T_L$, the torque balance equation can be stated as

$$T_m - J\dot{\omega}_m - B\omega_m - T_L = 0 \tag{7.18}$$

Solving this equation for speed and using the other equations, we can proceed as follows:

$$J\dot{\omega}_m + B\omega_m = T_m - T_L = K\left[\frac{G(E_d - h\omega_m) - (K\omega_m)}{R_o}\right] - T_L \tag{7.19}$$

$$\left[JD + B + \frac{K^2}{R_o} + \frac{GKh}{R_o}\right]\omega_m = \frac{GK}{R_o}E_d - T_L \tag{7.20}$$

Normalizing Eq. (7.20) and stating it in transfer function format, we can see that the speed is the following first-order function of the input and disturbance, with the static gain, disturbance sensitivity, and time constant as stated:

$$\omega_m = \frac{\dfrac{1/h}{\left(1 + \left(\dfrac{1 + \dfrac{B}{K^2/R_o}}{\dfrac{Gh}{K}}\right)\right)} E_d - \dfrac{R_o/K^2}{\left(1 + \dfrac{B}{K^2/R_o} + \dfrac{Gh}{K}\right)} T_L}{\dfrac{\dfrac{J}{K^2/R_o}}{\left(1 + \dfrac{B}{K^2/R_o} + \dfrac{Gh}{K}\right)} D + 1} \tag{7.21}$$

$$\text{Static gain} = \left.\frac{\partial \omega_m}{\partial E_d}\right|_{D=0} = \frac{G/K}{\left(1 + \dfrac{B}{K^2/R_o} + \dfrac{Gh}{K}\right)} = \frac{1/h}{\left(1 + \left(\dfrac{1 + \dfrac{B}{K^2/R_o}}{\dfrac{Gh}{K}}\right)\right)} \tag{7.22}$$

$$\text{Disturbance sensitivity} = \left.\frac{\partial \omega_m}{\partial T_L}\right|_{D=0} = -\frac{R_o/K^2}{\left(1 + \dfrac{B}{K^2/R_o} + \dfrac{Gh}{K}\right)} \tag{7.23}$$

$$\text{Time constant} = \tau = \frac{\dfrac{J}{K^2/R_o}}{\left(1 + \dfrac{B}{K^2/R_o} + \dfrac{Gh}{K}\right)} \tag{7.24}$$

It is interesting to compare this closed-loop control system to the previous open-loop control system in terms of their performance. Notice that the static gain, disturbance sensitivity, and time constant of the closed-loop system are each modified by the $(1 + B/(K^2/R_o) + (Gh)/K)$ term in the denominator. The effect of this term is to reduce each performance factor. A reduction in static gain, while not desirable, can be compensated for very easily by increasing the input. However, it is highly desirable to decrease the disturbance sensitivity and the time constant. A reduction in disturbance sensitivity means that the motor speed will not vary as much if a disturbance appears and thus will be held more nearly constant. A reduction in time constant means that the system will respond faster to an input change or a disturbance. Faster system response is usually a prime goal of a design engineer.

This system is used as an example for sizing and component selection in Chapter 10.

### 7.2.2 DC Motor Position Control

There are numerous applications in which a position control system is necessary. In this type of system, the output position $\theta$ of an actuator must have a linear response to an operator's input command $\theta_d$, as depicted in Figure 7.3. As in the previous two systems, a DC motor is used, except that a gear train is employed to reduce the speed of the motor $\omega_m$ as it moves the output inertial load $\theta$.

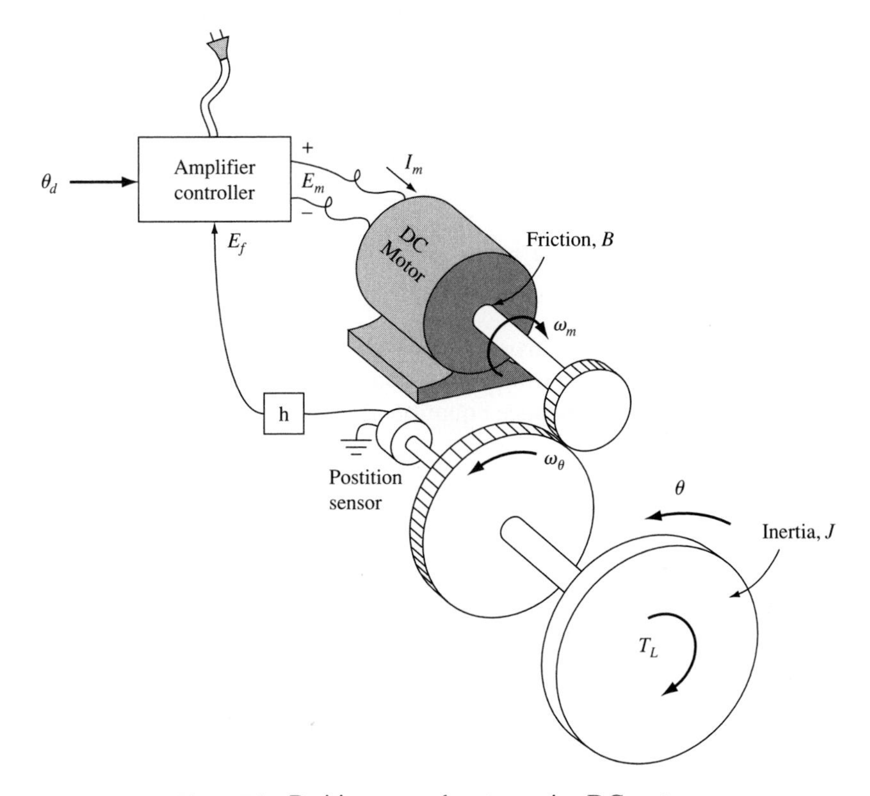


**Figure 7.3** Position control system using DC motor.

The motor equations are the same as used before:

$$T_m = KI_m \tag{7.25}$$

$$E_m = K\omega_m \tag{7.26}$$

The electronic amplifier or controller has inputs from the set point or desired position $\theta_d$ (with a gain of $c$) and from the feedback voltage $E_f$ (which represents the actual position.) The equation for the amplifier is

$$E_m = G(c\theta_d - E_f) - R_oI_m \tag{7.27}$$

The feedback position sensor could be a rotary potentiometer similar to the volume control on a radio. If a constant voltage is supplied across the potentiometer, then the output voltage can be linearly proportional to the angular position of the output shaft, $\theta$, with a sensitivity of $h$:

$$E_f = h\theta \tag{7.28}$$

The motor speed $\omega_m$ is reduced to a lower speed of the output shaft $\omega_\theta$ by a gear train with a speed ratio $R_s$. ($R_s$ is less than 1, which means the gear ratio is greater than 1.) Attendant upon the reduction in speed is an increase in torque from the motor $T_m$ to the output shaft $T_\theta$. We have:

$$\omega_\theta = R_s\omega_m \tag{7.29}$$

$$T_\theta = \frac{1}{R_s}T_m \tag{7.30}$$

Angular speed, of course, is the derivative of angular position:

$$\omega_\theta = \dot{\theta} \tag{7.31}$$

Neglecting the inertia of the gear train, and assuming some linear viscous damping $B$ in the drive train and inertia of the load with a disturbance torque $T_L$, we can state the torque balance equation as

$$T_\theta - J\ddot{\theta} - B\dot{\theta} - T_L = 0 \tag{7.32}$$

Combining the preceding equations, we can proceed as follows:

$$J\ddot{\theta} + B\dot{\theta} = \frac{T_m}{R_s} - T_L = \frac{1}{R_s}\left\{K\left[\frac{G(c\theta_d - h\theta) - \left(K\dfrac{D\theta}{R_s}\right)}{R_o}\right]\right\} - T_L \tag{7.33}$$

$$\left[JD^2 + \left(B + \frac{K^2}{R_oR_s^2}\right)D + \frac{GKh}{R_oR_s}\right]\theta = \frac{GKc}{R_oR_s}\theta_d - T_L \tag{7.34}$$

Normalizing Eq. (7.34) and stating it in transfer function format, we can see that the position is now a second-order function of the input and disturbance:

$$\theta = \frac{\dfrac{c}{h}\theta_d - \dfrac{1}{\left(\dfrac{GKh}{R_oR_s}\right)}T_L}{\dfrac{J}{\left(\dfrac{GKh}{R_oR_s}\right)}D^2 + \dfrac{(B + K^2/(R_oR_s^2))}{\left(\dfrac{GKh}{R_oR_s}\right)}D + 1} \tag{7.35}$$

The static gain of this system becomes a function of the input gain and the feedback gain and is independent of all other system coefficients. This is a very interesting result, in that the static system performance and accuracy are determined by the feedback element and could have nonlinear or inaccurate components in the forward loop, yet still be very accurate in the closed loop. The static gain is

$$G_s = \left.\frac{\partial \theta}{\partial \theta_d}\right|_{D=0} = \frac{c}{h} \tag{7.36}$$

The sensitivity of the position to a variation in torque is the inverse of a stiffness. Thus, even though a mechanical spring is not used in this system, the effect of position feedback is to create an artificial spring effect, or stiffness, of the system. The static ($D = 0$) response of the system with no input can be expressed as $T_L = k_s\theta$. (Don't worry about the negative sign in the transfer function, because the disturbance torque is defined in the direction opposite to that indicated by the sign convention.) The static stiffness is

$$k_s = -\left.\frac{\partial T_L}{\partial \theta}\right|_{D=0} = \frac{GKh}{R_o R_s} \tag{7.37}$$

The dynamic characteristics are (See Appendix E, Section E.3.1):

$$\omega_n = \sqrt{\frac{\left(\dfrac{GKh}{R_o R_s}\right)}{J}} \tag{7.38}$$

$$\zeta = \frac{B + K^2/(R_o R_s^2)}{2\sqrt{J\left(\dfrac{GKh}{R_o R_s}\right)}} \tag{7.39}$$

These systems are typically very responsive and very stiff.

## 7.3 FLUID-MECHANICAL SYSTEMS

Fluid control systems are used in a wide variety of applications involving precision motion control at high power levels. Fluid power control systems consist of hydraulic and pneumatic systems and are known for their ability to move heavy objects with very fast response. Fluid power has an inherent ability to provide high power actuation in a small volume, which gives rise to its high responsiveness [Refs. 2–5].

The fundamental configuration of a fluid power control system is a servovalve that is capable of precise control of the pressure and flow delivered to an actuator. For a servo system, the actuator output must be sensed and fed back to compare the output with the desired input. The error signal from this comparison is then used to actuate the valve.

Typical applications are flight controls (ailerons, rudder, and horizontal stabilizer), power steering on automobiles, and a vast variety of other automotive and industrial applications. The valves can be spool valves, flapper-nozzle, or poppet valves. The feedback can be mechanical or electrical.

Fluid control systems have numerous nonlinearities that are sometimes hard to avoid, so they offer a good opportunity to study nonlinear modeling and simulation. In this section, hydromechanical, pneumomechanical, and electrohydraulic servo systems are examined.

### 7.3.1 Hydraulic Position Servo

Shown in Figure 7.4 is a hydraulic position servo with mechanical feedback. This type of system was popular in flight control systems for military aircraft in the 1960s and is very similar to automobile power steering. The servovalve provides a metered pressure differential $\Delta P_o$ and flow $Q_o$ in response to the position $x$ of the spool valve. This pressure and flow move the double-acting, equal-area $A$ actuator to a position $z$. The actuator is moving a mass $M$ and experiences some unknown load disturbance force $F_L$. A rigid mechanical feedback lever is attached to the actuator and to the servovalve with frictionless pinned connections. As the input (or desired position) $u$ and the actuator position (or actual position) $z$ change, the feedback lever causes the servovalve to modulate the pressure and flow to the actuator, to reduce the error $x$ of the servo system.

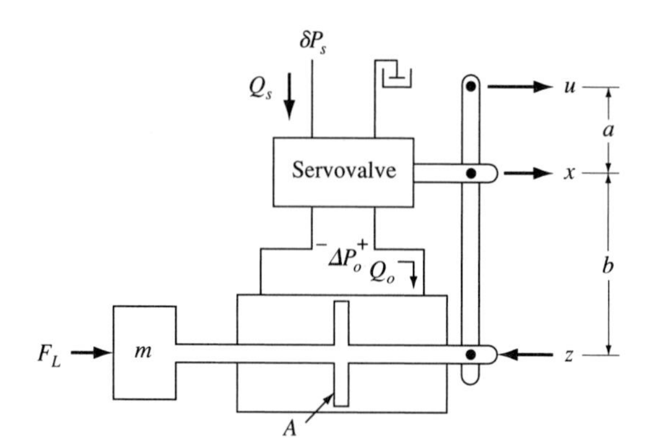


**Figure 7.4** Hydromechanical position servo system.

The servovalve is an underlapped spool valve that has linear characteristics for small-signal operation around the null position ($x$, $\Delta P_o$, and $Q_o$ are all equal to zero) and can be represented by a linear model using the valve pressure gain $G_p$ and output resistance $R$ that expresses the loss of pressure as a function of the flow delivered to the actuator:

$$\Delta P_o = G_p x - RQ_o \tag{7.40}$$

If we assume that the compressibility of the hydraulic fluid is small (i.e., the bulk modulus is large when considering the output resistance of the valve), the actuator has equal areas on both sides of the piston, and the flow entering one side of the actuator is instantaneously equal to the flow being pushed by the other side of the piston back into the valve, then the continuity equation for either side of the actuator yields the result

$$Q_o = \dot{V} = A\dot{z} \tag{7.41}$$

The force balance equation on the actuator takes into account the inertia, pressure forces, and load disturbance. In this system there are no spring forces, and frictional forces are neglected. We obtain

$$M\ddot{z} = A\Delta P_o - F_L \tag{7.42}$$

The kinematics of the feedback lever can be calculated using ratio proportioning and superposition:

$$x = \frac{b}{a+b}u \qquad \text{let } b' = \frac{b}{a+b} \tag{7.43}$$

Notice that the motion of $x$ is in the direction opposite that of the motion of $z$, so a negative sign must be used in the proportionality equation:

$$x = \frac{-a}{a+b}z \qquad \text{let } a' = \frac{a}{a+b} \tag{7.44}$$

Since these two equations are linear, they yield a linear combination when superposition is used:

$$x = b'u - a'z \tag{7.45}$$

Starting with the force balance equation and substituting the continuity equation, valve equation, and lever equation yields the following normalized transfer function:

$$z = \frac{\dfrac{b'}{a'}u - \dfrac{1}{k_s}F_L}{\dfrac{M}{k_s}D^2 + \dfrac{RA^2}{k_s}D + 1} \tag{7.46}$$

The static gain of this system, $G_s$, is the variation of the output $z$ with respect to variations in the input $u$ with no disturbances and in steady-state operation ($D = 0$). It is very interesting that the static gain is a function of the lever ratio and is independent of the characteristics of all of the other components:

$$G_s = \left.\frac{\partial z}{\partial u}\right|_{D=0} = \frac{b'}{a'} = \frac{b}{a} \tag{7.47}$$

In the preceding transfer function, there is a natural grouping of terms that can be recognized as a static stiffness $k_s$, giving the relation between force and displacement ($F_L = k_s z$ in the steady state, $D = 0$, and with no input):

$$k_s = -\left.\frac{\partial F_L}{\partial z}\right|_{D=0} = G_p A a' \tag{7.48}$$

The negative sign comes from the fact that the disturbance force is defined in the direction opposite that of the positive direction for $z$.

This system is second order with dynamic characteristics

$$\omega_n = \sqrt{\frac{k_s}{M}} \tag{7.49}$$

and

$$\zeta = \frac{RA^2}{2\sqrt{k_s M}} \tag{7.50}$$

The proper sizing of the valve, actuator, and feedback can result in a very responsive, extremely stiff servo system.

### 7.3.2 Pneumatic Position Servo

A position servo composed of a three-way valve connected to a spring-loaded actuator with a mechanical lever feedback is illustrated in Figure 7.5. There are two aspects of understanding the operation of this system. First, a decreasing motion of the valve position $y$ causes an increase in pressure, $\delta P_o$, which will in turn cause an increase in output position $\delta z$. If the input command $\delta u$ is held constant, the feedback lever will cause the valve position to be returned to its steady-state position. Second,

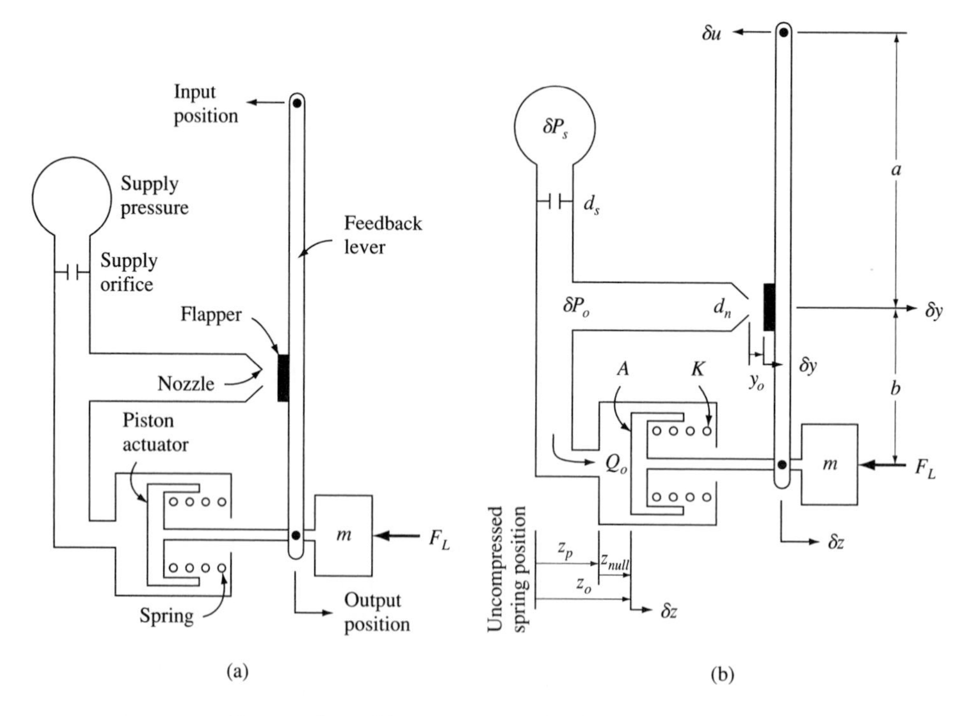


**Figure 7.5** Pneumatic position servo system. (a) Configuration of flapper-nozzle pneumatic servo. (b) Geometry of pneumatic servo.

if a load disturbance force $F_L$ causes a decrease in the output position $z$, the valve will be actuated, and the pressure to the actuator will increase to develop a force that will oppose the disturbance force. This will cause a stiffness effect that will tend to reject the disturbance. This system is used as an example for sizing and component selection in Chapter 10.

In this system, the actuator stroke starts at an unactuated position when the supply pressure is off and increases to a null position as the supply pressure is activated (assuming that there is no input or load disturbance). Thus, we can use a position-referencing scheme that considers the state when the supply pressure is off $(z_o, y_o, u_o)$ and then examine small variations $\delta z$, $\delta y$, and $\delta u$ in the positions around the null operation:

$$z = z_o + \delta z \tag{7.51}$$

$$y = y_o + \delta y \tag{7.52}$$

$$u = u_o + \delta u \tag{7.53}$$

We will first use a fully nonlinear model of this system and express the result in state-space format. Since it is very difficult to analyze the response characteristics of a nonlinear system based upon its coefficients, a linearized approximation of the nonlinearities of the system is used to derive a linear transfer function.

**Nonlinear Analysis.** The three-way pneumatic valve is a flapper-nozzle valve [Refs. 5, 6] with an upstream orifice of diameter $d_s$ and a nozzle of diameter $d_n$. By assuming incompressible flow in the fixed orifice and the variable-area nozzle, one can derive a nonlinear model for the output flow $Q_o$ of the flapper-nozzle valve as a function of the maximum supply flow $Q_s^*$, the pressure ratio $\delta P_o/\delta P_s$, and a normalized nozzle position variable, $\delta y/y_o$. We obtain

$$Q_o = Q_s^* \left\{ \sqrt{1 - \frac{\delta P_o}{\delta P_s}} - \alpha^* \left(1 + \frac{\delta y}{y_o}\right) \sqrt{\frac{\delta P_o}{\delta P_s}} \right\} \tag{7.54}$$

where

$$Q_s^* = C_{ds} \frac{\pi d_s^2}{4} \sqrt{\frac{2\delta P_s}{\rho}} \tag{7.55}$$

$$\alpha^* = 4 \frac{C_{dn}}{C_{ds}} \left(\frac{d_n}{d_s}\right)^2 \frac{y_o}{d_n} \tag{7.56}$$

and $C_{dn}$ and $C_{ds}$ are discharge coefficients for the nozzle and supply orifice, respectively.

Good design guidelines for flapper-nozzle valves dictate the following optimum or null values (for nominal discharge coefficients):

$$\alpha^* = 0.70 \tag{7.57}$$

$$\frac{y_o}{d_n} = 0.125 \tag{7.58}$$

$$\frac{d_s}{d_n} = \sqrt{\frac{4}{\alpha^*}\frac{C_{dn}}{C_{ds}}\frac{y_o}{d_n}} = 0.80 \tag{7.59}$$

$$\frac{\delta P_o^*}{\delta P_s} = 0.671 \tag{7.60}$$

Therefore, when we select the nozzle diameter and the supply pressure, the valve characteristics are completely determined.

This valve has nonlinear characteristics, as shown in Figure 7.6. The input variable shown in the figure is $\delta y/y_o$, and $\delta y$ is varied between $\pm y_o$.

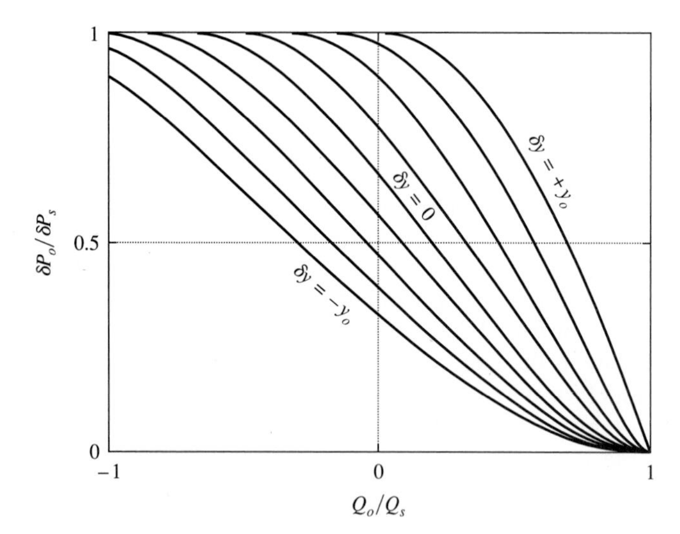


**Figure 7.6** Nonlinear characteristics of three-way flapper-nozzle valve.

Applying the continuity equation to the control volume $V$ of the actuator yields the differential equation

$$\frac{V}{\beta}\delta \dot{P}_o = Q_o - \dot{V} \tag{7.61}$$

The volume inside the actuator is a function of the "dead volume," or the volume of air when the actuator is fully retracted $V_o$ and the "swept volume," or the area times the stroke:

$$V = V_o + A(z_{null} + \delta z) \tag{7.62}$$

Differentiating, we obtain

$$\dot{V} = A\delta\dot{z} \tag{7.63}$$

Since air is compressible, the bulk modulus can be expressed as a function of the polytropic constant $n$ (where $1.0 < n < 1.4$, depending upon the speed of response of the system) and the pressure inside the actuator:

$$\beta = n(\delta P_o + P_{atm}) \tag{7.64}$$

Since we are expecting very fast response, we should use $n = 1.4$.

A force balance of the output motion of the actuator considers the inertia and spring forces along with the pressure force and disturbance force. A preload force in the spring is indicated by the term $k\, z_p$. While this is actually a nonlinearity, and the model would not be correct if the piston could retract beyond its retraction limit, it will be accurate for normal operation. Friction is not considered.

The force balance equation is

$$M\delta\ddot{z} + k(z_p + z_{null} + \delta z) = A\,\delta P_o - F_L \tag{7.65}$$

The feedback lever is a rigid member with frictionless pin connections, so the lever rotates and moves the valve input position $\delta y$ as a function of the input position $\delta u$ and the output position $\delta z$. The kinematics of the feedback lever can be calculated using ratio proportioning of the individual input and output motions and then applying superposition. First,

$$\delta y = \frac{-b}{a+b}\delta u \qquad \text{let } b' = \frac{b}{a+b} \tag{7.66}$$

$$\delta y = \frac{a}{a+b}\delta z \qquad \text{let } a' = \frac{a}{a+b} \tag{7.67}$$

Note that $a' + b' = 1$ or that $b' = 1 - a'$.

Since Eqs. (7.66) and (7.67) are linear, they can be combined using superposition:

$$\delta y = a'\delta z - (1 - a')\delta u \tag{7.68}$$

The preceding equations are the basic equations required for the modeling. Since there are three nonlinearities in these equations, the system should be represented in state-space form. However, we will retain the original system variables instead of converting to state variables $x_1$, $x_2$, and $x_3$. To do this, we will need to define the actuator velocity

$$v = \delta\dot{z} \tag{7.69}$$

Since the force balance equation is a second-order differential equation, we see that we need two state variables, $\delta z$ and $v$. Also, since the continuity equation is a first-order differential equation, we see that we need an additional state variable $\delta P_o$.

Solving all of these equations and reducing to the state-space format yields the following final form of derivatives of the state variables $\delta z$, $v$, and $\delta P_o$ as a function of the input $\delta u$ and the disturbance $F_L$:

$$\delta\dot{z} = v \tag{7.70}$$

$$\dot{v} = \frac{1}{M}[A\,\delta P_o - k(z_p + z_{null} + \delta z) - F_L] \tag{7.71}$$

$$\delta\dot{P}_o = \frac{n(\delta P_o + P_{atm})}{V_o + A(z_{null} + \delta z)} \tag{7.72}$$

$$\left[Q_s^*\left\{\sqrt{1 - \frac{\delta P_o}{\delta P_s}} - \alpha^*\left(1 + a'\frac{\delta z}{y_o} - (1 - a')\frac{\delta u}{y_o}\right)\sqrt{\frac{\delta P_o}{\delta P_s}}\right\} - Av\right]$$

The initial conditions at steady-state null are $\delta z = 0, v = 0$, and $\delta P_o = \delta P_o^* = 0.671\ \delta P_s$. We mention this because, without forethought, one might use all initial conditions equal to 0 for an input from the steady state. If one uses the wrong initial conditions, the simulation will be wrong.

**Linear Analysis.** There are three nonlinearities in the pneumatic servo system just discussed. First is the nonlinear characteristics of the 3-way valve, and the other two nonlinearities are the volume and bulk modulus when they are used as coefficients in equations.

The valve can be linearized by taking a Taylor series expansion of the nonlinear flow equation about the operating points $Q_o = 0$ and $\delta y/y_o = 0$. Evaluating the original nonlinear equations at these conditions reveals that $\delta P_o/\delta P_s = 0.671$. If we carry out the Taylor series expansion about the specified nominal operating conditions and rearrange the resulting terms, we get the following linear approximation to the valve equation:

$$\delta P_o = \delta P_o^* - G_p\delta y - R_o Q_o \tag{7.73}$$

In this equation,

$$\delta P_o^* = \frac{\delta P_s}{[1 + \alpha^{*2}]} = 0.671\ \delta P_s \tag{7.74}$$

$$G_p = \frac{\dfrac{\delta P_s}{y_o}}{\left[1 + \dfrac{\alpha^{*2} + \alpha^{*-2}}{2}\right]} = 0.441\ \frac{\delta P_s}{y_o} \tag{7.75}$$

$$R_o = \frac{\dfrac{2\delta P_s}{Q_s^*}}{\alpha^*\sqrt{1 + \alpha^{*2}}[1 + \alpha^{*-2}]} = 0.770\ \frac{\delta P_s}{Q_s^*} \tag{7.76}$$

These linearized characteristics are plotted in Figure 7.7. Notice that around the point $\delta y = 0$, $\delta P_o/\delta P_s = 0.671$, and $Q_o = 0$, the approximation is very good compared to the nonlinear characteristics shown in Figure 7.6.

The continuity equation (See Section 5.4) for the volume inside the actuator yields

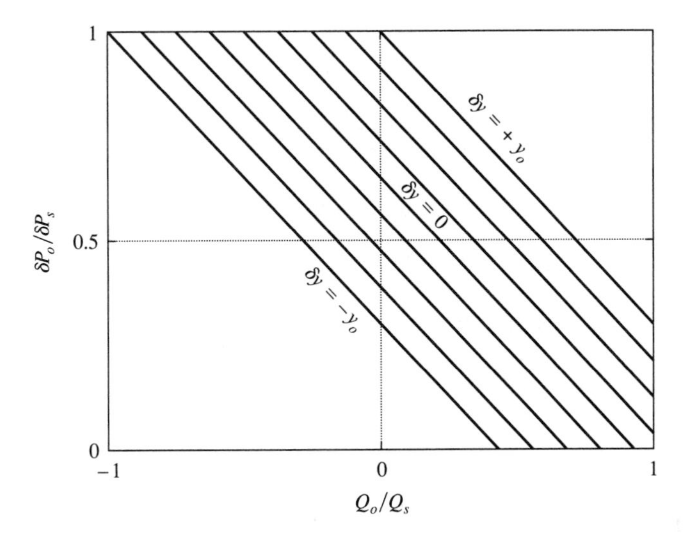


**Figure 7.7** Linearized pressure-flow characteristics of three-way flapper-nozzle valve.

$$Q_o = \frac{V}{\beta}\delta\dot{P}_o + \dot{V} \tag{7.77}$$

The actual volume is variable with the stroke of the actuator, and if we used this variable volume in Eq. (7.77), it would result in a nonlinearity. We can approximate this nonlinearity as a constant volume $V^*$, using an average or nominal volume. Accordingly, we assume that the volume in the actuator is constant and is equal to the total volume at the null position:

$$V \approx V_o + Az_{null} = V^* \tag{7.78}$$

Even though we use an average volume and assume it constant for the purposes of fluid capacitance, we must consider the rate of change of the volume with respect to time:

$$\dot{V} = A\,\delta\dot{z} \tag{7.79}$$

The bulk modulus of air varies with the absolute pressure. Again, this would give a variable coefficient in the continuity equation. If we use an average or nominal value of pressure $\delta P_o^*$ and consider it to be constant, then the bulk modulus is constant, and, the continuity equation is linear.

$$\beta = n(\delta P_o + P_{atm}) \approx n(\delta P_o^* + P_{atm}) = \beta^* \tag{7.80}$$

The force balance equation again considers the mass, the mechanical spring, the pressure force, and the disturbance force:

$$M\,\delta\ddot{z} + k(z_p + z_{null} + \delta z) = A\,\delta P_o - F_L \tag{7.81}$$

The feedback lever gives an input to the valve that is a linear difference of the desired position $\delta u$ and the actuator output $\delta z$:

$$\delta y = a'\delta z - (1 - a')\delta u \tag{7.82}$$

By combining the continuity equation with the linearized valve equation and the equation for the derivative of the volume, and then rearranging, we obtain an intermediate equation for the pressure in the actuator:

$$\delta P_o = \frac{\delta P_o^* - G_P\delta y - R_o A\delta\dot{z}}{R_o\dfrac{V^*}{\beta^*}D + 1} \tag{7.83}$$

If we place this output pressure equation in the force balance equation, rearrange terms, and normalize, we can derive the transfer function of the position response to the input command and the disturbance force:

$$\delta z = \frac{G_s\,\delta u + \delta z_{ss} - \dfrac{\left(R_o\dfrac{V^*}{\beta^*}D + 1\right)}{k_s}F_L}{R_o\dfrac{V^*M}{\beta^*k_s}D^3 + \dfrac{M}{k_s}D^2 + \dfrac{R_oA^2}{k_s}\left(1 + \dfrac{V^*/\beta^*}{A^2/k}\right)D + 1} \tag{7.84}$$

In deriving this equation, we recognize a natural grouping of terms as the static stiffness of the system $k_s$. This static stiffness, due to the mechanical spring stiffness $k$ and the artificial stiffness formed by the closed-loop feedback of the position, is

$$k_s = k + G_pAa' \tag{7.85}$$

where $G_pAa'$ is the stiffness of the system due to loop closure, and $k/(G_pAa')$ is the ratio of the mechanical (spring) stiffness to the loop closure stiffness. In most cases, this ratio is small.

In the derivation of this transfer function, there is a term relating to the derivative of $z_p + z_o$ which was omitted because $D(z_p + z_o) = 0$. Notice that there is also a constant term in the transfer function relating to the preload in the spring and the supply pressure and area. We would like this term to be zero in the steady state so that the actuator will be at the null position:

$$\delta z_{ss} = \frac{A\,\delta P_o^* - k(z_p + z_{null})}{k_s} = 0 \tag{7.86}$$

The static gain of the system, $G_s$, is basically $b'/a'$, but is reduced by the ratio of stiffnesses:

$$G_s = \frac{(1 - a')/a'}{\left(1 + \dfrac{k}{G_pAa'}\right)} \tag{7.87}$$

The system has third-order dynamics, as can be seen by the denominator of Eq. 7.84.

## 7.4 ELECTRO-HYDRAULIC POSITION SERVO

With the explosion of the use of electronics, electronic controls, and digital computer controls, there has been a corresponding explosion of the combination of electronics with hydraulic controls. The well known strength of electronics is its ability to perform computation and control functions; hydraulics is well known for its ability to operate at high power levels and to provide responsive systems; therefore, the combination is natural.

Shown in Figure 7.8 is an electro-hydraulic position servo system. An electro-hydraulic servovalve is connected to an actuator that drives an inertial load. The position of the actuator is measured with a position sensor, and the signal is fed back to an electronic amplifier that compares the actual position to the desired position and amplifies the error between the two.

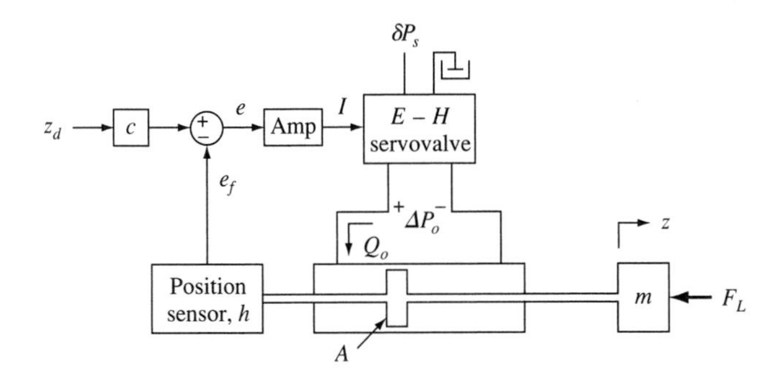


**Figure 7.8** Electro-hydraulic position servo control system.

An electro-hydraulic servovalve consists of an electromagnetic torque motor that converts current into a deflection of the input of the valve. Quite often, the valve has two stages: The first stage is a pilot valve that drives the second-stage main spool valve. The pilot valve is usually a flapper-nozzle valve. The reason a two-stage valve is used is to reduce the force required for the electromagnet to move the main valve. The small flapper-nozzle valve requires a very small force to move the flapper. The output pressure from the flapper-nozzle valve is used to position the main spool valve as if it were a ram-type actuator and thereby proportions the output flow from the valve.

The pressure-flow characteristics of the spool valve are almost linear for small-signal operation around the null, since the spool valve is slightly underlapped and can be represented by a linear model using the valve pressure gain $G_p$ to the input current $I$ and output resistance $R$. The model expresses the loss of pressure as a function of the flow delivered to the actuator:

$$\Delta P_o = G_p I - R Q_o \tag{7.88}$$

If we assume that the compressibility of the hydraulic fluid is small (i.e., the bulk modulus is large when considering the output resistance of the valve), the actuator has equal areas on both sides of the piston, and the flow entering one side of the actuator is instantaneously equal to the flow being pushed by the other side of the piston back into the valve, then the continuity equation for either side of the actuator yields

$$Q_o = \dot{V} = A\dot{z} \tag{7.89}$$

The force balance on the actuator considers the inertia, linear viscous friction, pressure forces, and load disturbance; in this example, there are no springs. The force balance equation is

$$M\ddot{z} + B\dot{z} = A\,\Delta P_o - F_L \tag{7.90}$$

The feedback is from a position sensor, which could be a potentiometer in which the slider is connected to the actuator motion. However, this approach has reliability problems with the sliding contact in the potentiometer. A more popular approach is to use a linear variable differential transformer (LVDT) with an iron core moving inside two transformers that results in a DC voltage $E_f$ which is linear with position after the AC voltages are converted to DC. The gain of the sensor is $h$ (volt/in), and the voltage is

$$E_f = hz \tag{7.91}$$

The amplifier is a summing junction that subtracts the feedback position signal $E_f$ from the input command signal to form an error signal which is then amplified to the current $I$ that drives the valve. In addition to the summation and amplification, a dynamic compensation is added to the circuit to improve the overall dynamics of the system. This dynamic compensation is a proportional-plus-derivative (P-D) controller or filter that causes the error signal to be summed with its derivative and then amplified. The P-D control is accomplished by the amp in Figure 7.8 (See Chapter 4). The mathematics for the circuit can be stated as follows:

$$I = G_a\left(1 + \frac{D}{\omega_d}\right)(cz_d - E_f) \tag{7.92}$$

Starting with the force balance equation and substituting the continuity equation, valve equation, and feedback-and-summing-junction equation yields the following normalized transfer function:

$$z = \frac{\left(\dfrac{D}{\omega_d} + 1\right)\dfrac{c}{h}z_d - \dfrac{1}{k_s}F_L}{\dfrac{M}{k_s}D^2 + \left(\dfrac{B + RA^2}{k_s} + \dfrac{1}{\omega_d}\right)D + 1} \tag{7.93}$$

In this transfer function, the static stiffness $k_s$ is the relation between force and displacement in the steady state with no input (the negative sign comes from the fact that the disturbance force is defined in the direction opposite that of the sign convention):

$$k_s = G_a G_p A h \tag{7.94}$$

The static gain of this system $G_s$ is the variation of the output $z$ with respect to variations in the input $u$, with no disturbances and in the steady state. It is very interesting that the static gain is a function of the input gain and the feedback gain and is independent of the characteristics of all of the other components:

$$G_s = \frac{c}{h} \tag{7.95}$$

The dynamic characteristics are:

$$\omega_n = \sqrt{\frac{k_s}{M}} \tag{7.96}$$

$$\zeta = \frac{B + RA^2 + k_s/\omega_d}{2\sqrt{k_s M}} \tag{7.97}$$

The proper sizing of the amplifier gain, valve, actuator, and feedback can result in a very responsive, extremely stiff servo system. With this high stiffness in mind, one can visualize the advantage of the use of the P-D controller to enhance the dynamics of the system. Notice that if $k_s$ is very high—say, infinity—then the numerator and denominator terms, $D/\omega_d + 1$, in the command transfer function cancel, so the system has no dynamics to the input command and therefore would be extremely fast. Of course, we wouldn't set the gain to infinity, but we could enhance the system dynamics; this is always a goal of good servo design.

## 7.5 SUMMARY

This chapter has presented the modeling of mixed discipline systems—systems involving more than one discipline. Electro-mechanical systems, hydraulic, pneumatic, and electro-hydraulic-mechanical systems were considered. Emphasis was placed upon the overall system performance and the interpretation of the static and dynamic performance characteristics, based upon the system parameters. With this knowledge, the designer can know what parameters or components to vary to obtain the desired system performance. A step upward in complexity has been taken from the chapters on basic modeling in that servo systems, nonlinear systems, and systems with multiple inputs were considered.

## REFERENCES

7.1 Say, M. G., and Taylor E. Openshaw. *Direct Current Machines*. John Wiley & Sons, New York, 1980.

7.2 McCloy, D., and Martin, H. *The Control of Fluid Power*. John Wiley & Sons, New York, 1980.

7.3 Merritt, Herbert E. *Hydraulic Control Systems*. John Wiley and Sons, New York, 1967.

7.4 Watton, John. *Fluid Power Systems*. Prentice Hall, New York, 1989.

7.5 Andersen, Blaine W. *The Analysis and Design of Pneumatic Systems*. John Wiley and Sons, New York, 1967.

7.6 Blackburn, J.R., Reethof, G., and Shearer, J.L. *Fluid Power Control*. The MIT Press, Cambridge, Massachusetts, 1960.

## NOMENCLATURE

$A$ area
$B$ linear viscous damping coefficient
$c$ input gain
$d_s$ diameter of supply orifice
$d_n$ diameter of nozzle
$E$ voltage
$E_f$ feedback voltage
$F_L$ load disturbance force
$G_q$ valve flow gain
$G_s$ static gain
$G_p$ valve pressure gain
$G, G_a$ amplifier gain
$h$ gain of the sensor
$I$ current
$J$ moment of inertia
$k$ mechanical spring stiffness
$k_s$ static stiffness
$K$ motor constant
$K_t$ proportionality constant for the torque-current equation
$K_v$ proportionality constant for the voltage-speed equation
$M$ mass
$n$ polytropic constant
$\Delta P_o$ pressure differential
$P_{atm}$ atmospheric pressure
$\delta P_s$ supply pressure
$\delta P_o$ output pressure
$Q_s^*$ maximum output flow
$Q_o$ output volume flow
$Q_s$ supply flow
$R, R_o$ output resistance
$R_s$ speed ratio of a gear train
$T$ torque
$T_L$ load disturbance torque

$T_m$ motor torque
$u$ input or desired position
$v$ velocity
$V$ control volume
$V_o$ volume of air when the actuator is fully retracted
$x$ error of the servo system
$\delta y$ valve input position or stroke
$y_o$ valve null position
$z$ output position
$z_{ss}$ output position in steady state
$\alpha$ normalized valve stroke
$\alpha^*$ null value of $\alpha$
$\beta$ bulk modulus
$\omega$ rotational speed
$\omega_d$ input speed (set-point speed)
$\omega_n$ natural frequency
$\omega_m$ motor speed
$\theta$ output position
$\theta_d$ input position (set-point or desired position)
$\zeta$ damping ratio

## PROBLEMS

**7.1** A hydraulic cylinder is to be used in an application in which the piston will move at a free speed until it bottoms out on a plate. During this extension, there are no forces acting on the piston (i.e., no friction, no inertia, no spring). When the piston hits the plate, it is to provide a maximum force to compress a part placed on the plate. At the time of the compression, the piston is not moving (i.e., its speed is zero). The hydraulic supply pressure is 1000 psi. The free speed of the actuator should be 5 inches per second, and the compression force should be 300 pounds. *Calculate* what size of actuator will satisfy the force requirement. *Calculate* what flow is required from a valve in order to achieve the specified speed using the selected actuator.

**7.2** Work Problem 7.1 using compressed air at 150 psi.

**7.3** A DC motor is connected to 125 volts DC and is used to spin a grindstone. The diameter of the grindstone disk is 200 mm, and the thickness is 15 mm. The density of the disk material is 3000 kg/m$^3$. The free speed of the motor (with maximum voltage applied and no torque loading) is 1000 rpm. The voltage source that is used to drive the motor has an output impedance of 2 Ω (i.e., $e_m = e_o - R_o i_m$).

Neglecting the inertia and friction of the motor itself, *derive* a transfer function for the dynamic response of the speed of the motor as a function of the input voltage.

*Calculate* the time response using Laplace transform techniques for a step input of 125 volts DC with zero initial conditions.

**7.4** *Work* Problem 7.3 using digital simulation.

**7.5** Shown in Figure P7.5 is a DC torque motor connected to a mechanical load through a gear reduction. The motor produces a torque in linear proportion to the current delivered from a constant-current source. The mechanical load is a disk with inertia $J$, and a translational spring and dashpot connected to the edge of the disk. *Write* the modeling equations for this system, and *derive* a transfer function for the angular position of the disk as a function of the input current.

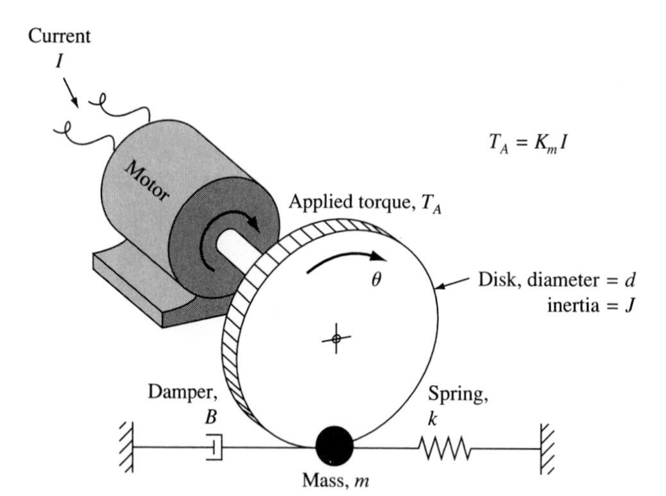


**Figure P7.5** Motor-driven mechanical load.

**7.6** A small DC motor is used to spin a disk with an op-amp as illustrated in Figure P7.6. In this application, we do not want the disk to spin up very fast, so a dynamic filter is added to the op-amp. The op-amp acts as an ideal voltage source with an output impedance $R_o$. *Write* the modeling equations for this system, and *derive* a state-space representation of the system using the natural dynamics ($e_o$ and $\omega$).

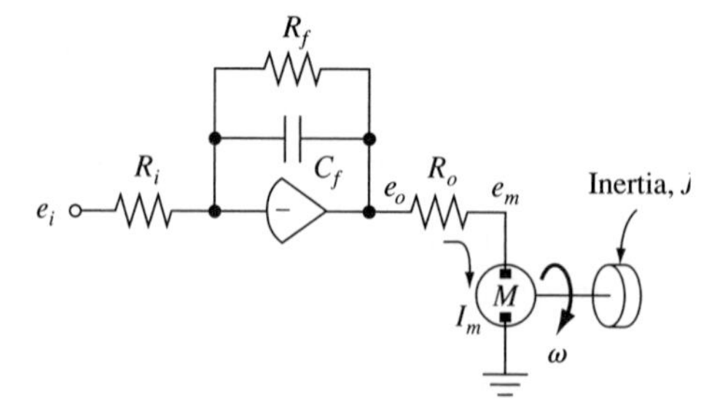


**Figure P7.6** Op-amp drive of small DC motor.

**7.7** Shown in Figure P7.7 is an ideal DC electric motor driving an inertial load $J$. The position of the output is fed back by a position sensor to an op-amp, which then sums

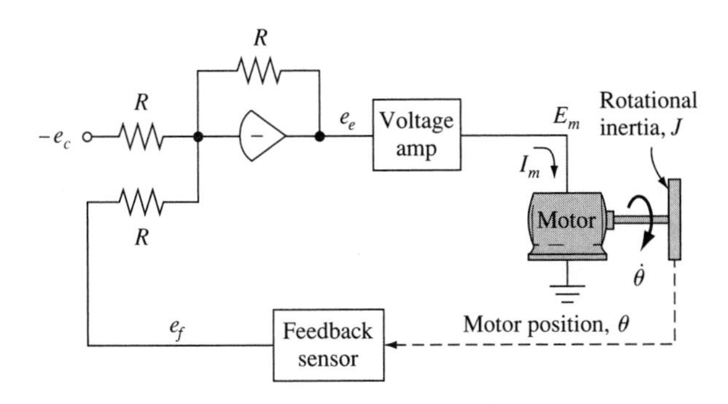


**Figure P7.7** Electronic position servo control system.

together the voltage command signal and the voltage from the position sensor. The output of the op-amp is further amplified by an ideal voltage amplifier. Write the modeling equations for this system, and derive the differential equation for the output position $\theta$ as a function of the input command $e_c$.

**7.8** In what follows, use the basic models for flows in the flapper-nozzle valve to *derive* the normalized model for the flapper-nozzle valve discussed in Section 7.3.2. *Plot* the output pressure $\delta P_o$ versus the normalized valve clearance $\alpha$ with no output flow. *Show* that the point $\alpha = 0.7$ is a good trade-off from among the maximum gain, the minimum mean output pressure, and the maximum linear modulation of the output pressure. *Show* that the null output pressure is $0.671\ \delta P_s$. *Linearize* this model to obtain the results given in Section 7.3.2 under the heading "Linear Analysis." The parameters for flow in the flapper-nozzle valve are:

Supply flow: $$Q_s = C_{ds}A_s\sqrt{\frac{2(\delta P_s - \delta P_o)}{\rho}}$$

Maximum supply flow: $$Q_s^* = C_{ds}A_s\sqrt{\frac{2\delta P_s}{\rho}}$$

Output flow: $$Q_o = Q_s - Q_n$$

Nozzle flow: $$Q_n = C_{dn}A_n\sqrt{\frac{2\delta P_o}{\rho}}$$

Supply orifice flow area: $$A_s = \frac{\pi}{4}d_s^2$$

Nozzle circumferential flow area: $$A_n = \pi d_n y$$

In your derivation of the output flow normalized to the maximum supply flow, you should find a natural grouping of terms such as the following:

Normalized valve stroke: $$\alpha = 4\frac{C_{dn}}{C_{cs}}\left(\frac{d_n}{d_s}\right)^2\frac{y}{d_n}$$

For symmetric operation, the null position of the valve is $y_o$, and the valve can stroke $\pm\delta y$.

Valve position: $y = y_o + \delta y$

Thus, the null value of $\alpha$: $\alpha^* = 4 \frac{C_{dn}}{C_{cs}} \left(\frac{d_n}{d_s}\right)^2 \frac{y_o}{d_n}$

For this model to be correct, the circumferential (or curtain) flow area $A_n$ of the nozzle should be smaller than the area $A_s$ of the supply orifice; that is,

$$\frac{y_o}{d_n} < 0.125$$

# Part 3

# System Dynamic Response Analysis

The US Army Tactical Missile System. (Photo courtesy of Lockheed Martin Vought Systems Corporation.)

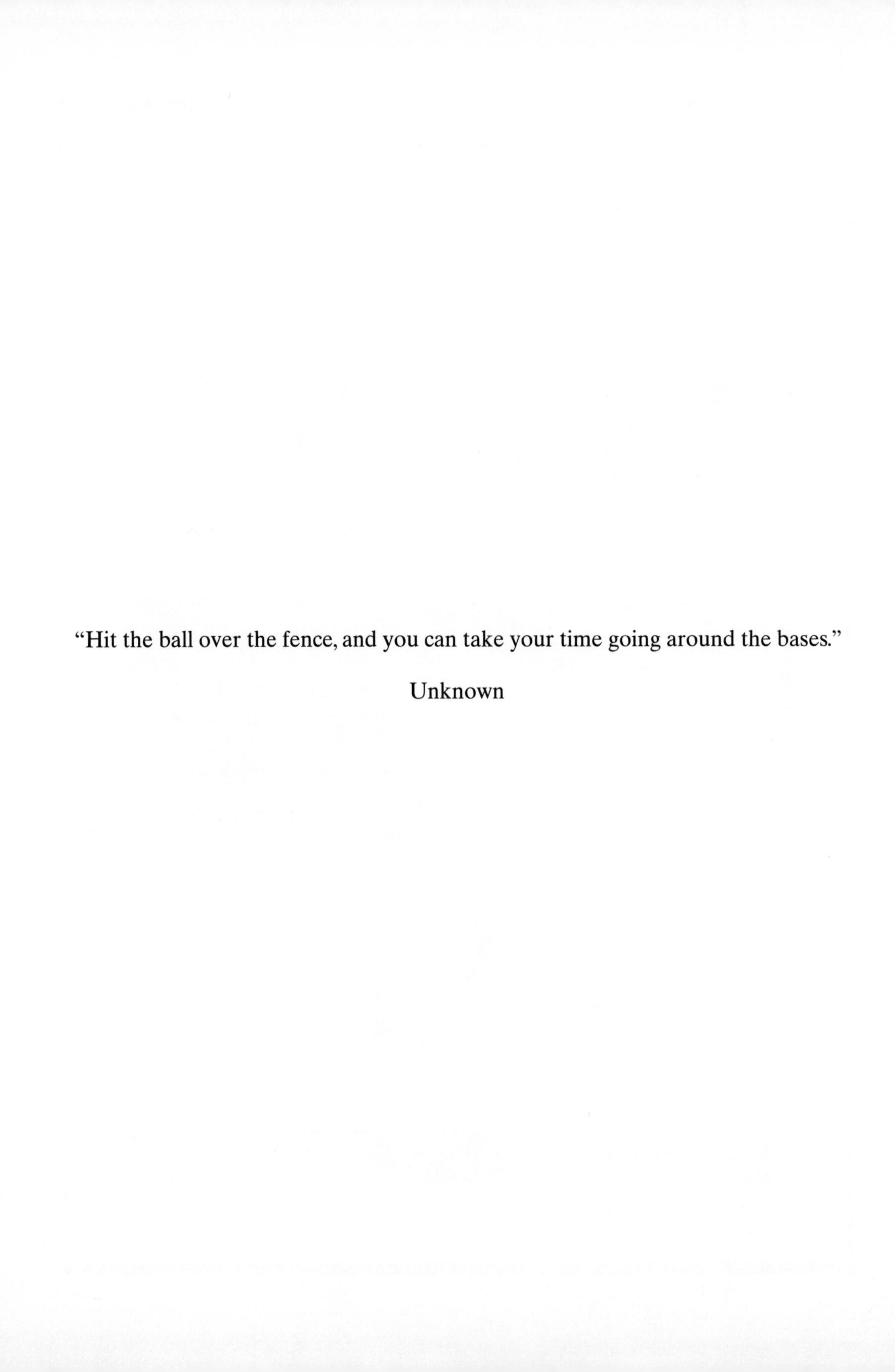

“Hit the ball over the fence, and you can take your time going around the bases.”

Unknown

# CHAPTER 8

# Frequency Response

## 8.1 INTRODUCTION

The previous chapters have been concerned with the modeling of dynamic systems—that is, the development of appropriate sets of differential and algebraic equations to describe a system's behavior. We now turn to the simulation phase, which treats the solution of the equations under specified input excitations and conditions. Normally, we are interested in three principal aspects of system behavior:

(a) inherent system characteristics and stability,
(b) response to harmonic inputs, and
(c) response to transient inputs.

In this chapter, we discuss the first two of these; in the next, we consider the third. (Response to random inputs is another area of common interest, but is not considered in this text.)

Many dynamic systems are subject to excitations that are harmonic in their nature. Harmonic inputs may occur, for example, in mechanical systems or subsystems that include rotating components or in electronic systems that incorporate oscillator circuits of various types. **Frequency response** describes the steady-state behavior of the system to harmonic excitations over a range of input frequencies.

It is also possible to determine important behavioral characteristics of dynamic systems by subjecting them to harmonic inputs and observing the response. This can be done experimentally, analytically, or numerically. In this chapter, we consider the analytical and numerical determination of system frequency response characteristics. Initially, we draw on the closed-form results found in Appendix E and the Laplace transform methods discussed in Appendix F. (If necessary, you might want to review that material before proceeding with the discussion that follows.) We then consider computer methods designed to evaluate system frequency response numerically and show their application to typical problems of interest to the systems engineer.

## 8.2 PRELIMINARIES

Before discussing specific systems, we set out some preliminary general concepts that will aid our understanding of the material that follows. The behavior of many dynamic systems can be described by ordinary, linear, constant-coefficient differential equations; the general classical form is

$$a_n \frac{d^n x}{dt^n} + a_{n-1} \frac{d^{n-1} x}{dt^{n-1}} + \cdots + a_1 \frac{dx}{dt} + a_0 x = Gu(t) \tag{8.1}$$

Solutions to the general form can be expressed as

$$x(t) = x_h(t) + x_p(t) \tag{8.2}$$

The first term on the right of this equation is the solution to the **homogeneous equation**

$$a_n \frac{d^n x}{dt^n} + a_{n-1} \frac{d^{n-1} x}{dt^{n-1}} + \cdots + a_1 \frac{dx}{dt} + a_0 x = 0 \tag{8.3}$$

The second term in Eq. (8.2) is a solution particular to the type of excitation function $Gu(t)$. The homogeneous equation expresses the inherent characteristics of the system, which are independent of what kind of excitation is applied to it. The solution can be written in the form

$$x_h(t) = Ce^{\lambda t} \tag{8.4}$$

where $\lambda$ is determined by substitution into the homogeneous equation:

$$a_n C\lambda^n e^{\lambda t} + a_{n-1} C\lambda^{n-1} e^{\lambda t} + \cdots + a_1 C\lambda e^{\lambda t} + a_0 Ce^{\lambda t} = 0 \tag{8.5}$$

Simplification reduces this expression to a polynomial in $\lambda$:

$$a_n \lambda^n + a_{n-1} \lambda^{n-1} + \cdots + a_1 \lambda + a_0 = 0 \tag{8.6}$$

This expression is called the **characteristic equation,** since important behavior characteristics of the system are reflected in the nature of the roots of the equation. If the roots of the equation are distinct, there are $n$ different solutions to the characteristic equation, and since the differential equation is linear, we can superpose these individual solutions to obtain

$$x_h(t) = C_n e^{\lambda_n t} + C_{n-1} e^{\lambda_{n-1} t} + \cdots + C_1 e^{\lambda_1 t} \tag{8.7}$$

The roots $\lambda_i$ of the homogeneous equation give us information about the stability of the system and the speed with which it responds to external inputs. The roots can be real or complex, positive or negative, and they determine the system's behavior in ways discussed subsequently.

It is also useful to define **transfer functions** for linear systems. For systems with a single input, the transfer function expresses the ratio of the output response to the input excitation, as determined from the Laplace transform of the system equation if the initial conditions pertaining to the dependent variable are zero. The Laplace transform of Eq. (8.1) for zero initial conditions gives

$$[a_n s^n + a_{n-1} s^{n-1} + \cdots + a_1 s + a_0]X(s) = GU(s) \tag{8.8}$$

Solving for the ratio of the output to the input, we obtain

$$\mathrm{TF}(s) = \frac{X(s)}{U(s)} = \frac{G}{a_n s^n + a_{n-1} s^{n-1} + \cdots + a_1 s + a_0} \tag{8.9}$$

If derivatives are present in the input, the Laplace transform of the input is represented as:

$$\mathcal{L}(f(t)) = G(\beta_k s^k + \beta_{k-1} s^{k-1} + \cdots + \beta_1 s + \beta_0)U(s) \tag{8.10}$$

and the transfer function takes the form

$$\frac{X(s)}{U(s)} = \frac{G(\beta_k s^k + \beta_{k-1} s^{k-1} + \cdots + \beta_1 s + \beta_0)}{a_n s^n + a_{n-1} s^{n-1} + \cdots + a_1 s + a_0} \tag{8.11}$$

where $k \leq n$.

We next consider first-, second-, third-, and higher order systems to determine their inherent characteristics and frequency response.

## 8.3 FIRST-ORDER SYSTEMS

### 8.3.1 Introduction

The RC circuit of Section 4.3.3 is a good example of a first-order system, so let us consider its frequency response characteristics. Figure 8.1 shows the circuit; the differential equation that governs it is Eq. (8.12)

$$RC\dot{e}_1 + e_1 = e_0(t). \tag{8.12}$$

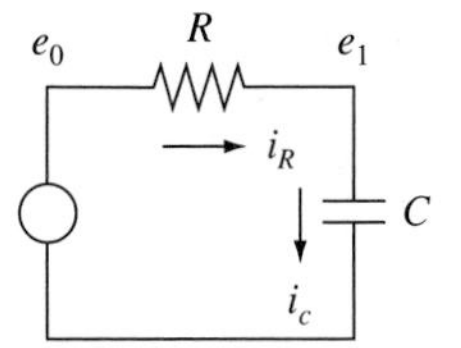


**Figure 8.1** Resistor-capacitor circuit.

Suppose the excitation voltage in the circuit is a harmonic function of time:

$$e_0(t) = E \cos \omega t \tag{8.13}$$

In this case, the governing equation takes the form

$$RC\dot{e}_1 + e_1 = E \cos \omega t \tag{8.14}$$

where $E$ is the amplitude of the applied harmonic voltage.

### 8.3.2 Classical Solution

The standard form for the first-order equation, as described in Appendix E, is

$$\tau\dot{x} + x = Gu(t) \tag{8.15}$$

The corresponding homogeneous equation is

$$\tau\dot{x} + x = 0 \tag{8.16}$$

with solution

$$x_h(t) = Ae^{\lambda t} \tag{8.17}$$

Substitution into the characteristic equation gives

$$\lambda = -1/\tau \tag{8.18}$$

For a harmonic excitation,

$$Gu(t) = Gu_0 \cos \omega t \tag{8.19}$$

The classical solution to this equation is

$$x(t) = \underbrace{Ae^{\lambda t}}_{\substack{\text{homogeneous} \\ \text{(transient)}}} + \underbrace{B\cos(\omega t + \phi)}_{\substack{\text{particular} \\ \text{(steady state)}}} \tag{8.20}$$

where

$$B = \frac{Gu_0}{\sqrt{1 + \omega^2\tau^2}} \tag{8.21}$$

$$\phi = \tan^{-1}(-\omega\tau) \tag{8.22}$$

and $A$ is found from the initial conditions.

In considering frequency response behavior, we are concerned with the steady-state nature of the response. That is, we wait for the transient response to die away and determine the amplitude of the steady response to the harmonic input. This assumes that the root of the characteristic equation for the first-order system is negative, so that the exponential term in the solution gets smaller with increasing time. If this is not so, the exponential term grows larger and larger with time, and the system is **unstable.** Consequently, it does not ever achieve a steady-state condition.

Normally, our first consideration in dynamic systems design is to ensure stability, and we therefore are obliged to examine the roots of the characteristic equation of the system. For the electrical circuit under consideration, we note that

$$\tau = RC \quad \text{and} \quad Gu_0 = E \tag{8.23}$$

Stability for this system requires only that $R$ and $C$ have positive values to guarantee a negative coefficient on the transient exponential term in the homogeneous solution.

The amplitude and the phase of the steady-state response of the $RC$ circuit are, respectively,

$$B = \frac{E}{\sqrt{1 + \omega^2\tau^2}} \tag{8.24}$$

and

$$\phi = \tan^{-1}(-\omega\tau) \tag{8.25}$$

In examining these results, it is helpful to plot the amplitude of the response divided by the amplitude of the input function, which gives the normalized output response amplitude for the circuit:

$$\frac{\dfrac{E}{\sqrt{1 + \omega^2\tau^2}}}{E} = \frac{1}{\sqrt{1 + \omega^2\tau^2}} \tag{8.26}$$

We also normalize the frequency scale by multiplying the excitation frequency $\omega$(rad/sec) by the time constant $\tau$ (sec). (*Note:* In this chapter, we use "sec" as the abbreviation for "second" to avoid confusion with the Laplace operator $s$; other chapters use the SI abbreviation "s" for "second".) Normalizing in this way produces a plot with dimensionless axes and extends the use of the plot beyond a single numerical example. Figure 8.2 shows the amplitude and phase expressions plotted over a range of frequencies of interest.

It is convenient to represent the response amplitude on a base-10 logarithmic scale:

$$\text{dB} = 20 \log_{10}(B/Gu_0) \tag{8.27}$$

Figure 8.3 shows the circuit response plotted in this way. Here the normalized amplitude is plotted in decibels (dB), and the normalized frequency is plotted on a log scale. Henrik Wade Bode (American Engineer, 1905– ) [Ref. 8.1] observed that natural symmetries and asymptotic behaviors are best displayed when logarithmic scales are used in this way, and such graphs are therefore called **Bode plots**.

### 8.3.3 Frequency Response

We next consider the solution of the $RC$ circuit equation using the Laplace transform methods of Appendix F. Assuming that the initial value of the voltage $e_1$ is zero, applying the Laplace transform to Eq. (8.13) yields

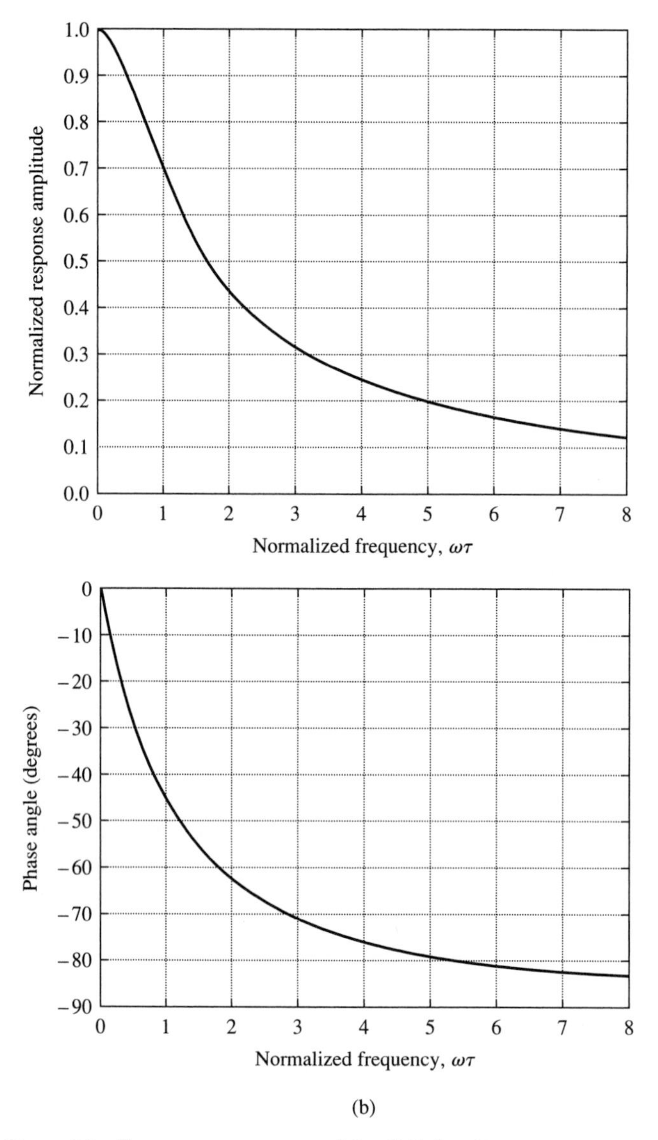


**Figure 8.2** Frequency response of the *RC* circuit.

$$\tau s E_1(s) + E_1(s) = EU(s) \tag{8.28}$$

We solve this algebraic equation for $E_1(s)/U(s)$ to obtain

$$\frac{E_1(s)}{U(s)} = \frac{E}{s\tau + 1} \tag{8.29}$$

Applying the methods of Appendix F allows us to determine the solution for $e_1(t)$. The result is the same as the classical solution given in Eq. (8.2). Our purpose here is not to develop another way of finding the same time-history solution as before, but

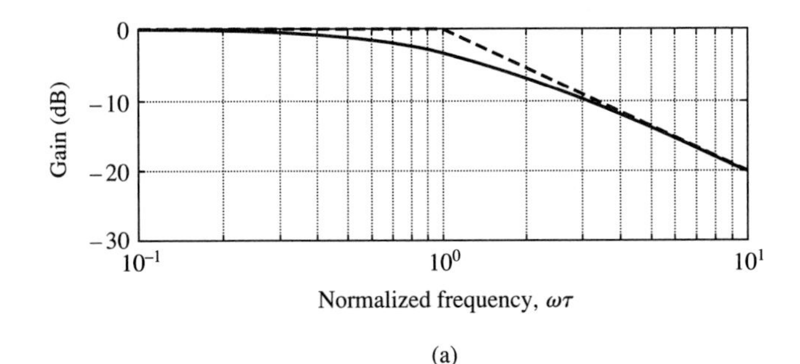


(a)

0
−30
−60
−90
Phase (degrees)
$10^{-1}$
$10^{0}$
$10^{1}$
Normalized frequency, $\omega\tau$

(b)

**Figure 8.3** *RC* circuit Bode plot.

to provide an alternative means of obtaining the frequency response solution. Suppose we now substitute $j\omega$ for $s$ in the previous equation. Here $j = \sqrt{-1}$ is a complex number, and $\omega$ is the excitation frequency. The resulting transfer function $(\overline{\text{TF}})$ is now written as a complex number:

$$\overline{\text{TF}} = \frac{E_1(j\omega)}{U(j\omega)} = \frac{E}{j\omega\tau + 1} \tag{8.30}$$

Let us calculate the magnitude of the complex number in this equation:

$$\text{TF} = \text{MAG}\,(\overline{\text{TF}}) = \frac{\text{MAG}(E)}{\text{MAG}(j\omega\tau + 1)} \tag{8.31}$$

$$\text{TF} = \frac{E}{\sqrt{1 + \omega^2\tau^2}} \tag{8.32}$$

The result, we see, is the same as is obtained from the time history, but is found quite quickly using the transfer function form. The phase of this complex number is defined by the angle whose tangent is the ratio of its complex part to its real part and is found to be

$$\phi = \tan^{-1}(-\omega\tau) \tag{8.33}$$

We see that this result is also the same as was previously determined using classical methods.

Equation (8.30) is the complex form of the transfer function of the system and provides a convenient method of determining the system's frequency response characteristics. We will find this approach very helpful, especially in the treatment of higher order systems.

### 8.3.4 Frequency Response Simulation

Frequency response simulation calculations are easily carried out using the transfer function form of representing the system. We need only to sweep through the desired range of input frequencies and, at each frequency, compute the value of the complex form of the transfer function, its magnitude, and its phase. Computational systems that automatically treat operations on complex numbers are very helpful in this regard. The resulting data may then be tabulated or plotted to provide useful information about the system's behavior. Examples of this technique follow in the next two subsections.

### 8.3.5 Low-Pass Filter

Sinusoidal input signals passing through a dynamic system are altered by the dynamic characteristics of the system. This is shown in Figure 8.4, in which both the amplitude and phase of the output $x(t)$ are different from those of the input $u(t)$. The frequency of the output signal, however, is the same as the frequency of the input signal, as is shown by Eqs. (8.24) and (8.25).

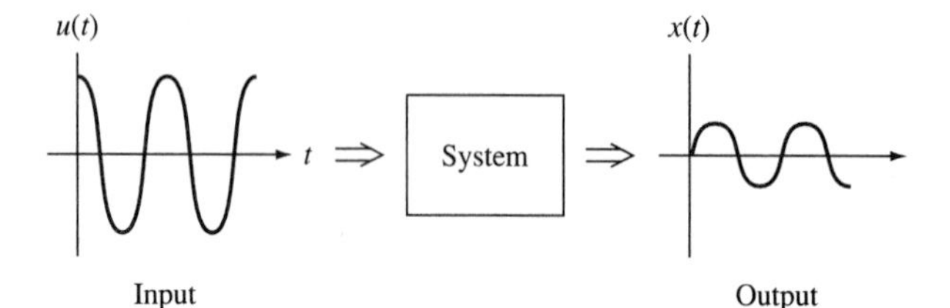


**Figure 8.4** Typical system response.

We note from Figure 8.3 that as the frequency of the input signal to the *RC* circuit increases, the output signal is reduced in amplitude and shifted in phase. Thus, the *RC* circuit can be considered a **low-pass filter,** in the sense that below a certain input frequency, the input amplitude is reproduced almost unchanged at the output, but above that frequency, the output amplitude is much smaller than the input amplitude to the system. Thus, the system only reproduces (without significant amplitude or phase modification) signals with frequencies below a particular value. Observe also in Figure 8.3 that the frequency response characteristics of the *RC* circuit can be approximated by two straight lines on the scale of dB vs. log $\omega\tau$. The knee of the curve is called the **corner frequency** or **break frequency**. Signals with frequencies above the corner frequency are attenuated and shifted in phase on output.

**Example 8.1 RC Circuit**

An *RC* circuit contains a resistor $R = 5000\ \Omega$ and a capacitor $C = 1\ \mu$f. Find the frequency response characteristics of this circuit.

**Solution** The time constant is

$$\tau = RC = 5000 \frac{\text{volt}}{\text{amp}} (1 \times 10^{-6}) \frac{\text{amp sec}}{\text{volt}} = 0.005 \text{ sec} \tag{8.34}$$

From Figure 8.3, the corner frequency is $\omega\tau = 1.0$; thus,

$$\omega = \frac{1}{\tau} = 200 \text{ rad/sec} = 31.8 \text{ Hz} \tag{8.35}$$

which means that the circuit passes frequencies below about 30 Hz with little amplitude attenuation or phase distortion. This design is easily modified by changing the values of $R$ and $C$ to move the corner frequency up or down. Suppose, for example, that we wish to pass frequencies up to 60 Hz. Then the product $RC$ must be reduced by a factor of two to increase the corner frequency appropriately. A change in either $R$ or $C$ or both consistent with other system constraints could be employed to accomplish the modified design goal.

### 8.3.6 A First-Order Mechanical System

Figure 8.5 shows a mass coupled to a moving platform with viscous friction at the interface between the two. The position $y(t)$ of the platform is a specified function of time, and the desired system output is $v(t)$, the velocity of the mass.

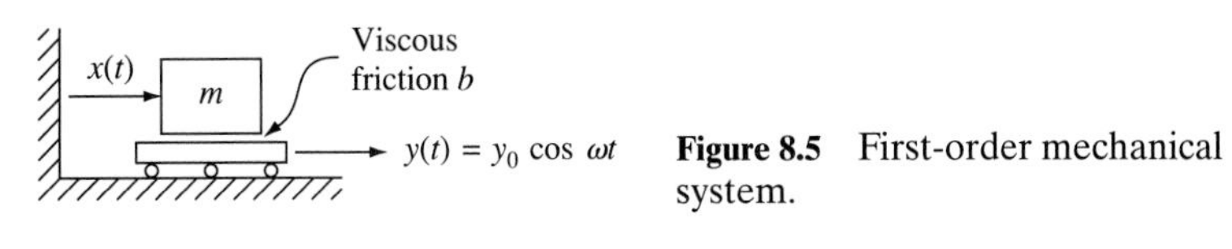


**Figure 8.5** First-order mechanical system.

The equation of motion of the system can be shown to be

$$m\ddot{x} + b(\dot{x} - \dot{y}) = 0 \tag{8.36}$$

or

$$m\dot{v} + bv = b\dot{y} \tag{8.37}$$

We want to know the response as a function of a harmonic input, so we take the Laplace transform of Eq. (8.37) (with zero initial conditions):

$$msV(s) + bV(s) = bsY(s) \tag{8.38}$$

$$\frac{V(s)}{Y(s)} = \frac{bs}{ms + b} = \frac{s}{\tau s + 1} \tag{8.39}$$

Substituting $j\omega$ gives

$$\frac{V(j\omega)}{Y(j\omega)} = \frac{j\omega b}{mj\omega + b} \tag{8.40}$$

The amplitude and phase of the response are, respectively,

$$A = \frac{b\omega}{\sqrt{m^2\omega^2 + b^2}} \tag{8.41}$$

and

$$\phi = 90^\circ - \tan^{-1}\left(\frac{m\omega}{b}\right) \tag{8.42}$$

**Example 8.2 Viscous-Coupled Mechanical System**

The mechanical system of Figure 8.5 has a mass of 2.0 kg and a viscous damping constant of 3.5 N sec/m. Find the frequency response of this system using MATLAB.

**Solution** Referring to the Laplace form of the transfer function, we determine the array of coefficients of the numerator (`num`) polynomial and the array of coefficients of the denominator (`den`) polynomial. For the case at hand,

$$\frac{V(s)}{Y(s)} = \frac{bs}{ms + b} = \frac{3.5s}{2.0s + 3.5} \tag{8.43}$$

The polynomial coefficients are expressed as row vectors in order of *descending powers* of $s$:

```
num = [3.5  0]        (8.44)
den = [2.0  3.5]      (8.45)
```

We employ the "MATLAB Systems and Signals Toolbox" function `bode` to perform the required calculations that were discussed in Section 8.3.4. The interactive command is simply

```
bode(num,den)         (8.46)
```

This produces a Bode plot directly on the computer screen, and the results are shown in Figure 8.6. The excitation frequency range may be selected automatically by the `bode` routine, as was done in this case, or may be supplied as an input array.

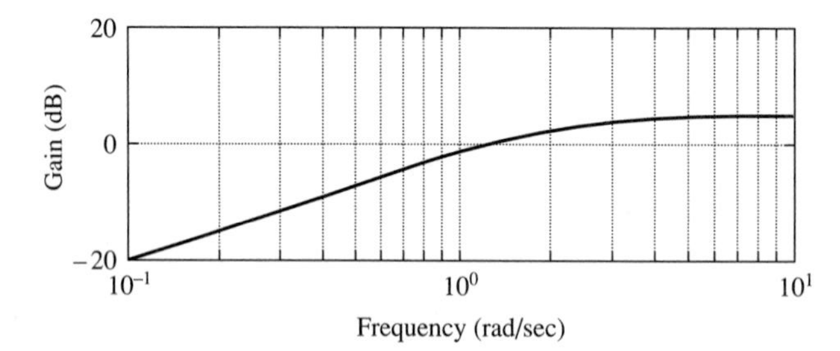


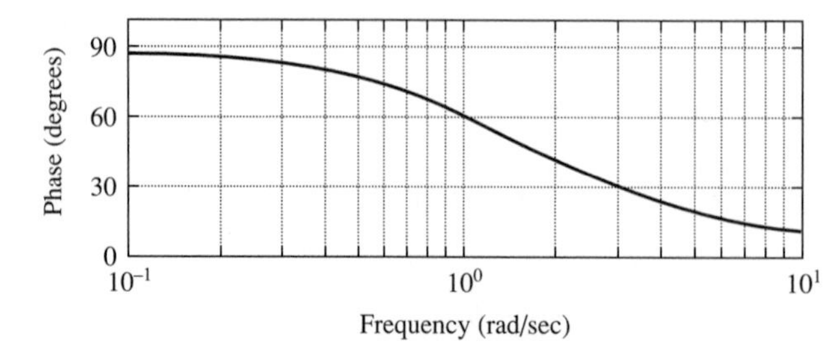


**Figure 8.6** Frequency response of first-order mechanical system.

Notice in the foregoing example that at low frequencies the coupling between the mass and the input platform is low; the coupling increases with frequency up to a frequency of $1/\tau = 1.75$ rad/sec. At higher frequencies, the ratio of the magnitudes of the output to the input is essentially constant. This means that the viscous coupling is so high that the two masses move as one, which is similar to the behavior of a torque converter in an automatic transmission.

## 8.4 SECOND-ORDER SYSTEMS

### 8.4.1 Classical Solution

The spring-mass-damper system of Chapter 3 is an example of a frequently occurring mechanical system that is of second order. Figure 8.7 shows such a system subjected to a force that varies harmonically with time.

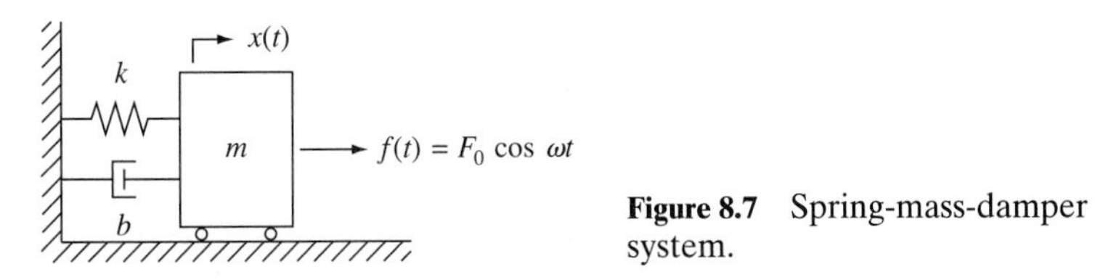


**Figure 8.7** Spring-mass-damper system.

The equation of motion of the system is

$$m\ddot{x} + b\dot{x} + kx = F_0 \cos \omega t \tag{8.47}$$

To put this system into the standard form described in Appendix E, we divide each term by the spring stiffness $k$:

$$\frac{m}{k}\ddot{x} + \frac{b}{k}\dot{x} + x = \frac{F_0}{k}\cos \omega t \tag{8.48}$$

Using this notation, we rewrite the equation in the form

$$\frac{1}{\omega_n^2}\ddot{x} + \frac{2\zeta}{\omega_n}\dot{x} + x = x_0 \cos \omega t \tag{8.49}$$

where $x_0 = F_0/k$ is the displacement that would result if the force were applied statically. The undamped natural frequency is

$$\omega_n = \sqrt{k/m} \tag{8.50}$$

and the damping ratio is

$$\zeta = \frac{b}{b_c} \tag{8.51}$$

where

$$b_c = 2\sqrt{km}$$

is the critical damping factor.

The characteristic equation is

$$\frac{1}{\omega_n^2}\lambda^2 + \frac{2\zeta}{\omega_n}\lambda + 1 = 0 \tag{8.52}$$

This equation is quadratic in $\lambda$ with roots

$$\lambda_{1,2} = \frac{\omega_n^2}{2}\left(-\frac{2\zeta}{\omega_n} \pm \sqrt{\frac{4\zeta^2}{\omega_n^2} - \frac{4}{\omega_n^2}}\right) \tag{8.53}$$

or

$$\lambda_{1,2} = \omega_n\left(-\zeta \pm \sqrt{\zeta^2 - 1}\right) \tag{8.54}$$

The homogeneous solution is

$$x_h(t) = C_1e^{\lambda_1 t} + C_2e^{\lambda_2 t} \tag{8.55}$$

The nature of the solution depends on the amount of viscous damping attributed to the system, and stability is determined by the roots of the characteristic equation. The system is stable if the real parts of the roots are negative. Stable systems have transients that die away with time and, when subjected to steady harmonic inputs, display steady-state responses. Note from the preceding equations that linear second-order systems are guaranteed to be stable if the damping ratio $\zeta$ and the frequency $\omega_n$ are positive constants.

If $\zeta$ is less than 1, the system is said to be underdamped. In that case, the roots are complex conjugates, and the homogeneous solution is harmonic. Thus,

$$\lambda_{1,2} = \omega_n\left(-\zeta \pm j\sqrt{1 - \zeta^2}\right) \tag{8.56}$$

The steady-state solution to the second-order system is given next, while its frequency response is shown in Figure 8.8 for a range of damping ratios of interest. The response is characterized by its amplitude and by its phase with respect to the input. We have

$$x(t) = \frac{x_0\cos(\omega t + \phi)}{\sqrt{(1 - \overline{\omega}^2)^2 + (2\zeta\overline{\omega})^2}} = A\cos(\omega t + \phi) \tag{8.57}$$

where $\overline{\omega} = \omega/\omega_n$ and the amplitude is

$$A = \frac{x_0}{\sqrt{(1 - \overline{\omega}^2)^2 + (2\zeta\overline{\omega})^2}} \tag{8.58}$$

The phase angle is

$$\phi = \tan^{-1}[-2\zeta\,\overline{\omega}/(1 - \overline{\omega}^{\,2})] \tag{8.59}$$

Several interesting characteristics of second-order systems are revealed in the plots of Figure 8.8. Notice that when the excitation frequency is equal to the natural frequency of the system, the output response can be significantly amplified. This

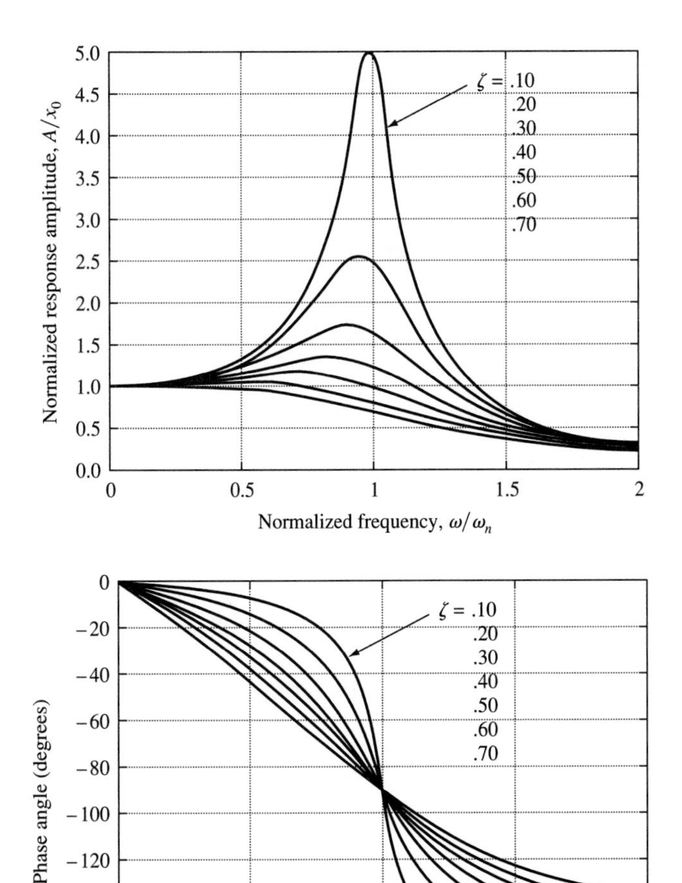


**Figure 8.8** Frequency response of second-order system.

condition is referred to as **resonance.** In mechanical systems, resonance can be either destructive or useful. In an electrical circuit, the maximum output occurs when the circuit parameters are tuned to the input frequency. Thus, resonance may be the desired operating state.

Observe also the strong dependence of the response on the amount of damping present. Lightly damped systems respond very strongly to frequency changes near resonance, while heavily damped systems do not. If the frequency ratio is kept below 0.5 and the damping is around 0.7 or greater, the output amplitude differs very little from the input amplitude, and thus there is almost no amplification whatsoever of the signal. The phase shift is limited to 40 degrees under these conditions.

**Example 8.3 Transmission System**

A portion of a helicopter transmission support system is represented by a spring-mass-damper system such as the one shown in Figure 8.7. Find the inherent dynamic characteristics. If a harmonic excitation is applied at 300 rpm = 5 Hz with a force amplitude of 2450 lbf, find the amplitude of the steady-state displacement response and the response phase with respect to the input. The part in question weighs 725 lbf, and the damping is estimated to be around 15 percent of the critical value (i.e., $\zeta = 0.15$).

The physical parameters are

$$k = 9950 \text{ lbf/in} \quad \text{and} \quad m = \frac{725 \text{ lbf}}{386 \text{ in/sec}^2} \tag{8.60}$$

**Solution** The static response amplitude is

$$x_0 = \frac{F_0}{k} = \frac{2450 \text{ lbf}}{9950 \text{ lbf/in}} = 0.246 \text{ in} \tag{8.61}$$

First, we find the natural frequency:

$$\omega_n = \sqrt{\frac{k}{m}} = \sqrt{\frac{9950 \text{ lbf/in}}{725 \text{ lbf}/(386 \text{ in/sec})^2}} = 72.8 \text{ rad/sec} = 11.6 \text{ Hz} \tag{8.62}$$

Then the frequency ratio is $\overline{\omega} = \omega/\omega_n = 5/11.6 = 0.43$. This operating point is shown on the frequency response plot of Figure 8.9.

We now compute the denominator of the response amplitude equation:

$$\sqrt{(1 - \overline{\omega}^2)^2 + (2\zeta\overline{\omega})^2} = \sqrt{(1 - .43^2)^2 + (2(.15)(.43))^2} = 0.825 \tag{8.63}$$

The amplitude of the steady-state response is found to be

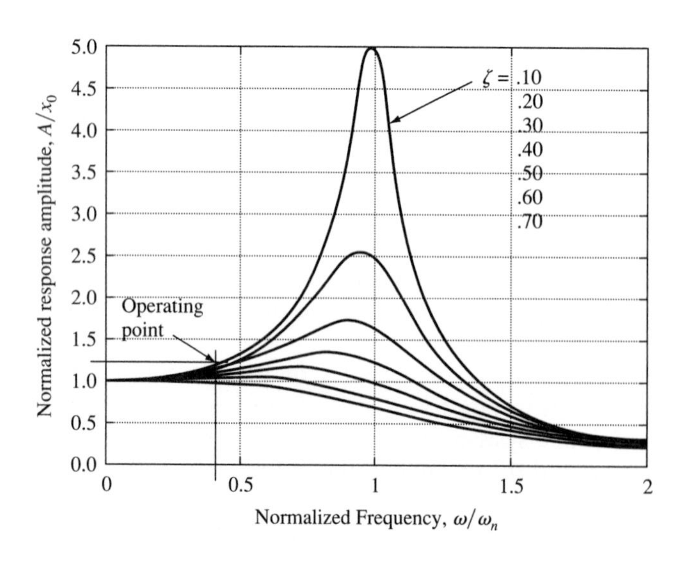


**Figure 8.9** Frequency response for Example 8.3.

$$\text{Amplitude} = \frac{0.246 \text{ in}}{0.825} = 0.298 \text{ in} \approx 0.3 \text{ in} \tag{8.64}$$

Since $1/(0.825) = 1.21$, we obtain a 21 percent amplification in the response due to the dynamics of the system. The response phase is

$$\phi = \tan^{-1}\left(\frac{-2\zeta}{1 - \overline{\omega}^2}\right) = -20.2 \text{ degrees} \tag{8.65}$$

It is often desirable to operate a second-order system at a resonance state. The next example illustrates this.

**Example 8.4 Tuning Circuit**

An electronic tuning circuit component is represented by the *RLC* circuit of Figure 8.10. The governing differential equation was developed in Section 4.3.5 and was found to be

$$LC\ddot{e}_2 + RC\dot{e}_2 + e_2 = e_0(t) \tag{8.66}$$

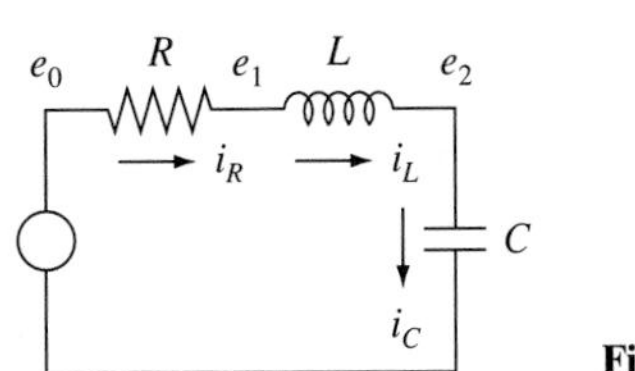


**Figure 8.10** *RLC* circuit.

For the given application, the parameter values are $L = 0.1$ mh, $C = 10$ $\mu$f, and $R = 1\ \Omega$. At what frequency is the circuit tuned, and how much damping is inherent in the design?

**Solution** The natural frequency is given by

$$\omega_n^2 = \frac{1}{LC} = \frac{1}{(0.1 \times 10^{-3} \text{ ohm sec})(10 \times 10^{-6} \text{ sec/ohm})} = 10^9 \frac{1}{\text{sec}^2} \tag{8.67}$$

$$\omega_n = 3.16 \times 10^4 \text{ rad/sec} = 5033 \text{ Hz} \tag{8.68}$$

The circuit damping is

$$\zeta = \frac{RC}{2}\omega_n \frac{(1 \text{ ohm})(10 \times 10^{-6} \text{sec/ohm})}{2}(3.16 \times 10^4 \text{ rad/sec}) = 0.158 \tag{8.69}$$

The amplification of the output signal of a second-order system excited at resonance is proportional to

$$Q = \frac{1}{2\zeta} \tag{8.70}$$

For this example, we find that $Q = 3.16$. Design considerations for a circuit of this type are quite straightforward. To shift the natural or tuning frequency, the product $LC$ needs to be adjusted. To increase $Q$, the quantity $R\sqrt{C/L}$ has to be reduced.

### 8.4.2 Frequency Response

For the situation in which the initial conditions are zero, the Laplace transform applied to Eq. (8.49) gives

$$\left[\frac{s^2}{\omega_n^2} + \frac{2\zeta s}{\omega_n} + 1\right] X(s) = x_0 U(s) \tag{8.71}$$

Now we compute the transfer function

$$\frac{X(s)}{U(s)} = \frac{x_0}{\dfrac{s^2}{\omega_n^2} + \dfrac{2\zeta s}{\omega_n} + 1} \tag{8.72}$$

and substitute $j\omega$ for $s$, just as we did for the first-order system, to obtain

$$\frac{X(j\omega)}{U(j\omega)} = \frac{x_0}{1 - \overline{\omega}^2 + j2\zeta\overline{\omega}} \tag{8.73}$$

The magnitude of the transfer function is found to be

$$A = \frac{x_0}{\sqrt{(1 - \overline{\omega}^2)^2 + (2\zeta\overline{\omega})^2}} \tag{8.74}$$

and its phase is

$$\phi = \tan^{-1}[-2\zeta\overline{\omega}/(1 - \overline{\omega}^2)] \tag{8.75}$$

These results are the same as we found earlier using the classical solution and again point to the ease with which the frequency response is obtained directly from the transfer function.

**Example 8.5 Tuning Circuit Stability**

Find the roots of the denominator polynomial of the transfer function for the circuit of Example 8.4 using MATLAB, and evaluate the stability of the circuit.

**Solution** The denominator polynomial is given by

$$\mathrm{den}(s) = \frac{s^2}{\omega_n^2} + \frac{2\zeta s}{\omega_n} + 1 = 10^{-9}s^2 + 10^{-5}s + 1 \tag{8.76}$$

The row vector of the denominator polynomial coefficients entered in MATLAB is

$$\texttt{den} = [10^{-9}\ 10^{-5}\ 1] \tag{8.77}$$

The roots of this polynomial are found to be

$$\lambda_{1,2} = -500 \pm j31225 \tag{8.78}$$

The circuit is stable, since the real part of each root is negative.

## 8.5 HIGHER ORDER SYSTEMS

### 8.5.1 Introduction

A large number of practical engineering systems can be modeled using the first- or second-order linear systems considered in this chapter. The solutions discussed in previous sections are very useful in examining the behavior of such systems. There are also a great number of systems and devices that require third-, fourth-, and higher-order models to describe their physical behavior adequately. These systems are considered next.

### 8.5.2 Pneumatic servo system

In Chapter 7, the model of a pneumatic servo system was developed. The system contains pneumatic as well as mechanical elements and is shown in Figure 8.11.

The following equation provides us with a description of this third-order system:

$$\alpha_3 \dddot{z} + \alpha_2 \ddot{z} + \alpha_1 \dot{z} + z = \gamma u \tag{8.79}$$

The system input is $u(t)$ and the output is $z(t)$.

The constants in Eq. (8.78) have the following values for a specific case:

$$\begin{aligned} \alpha_3 &= 7.8068 \times 10^{-6}\,\text{sec}^3 \\ \alpha_2 &= 2.9107 \times 10^{-4}\,\text{sec}^2 \\ \alpha_1 &= 3.0972 \times 10^{-2}\,\text{sec} \\ \gamma &= 20.76\ \text{in/in} \end{aligned} \tag{8.80}$$

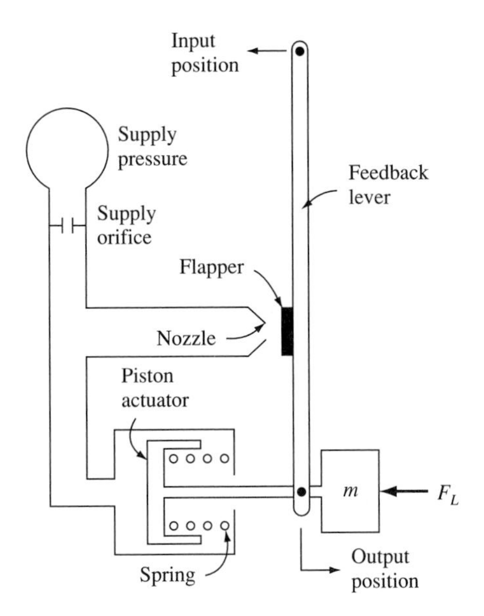


**Figure 8.11** Pneumatic servo system.

In order to determine the frequency response characteristics, we proceed directly to the transfer function representation of the system:

$$\frac{Z(s)}{U(s)} = \frac{\gamma}{[\alpha_3 s^3 + \alpha_2 s^2 + \alpha_1 s + 1]} \tag{8.81}$$

We substitute $j\omega$ for $s$ and determine the amplitude and phase of this complex number to determine the frequency response. This gives a complex number representing the transfer function of the system. The magnitude of the number represents the system response amplitude, and the phase of the number represents the phase of the output with respect to the input.

**Example 8.6 Servo Frequency Response**

Determine the frequency response of the pneumatic servo just described, using MATLAB.

**Solution** The numerator and denominator polynomial coefficients are expressed, respectively, as

$$num = [\gamma] \quad \text{and} \quad den = [\alpha_3 \ \alpha_2 \ \alpha_1 \ 1] \tag{8.82}$$

Figure 8.12 shows the results obtained by using the `bode` function to find the frequency response of this system. To accommodate the wide variation in the magnitudes of the coefficients, we first use the MATLAB output format command to establish an exponential form for the values entered; `format short e` will specify the five-digit mantissa in exponential form.

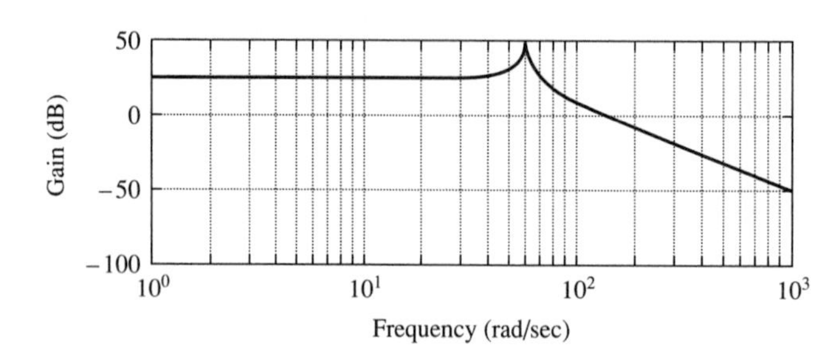


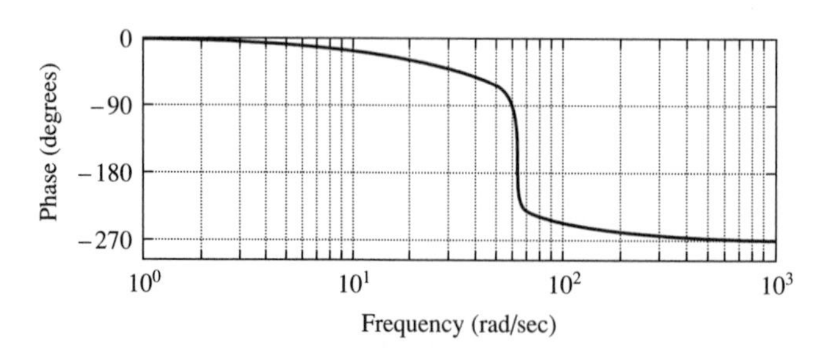


**Figure 8.12** Pneumatic servo Bode plot.

The peak in the first plot in the figure indicates a system natural frequency just above 60 rad/sec. We can check this observation by using the MATLAB function `damp` to calculate the eigenvalues, together with the corresponding natural frequencies, as well as the damping of the system.

### 8.5.3 Torsional System

A torsional transmission system uses a viscous friction device to transmit motion. The schematic of such a system is shown in Figure 8.13.

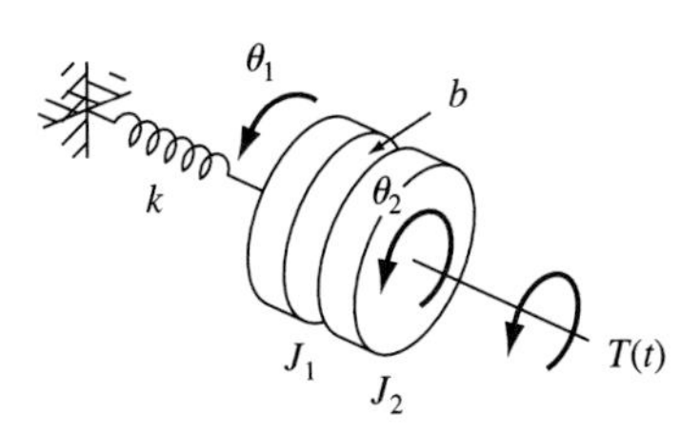


**Figure 8.13** Torsional system.

The translational counterpart of this system was considered in Section 3.3.4 of Chapter 3. Substitution of $J$ for $m$, and $\theta$ for $x$ into the governing equation gives the following fourth-order equation of motion for the torsional system:

$$[J_1J_2D^4 + (J_1 + J_2)bD^3 + J_2kD^2 + bkD]\theta_2 = [J_1D^2 + bD + k]T \tag{8.83}$$

Note that while this equation contains a fourth-order derivative, the system is really third order, since there is no constant term on the left-hand side of the equation. In the next example, we evaluate the system's inherent characteristics and determine the frequency response behavior for a specific application.

**Example 8.7 Torsional System**

Determine the frequency response characteristics of $\dot{\theta}_2$ in the torsional system just described if the system properties are $J_1 = 1.5$ kg m$^2$, $J_2 = 1.75$ kg m$^2$, $k = 1200$ N m/rad, and $b = 40$ N m/(rad/sec).

**Solution** We first rewrite the equation of motion to show $\dot{\theta}_2 = \omega_2$ and the input derivative terms:

$$[J_1J_2D^3 + (J_1 + _{J2})bD^2 + J_2kD + bk]\dot{\theta}_2 = [J_1D^2 + bD + k]T \tag{8.84}$$

Now we find the Laplace transform of the transfer function and then substitute the numerical values into the numerator and denominator polynomials of the resulting expression:

$$\frac{\Omega_2(s)}{T(s)} = \frac{[J_1s^2 + bs + k]}{[J_1J_2s^3 + (J_1 + J_2)bs^2 + J_2ks + bk]} \tag{8.85a}$$

$$\frac{\Omega_2(s)}{T(s)} = \frac{[1.5s^2 + 40s + 1200]}{[2.625s^3 + 130s^2 + 2100s + 4800]} \tag{8.85b}$$

Thus,

$$\texttt{num} = [1.5\ \ 40\ \ 1200] \tag{8.86}$$

$$\texttt{den} = [2.625\ \ 130\ \ 2100\ \ 4800] \tag{8.87}$$

To evaluate the system's inherent characteristics, we find the roots of the characteristic equation, which is expressed by the denominator polynomial. Using the MATLAB routine `roots(den)`, we obtain

$$\lambda_1 = -23.403 + j11.184 \tag{8.88}$$

$$\lambda_2 = -23.403 - j11.184 \tag{8.89}$$

$$\lambda_3 = -2.718 \tag{8.90}$$

The solution to the homogeneous equation is

$$x_h(t) = C_1 e^{\lambda_1 t} + C_2 e^{\lambda_2 t} + C_3 e^{\lambda_3 t} \tag{8.91}$$

The real part of all three roots is negative, so there are no growing exponential terms, and the system is stable. The complex roots indicate an underdamped component; thus, the complex conjugates $\lambda_1$, $\lambda_2$ represent a second-order component and can be expressed in terms of the damping ratio and natural frequency of a second-order system.

From the discussion in Section 8.3.1,

$$\lambda = a + jb = -\zeta\omega_n + j\omega_n\sqrt{1 - \zeta^2} \tag{8.92}$$

Equating real and imaginary terms on each side of the equation gives

$$a = -\zeta\omega_n \quad \text{and} \quad b = \omega_n\sqrt{1 - \zeta^2} \tag{8.93}$$

Solving these for the corresponding damping ratio and natural frequency results in

$$\zeta = \frac{-a}{\sqrt{a^2 + b^2}} \tag{8.94}$$

$$\omega_n = \sqrt{a^2 + b^2} \tag{8.95}$$

For the torsional system, we find that

$$\zeta = 0.9023 \tag{8.96}$$

$$\omega_n = 25.9 \text{ rad/sec} \tag{8.97}$$

We next determine the frequency response of this system. Using the MATLAB function `bode(num,den)` gives the plot shown in Figure 8.14.

## 8.6 ALTERNATIVE FORMS OF SYSTEM DESCRIPTION

The mathematical representation of a system model can be expressed in a number of ways, and each may have a specific benefit for a given task. We saw this in the discussion of the transfer function and its relation to frequency response calculations.

For a linear system, the transfer function in the Laplace variable $s$ can be expressed as the ratio of two polynomials:

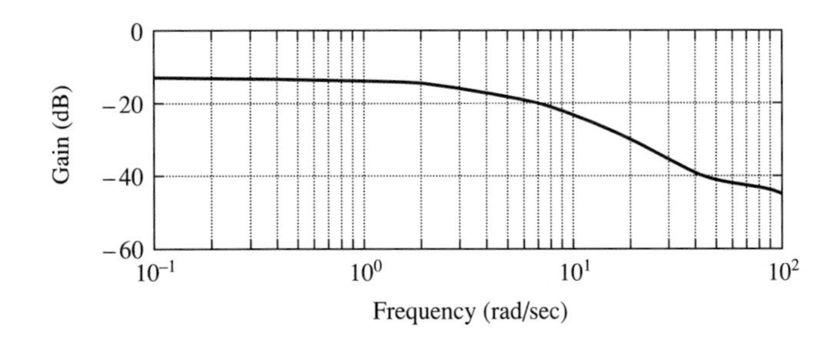


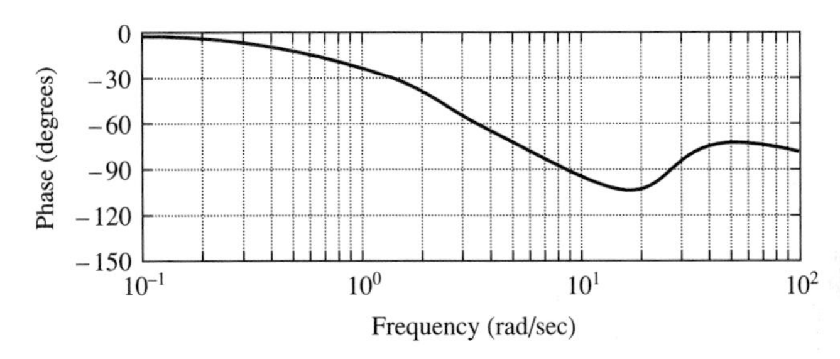


**Figure 8.14** Frequency response of torsional system.

$$\mathrm{TF}(s) = \frac{\beta_k s^k + \cdots + \beta_1 s + \beta_0}{\alpha_n s^n + \cdots + \alpha_1 s + \alpha_0} = \frac{\mathrm{num}(s)}{\mathrm{den}(s)} \tag{8.98}$$

The $\beta$'s are the constants in the numerator polynomial, and the $\alpha$'s are the constants in the denominator polynomial.

The transfer function could also be written in terms of the roots of these two polynomials. The **zeros** $-z_i$ are the roots of the numerator polynomial. Then the numerator can be expressed as a constant $K_1$ times the product of the polynomial factors:

$$\beta_k s^k + \cdots + \beta_1 s + \beta_0 = K_1(s + z_1)(s + z_2)\cdots(s + z_k) \tag{8.99}$$

Likewise, the **poles** $-p_i$ are the roots of the denominator polynomial, and $K_2$ is a constant. We may write

$$\alpha_n s^n + \cdots + \alpha_1 s + \alpha_0 = K_2(s + p_1)(s + p_2)\cdots(s + p_n) \tag{8.100}$$

The transfer function may then be expressed in terms of the zeros and poles as

$$\mathrm{TF}(s) = K\frac{(s + z_1)(s + z_2)\cdots(s + z_k)}{(s + p_1)(s + p_2)\cdots(s + p_n)} \tag{8.101}$$

where $K = K_1/K_2$ is a constant.

**Example 8.8 Pole-Zero Transfer Function Form**

Convert the transfer function representation of the torsional system in Example 8.7 into the pole-zero form.

**Solution** The roots of the numerator and denominator polynomials are found to be

$$-z_{1,2} = -13.333 \pm j24.944 \tag{8.102}$$

$$-p_{1,2} = -23.403 \pm j11.184 \tag{8.103}$$

$$-p_3 = -2.718 \tag{8.104}$$

The zeros (o) and the poles (x) are plotted on the complex plane in Figure 8.15 . The MATLAB function **rlocus**(num,den) can be used to generate this plot automatically.

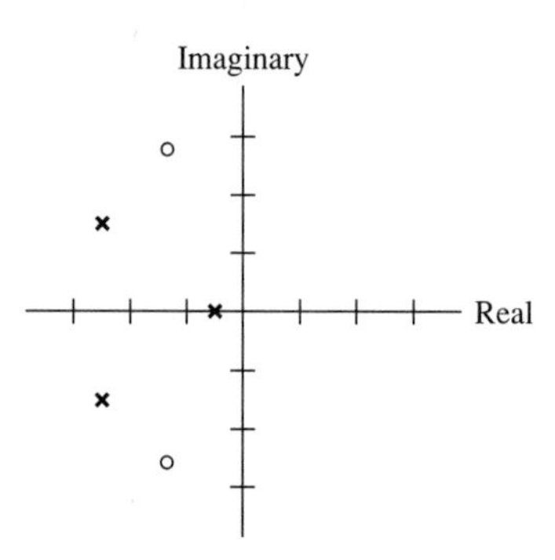


**Figure 8.15** Torsional system poles and zeros.

The corresponding pole-zero form of the transfer function is

$$\mathrm{TF}(s) = \frac{(s + 13.333 + j24.944)(s + 13.333 - j24.944)}{(s + 23.403 + j11.184)(s + 23.403 - j11.184)(s + 2.718)} \tag{8.105}$$

At times, it is convenient to represent the system in terms of its state-space form, in which an $n$th order system is represented as an equivalent system of $n$ first-order differential equations. Appendix G discusses the conversion from one form to another. The standard state-space form for linear systems is

$$\dot{\mathbf{x}} = \mathbf{Ax} + \mathbf{Bu} \tag{8.106}$$

where **A** and **B** are matrices of constants and **x** and **u** are vectors. Examples of the state-space representation for specific systems can be found in Appendix G and in previous chapters. (See Example 3.7, for instance.) Also see Appendix H.

## 8.7 SUMMARY

In this chapter, we have discussed inherent characteristics for linear, constant-coefficient systems and calculated the frequency response characteristics of those systems. Both classical closed-form methods and numerical methods using MATLAB procedures were introduced. These concepts find wide application in engineering, and a vast number of dynamic system applications owe their success to design methods based on those principles.

However, if linear, constant-coefficient equations do not provide a satisfactory system model, the methods discussed here are not applicable. It then may be necessary to perform a complete time response simulation of the system's behavior, including any nonlinearities. Time response and digital simulation are discussed in the next chapter.

## REFERENCES

8.1 Bode, Henrik Wade. *Network Analysis and Feedback Amplifier Design*. Van Nostrand, New York, 1945.

8.2 Cannon, Robert H., Jr. *Dynamics of Physical Systems*. McGraw-Hill, Inc., New York, 1967.

8.3 Goode, Stephen W. *An Introduction to Differential Equations and Linear Algebra*. Prentice Hall, Englewood Cliffs, NJ, 1991.

8.4 Ogata, Katsuhiko. *Solving Control Engineering Problems with MATLAB*. Prentice Hall, Inc., Englewood Cliffs, NJ, 1994.

8.5 Ogata, Katsuhiko. *System Dynamics,* 2nd ed., Prentice Hall, Inc., Englewood Cliffs, NJ, 1992.

8.6 Palm, William J., III. *Modeling, Analysis and Control of Dynamic Systems*. John Wiley & Sons, Inc., New York, 1983.

## NOMENCLATURE

$a$ constant
$A$ constant
$\mathbf{A}$ matrix
$b$ viscous damper constant
$B$ amplitude
$\mathbf{B}$ matrix
$C$ capacitance, constant
dB decibel
$e$ voltage
$E$ voltage amplitude
$f$ input function, farad
$F$ force
$F_0$ force amplitude
$G$ gain
$h$ henry
$i$ current
$j$ complex number $\sqrt{-1}$
$J$ mass moment of inertia
$k$ spring constant
$L$ inductance
$\mathcal{L}$ Laplace transform operator
$m$ mass
MAG magnitude

$Q$ amplification factor
$R$ resistance
$t$ time
$T$ torque
$TF$ transfer function
$s$ Laplace variable
$u$ input function
$U(s)$ Laplace transform of input function
$v$ velocity
$V(s)$ Laplace transform of velocity
$x$ displacement
$X(s)$ Laplace transform of displacement
$x_h$ homogeneous solution
$x_p$ particular solution
$y$ input function
$Y_0$ amplitude
$z$ output variable
$\alpha$ constant
$\beta$ constant
$\gamma$ constant
$\omega$ excitation frequency
$\omega_n$ natural frequency
$\overline{\omega}$ frequency ratio $\omega/\omega_n$
$\lambda$ root of characteristic equation
$\zeta$ damping ratio
$\theta$ angle of twist
$\Theta(s)$ Laplace transform of angle
$\tau$ time constant
$\phi$ phase angle
$\Omega$ ohm, Laplace transform of angular velocity

## PROBLEMS

Where appropriate, use MATLAB or other suitable computational tools to solve the following problems. Remember to correctly convert to and from frequencies in Hz and in rad/sec.

**8.1** *Find* the time constant for an *RC* circuit (Figure 8.1) that has parameters $R = 8200\ \Omega$ and $C = 2.2\ \mu$f. *Sketch* the amplitude frequency response and *determine* the corner frequency. Approximately how long does it take the transient part of the solution to this system to die out?

**8.2** The first-order mechanical system of Section 8.3.6 has a mass of $m = 4$ kg and a damping constant of $b = 3.2$ N sec/m. At *what* frequency does the output response amplitude become essentially constant? That is, at *what* frequency is the output amplitude within 2 percent of the value that would occur if the input frequency were infinite?

**8.3** A harmonic signal of amplitude 1 and frequency 70 Hz is the input to a linear first-order system whose time constant is 0.5 second. *What* is the amplitude of the output? *What* is the phase of the output with respect to the input? *Does* the output lag behind the input or lead it?

**8.4** *Answer* the same questions as in the previous problem, except that now the system is a second-order linear system with a natural frequency of 21 rad/sec and a damping ratio of 0.25.

**8.5** The spring-mass-damper system of Example 8.3 has an input excitation with a frequency that is equal to its natural frequency (resulting in resonance). *Redesign* the system so that resonance is avoided and so that the output displacement is no greater than 1.5 times the value that would occur if the load were applied statically.

The new natural frequency can be above or below the frequency of the input signal. If it is below, the system will pass through resonance when coming up to operating conditions.

**8.6** A linear second-order system has a natural frequency of 110 Hz and a damping ratio of 0.37. *What* is the acceptable range of input signal frequencies if the amplitude of the output is to be practically constant?

**8.7** It is proposed to attach an additional spring-mass-damper system to a primary spring-mass-damper system as shown in Figure P8.7. *Find* the steady-state amplitude of the displacement response of the primary mass, and plot it with respect to the input frequency for the cases with and without the attached system. Such attached systems can be used to absorb unwanted vibrations. *Comment* on the effectiveness of reducing vibration for this system. The parameters of the problem are:

$w_1 = 200$ lbf $\quad w_2 = 50$ lbf

$k_1 = 18{,}500$ lbf/in $\quad k_2 = 4600$ lbf/in

$b_1 = 20$ lbf sec/in $\quad b_2 = 5$ lbf sec/in

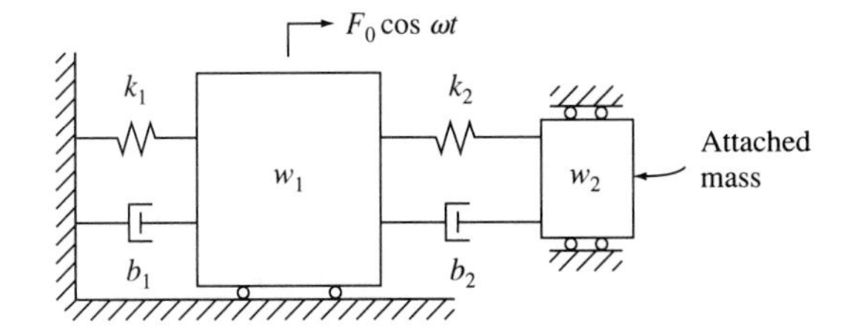


**Figure P8.7** Vibration absorber system.

**8.8** *Derive* the transfer function relating the output $e_1$ to the input $e_0$ for the circuit shown in Figure P8.8.

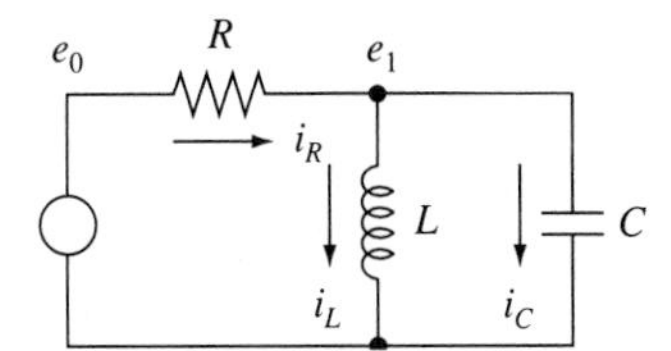


**Figure P8.8** *RLC* circuit.

**8.9** A spring-mass-damper system has values of $m = 4$ kg, $k = 320$ N/m, and $b = 6.2$ N sec/m. An input force of $F = 5$ N is applied at an input frequency of 2.5 Hz. *Find* the undamped natural frequency, the damping ratio, and the damped natural frequency of the system. *What* is the steady-state displacement amplitude of the mass?

**8.10** *Determine* the transfer function relating the output voltage $e_2$ to the input voltage $e_0$ for the *RLC* circuit of Figure 4.11. *Develop* the corresponding Bode plot using data from Example 4.3. What is the natural frequency of the system? *Redesign* the circuit so as to increase its natural frequency by 30 percent, but keep the damping ratio at 0.707.

**8.11** *Find* the roots of the characteristic equation of the following systems. *What* are the time constants and/or natural frequencies? *What* are the damping ratios? *Comment* on the stability of each. *Solve* by hand and *check* by digital computation.

**a.** $2\ddot{x} + 8\dot{x} + 32x = 15u(t)$

**b.** $0.01\ddot{x} + 0.02\dot{x} + 1 = u(t)$

**c.** $\ddot{x} + 20\dot{x} + 25 = 134u(t)$

**d.** $10\ddot{x} + 600\dot{x} + 1000 = u(t)$

**e.** $100\ddot{x} + 400\dot{x} + 1 = u(t)$

**f.** $7\dddot{z} + 5\ddot{z} + 2\dot{z} + 3z = 2.5f(t)$

**g.** $5\ddot{z} + 2\dot{z} + 4z - 3\dot{v} - 4v = 8f(t)$

$\ddot{v} + 3\dot{v} + 12v - 3z = 0$

**h.** $\dddot{z} + 2\ddot{z} + 4\dot{z} + 5z = 4f(t)$

**8.12** *Find* the transfer function for each of the systems of the previous problem, relating the output $x$ or $z$ to the input $u$ or $f$.

**8.13** *Find* the set of state-space equations for each of the systems of Problem 8.11.

**8.14** A mechanical system with an input $f(t)$ and three displacement variables has the following equations:

$$\ddot{z}_1 + z_1 = f(t) + 2z_2 + 2z_3$$

$$\ddot{z}_2 + 2z_2 = 4z_1 - 2z_3$$

$$\dot{z}_3 + 5z_3 = 5z_1 + \dot{z}_2$$

*Find*:

**a.** The transfer function for $f(t)$ as the input and $z_1(t)$ as the output.

**b.** The poles and zeros, and write the transfer function in pole-zero form.

**c.** The time constants, natural frequencies, and damping ratios.

**d.** The state-space form of the system equations and the eigenvalues of the system.

**8.15** The torsional system shown in Figure P8.15 represents an electric motor, shafting, and an inertial load. *Find* the frequency response of the angular velocity $\dot{\theta}_2$ of the load in response to a harmonic torque input $T_1$ at the electric motor. *Present* a Bode plot for the system. *Find* the roots of the characteristic equation. The parameters of the problem are:

$J_1 = 3$ kg m$^2$ $\quad J_2 = 4.25$ kg m$^2$

$b_1 = 26$ N m/(rad/sec) $\quad b_2 = 30$ N m/(rad/sec) $\quad k = 1500$ N m/rad

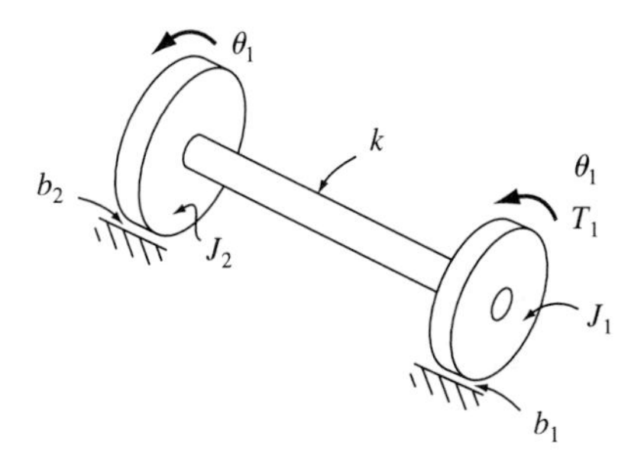


**Figure P8.15** Torsional system.

**8.16** *Determine* the transfer function coefficients for the circuit of Problem 4.14 if:

$$R_1 = 10\ \Omega \qquad R_2 = 100\ \Omega$$

$$C_1 = 1\ \mu\text{f} \qquad C_2 = 10\ \mu\text{f}$$

$$L = 5\ \text{mh}$$

*Find* the eigenvalues, and check the stability of the system. If the system is stable, *determine* its Bode plot.

**8.17** *Determine* the transfer function coefficients for the circuit of Problem 4.16 if:

$$R_1 = 5\ \Omega \qquad R_2 = 10\ \Omega$$

$$C_1 = 10\ \mu\text{f} \qquad C_2 = 22\ \mu\text{f}$$

$$L = 10\ \text{mh}$$

*Convert* the system representation from transfer function to state-space form. *Find* the eigenvalues, and *check* the stability of the system. If the system is stable, *determine* its Bode plot.

**8.18** *Calculate* the eigenvalues, time constants and/or natural frequencies, as well as the damping ratios, for each of the systems whose transfer functions are as follows:

**a.** $\text{TF}(s) = \dfrac{2(s^2 + 1)}{4s^2 + 9s + 6}$

**b.** $\text{TF}(s) = \dfrac{s^2 + 2s + 2.5}{2s^3 + 2.5s^2 + 1s + 2.75}$

**c.** $\text{TF}(s) = \dfrac{s^2 + 2s + 3}{s^3 + 3s^2 + s + 2}$

**d.** $\text{TF}(s) = \dfrac{s + 2.5}{s^3 + 1.75s^2 + 1.25s + 3}$

**e.** $\text{TF}(s) = \dfrac{2s^2 + 1}{s^3 + 2s^2 + 5s}$

**f.** $\text{TF}(s) = \dfrac{2s + 1}{s^4 + 2s^3 + 5s^2 + 8s + 4}$

**g.** $\text{TF}(s) = \dfrac{s^2 + 2s + 1}{2s^4 + 3s^3 + 9s^2 + 7s + 5}$

**h.** $\text{TF}(s) = \dfrac{2s^2 + 1}{s^4 + 2s^3 + 7s^2 + 8s + 4}$

**8.19** *Express* each of the systems of the previous problem in pole-zero form.

**8.20** *Convert* each of the systems of Problem 8.18 to state-space form using

**a.** Hand calculations

**b.** The MATLAB function `tf2ss`.

**8.21** *Determine* the transfer function for each of the systems whose governing equations are as follows:

**a.** $\dot{z} + 8z = G$ ($G$ is a constant.)

**b.** $2\dot{z} + 3z = Gt$

**c.** $0.25\dot{z} + 7.5z = G \sin \omega t$

**d.** $\ddot{z} + 5\dot{z} + 4z = \dot{y}$

**e.** $2.5\ddot{z} + 2\dot{z} + 3.75z = G$

**f.** $\ddot{z} + 0.25\dot{z} + 0.5z = G \cos \omega t$

**8.22** *Convert* the equations of Problem 8.21 to state-space form.

**8.23** *Determine* the frequency response characteristics of the output voltage for the electrical circuit shown in Figure P4.25.

**8.24** An *RC* circuit such as the one shown in Figure 8.1 is to have a time constant of 0.5 $\mu$sec and a corner frequency of 10 kHz. Design a 5 volt system that meets this requirement.

**8.25** A second-order system with a natural frequency of 120 Hz is subject to constant amplitude harmonic input and must have a dynamic response that is no more than 35% greater than the value that would occur if the input were static. *What* is the smallest damping ratio that can be employed when building this system?

**8.26** An automobile is traveling over a wavy stretch of highway. The input displacement to the tires is assumed to be a harmonic function. As the car travels at different speeds, the frequency of the input to the system changes. Using the model developed in Example 3.7 and the following numerical data, *find* the input frequency that causes the largest steady-state response of the auto body:

$$k_1 = 1000 \text{ lbf/in} \qquad w_1 = 20 \text{ lbf}$$

$$k_2 = 100 \text{ lbf/in} \qquad w_2 = 150 \text{ lbf} \quad b = 5 \text{ lbf sec/in}$$

# CHAPTER 9

# Time Response and Digital Simulation

## 9.1 INTRODUCTION

In this chapter, we discuss the determination of the time response behavior of dynamic systems by both classical analytical solution methods and digital simulation. We seek to determine the response to inputs that are not steady with time and to plot the responses of system variables as functions of time so that we may better understand the behavior of a system through visualizing its response. When applicable, analytical methods give closed-form solutions and provide a convenient means of evaluating the effect of variations in system parameters on the response of the system. Thus these methods are valuable tools for system design considerations. Closed-form solutions for the frequently used linear first- and second-order systems are discussed in Appendix E.

Digital simulation methods, on the other hand, provide convenient procedures for quickly examining the response of linear or nonlinear systems to proposed design modifications and are an integral part of many widely used, commercially distributed computer codes. These methods are capable of rapidly determining and plotting the numerical representation of the system response to a particular set of parameters. Proposed changes to the system also can be evaluated rapidly.

## 9.2 ANALYTICAL METHODS

### 9.2.1 First-Order Systems

Consider the $RC$ circuit of Section 4.3.3, shown in Figure 9.1. The governing equation for the voltage $e_1(t)$ is

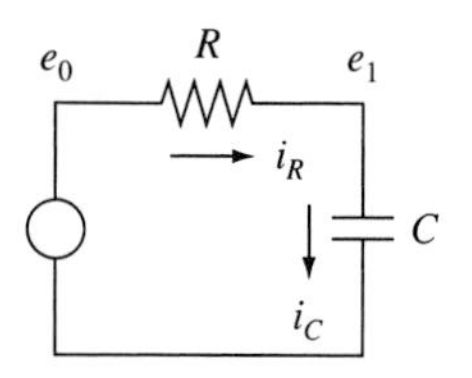


**Figure 9.1** *RC* circuit.

$$RC\dot{e}_1 + e_1 = e_0(t) \tag{9.1}$$

A good measure of the system behavior is the way in which the circuit responds to a step input—a sudden excitation that commands the system to move to a new state. (See Figure 9.2.)

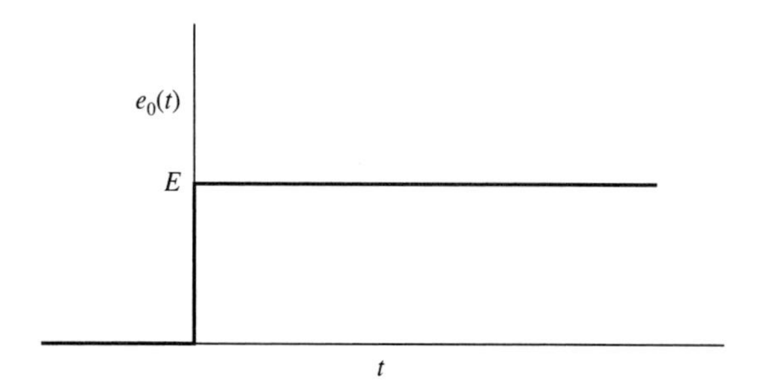


**Figure 9.2** Step input.

We consider the case in which the capacitor is initially uncharged and the voltage $e_0(t)$ is applied or switched on suddenly. The input voltage changes from zero to a constant value $E$. The analytical solution to this input can be found by combining the homogeneous solution with the particular solution corresponding to the step input. The solution, developed in Appendix E using the standard form of the first-order equation, is

$$x(t) = x_0 e^{-t/\tau} + Gu_0(1 - e^{-t/\tau}) \tag{9.2}$$

Here $x_0$ is the initial condition on the response variable, and $Gu_0$ is the amplitude of the step. If the initial voltage is zero, the circuit's response is

$$e_1(t) = E(1 - e^{-t/\tau}) \tag{9.3}$$

**Example 9.1** ***RC*** **Circuit Response**

Determine the response of an *RC* circuit whose parameters are $R = 5000\ \Omega$ and $C = 1\ \mu f$ if the circuit is subjected to a step input of $E = 5$ volts. How long does it take $e_1(t)$ to reach a steady value?

**Solution** The time constant for the system is the product *RC*, as we have noted before:

$$\tau = RC = 5000 \text{ ohm } (1 \times 10^{-6}) \frac{\text{s}}{\text{ohm}} = 0.005 \text{ s} = 5 \text{ ms} \tag{9.4}$$

The time response plot from Appendix E, Figure E.3, is reproduced in Figure 9.3, and Table 9.1 shows the output response as a function of multiples of the system time constant.

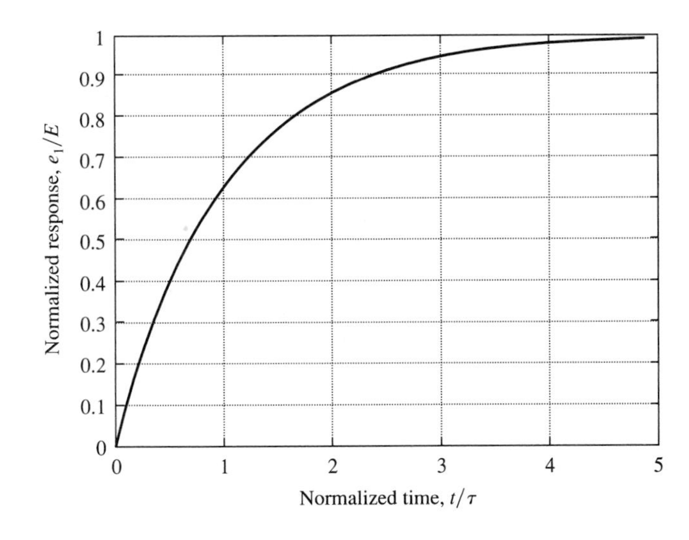


**Figure 9.3** RC circuit response to a step input.

From these closed-form results, we see that the response is 98.2 percent of its ultimate value at an elapsed time equal to four time constants, or $t = (4)(5 \text{ ms}) = 20$ ms. If we require a circuit with a faster response, we need to reduce the time constant. For example, if we want a 10 ms response time, the product $RC$ must be halved by making suitable changes in the circuit parameters.

**TABLE 9.1** RESPONSE OF THE RC CIRCUIT TO A STEP INPUT

| $t$ | $e_1(t)/E$ |
|---|---|
| 0 | 0.000 |
| $1\tau$ | 0.632 |
| $2\tau$ | 0.865 |
| $3\tau$ | 0.950 |
| $4\tau$ | 0.982 |
| $5\tau$ | 0.993 |
| $6\tau$ | 0.998 |

### 9.2.2 Second-Order Systems

The time response of second-order systems to step inputs is an important measure of their suitability to a given purpose. Analytical solutions for the step response of a second-order system are derived in Appendix E, Section E.3.2.

**Example 9.2 Second-Order Mechanical System**

A component of a computer disk drive is modeled as a spring-mass-damper system as shown in Figure 9.4. The position of the end of the spring, $x_s(t)$, is given a step input $X_s$, and we want to know the response characteristics of the mass, which must be at the new position within 1 ms.

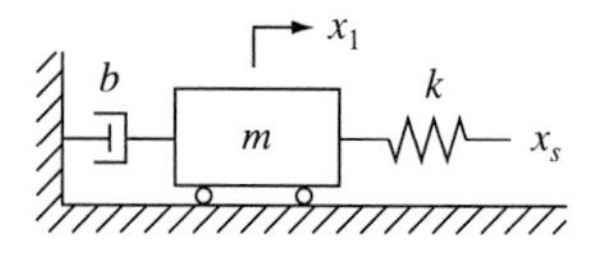


**Figure 9.4** Second-order mechanical system.

**Solution** The equation of motion for this system is

$$m\ddot{x}_1 + b\dot{x}_1 + kx_1 = kx_s(t) \tag{9.5}$$

We know that the damping is small, and suspect that $\zeta < 1$. The analytical solution to Eq. (9.5) is derived in Appendix E. The solution corresponding to an underdamped system with zero initial conditions is

$$x_1(t) = Gu_0\left\{1 - e^{-\zeta\omega_n t}\left[\cos \omega_d t + \frac{\zeta}{\sqrt{1 - \zeta^2}} \sin \omega_d t\right]\right\} \tag{9.6}$$

where $Gu_0 = X_s$. A plot of the response is reproduced in Figure 9.5.

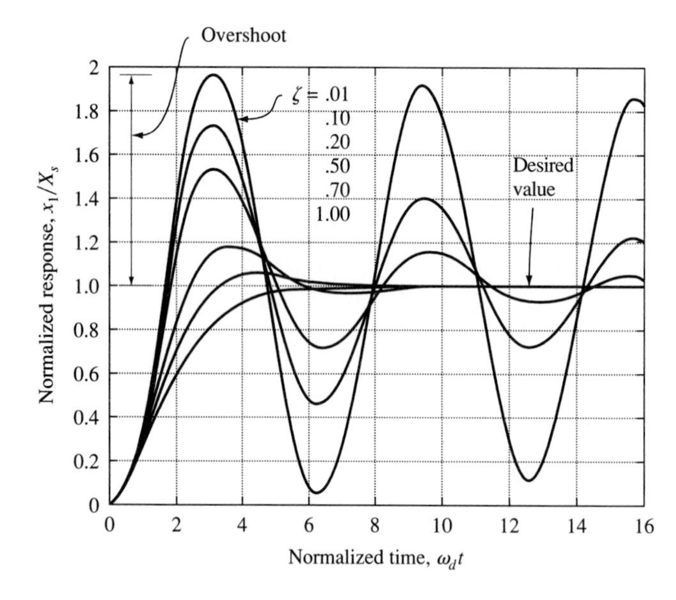


**Figure 9.5** Response of an underdamped second-order system to a step input.

Damping is a critical parameter in determining whether the response is oscillatory. From the discussion in Section 8.4 and the results of Appendix E, we note that

$$\omega_n = \sqrt{k/m} \qquad \omega_d = \omega_n\sqrt{1 - \zeta^2} \tag{9.7}$$

The damping ratio $\zeta$ is

$$\zeta = \frac{b}{b_c} \quad \text{where} \quad b_c = 2\sqrt{km} \tag{9.8}$$

is the critical damping factor. If the device is underdamped, the response will overshoot the desired value, as shown in Figure 9.5. For the mechanism in question, it is required that the position be achieved within 1 ms and that overshoot not exceed 20 percent. The first requirement places a condition on the system settling time. From the response plot, we note that the peak value of position occurs at about $\omega_d t = 3$ rad. Thus,

$$\omega_d = \frac{3 \text{ rad}}{0.001 \text{ s}} = 3000 \text{ rad/s} \approx 477 \text{ Hz} \tag{9.9}$$

The system parameters would have to be selected so as to achieve this frequency value.

**Overshoot** in a system is defined as the amount the transient response exceeds the steady-state value produced by a step input. The amount of overshoot is directly related to the system damping. The analytic relationship between the overshoot and the damping can be found by determining the first point in time at which the velocity of the system becomes zero in response to a step input. Substituting this value of time into the position response of the system yields the peak value of the response. The resulting expression is

$$\text{OS} = e^{-\frac{\zeta\pi}{1-\zeta^2}} \tag{9.10}$$

For $\zeta = 0.20$ (20 percent damping), this equation gives 0.52, which corresponds to 52 percent overshoot. To achieve the stated goal, the damping will have to be larger. Solving Eq. (9.10) for the damping corresponding to a 20 percent overshoot gives $\zeta = 0.42$, the required system value to meet our objectives.

The important thing about Example 9.2 and the previous one is that significant system design decisions can be made using closed-form solutions. No specific system parameters need be selected in order to determine behavior trends, since they are known in terms of the time constant or natural frequency and damping ratio parameters. This illustrates the value of using analytical solutions when they are available. Closed-form solutions for first- and second-order systems are summarized in Tables E.2 and E.3, respectively, in Appendix E for easy reference.

### 9.2.3 Higher Order Systems

In treating linear systems of higher order than second, we can put the system into state-space form:

$$\dot{\mathbf{x}} = \mathbf{A}\mathbf{x} + \mathbf{B}\mathbf{u}(t) \tag{9.11}$$

The homogeneous solution can be written in terms of the **state transition matrix**: (Ref. 9.11)

$$\mathbf{x}_h(t) = \mathbf{S}(t)\mathbf{x}(0) \tag{9.12a}$$

Here $\mathbf{x}(0)$ is the initial-condition vector, and the **matrix exponential** is defined by the equation

$$\mathbf{S}(t) = e^{\mathbf{A}t} \tag{9.12b}$$

The particular solution depends upon the form of the excitation and can be written for the general case using a convolution integral. The total solution is the sum of the homogeneous solution and the particular solution and is

$$\mathbf{x}(t) = \mathbf{S}(t)\mathbf{x}(0) + \int_0^t \mathbf{S}(t-\eta)\mathbf{B}\mathbf{u}(\eta)d\eta \tag{9.12c}$$

where $\eta$ is a dummy variable used for integration.

The exponential operations indicated in Eq. (9.12b) may not be practical in closed form for systems of any significant order, but when the input is of simple form (impulse, step, etc.), the procedure can be implemented numerically, so that the response of higher order systems is easily evaluated.

**Example 9.3 Pneumatic Servo System**

Find the response of the third-order pneumatic servo system discussed in Chapters 7 and 8 to a step input.

**Solution** The system transfer function and its coefficients are given in Section 10.5 and Section 8.5.2 and are as follows:

$$\frac{Z(s)}{U(s)} = \frac{\gamma}{[\alpha_3 s^3 + \alpha_2 s^2 + \alpha_1 s + 1]} \tag{9.13}$$

$$\begin{aligned} \alpha_3 &= 7.8068 \times 10^{-6} \quad \text{sec}^3 \\ \alpha_2 &= 2.9107 \times 10^{-4} \quad \text{sec}^2 \\ \alpha_1 &= 3.0972 \times 10^{-2} \quad \text{sec} \\ \gamma &= 20.76 \end{aligned} \tag{9.14}$$

The numerator and denominator polynomial coefficients for this problem may be expressed as the elements of one-dimensional arrays as follows:

$$\texttt{num} = [\gamma] \qquad \texttt{den} = [\alpha_3 \ \alpha_2 \ \alpha_1 \ 1] \tag{9.15}$$

Here, `num` and `den` are row arrays containing the polynomial coefficients arranged in descending order of powers of $s$.

The MATLAB function `step` employs the matrix exponential method of Section 9.2.3 to generate a numerical solution for the response of a system to a step function of unit amplitude. We use the command `step(num,den)` to carry out the calculation of the response of the pneumatic servo to a step input; the results are shown in Figure 9.6. (Invoking `step` without any left-hand arguments produces a plot of the response directly on the computer screen. Note that when using this function, it is not necessary to convert to state-space form; an option is provided to enter the polynomial coefficients directly. See the MATLAB user's guide for further information.)

Notice in Figure 9.6 that about 9.7 cycles of motion occur in the first second. This corresponds to a frequency of

$$\omega_n = (9.7 \text{ cycles/s})(2\pi \text{ rad/cycle}) = 60.9 \text{ rad/s} \tag{9.16}$$

which is the natural frequency identified in the frequency response plot of Figure 8.12. Notice also that the step response does not settle very rapidly, indicating a system component with light damping. We also observe that the static amplitude value ultimately achieved is equal to the static gain, 20.76, as would be expected.

The methods presented so far in this chapter apply to linear systems that can be easily described in analytic form. In cases where analytical methods are not appropriate, or where it is simply more convenient to obtain a numerical solution, the digital simulation approach, discussed in the next section, is used.

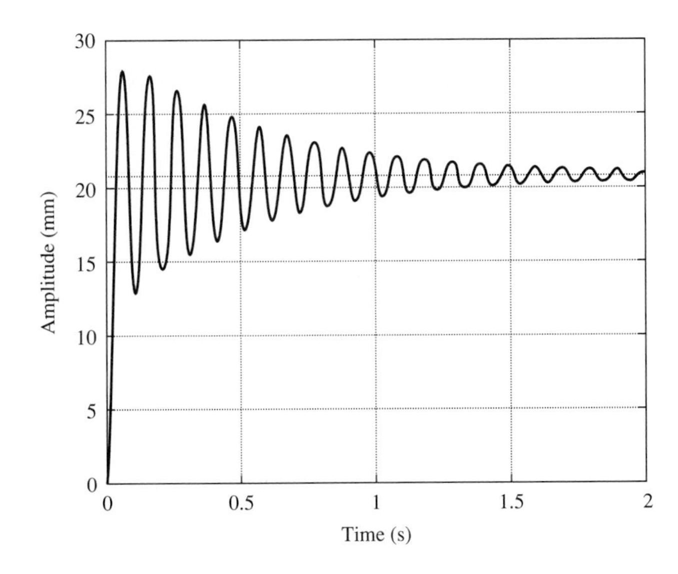


**Figure 9.6** Step response of a pneumatic servo.

## 9.3 DIGITAL SIMULATION

### 9.3.1 Introduction

While many low-order linear systems can be solved using standard differential equation solution procedures as discussed in Appendix E and Appendix F, numerical methods are advantageous when a large number of differential equations is necessary to describe the system or when the physical situation must be represented by nonlinear differential equations. In these cases, digital simulation is an indispensable tool. It is assumed that higher order systems can be transformed into an equivalent set of first-order, state-space equations, and only first-order equations are discussed here.

Figure 9.7 shows the solution to a differential equation (solid line) and an approximate solution to the same equation obtained by a numerical method (dots). While a closed-form solution consists of an analytic function relating the dependent and independent variables ($x$ and $t$, respectively), a numerical solution consists of a series of points $x_j$ obtained at a discrete set of points $t_j$ of time. The distance between the time points (the time step) is denoted by $h$. The figure makes clear the strong dependence of the quality of the numerical solution on the size of $h$. That is, toward the end of the record of time given, the analytic solution begins an oscillation that the discrete time point spacing is unable to represent. A shorter time step is obviously needed in this region of the solution.

### 9.3.2 Euler's Method

Suppose a single first-order differential equation is given in which $f(x,t)$ may be a linear or a nonlinear function, depending on the problem at hand:

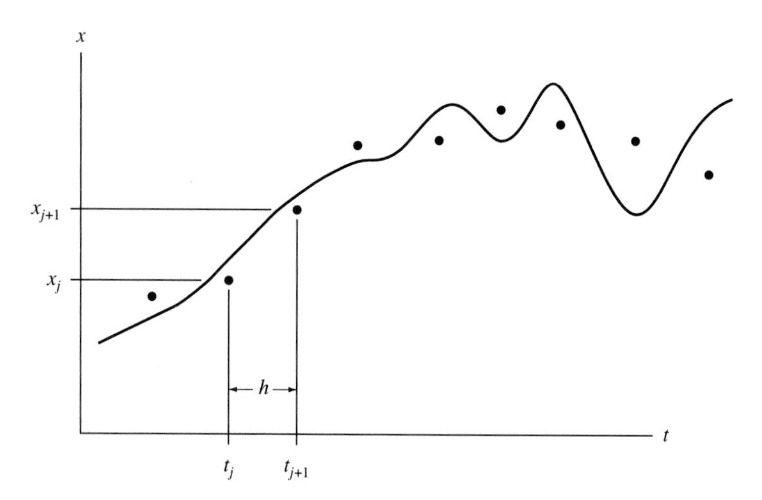


**Figure 9.7** Numerical integration of a dynamic system.

$$\frac{dx}{dt} = f(x, t) \quad \text{with initial condition } x(t_0) = x_0 \tag{9.17}$$

This is an initial-value problem, since $x(t)$ is given at some initial time $t_0$, and we desire the solution values at succeeding times $t$. Equation (9.17) may be written as

$$\int_{x_j}^{x_{j+1}} dx = \int_{t_j}^{t_{j+1}} f(x,t)dt \tag{9.18}$$

Suppose the time step $h = t_{j+1} - t_j$ is small enough so that the slope $\dot{x} = f(x,t)$ may be taken to be the constant $f(x_j, t_j)$ over the interval $h$. Then

$$x_{j+1} - x_j = f(x_j,t_j)\int_{t_j}^{t_{j+1}} dt = hf(x_j,t_j) = hf_j \tag{9.19}$$

Thus, the new value of $x$, $(x_{j+1})$ may be computed from previous values by the relation

$$x_{j+1} = x_j + hf_j \tag{9.20}$$

This simple technique of integration is called **Euler's method** (after Leonard Euler, Swiss mathematician, 1707–83), or the slope projection method, and is shown in Figure 9.8.

Geometrically, the Euler process assumes that the slope $dx/dt = f(x,t)$ is constant over the interval $h$. This assumption has obvious limitations. The error involved in using Euler's method can be examined by considering a Taylor series expansion of $x(t)$ (Brook Taylor, English mathematician, 1685–1731). Recalling the Taylor series expansion discussed in calculus, we write

$$x(t_{j+1}) = x(t_j) + h\left.\frac{dx}{dt}\right|_j + \left.\frac{h^2}{2!}\frac{d^2x}{dt^2}\right|_j + \dots \tag{9.21}$$

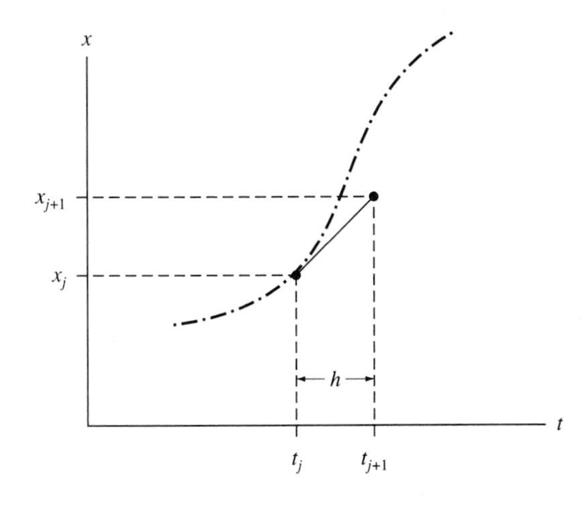


**Figure 9.8** Euler integration technique.

Replacing $dx/dt$ with $f(x,t)$ we obtain

$$x_{j+1} = x_j + hf_j + \frac{h^2}{2!}\frac{d^2x}{dt^2}\bigg|_j + \ldots \tag{9.22}$$

In Euler's method, we neglect terms beyond $hf_j$; hence, the error caused at each step is proportional to $h^2$, and is called the truncation error.

**Example 9.4 Euler's Method**

Use Euler's method to solve the following two first-order equations. The first displays exponential behavior; the second has a harmonic solution.

$$\dot{x} = -x, \text{ with inital condition } x(0) = 1, \quad 0 \le t \le 5 \text{ s}, \tag{9.23}$$

and

$$\dot{y} = -3y + 10\sin t, \text{ with initial condition } y(0) = -1, \quad 0 \le t \le 12 \text{ s} \tag{9.24}$$

**Solution** For Eq. (9.23) we select a step size of $h = 0.5$ s then

$$t = 0.5: \qquad x_1 = x_0 + 0.5(-1) \;\; = 1 - 0.5 = 0.5 \tag{9.25}$$

$$t = 1.0: \qquad x_2 = x_1 + 0.5(-0.5) = 0.5 - 0.25 = 0.25 \tag{9.26}$$

$$\vdots \qquad\qquad \vdots$$

This recursive process is continued until we reach the desired final time point, $t = 5$.

The exact solution to Eq. (9.23) can be found using the methods of Appendix E. We find:

$$x_e(t) = e^{-t} \tag{9.27}$$

Notice that the exact solution approaches zero with increasing time.

Table 9.2 gives a comparison of the Euler solution with the exact solution to this problem for step sizes, $h = 0.5$, $h = 0.1$, and $h = 0.0125$. See also Figure 9.9(a). The error in the calculated result at some specific time $t$ is called the *global error*. Consider the global error (at $t = 2$ s for example) as a function of the step size used in the calculation.

**TABLE 9.2** SOLUTIONS TO $\dot{x} = -x, x(0) = 1$

| $h = 0.5$ | | | |
|---|---|---|---|
| $t$ | $x$ (Euler) | $x$ (exact) | error (%) |
| 0.5 | 0.500000 | 0.606531 | 17.56 |
| 1.0 | 0.250000 | 0.367879 | 32.04 |
| 1.5 | 0.125000 | 0.223130 | 43.98 |
| 2.0 | 0.062500 | 0.135335 | 53.82 |
| 2.5 | 0.031250 | 0.082085 | 61.93 |
| 3.0 | 0.015625 | 0.049787 | 68.62 |
| 3.5 | 0.007813 | 0.030197 | 74.13 |
| 4.0 | 0.003906 | 0.018316 | 78.67 |
| 4.5 | 0.001953 | 0.011109 | 82.42 |
| 5.0 | 0.000977 | 0.006738 | 85.51 |
| $h = 0.1$ | | | |
| $t$ | $x$ (Euler) | $x$ (exact) | error (%) |
| 0.5 | 0.590490 | 0.606531 | 2.64 |
| 1.0 | 0.348678 | 0.367879 | 5.22 |
| 1.5 | 0.205891 | 0.223130 | 7.73 |
| 2.0 | 0.121577 | 0.135335 | 10.17 |
| 2.5 | 0.071790 | 0.082085 | 12.54 |
| 3.0 | 0.042391 | 0.049787 | 14.86 |
| 3.5 | 0.025032 | 0.030197 | 17.11 |
| 4.0 | 0.014781 | 0.018316 | 19.30 |
| 4.5 | 0.008728 | 0.011109 | 21.43 |
| 5.0 | 0.005154 | 0.006738 | 23.51 |
| $h = 0.0125$ | | | |
| $t$ | $x$ (Euler) | $x$ (exact) | error (%) |
| 0.5 | 0.604622 | 0.606531 | 0.31 |
| 1.0 | 0.365568 | 0.367880 | 0.63 |
| 1.5 | 0.221031 | 0.223130 | 0.94 |
| 2.0 | 0.133640 | 0.135335 | 1.25 |
| 2.5 | 0.080802 | 0.082085 | 1.56 |
| 3.0 | 0.048855 | 0.049787 | 1.87 |
| 3.5 | 0.029539 | 0.030197 | 2.18 |
| 4.0 | 0.017860 | 0.018315 | 2.49 |
| 4.5 | 0.010798 | 0.011109 | 2.80 |
| 5.0 | 0.006529 | 0.006738 | 3.10 |

We note from Table 9.3 that as the step size is halved, the global error of Euler's method is approximately halved also. Thus, even though the per-step error is proportional to $h^2$, the global error is approximately proportional to $h$. This difference of one power of $h$ between the per-step error and the global error is characteristic of numerical solution methods, and a knowledge of that relationship is quite useful for extrapolation purposes.

It should be noted that these step sizes are extremely large in order to illustrate the concepts. One would not expect to obtain useable results with such large step sizes and the attendant errors.

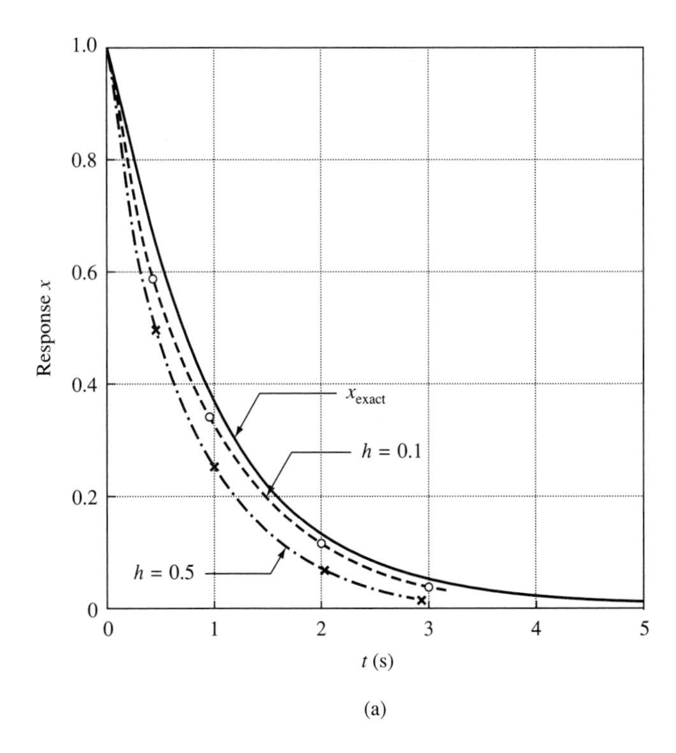


(a)

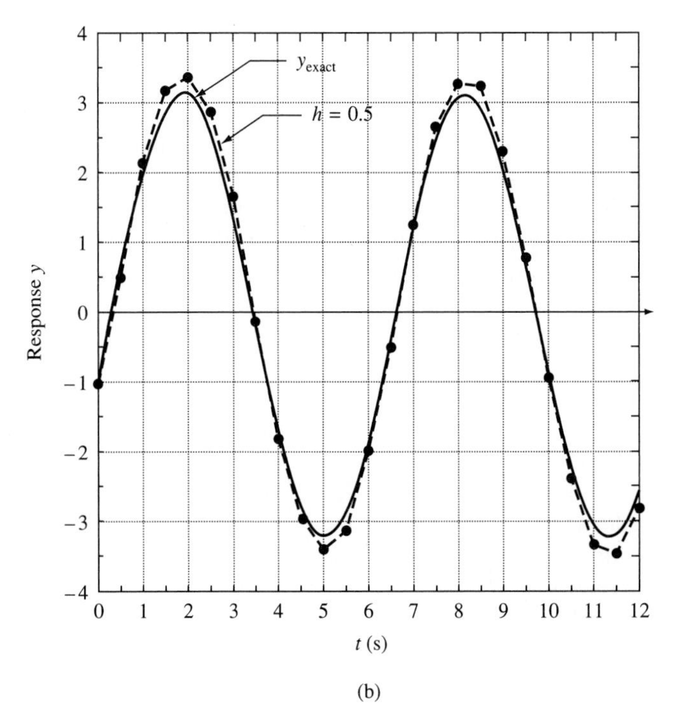


(b)

**Figure 9.9** Solutions to the first-order systems of Example 9.4. (a) $\dot{x} = -x$, (b) $\dot{y} = -3y + 10\sin t$.

**TABLE 9.3** FINAL GLOBAL ERRORS USING EULER'S METHOD TO SOLVE $\dot{x} = -x$, $x(0) = 1$

| Step size $h$ (s) | 0.5 | 0.25 | 0.1 | 0.05 | 0.025 | 0.0125 |
|---|---|---|---|---|---|---|
| Global error in $x(t = 2)$, (percent) | 53.8 | 26.0 | 10.2 | 5.04 | 2.51 | 1.25 |

Equation (9.24) was solved using the Euler method and a time step $h = 0.5$ s; Figure 9.9(b) shows the result of that calculation.

### 9.3.3 More Accurate Methods

Euler's method uses only two items of information to propagate the solution from $t_j$ to $t_{j+1}$: the current solution estimate $x_j$ and the function value or slope $f_j$. If more information concerning the character of the solution is used, the numerical results will obviously be more accurate for a given step size; that is, they will have smaller per-step truncation errors. Many such methods have been devised over the years. Two are discussed in this chapter: the Adams method and Runge-Kutta methods.

The Adams method (J. C. Adams, English mathematician, 1819–92) is typical of methods employing three pieces of information to advance the solution. It uses the equation

$$x_{j+1} = x_j + \frac{h}{2}(3f_j - f_{j-1}) \tag{9.28}$$

which employs a weighted combination of the slopes at time $t_j$ and time $t_{j-1}$, to propagate the solution. With the Adams method, the per-step error term is proportional to $h^3$, and hence, the global error is proportional to $h^2$. Note that the method is not a **self-starting** process, since solutions must be known at $t_j$ and at $t_{j-1}$ before the values at $t_{j+1}$ can be determined. When starting the process, the initial value is used for the $t_0$ point but a self-starting technique such as the Euler method would have to be used to generate an estimate of the solution at $t_1$ before the Adams method could be initiated.

### 9.3.4 Runge-Kutta Methods

A technique that employs five information values and is also self-starting is the Runge-Kutta scheme (after C. D. T. Runge, German mathematician, 1856–1927, and W. Kutta, German mathematician, 1867–1944). The Runge-Kutta approach is based on retaining higher order terms in the Taylor series expansion of the dependent variable. The fourth-order method, presented shortly, utilizes information about the derivative at four points within the time-step interval. This provides a much more accurate estimate of the solution than the methods discussed previously, since curvature of the solution over the time step is now accounted for. (Recall that we simply used a straight-line projection in the Euler method.) Runge-Kutta methods are widely used in system simulation studies because of their accuracy and ease of implementation.

A frequently used fourth-order Runge-Kutta recursion process is defined by the equation

$$x_{j+1} = x_j + \frac{h}{6}(A + 2B + 2C + D) \tag{9.29}$$

where

$$A = f(x_j, t_j) \tag{9.30}$$

$$B = f\left(x_j + \frac{h}{2}A, t_j + \frac{h}{2}\right) \tag{9.31}$$

$$C = f\left(x_j + \frac{h}{2}B, t_j + \frac{h}{2}\right) \tag{9.32}$$

$$D = f(x_j + hC, t_j + h) \tag{9.33}$$

Note that the arguments of the functions defined in *B*, *C*, and *D* are not necessarily points on the solution curve $x(t)$. The per-step and global errors of this method are proportional to $h^5$ and $h^4$, respectively.

**Example 9.5 Runge-Kutta Method**

Solve the first-order problem of Eq. (9.23) using the fourth-order Runge-Kutta method.

**Solution** The problem is solved using the Runge-Kutta method as just outlined, and the results are presented in Table 9.4. Note that for comparable step sizes, the Runge-Kutta method produces results that are much more accurate than does the Euler method.

**TABLE 9.4** RUNGE-KUTTA (R–K) SOLUTION TO $\dot{x} = -x$, $x(0) = 1$

| $h = 0.5$<br>$t$ | $x$ (R-K) | $x$ (exact) | error (%) |
|---|---|---|---|
| 0.5 | 0.606771 | 0.606531 | − 0.040 |
| 1.0 | 0.368171 | 0.367879 | − 0.079 |
| 1.5 | 0.223395 | 0.223130 | − 0.119 |
| 2.0 | 0.135550 | 0.135335 | − 0.158 |
| 2.5 | 0.082248 | 0.082085 | − 0.198 |
| 3.0 | 0.049905 | 0.049787 | − 0.238 |
| 3.5 | 0.030281 | 0.030197 | − 0.278 |
| 4.0 | 0.018374 | 0.018316 | − 0.317 |
| 4.5 | 0.011149 | 0.011109 | − 0.357 |
| 5.0 | 0.006765 | 0.006738 | − 0.397 |

We solved this problem several times using the Runge-Kutta method with different step sizes, and the effect of the step size on the global error is shown in Table 9.5. Note that as the step size is halved, the error is reduced by a factor of about 16.

**TABLE 9.5** RUNGE-KUTTA SOLUTION TO $\dot{x} = -x$, $x(0) = 1$

| | | | |
|---|---|---|---|
| Step size $h$ (s) | 0.5 | 0.25 | 0.125 |
| Global error in $x(t = 2)$, (percent) | 0.158 | 0.008 | 0.00046 |

## 9.4 SYSTEMS OF EQUATIONS

The preceding methods may be used for more than one equation if we regard the symbols in Eq. (9.17) as vectors. That is, **x** and **f** are vectors, and we have a set of equations of the state-space form:

$$\frac{d\mathbf{x}}{dt} = \dot{\mathbf{x}} = D\mathbf{x} = \mathbf{f}(\mathbf{x},t), \text{ with initial-condition vector } \mathbf{x}(t_0) = \mathbf{x}_0 \tag{9.34}$$

There is a separate definition of the derivative for each variable; that is, for $n$ equations,

$$\begin{aligned} \dot{x}_1 &= f_1(x_1, x_2, \ldots, x_n, t) \\ \dot{x}_2 &= f_2(x_1, x_2, \ldots, x_n, t) \\ &\vdots \\ \dot{x}_n &= f_n(x_1, x_2, \ldots x_n, t) \end{aligned} \tag{9.35}$$

The functions $f_i$ express the coupling between the solution variables $x_i$. In applying the fourth-order Runge-Kutta method to a system of equations, we obtain the numerical solution for the vector of dependent variables, **x**, as given by the expression

$$\mathbf{x}_{j+1} = \mathbf{x}_j + \frac{h}{6}(\mathbf{A} + 2\mathbf{B} + 2\mathbf{C} + \mathbf{D}) \tag{9.36}$$

where

$$\mathbf{A} = \mathbf{f}(\mathbf{x}_j, t_j) \tag{9.37}$$

$$\mathbf{B} = \mathbf{f}\left(\mathbf{x}_j + \frac{h}{2}\mathbf{A}, t_j + \frac{h}{2}\right) \tag{9.38}$$

$$\mathbf{C} = \mathbf{f}\left(\mathbf{x}_j + \frac{h}{2}\mathbf{B}, t_j + \frac{h}{2}\right) \tag{9.39}$$

$$\mathbf{D} = \mathbf{f}(\mathbf{x}_j + h\mathbf{C}, t_j + h) \tag{9.40}$$

Here **A, B, C,** and **D** are vectors corresponding to the function evaluations shown in Eqns. (9.37) through (9.40) above. A fixed step size implementation of the fourth-order Runge-Kutta method in the MATLAB language is presented in Appendix H in the file `rk4.m`, and an example of its use for systems of equations is given later in this chapter.

## 9.5 SELECTION OF THE STEP SIZE

It is clear from Figure 9.7 that the accuracy with which the numerical solution is computed depends upon the step size employed. A fixed step size that is too large can give results either with large errors or cause the numerical process to become unstable. On the other hand, too small an $h$ causes an inordinately large number of steps to be taken, which, in the extreme, can cause large computational errors due to round-off. This happens because computations are normally performed with a limited number of digits. For example, the fraction one-third may be represented by 0.33333333 on a computer with an eight-digit floating-point word length. Multiplying this representation by 3 gives the result 0.99999999, an approximation to 1.0. Repeated rounding off of results in this way, however, can cause a significant error when a very large number of calculations is performed. Hence, as $h \to 0$, the number of calculations required to reach a given time $t_j$ approaches infinity, and serious rounding effects can occur.

The preceding observations are illustrated in Figure 9.10, in which the global error in a numerical solution is plotted against $h^p$, where $p$ is the order of the method. That is, $p = 1$ for the Euler method; $p = 4$ for the fourth-order Runge-Kutta method, etc.

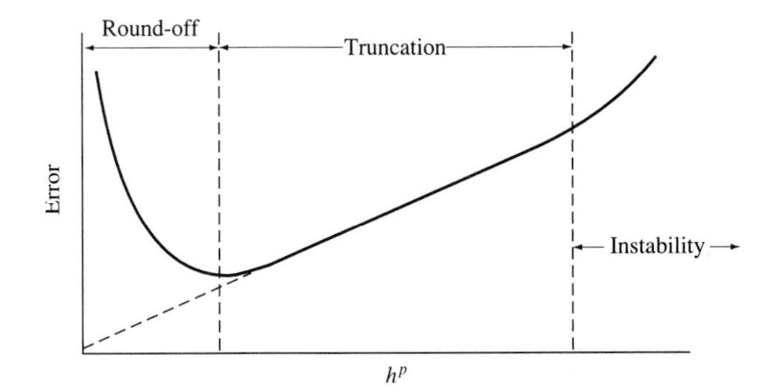


**Figure 9.10** Global error vs. $h^p$.

### 9.5.1 Selection of a Fixed Step Size

In selecting the proper step size for a numerical method of solving a system equation, the primary goal is to divide the resulting waveform into enough pieces to be able to resolve the character of the wave accurately, without resorting to the use of an excessive number of points and thereby incurring significant round-off errors. Two different dynamic effects must be considered in the selection of the step size:

the speed of the response, determined by the **natural dynamics** of the system, and the waveform of the **input signal**.

**Step Size Based on Natural Dynamics.** The dynamic response characteristics of a system are determined by the roots of its characteristic equation. The roots can be real or complex, and we can think of the system as being composed of first- and second-order components. Complex roots occur in conjugate pairs for polynomials with real coefficients. Real roots correspond to first- or second-order system components with a nonoscillatory response to initial conditions, while complex conjugate roots correspond to second-order system components with an oscillatory response to initial conditions.

The characteristic period for components with real roots is the exponent or time constant. The characteristic period for an oscillatory system is the period corresponding to its oscillatory natural frequency.

For the best results in determining the system response numerically, it is recommended that a fourth-order method such as Runge-Kutta be used and that the smallest time constant, or the period associated with the highest frequency of the dynamic system, be divided into about 10 parts. For real roots,

$$\tau = \frac{1}{\lambda} \tag{9.41}$$

For complex conjugate roots,

$$\lambda_{1,2} = a \pm jb \tag{9.42}$$

Recall from Section 8.5.3 that

$$\omega_n = \sqrt{a^2 + b^2} = |\lambda| \tag{9.43}$$

Thus,

$$\tau = \frac{1}{\omega_n} = \frac{1}{|\lambda|} \tag{9.44}$$

Since the system characteristic time periods are related to the inverses of the eigenvalues (roots of the characteristic equation), the step size should be about one-tenth of the inverse of the magnitude of the largest eigenvalue, $\lambda_{max}$:

$$h = \frac{1}{10\,|\lambda_{max}|} \tag{9.45}$$

The largest eigenvalue of the system is used because it corresponds to the response with the highest frequency (shortest wavelength), which requires the smallest step size in order to resolve the wave.

The foregoing method can be used to select step sizes for a system whose eigenvalues are known; however, for high order systems, the determination of eigenvalues may be as time consuming as calculating the simulation response itself. Hence, it may be more convenient to estimate the proper step size based on the coefficients of the characteristic equation, as found from the classical form of the governing differential equation. The average of the eigenvalues can then be related to these coefficients. The characteristic equation can be written as

$$\alpha_n \lambda^n + \alpha_{n-1} \lambda^{n-1} + \cdots + \alpha_1 \lambda + \alpha_0 = 0 \tag{9.46}$$

It can be shown that the lowest-order coefficient $\alpha_0$, normalized by the highest order coefficient $\alpha_n$, is the product of all eigenvalues and that the $n$th root of the lowest-order normalized coefficient is the logarithmic mean of the eigenvalues,

$$\bar{\lambda}_L = \sqrt[n]{\frac{\alpha_0}{\alpha_n}} \tag{9.47}$$

This result does not provide the largest eigenvalue of the system, but the step size found using the approach gives very good simulation results and is very easy to calculate if the characteristic equation is known. The step size is

$$h = \frac{1}{10\,\bar{\lambda}_L} = \frac{1}{10} \frac{1}{\sqrt[n]{\dfrac{\alpha_0}{\alpha_n}}} \tag{9.48}$$

Thus, based on the natural dynamics of the system, the step size should be selected according to the largest actual eigenvalue or according to the mean eigenvalue predicted by the coefficients of the characteristic equation. Since we have allowed some latitude on the selection, either approach will suffice.

**Step Size Based on the Input Signal.** In some cases, the input signal may contain higher frequencies (smaller periods) than the system natural dynamics. Just as before, we want to use a step size that will divide the shortest dynamic period into about 10 steps, and if the input frequency $\omega_{in}$ is higher than the largest frequency found from the system eigenvalues, the step size will be governed by the input. If the input signal is a complex waveform, the smallest recognizable period $\tau_{in}$ should be divided into about 10 steps; that is,

$$h = \frac{1}{10\omega_{in}} \quad \text{or} \quad h = \frac{\tau_{in}}{10} \tag{9.49}$$

For a fixed-step solution method, we can select $h$ from Eq. 9.45, Eq. 9.48, or Eq. 9.49. Of course, the *smallest* value as of the three should be used.

Further, it is usually best to use a convenient multiple of 1, 2, 2.5, 5, etc., multiplied by some power of 10. For example, if a value of 0.03876 were calculated as the smallest recommended step size, a time step of 0.025 or perhaps 0.05 would be suitable.

The preceding guidelines for selecting a step size are based upon a fourth-order method such as Runge-Kutta. Lower order methods, such as the Euler method, would require much smaller step sizes to achieve comparable accuracy.

**Step Size for Nonlinear Systems.** Eigenvalues are properties of linear systems. If a system is nonlinear, it has no eigenvalues in the strict sense. However, the governing state-space equations may be reduced to a linearized equation using suitable approximations, and estimates of the step size may be obtained from this linear model. In linearizing the system, the parameters should be selected to give the largest values for the coefficients in the characteristic equation. Notice that, for the purpose of selecting a step size, we need only the lowest-ordered normalized coefficient in the characteristic equation, so only a few terms in the linearized system may need to be evaluated.

If linearization is too difficult or time consuming, we can resort to empirical methods. The step size can always be selected by intuition and the simulation performed with successively smaller and larger step sizes, until no change in the computed response is observed. However, remember that a step size that is too large can cause numerical instabilities which could be confused with system instabilities. If the result is reasonable in light of the physics of the problem, we can probably move on to other things. Alternatively, we can use methods that automatically select and adjust the step size as the solution progresses. Automatic selection of the step size is employed in many commercially available simulation codes and is discussed later in the chapter.

### 9.5.2 Selection of Output Interval and Final Time

In most simulations, it is neither necessary nor desirable to print or plot the results after every time step. To avoid the generation of unnecessary output, results can be printed or plotted only at specified intervals. The output interval is the time between two distinct recordings of data, and of course, it must be an integer multiple of the step size. Although there are no definite rules, suggested output intervals are as follows:

| | |
|---|---|
| $h$ | for a first-order system |
| $2h$ to $5h$ | for a second-order system |
| $4h$ to $10h$ | for a third-order system |
| $10h$ to $20h$ | for a high-order system |

The final time is the ending time of the simulation and may also generally be a function of the natural dynamics of the system. Thus, it, too, may be based on the solution step size. The time required to achieve a desired response or reach a steady state is generally different for each simulation; however in the absence of other information, the following guidelines may be used for estimating the required final time:

| | |
|---|---|
| $50h$ | for a first-order system |
| $50h$ to $200h$ | for a second-order system |
| $100h$ to $500h$ | for a third-order system |
| $200h$ to $1000h$ | for a high-order system |

These final time estimates are based on the **natural dynamics** of the system; if a variable input is used, then they should be added to the time at which the input function reaches an approximate steady state. In calculating the final time, use the step size based upon the natural dynamics.

**Example 9.6 Response of a Second-Order System**

Figure 9.11 illustrates a spring-mass-damper system which is given an initial displacement of 2 in and is released from rest. The system parameters give $b/m = 0.5$ and $k/m = 4$. Find the displacement and velocity response from $t = 0$ to $t = 10$ s.

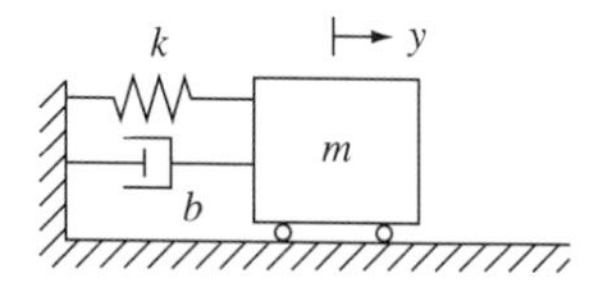


**Figure 9.11** Spring-mass-damper system.

**Solution** For a linear spring and damper, the equation of free motion is

$$m\ddot{y} + b\dot{y} + ky = 0 \tag{9.50}$$

We convert this second-order equation into two first-order equations by substituting the following definitions into the equation of motion:

$$\text{Let } x_1 = \dot{y} \tag{9.51}$$

$$x_2 = y \tag{9.52}$$

$$\text{Then } \dot{x}_1 = \ddot{y} \tag{9.53}$$

$$\dot{x}_2 = \dot{y} = x_1 \tag{9.54}$$

The equation of motion becomes

$$\dot{x}_1 + \frac{b}{m}x_1 + \frac{k}{m}x_2 = 0 \tag{9.55}$$

Equations (9.55) and (9.54) can be expressed as the derivatives of $x_1$ and $x_2$, respectively, to obtain two first-order differential equations:

$$\dot{x}_1 = -\frac{b}{m}x_1 - \frac{k}{m}x_2 \tag{9.56}$$

$$\dot{x}_2 = x_1 \tag{9.57}$$

Following the notation of Section 9.4, we have

$$\dot{x}_1 = f_1(x_1, x_2, t) = -\frac{b}{m}x_1 - \frac{k}{m}x_2 \tag{9.58}$$

$$\dot{x}_2 = f_2(x_1, x_2, t) = x_1 \tag{9.59}$$

Now $b/m = 0.5$ (1/s) and $k/m = 4.0$ (1/s$^2$). The two equations of motion and their initial conditions are then

$$\dot{x}_1 = -0.5x_1 - 4.0x_2 \quad \text{with} \quad x_1(0) = 0 \tag{9.60}$$

and

$$\dot{x}_2 = x_1 \quad \text{with} \quad x_2(0) = 2 \tag{9.61}$$

The natural frequency is

$$\sqrt{k/m} = \sqrt{4} = 2 \text{ rad/s} \tag{9.62}$$

Thus,

$$h = \frac{1}{10\sqrt{4}} = 0.05 \text{ s} \tag{9.63}$$

The system response was found using the fourth-order, fixed-step-size Runge-Kutta method discussed in Section 9.4 and implemented using the MATLAB language in the function `rk4.m`. (See Appendix H.) Two files are needed to solve the problem: one to describe the differential equations to be integrated and one to initialize the parameters, invoke the integration routine, and plot the results.

The differential equations are described by the file `smddx.m` (**s**pring-**m**ass-**d**amper **dx**), which is as follows (note that the M-file name and the function name should be the same):

```
function xdot = smddx(t,x)
% file smddx.m
xdot(1) = -0.5 * x(1) -4 * x(2);
xdot(2) = x(1);
```

(% indicates a comment in a MATLAB M-file.)

The file used to initialize variables, call the integration procedure, and plot the results is `smd4.m` and is as follows:

```
% driver file for spring-mass-damper using rk4.m
% initial time
t0 = 0;
% final time
tf = 10.;
% column vector of initial conditions
x0 = [0 2]';
% step size
h = .05;
[t,x]=rk4('smddx', t0, tf, x0, h);
```

```
plot(t,x)
grid
xlabel('Time (s)')
ylabel('Displacement (in) and Velocity (in/s)')
```

Here, `'smddx'` is the name of the M-file that defines the governing equations. We carry the solution out to a time of $200h = 10$ s and get a little over three cycles of the response. The results for the displacement and velocity of the mass are shown in Figure 9.12.

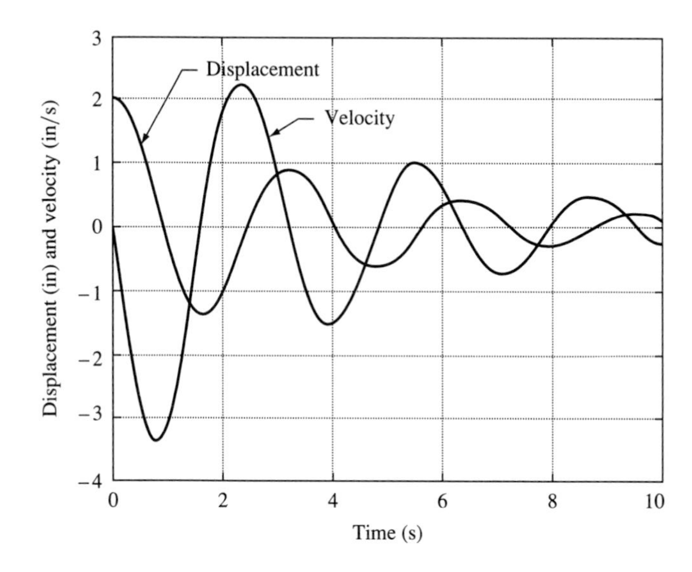


**Figure 9.12** Spring-mass-damper response to initial conditions.

## 9.6 VARIABLE STEP SIZE METHODS

In some instances, the step size dictated by the requirements outlined in Section 9.5 may not be required, except over a portion of the period during which the solution is found. As time advances, the frequency content of the input function may change, or, in nonlinear systems, the system dynamic characteristics may change. For example, a space vehicle on a trip from the earth to loop around the moon and return has a very different set of forces applied to it when it passes around the moon than when it is midway between the earth and the moon. Thus, it is easy to imagine the desirability of using a small step size during the high acceleration gradient portions of the solution and using a large step size when accelerations vary slowly.

In order to be able to change the integration time step as the solution progresses, some estimate of the current accuracy of the calculated results must be made, and different methods of doing this are available, depending on the particular numerical integration scheme that is being employed. One of the most straightforward approaches involves calculating the response at the next point in time in two ways.

First, the new solution point is calculated in the usual way by advancing from $t$ to $t + h$ using a step of size $h$. Let this solution be $x_a$. Then another approximation to the solution at $t + h$ is found by taking two small steps of size $h/2$. Call this second solution estimate $x_c$.

The truncation error (per-step error) can be estimated by means of a Taylor series expansion for both cases. We use the Euler method to illustrate the procedure. From Eq. (9.22), a Taylor series expansion gives

$$x(t + h) = x(t) + hf(t) + \frac{h^2}{2!}\frac{d^2x}{dt^2}\bigg|_t + \ldots \tag{9.64}$$

Employing Euler's method, we find that the approximate solution for the new point is

$$x_a = x(t) + hf(t) \tag{9.65}$$

The true solution can be expressed in terms of the approximate solution and the part that is truncated, which we call $h^2Q_a$:

$$x(t + h) = x_a + \frac{h^2 d^2x}{2!\, dt^2}\bigg|_t + \ \ldots = x_a + h^2Q_a \tag{9.66}$$

We repeat the solution, but now taking two steps of size $h/2$. The first step gives

$$x\left(t + \frac{h}{2}\right) = x(t) + \frac{h}{2}f(t) + \frac{h^2}{2^2}Q_b + \ldots \tag{9.67}$$

Let the intermediate point be $x_b$. The Euler method solution for the intermediate point is

$$x_b = x(t) + \frac{h}{2}f(t) \tag{9.68}$$

The true solution at the intermediate point can be written as

$$x\left(t + \frac{h}{2}\right) = x_b + \frac{h^2}{2^2}Q_b + \ldots \tag{9.69}$$

For the second step, the true value at $t + h$ can be expressed as

$$x(t + h) = x_b + \frac{h^2}{2^2}Q_b + \frac{h}{2}f\left(t + \frac{h}{2}\right) + \frac{h^2}{2^2}Q_c + \ldots \tag{9.70}$$

The approximation to the solution at $t + h$ is found by taking two steps using Euler's method:

$$x_c = x_b + \frac{h}{2}f\left(t + \frac{h}{2}\right) \tag{9.71}$$

The true value expressed in terms of the approximate solution and the portion truncated would be

$$x(t + h) = x_c + \frac{h^2}{2^2} Q_b + \frac{h^2}{2^2} Q_c \tag{9.72}$$

The terms $Q_a$, $Q_b$, and $Q_c$ are proportional to the second derivative of the function evaluated within the step. If we assume that the second derivative is essentially constant in the interval, then $Q_a = Q_b = Q_c = Q$, and the two separate expressions for $x(t + h)$ may be combined as

$$x(t + h) = x_c + \frac{h^2}{2^2} Q + \frac{h^2}{2^2} Q = x_c + 2 \frac{h^2}{2^2} Q \tag{9.73}$$

Also,

$$x(t + h) = x_a + h^2 Q \tag{9.74}$$

Subtracting Eq. (9.74) from Eq. (9.73) gives

$$0 = x_c - x_a + 2Q \frac{h^2}{2^2} (1 - 2) \tag{9.75}$$

or

$$2Q \frac{h^2}{2^2} = \frac{x_c - x_a}{(2^1 - 1)} \tag{9.76}$$

Therefore, an estimate of the truncation error that occurs during the step can be computed from the two separate solutions calculated for the value of $x(t + h)$:

$$E_T \cong \frac{x_c - x_a}{2^1 - 1} \tag{9.77}$$

Since the Euler method is a first-order method, the constant 2 appears to the first power in this equation. If the analysis were repeated for a method of order $p$, we would obtain

$$E_T \cong \frac{x_c - x_a}{2^p - 1} \tag{9.78}$$

for an estimate of the per-step truncation error.

For the fourth-order Runge-Kutta method, we could estimate the per-step error by

$$E_T \cong \frac{x_c - x_a}{2^4 - 1} = \frac{x_c - x_a}{15} \tag{9.79}$$

If, during the solution process, it is found that the per-step error estimate is larger than some preselected value, the step size may be reduced at that point in the computation and the time advance continued. On the other hand, if the error is very small, the step size may safely be increased. A more detailed discussion of this process may be found in Refs. [9.3] and [9.6].

An alternative method of estimating the error involves using integration methods of different orders to calculate the estimate over a given time step. For example, we might use a fourth-order Runge-Kutta method to advance the solution one time step and then repeat the calculation for that step using a fifth-order Runge-Kutta method. An estimate of the error can then be found from the difference in the two calculated values at the endpoint and a decision made about whether to change the step size. Such a procedure is implemented in the MATLAB ordinary differential equation routines, `ode23` and `ode45` (Ref. [9.12]).

The preceding approach or something similar to it is an option in many simulation programs. Automatic selection and control of the step size removes the need to work through the step size requirements discussed previously and provides some measure of protection against abortive or unnecessarily expensive computations. If the time step requirements vary drastically over the range in which the solution is valid, the extra expense required to calculate an estimate of the error in the solution at each step may pay for itself many times over. It all depends upon the characteristics of the particular problem at hand.

**Example 9.7 Variable-Step Solution to a Second-Order System**

If we increase the damping constant in Example 9.6 so that $b/m = 5$ 1/s, the system is overdamped, and the response to an initial condition becomes nonoscillatory. At the beginning of the motion a small step size is needed, but as the response approaches the equilibrium position, much larger step sizes are sufficient. Calculate the displacement and velocity response for this system.

**Solution** The equations of motion for the system are:

$$\dot{x}_1 = -5.0x_1 - 4.0x_2 \quad \text{with} \quad x_1(0) = 0 \tag{9.80}$$

$$\dot{x}_2 = x_1 \quad \text{with} \quad x_2(0) = 2 \tag{9.81}$$

The response was found using the MATLAB variable-order error estimation method implemented in the procedure `ode45.m`. The routine uses Runge-Kutta fourth- and fifth-order methods to estimate the error at each step and automatically adjusts the step size to maintain a specified error. The default value for the desired accuracy of the solution, $10^{-6}$, was used for this and other examples in the chapter.

As before, two files are needed to solve the problem: one to describe the differential equations to be solved and one to initialize the parameters, call the integration routine, and plot the results.

The differential equations are described by the file `smd2dx.m`:

```
function xdot = smd2dx(t,x)
% file smd2dx.m
xdot(1) = -5 * x(1) -4 * x(2);
xdot(2) = x(1);
```

The file used to initialize variables, call the integration procedure, and perform the plotting is `smd45.m` and is as follows:

```
% driver file for spring-mass-damper using ode45
t0 = 0;
tf = 8.;
x0 = [0 2]';
[t,x]=ode45('smd2dx', t0, tf, x0);
plot(t,x(:,1),'o')
hold
plot(t,x(:,2),'+')
grid
xlabel('Time (s)')
ylabel('Displacement (in)and velocity (in/s)')
```

Figure 9.13 shows the results of this calculation, and the spacing of the points shown indicates the adjustment of the step size over the course of the solution. Notice that nearly 20 time steps per second are required at the beginning of the simulation, while only 2 steps per second are needed near the end of the calculation.

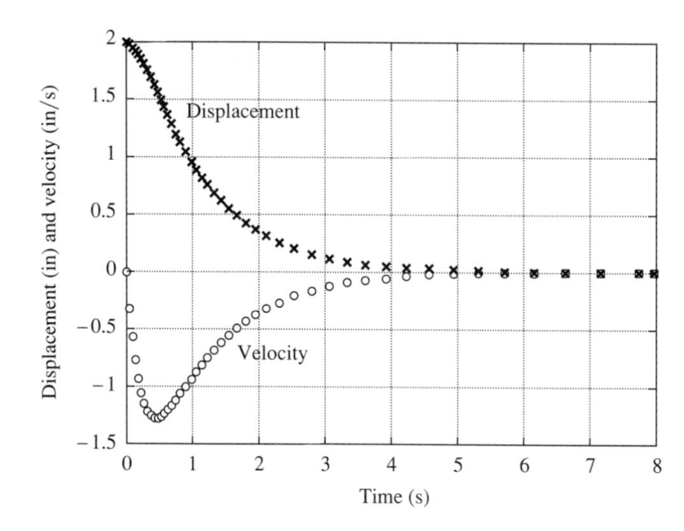


**Figure 9.13** Variable-step-size solution to an overdamped second-order system.

## 9.7 SOLUTION OF NONLINEAR DIFFERENTIAL EQUATIONS

The behavior of many physical systems is adequately described by linear differential and algebraic equations, and the solution for the simulation response is usually a straightforward procedure using the methods previously discussed. However, systems whose response must be described by nonlinear equations may present special difficulties for the engineer. Analytical solutions are hard to find or are nonexistent for these systems. Different classes of nonlinear equations can behave in markedly different ways. Small changes in initial conditions can sometimes produce disproportionately large changes in response, and since a nonlinear differential system has no fixed eigenvalues, the proper step size for numerical integration may not be obvious. Therefore, solving these systems must be approached with more than the usual care. Examples 9.8 and 9.9 illustrate the solution of nonlinear systems.

**Example 9.8 Nonlinear Electrical Circuit**

Figure 9.14 shows an electrical circuit consisting of a 175 volt source $e_0$, a 5 henry inductor $L$, and a resistor $R$. As the circuit comes up to operating conditions, the resistance varies as a function of the square of the current because of temperature changes. That is, $R = c_1 + c_2 i^2$, where $c_1 = 125$ ohms and $c_2 = 40$ ohms/amp$^2$. The equation governing the current flowing after the switch is closed is given in Section 4.3.4 of Chapter 4. Determine the response of the electrical current in the circuit.

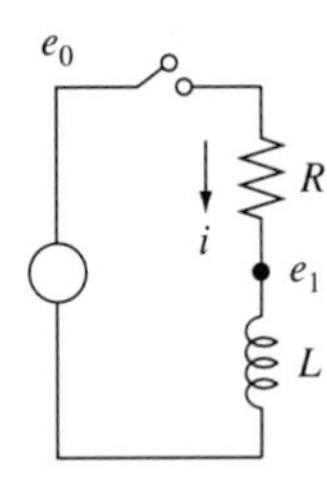


**Figure** 9.14 *RL* circuit.

**Solution** The governing equations are:

$$e_0 - L\frac{di}{dt} - Ri = 0 \tag{9.82}$$

$$\frac{di}{dt} = \frac{1}{L}[e_0 - c_1 i - c_2 i^3] \tag{9.83}$$

$$\frac{di}{dt} = 35 - 25i - 8i^3 \tag{9.84}$$

Initially there is no current, so $i(0) = 0$. The methods discussed in this chapter may be used to find the time response of the circuit. Selection of an integration step size can be based upon the response of the linear system. If $c_2$ were zero, the equivalent linear equation would have a time constant of 0.04 s; therefore, a step size of 0.004 would be adequate for use in a fixed-step-size digital simulation of the linear model. The files used to solve this problem are shown next.

The file used to initialize variables, call the integration procedure, and perform the plotting is as follows:

```
% driver file for nonlinear circuit
t0 = 0;
tf = 0.2;
x0 = 0;
h = .004;
[t,x]=rk4('cirdx', t0, tf, x0, h);
plot(t,x)
grid
xlabel('Time (s)')
ylabel('Current (amp)')
```

The file defining the governing equation is:

```
function xdot = cirdx(t,x)
% file cirdx.m
% nonlinear circuit example
xdot = 35 - 25*x(1) - 8*x(1)*x(1)*x(1);
```

The calculated current response of the nonlinear system is shown in the plot of Figure 9.15. The circuit reaches a steady state of a little more than 1 amp in about 0.1 s. The response is also found using the linear model, and it predicts a steady operational current of 1.4 amp, which is an error in amplitude of almost 40 percent. There is also a significant error in the time to maximum response with the linear model.

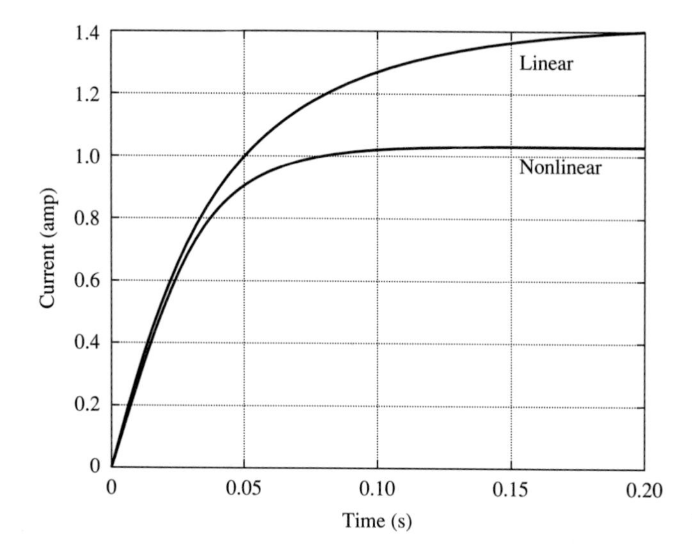


**Figure 9.15** Circuit response.

**Example 9.9 Response of a Pendulum**

The equation governing the motion of a freely swinging pendulum bob (see Figure 9.16) acted on by the force of gravity (assuming that we neglect the effect of air resistance) is

$$\frac{d^2\phi}{dt^2} + \frac{g}{L}\sin(\phi) = 0 \tag{9.85}$$

Using

$$x_1 = \dot{\phi}, \quad \dot{x}_1 = \ddot{\phi} \tag{9.86}$$

$$x_2 = \phi \tag{9.87}$$

we convert Eq. (9.85) to state-space form, obtaining

$$\dot{x}_1 = -\frac{g}{L}\sin(x_2) \tag{9.88}$$

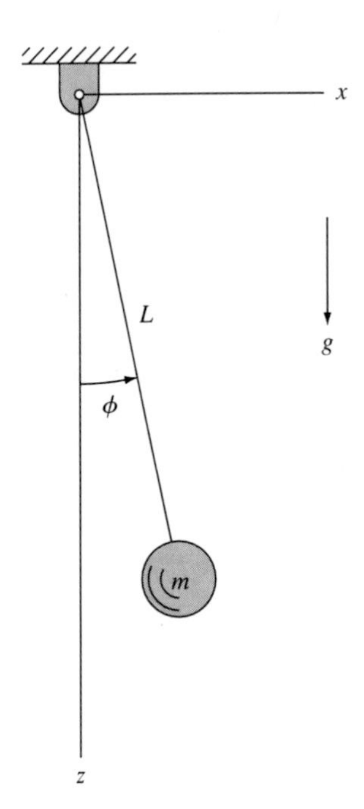


**Figure 9.16** Pendulum.

$$\dot{x}_2 = x_1 \tag{9.89}$$

If we pull the pendulum back, release it with zero initial velocity, and let it swing, the initial conditions on the problem are

$$x_1(0) = 0 \quad \text{initial velocity} \tag{9.90}$$

and

$$x_2(0) = \phi_0 \quad \text{initial angle} \tag{9.91}$$

Suppose the length $L$ of the pendulum is 40 inches. The acceleration of gravity is $g = 386$ in/s$^2$. We consider two initial angles, of 0.4 radian (about 23 degrees) and $\pi/2$ radians (90 degrees). Determine the angular position response of the pendulum to each of these initial conditions over the first 3 seconds of motion.

**Solution** The solutions to the foregoing equations were computed for the two initial release angles using the fourth-order Runge-Kutta method. The MATLAB routine `ode45.m` was employed to solve this problem, and the associated files are as follows:

```
% driver file for pendulum using ode45
t0 = 0;
tf = 3.;
% initial condition 'a'
x0 = [0 0.4]';
```

```
[t,x]=ode45('pendx', t0, tf, x0);
plot(t,57.3 *x(:,2),'w')
hold
% initial condition 'b'
x0 = [0 pi/2]';
[t,x1]=ode45('pendx', t0, tf, x0);
plot(t,57.3 *x1(:,2),'w')
grid
xlabel('Time (s)')
ylabel('Angular displacement (degrees)')

function xdot = pendx(t,x)
% file pendx.m
% pendulum problem
xdot(1) = -386./40. *sin(x(2));
xdot(2) = x(1);
```

The results are shown in Figure 9.17. For small angular motions, sin $\phi$ may be replaced with $\phi$ in the equations of motion, and they become a linear set of equations. For a 40 inch pendulum, the natural period of the linear system is about 2 seconds.

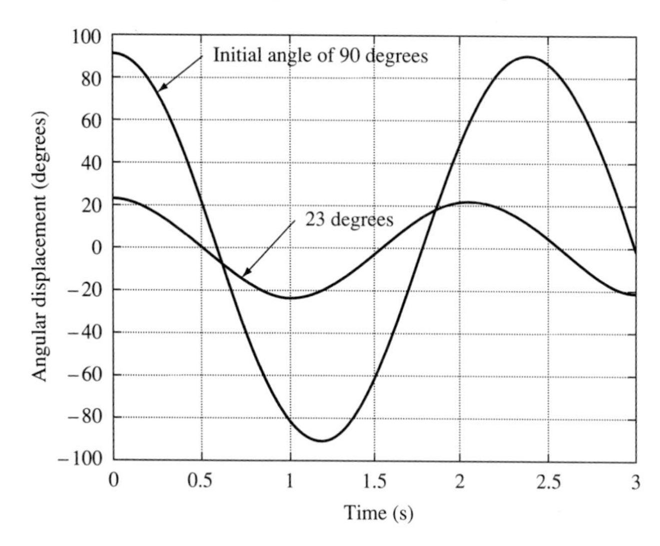


**Figure 9.17** Response of pendulum.

Notice that the period of the nonlinear pendulum system depends upon the initial conditions. For small angles, the angle and its sine are almost the same. For large angles of motion, however, they are quite different. The calculated period is approximately 2 seconds for the small initial angle, as predicted by linear theory, but increases to 2.35 seconds for the larger initial angle. This illustrates an important characteristic of nonlinear problems, namely, that response characteristics may vary with the initial conditions in a nonlinear manner. Figure 9.17 shows the response of the pendulum to the two different initial angles; the elongation of the period caused by the nonlinearity is plainly visible.

## 9.8 STIFF EQUATIONS

System-governing equations which have eigenvalues that differ by several orders of magnitude are identified as **stiff equations**. Stiff equations differ from the equations discussed previously in that they impose widely differing time step requirements on a numerical solution procedure and hence may present difficulties for the methods we have examined thus far. An example is given next. (See Ref. [9.3].)

**Example 9.10 Stiff Equations**

Determine the time response solution to the following equations:

$$\dot{x}_1 = 998x_1 + 1998x_2 \qquad x_1(0) = 1 \tag{9.92}$$

$$\dot{x}_2 = -999x_1 - 1999x_2 \qquad x_2(0) = 0 \tag{9.93}$$

**Solution** In state-space form, the preceding equations become

$$\dot{\mathbf{x}} = \mathbf{A}\mathbf{x} + \mathbf{B}\mathbf{u} \tag{9.94}$$

$$\mathbf{A} = \begin{bmatrix} 998 & 1998 \\ -999 & -1999 \end{bmatrix} \tag{9.95}$$

$$\mathbf{B} = \begin{bmatrix} 0 \\ 0 \end{bmatrix} \tag{9.96}$$

To compute the eigenvalues, we use the MATLAB procedure `eig(A)`, which gives the result

$$\lambda_1 = -1 \qquad \lambda_2 = -1000 \tag{9.97}$$

The exact solution is

$$x_1 = C_1 e^{-t} + C_2 e^{-1000t} \tag{9.98}$$

$$x_2 = C_3 e^{-t} + C_4 e^{-1000t} \tag{9.99}$$

We see that the system has two first-order components and that their time constants differ by three orders of magnitude. The step size required for numerical integration is governed by the larger of the two components. Thus, we will need a step size of about one-tenth of the reciprocal of $|\lambda_2| = 1000$:

$$h = \frac{1}{10}\frac{1}{|\lambda_2|} = 0.0001 \text{ s} \tag{9.100}$$

The contribution of the $\lambda_2$ component diminishes rapidly with time, and the solution becomes mainly composed of the $\lambda_1$ term.

Figure 9.18 gives the response of the system to the indicated initial conditions, as determined by using the variable-step-size procedure `ode45.m` discussed previously. An expanded view of the first 0.1 second of the response is shown in Figure 9.19. Observe that

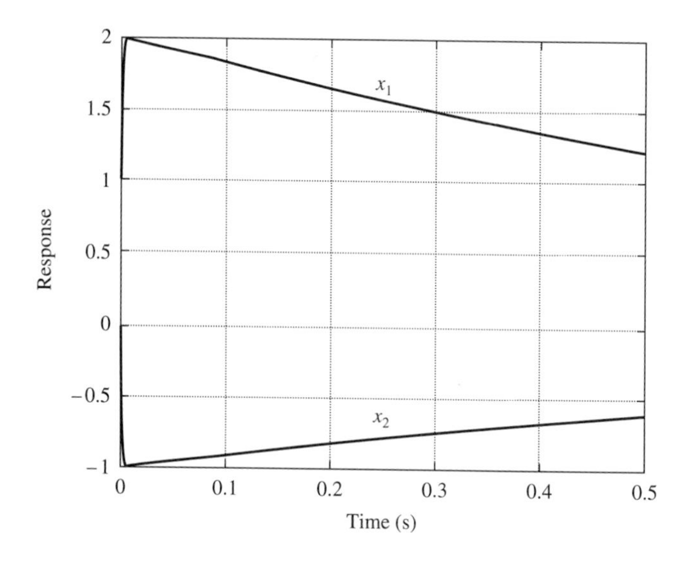


**Figure 9.18** Calculated response for stiff equations.

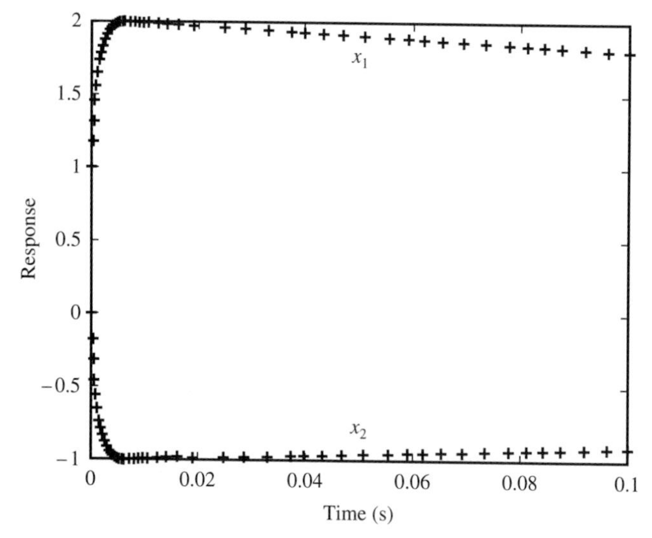


**Figure 9.19** Expanded time scale for response of stiff equations.

that the time from 0.0 s to 0.05 s required 36 time steps, while that from 0.05 s to 0.1 s took only 14 steps.

Special methods have been devised to treat stiff equations, since the usual methods may not perform satisfactorily. You should compute the time constants for the system, and if you find widely varying time step requirements, you may need to consider using procedures other than those we have discussed. References [9.3] and [9.10] have additional details concerning suitable computational methods, and algorithms are available from *netlib.att.com* on the Internet. In addition, MATLAB SIMULINK routines include methods suitable for stiff equations.

## 9.9 SUMMARY

This chapter has discussed determining the time response solution to dynamic system equations by using analytical methods as well as numerical methods. Digital simulation was emphasized, and MATLAB procedures for integrating ordinary differential equations were introduced through the solution of a number of sample problems.

## REFERENCES


9.1 Acton, Forman S. *Numerical Methods That Work*. Harper & Row, New York, 1970.

9.2 Burden, Richard L., and Faires, J. Douglas. *Numerical Analysis.* 5th ed. PWS-Kent Publishing Co., Boston, 1993.

9.3 Gear, C. William. *Numerical Initial Value Problems in Ordinary Differential Equations*. Prentice Hall, Inc., Englewood Cliffs, NJ, 1971.

9.4 Hornbeck, Robert W. *Numerical Methods*. Quantum Publishers, Inc., New York, 1975.

9.5 Jaluria, Yogesh. *Computer Methods for Engineering*. Allyn and Bacon, Inc., Boston, 1988.

9.6 James, M. L., Smith, G. M., and Wolford, J. C. *Applied Numerical Methods for Digital Computation*. 3rd ed. Harper & Row, New York, 1985.

9.7 Ogata, Katsuhiko. *Solving Control Engineering Problems with MATLAB*. Prentice Hall, Inc., Englewood Cliffs, NJ, 1994.

9.8 Parker, Thomas S., and Chua, Leon O. *Practical Numerical Algorithms for Chaotic Systems*. Springer-Verlag, New York, 1989.

9.9 Press, William, Flannery, Brian P., Teukolsky, Saul A., and Vetterling, William T., *Numerical Recipes in C*. Cambridge University Press, Cambridge, 1988.

9.10 Shampine, Lawrence F. *Numerical Solution of Ordinary Differential Equations*. Chapman & Hall, New York, 1994.

9.11 Takahashi, Yasundo, Rabins, Michael J., and David M. Auslander. *Control and Dynamic Systems.* Addison-Wesley Publishing Co., Reading, MA, 1970.

9.12 *The Student Edition of MATLAB, Version 4, User's Guide*. Prentice Hall, Inc., Englewood Cliffs, NJ, 1995.


## NOMENCLATURE

**A, B** matrices
A, B, C, D terms in Runge-Kutta method
$b$ viscous damper dissipation constant
$b_c$ critical damping factor
$C$ capacitance, constant

$e, E$ voltage
$E_T$ truncation error
$g$ gravitational acceleration
$G$ gain
$h$ time step
$i$ electric current
$k$ spring constant
$L$ inductance, length
$m$ mass
$Q, Q_a$ constants
$Q_b, Q_c$
$R$ resistance
$t$ time
$u$ input function
$x$ displacement
$x_e$ exact solution
$y$ displacement
$\alpha$ constant
$\gamma$ constant
$\lambda$ eigenvalue
$\overline{\lambda}_L$ logarithmic mean of eigenvalues
$\omega_{in}$ input frequency
$\omega_n$ natural frequency
$\omega_d$ damped natural frequency
$\phi$ angle
$\tau_{in}$ smallest period
$\tau$ time constant
$\zeta$ damping ratio

## PROBLEMS

Where appropriate, use MATLAB or other suitable computational tools to solve the following problems. System initial conditions may be assumed to be zero unless indicated otherwise. Identify any unstable systems.

**9.1** *What* must the time constant of a first-order system be if the system is to respond to a step input in 5 ms? (The system should be within 98 percent of its final value at 5 ms.) *Design* a DC electrical circuit that meets this requirement if the input voltage is 5 volts.

**9.2** *What* peak output voltage $e_2$ occurs in the circuit of Example 8.4 if it is subjected to a step input when connected to a 5 volt DC voltage source? *Use* analytical methods.

**9.3** The circuit shown in Figure P9.3 is given a step input when the switch is closed. The capacitor has no initial charge. *Derive* the differential equation describing the behavior of the system. Use analytical methods to *find* the solution for the output voltage as a function of time. *What* is the eventual steady-state value of the output voltage?

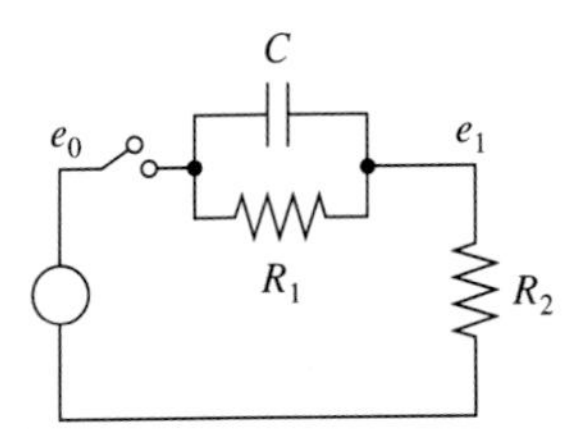


**Figure P9.3** Electric circuit.

**9.4** Use analytical methods to determine the maximum displacement and the maximum velocity of the spring-mass-damper system described in Example 8.3 (see Figure 8.7) if the input is a unit step function force.

**9.5** A second-order spring-mass-damper system is to have a natural frequency of 34 rad/s. *Design* a system that will respond to a step input with an overshoot of about 15 percent. What is the time at which the system reaches its maximum displacement?

**9.6** *Find* the step response of the voltage $e_2$ in the circuit of Problem 4.14 if the input is 5 volts DC, the capacitors have no initial charge, and the component data are as follows:

$$R_1 = 10\ \Omega \quad R_2 = 100\ \Omega$$

$$C_1 = 1\ \mu\text{f} \quad C_2 = 10\ \mu\text{f}$$

$$L = 5\ \text{mh}$$

**9.7** *Find* the step response of $e_2$ in the circuit of Problem 4.16 if the input is 5 volts DC, the capacitors have no initial charge, and the component data are as follows:

$$R_1 = 5\ \Omega \quad R_2 = 10\ \Omega$$

$$C_1 = 10\ \mu\text{f} \quad C_2 = 22\ \mu\text{f}$$

$$L = 10\ \text{mh}$$

**9.8** The equation describing a dynamic system is

$$\dot{z} + 2z = 10 \quad z(0) = 2.0$$

Use hand calculations and the Euler method to *estimate* the time response of this system. Take three steps only. *Explain* how you selected the step size, and *justify* the value you used.

**9.9** *Write* and successfully *execute* a program utilizing the Euler method for the numerical integration of a single ordinary first-order differential equation. As a test problem, use $\dot{x} = -2x$, $x(0) = 1.0$. Integrate this equation from $t = 0.0$ to $t = 5.0$ with $h = 0.1$. Tabulate the calculated $x_i$, the exact solution $x_e(t) = e^{-2t}$, and the percent error. Repeat with $h = 0.05$. Comment on the effect of the step size on the accuracy of the solution.

**9.10** *Write* and *execute* a digital computer program to numerically integrate a single first-order differential equation. Use the Runge-Kutta integration formulas as given in Section 9.3.4.

*Find* the numerical approximation to the solution of $\dot{x} = x$, with initial condition $x(0) = 1.0$. Integrate from $t = 0.0$ to $t = 5.0$. Use step sizes of $h = 0.01$, 0.1, 0.5, and 1.0 s, and compare your results to the exact solution, $x_e(t) = e^t$ at the point $t = 5$.

**9.11** Use digital simulation methods to *solve* the mechanical system problem described in Example 3.9 from Chapter 3.

**9.12** Use digital simulation methods to *solve* the transient thermal problem described in Example 6.4 from Chapter 6.

**9.13** The equation of motion for a linear spring-damper system subject to a step displacement input $u_0$ is

$$b\dot{x} + kx = ku(t) \qquad u(t) = u_0$$

Use digital simulation to *find* the response in the interval $0 < t < 4(b/k)$ if $x(0) = 0$, $b = 2$ Ns/cm, $k = 100$ N/cm, and $u_0 = 1$ cm. Plot $x$ vs. $t$.

**9.14** *Determine* the analytical expression for the unit step response of the system described by each of the following equations. *Find* the time constant $\tau$ of each system. *What* is the value of the response at $t = 3\tau$? *Sketch* the time response curve. Finally, *solve* the problem using digital simulation methods, and *compare* your results with the analytical solution.

**a.** $\dot{z} + 8z = u(t) \qquad z(0) = 0$

**b.** $2\dot{z} + 3z = u(t) \qquad z(0) = 1.2$

**c.** $0.25\dot{z} + 7.5z = u(t) \qquad z(0) - 0.5$

(Note that the MATLAB function `step` assumes zero initial conditions for the system.)

**9.15** *Determine* the analytical expression for the unit step response of the system described by each of the following equations. *Find* the eigenvalues, natural frequencies, and damping ratios for each system. *What* is the value of the response at $t = 3\tau$ where $\tau$ is the undamped natural period or largest time constant of the system? *Sketch* the time response curve. Finally, *solve* the problem using digital simulation methods, and *compare* your results with the analytical solution.

**a.** $\ddot{z} + 5\dot{z} + 4z = u(t) \qquad z(0) = 0 \qquad \dot{z}(0) = 0$

**b.** $2.5\ddot{z} + 2\dot{z} + 3.75z = u(t) \qquad z(0) = -1.0 \qquad \dot{z}(0) = 0$

**c.** $\ddot{z} + 0.25\dot{z} + 0.5z = u(t) \qquad z(0) = 0 \qquad \dot{z}(0) = -1.0$

**9.16** An 8 lbf weight freely falling under the influence of gravity strikes a spring of stiffness 7 lbf/in with an initial velocity of 10 in/s, as shown in Figure P9.16. Use both analytic

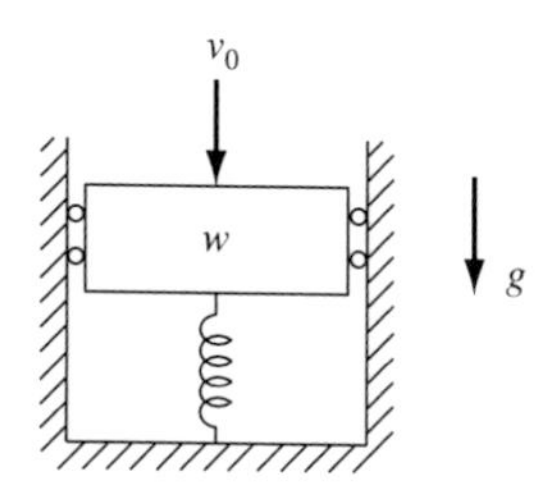


**Figure P9.16** Suspended weight.

and simulation methods to *find* the maximum downward deflection of the weight that will occur. Be sure to *include* the force due to the weight in your analysis.

**9.17** *Repeat* the digital simulation portion of Problem 9.14, except let the input now be a ramped step input with a ramp time of $T_1 = 1.25\tau$. (See Figure P9.17.)

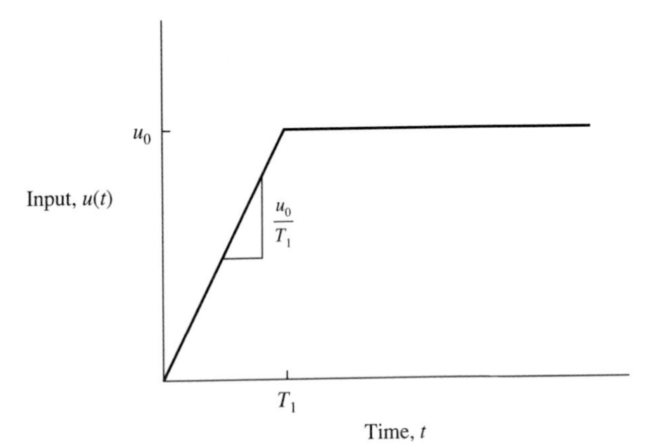


**Figure P9.17** Ramped step function.

**9.18** *Repeat* the digital simulation portion of Problem 9.15, except let the input now be a ramped step input with a ramp time of $T_1 = 1.25\tau$. (See Figure P9.17.)

**9.19** *Find* the unit step response of the system described by each of the following transfer functions. *Plot* your results. (Here $s$ is the Laplace transform variable.)

**a.** $\mathrm{TF}(s) = \dfrac{2(s^2 + 1)}{4s^2 + 9s + 6}$

**b.** $\mathrm{TF}(s) = \dfrac{s^2 + 2s + 2.5}{2s^3 + 2.5s^2 + 1s + 2.75}$

**c.** $\mathrm{TF}(s) = \dfrac{s^2 + 2s + 3}{s^3 + 3s^2 + s + 2}$

**d.** $\mathrm{TF}(s) = \dfrac{s + 2.5}{s^3 + 1.75s^2 + 1.25s + 3}$

**e.** $\mathrm{TF}(s) = \dfrac{2s^2 + 1}{s^3 + 2s^2 + 5s}$

**f.** $\mathrm{TF}(s) = \dfrac{2s + 1}{s^4 + 2s^3 + 5s^2 + 8s + 4}$

**g.** $\mathrm{TF}(s) = \dfrac{s^2 + 2s + 1}{2s^4 + 3s^3 + 9s^2 + 7s + 5}$

**h.** $\mathrm{TF}(s) = \dfrac{2s^2 + 1}{s^4 + 2s^3 + 7s^2 + 8s + 4}$

**9.20** Calculate the poles and zeros of each of the systems whose transfer functions are given in Problem 9.19. (See Section 8.6.)

**9.21** *Derive* the system equations for the circuit shown in Figure P9.21, and *find* the transfer function between the input voltage $e_0$ and the output voltage $e_1$. Determine the step re-

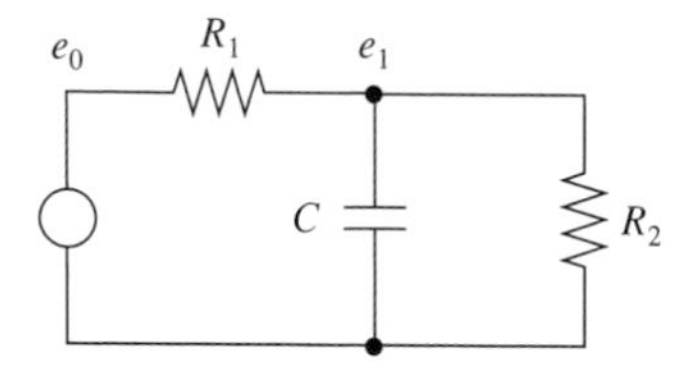


**Figure P9.21** Electric circuit.

sponse and plot it. *What* is the eventual value of the output voltage if the input is 5 v? Take $R_1 = 5\ \Omega$, $R_2 = 12\ \Omega$, and $C = 10\ \mu$f. Use digital simulation methods.

**9.22** A linear second-order system is described by the equation

$$\ddot{z} + 2\zeta\omega_n\dot{z} + \omega_n^2 z = u(t)$$

If $\omega_n$ is a known value, *overlay* on a single plot *sketches* of the unit step responses of the system for damping ratios $\zeta$ of 0.0, 0.5, and 1.0.

**9.23** Determine the maximum angular velocity $\dot{\theta}_2$ that occurs in the torsional system of Example 8.7 if the system experiences a unit step input torque. (See Figure 8.13.)

**9.24** Suppose the spring in Problem 9.13 is nonlinear such that the equation of motion of the system is subject to a step input

$$b\dot{x} + kx + qx^2 = ku(t)$$

*Find* and *plot* the response in the interval $0 < t < 4(b/k)$ if $q = 100$ N/cm$^2$ and

**a.** $u_0 = 1$ cm

**b.** $u_0 = 2$ cm

*Compare* the results found in (a) and (b), and comment on them.

**9.25** *Use* digital simulation to examine the response of the system

$$\begin{bmatrix} \dot{x}_1 \\ \dot{x}_2 \end{bmatrix} = \begin{bmatrix} 0 & 1 \\ -25 & -2 \end{bmatrix} \begin{bmatrix} x_1 \\ x_2 \end{bmatrix} + \begin{bmatrix} 0 \\ 25 \end{bmatrix} u_0$$

with initial conditions

$$x_1(0) = -1.0 \text{ cm}, \quad x_1 = \text{position (cm)}$$

and

$$x_2(0) = 0.0 \text{ cm/sec}, \quad x_2 = \text{velocity (cm/sec)}$$

*Find* the solution for a step input $u_0 = 1.0$ cm for $t > 0$.

**9.26** *Select* the step size and the final time for numerical integration (using a fixed-step fourth-order Runge-Kutta integration procedure) for the following systems:

**a.** $$\ddot{z} + 51z + 50 = 0$$

**b.** $$\dddot{z} + 3\ddot{z} + 403z + 401 = 0$$

**c.** $$\frac{d^4z}{dt^4} + 26\frac{d^3z}{dt^3} + 325\frac{d^2z}{dt^2} + 1300\frac{dz}{dt} + 1000z = 0$$

**9.27** *Calculate* the response of the following third-order system to a triangular pulse input of unit magnitude and duration $T_1$. (See Figure P9.27). Here, $T_1$ is the period corresponding to the smallest eigenvalue of the system. The initial conditions are zero.

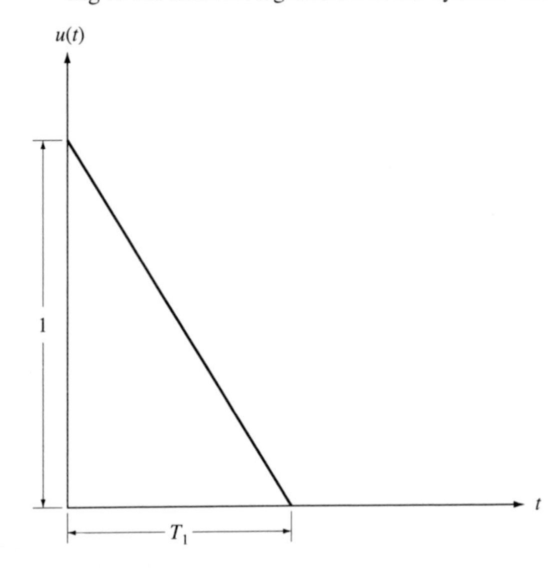


**Figure P9.27**

$$\dddot{z} + 3\ddot{z} + 28\dot{z} + 26z = 52u(t)$$

**9.28** A spring-mass-damper system subject to a step input $u_0$ has a nonlinear spring. The equation of motion of the system is

$$m\ddot{y} + b\dot{y} + ky + qy^2 = ku(t)$$

The initial conditions are $y(0) = \dot{y}(0) = 0$. If $m = 20$ kg, $b = 200$ Ns/m, and $k = 3500$ N/m, *determine* the response in the interval $0 < t < 1.5$ s for the following cases:

**a.** $q = 0$, linear case $ku_0 = 350\text{ N}$

**b.** $q = 5000\text{ N/m}^2$, $ku_0 = 350\text{ N}$

**c.** $q = 5000\text{ N/m}^2$, $ku_0 = 700\text{ N}$

*Plot* $y(t)$ in each case and compare the results.

**9.29** The equation

$$\ddot{\theta} + 0.2\theta + 10\sin(\theta) = 100\sin(2t)$$

is used to represent the motion of a damped pendulum excited by a harmonic torque. The system starts from rest. *Find* the response in the 0- to 3-second interval. Plot $\theta$ and $\dot{\theta}$ as functions of time.

**9.30** *Duplicate* the time response calculations for the electric circuit of Example 9.8.

**9.31** *Repeat* the pendulum response problem of Example 9.9, but let the second initial condition be 135 degrees. *Find* the period of oscillation for initial conditions of 23 degrees and 135 degrees, and *compute* the difference in the two periods.

**9.32** *Design* a pendulum that has a period of 4 s when released from an initial angle of 175 degrees.

**9.33** *Add* aerodynamic damping to the pendulum model of Example 9.9, and *compute* the response for release from an initial angle of 90 degrees. The pendulum mass is an aluminum sphere 1.5 inches in diameter. *Plot* the pendulum angle as a function of time for three cycles of oscillation. *Does* the damping cause a significant reduction in the amplitude of oscillation of the pendulum?

**9.34** If the mass in Problem 9.16 is 4 kg and its initial velocity is 0.5 m/s, *select* a linear spring so that the maximum deflection of the falling mass does not exceed 0.1 m. Could a nonlinear spring be used to minimize the deceleration of the mass, but still bring it to rest in the same distance? Use simulation methods to *try* out your ideas.

**9.35** *Solve* Problem 6.24.

**9.36** *Repeat* Problem 9.35 but let the wall material now be copper. *Consider* two models, one allowing for the variation in thermal conductivity of copper with temperature (see Figure 6.2) and the other using an average value for the conductivity of a material. *Comment* on the significance of the dependence of a material property on temperature in this problem.

**9.37** *Determine* the minimum acceptable amount of damping for the problem described in Example 9.2 if the input is a ramped step with a rise time $T_1$ of 0.5 ms instead of a step function. (See Figure P9.17.)

**9.38** A small automobile is traveling along a highway when it suddenly encounters a 1.0 in change in the height of the pavement due a resurfacing operation. The input displacement to the tires is assumed to be a step function. Using the model developed in Example 3.7 and the following numerical data, *find* and *plot* the response of the auto body to the step function input:

$$k_1 = 1000 \text{ lbf/in} \qquad w_1 = 20 \text{ lbf}$$

$$k_2 = 100 \text{ lbf/in} \qquad w_2 = 150 \text{ lbf} \qquad b = 5 \text{ lbf s/in}$$

# Part 4

# ENGINEERING APPLICATIONS

High-Performance Formula SAE® Race Car. (Photo courtesy of UTA Formula Team)

A Bachelor of Arts [or Science] Degree is a
license to acquire an education.

Gerald W. Johnson, 1957.

# CHAPTER 10

# System Design and Selection of Components

## 10.1 INTRODUCTION

Previous chapters have illustrated how to obtain a model for a dynamic system and how to analyze the response of the system once the coefficients of the governing equations are known; however, in many circumstances, it is more difficult to determine the best values of the physical parameters than it is to model the system. This determination of the coefficients is related to the engineering sizing and selection of components for the system to meet certain performance requirements.

In any system, there are some parameters that are fixed, and we cannot change their values. This is typical of loads and things to which we might attach control components. Other parameters, such as power supplies, may be limited according to their availability. Some parameters are parasitic in that they are a part of the system that we cannot eliminate or whose coefficients we have no control over. Some parameters directly affect the static and dynamic performance of the system over which we have complete control; however, even when we think that we have complete control of the selection of certain parameters, the availability of the components they represent may be limited to discrete sizes from various manufacturers (e.g., the diameters of pistons, the sizes of valves, the values of resistances, etc.).

In this chapter, we emphasize how to select the parameters and components for engineering systems. To do this, we must first have a model that accurately describes the system behavior. Then we must develop operational constraints, ranges, limits, performance goals, or desired characteristics to help us select the values of the components. We would like to have the same number of criteria as the required number of parameters. In this regard, it is extremely helpful to normalize and rearrange the terms in the system equations to make them significant recognizable factors (such as the static gain, loop gain, impedance ratio, time constant, ratio of open-loop to closed-loop factors, etc.). It is wise to initially idealize the system by neglecting friction, higher-order dynamic effects, and minor degradation factors in order to determine which coefficients have a predominate effect upon performance factors. Once this is done, the coefficients can be selected on the basis of the performance factor over which they have the most influence (e.g., feedback gain should be selected on the basis of static gain, output impedance of an amplifier should be selected on the basis of the system time constant or system dynamics, etc.).

At first glance, this will seem like an open-ended process; and indeed, a good spreadsheet relating the basic component values to the system static and dynamic performance characteristics and behavior is very useful and often necessary. However, in this chapter, we consider several examples and illustrate how to use certain performance factors and characteristics to help select the required component parameter values even when the performance is not stated or very obvious.

## 10.2 VOLTAGE POWER SUPPLY FILTER

An electronic power supply system, shown in Figure 10.1a, is used to provide a voltage source for a circuit. In this system, a transformer is used to convert 60 Hz, 120 VAC (rms), to 12 volts peak-to-peak. This reduced voltage is then rectified with a half-wave rectifier. A capacitor is used to smooth the rectified AC voltage to a DC voltage with some ripple (or variation in voltage). The DC voltage is then used to drive the load circuit that is represented by the equivalent load impedance $R_L$.

We know the characteristics of the transformer and the equivalent load impedance; our goal is to select the capacitor so as to provide a specified amount of ripple in the voltage supplied to the load. The larger the capacitor, the less will be the ripple (which is desirable); but at the same time, we want to minimize the capacitance to save circuit board space and for cost considerations. Therefore, what is the minimum value of capacitance that will give the allowable ripple?

For this application, we will allow approximately 5% ripple. Admittedly, this is a substantial amount of ripple for most applications in electronics; however, it does illustrate the procedure and makes a more visual graph when the output is plotted against time.

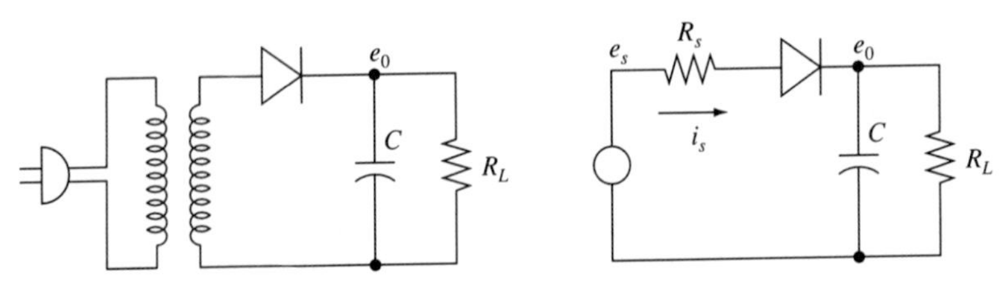


**Figure 10.1** Electronic filter circuit for transformer power supply. (a) Transformer circuit. (b) Equivalent ideal circuit.

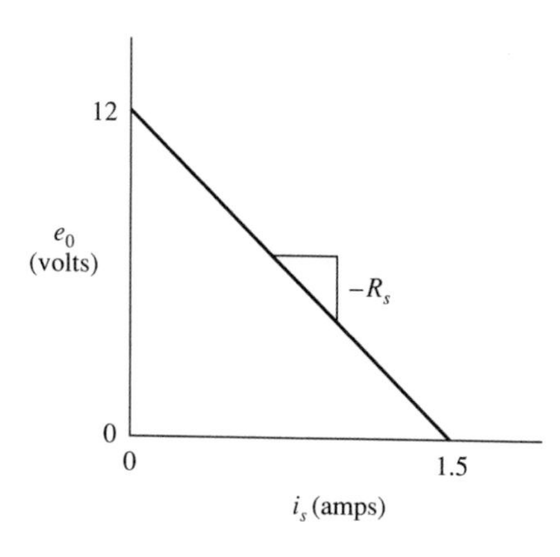


**Figure 10.2** Output characteristics of transformer power supply.

The transformer has limited current capability, and the output voltage $e_0$ degrades with the output current $i_s$, as shown in Figure 10.2. Thus, the equivalent circuit for this system is an ideal voltage source $e_s$ with an output resistance of $R_s$ as shown in Figure 10.1b. This transformer can produce 12 volts peak-to-peak with no load placed on the transformer ($R_L = \infty$) or can deliver 1.5 amp with a short circuit ($R_L = 0$). Thus, the effective output impedance ($\Delta e_0/\Delta i_s = 12$ volt/1.5 amp) is 8 ohms. The equivalent load impedance is 48 ohms.

When the rectifier is forward biased (i.e., when $e_s > e_0$), it allows current to pass with very little resistance; when the rectifier is reversed biased (i.e., when $e_s < e_0$), it allows no (or at least very litle) current to pass. Thus, the ideal component equation for the rectifier and source resistance is

$$i_s = \frac{e_s - e_0}{R_s} \qquad \text{if } e_s > e_0 \tag{10.1}$$

$$i_s = 0 \qquad \text{if } e_s < e_0 \tag{10.2}$$

The other component equations and the node equation are

$$i_c = C\,\dot{e}_0 \tag{10.3}$$

$$i_L = \frac{e_0}{R_L} \tag{10.4}$$

$$i_s = i_c + i_L \tag{10.5}$$

These equations can be combined to yield the following differential equation for the two operational modes of the circuit (i.e., $e_s > e_o$ and $e_s < e_0$):

$$\dot{e}_0 = \frac{e_s - \left(1 + \dfrac{R_s}{R_L}\right) e_0}{R_s C} \qquad \text{if } e_s > e_0 \tag{10.6}$$

$$\dot{e}_0 = \frac{-e_0}{R_L C} \qquad \text{if } e_s < e_0 \tag{10.7}$$

We could simply try different values of $C$ in a transient response simulation until we achieved the allowable ripple; however, we can get an estimate of the order of magnitude of the value of $C$ from a frequency response analysis. (See Chapter 8.)

The time constant of the system when it is charging the capacitor is

$$\tau = \frac{R_s C}{\left(1 + \frac{R_s}{R_L}\right)} \tag{10.8}$$

The frequency ratio (the excitation frequency, 60 Hz, divided by the circuit break frequency, $\omega_1 = 1/\tau$) is

$$\frac{\omega}{\omega_1} = N = \frac{R_s C}{\left(1 + \frac{R_s}{R_L}\right)}\omega \tag{10.9}$$

Solving this frequency ratio equation for the capacitance $C$ yields the following criterion:

$$C = \frac{\left(1 + \frac{R_s}{R_L}\right)}{R_s}\frac{N}{\omega} \tag{10.10}$$

If the system were linear, excitation of the circuit at a frequency ratio of 10 would produce an output of $-20$ dB, which is equivalent to 9.95% ripple (refer to Chapter 8). Since the system is nonlinear (due to the rectifier), the driving frequency is more closely approximated by a 30-Hz voltage signal instead of 60 Hz; thus, an actual frequency ratio of 10 will appear to respond more like 20 in this nonlinear system. A frequency ratio of 20 in a linear system would produce an expected ripple of 4.99%, which is close to our allowable value of 5%. Using the circuit values given earlier, we can calculate the ideal capacitance with a frequency ratio of 10:

$$C = \frac{\left(1 + \frac{R_s}{R_L}\right)}{R_s}\frac{N}{\omega} = \frac{\left(1 + \frac{8}{48}\right)}{8 \text{ ohm}} \frac{10}{60 \times 2\pi \frac{\text{rad}}{\text{s}}} = 3868\ \mu\text{f} \tag{10.11}$$

Of course, a capacitor of 3868 $\mu$f cannot be purchased. As a matter of fact, capacitors normally come in relatively coarse increments (1, 2.2, 3.3, 4.7, 6.8, etc.), simply because the capacitors themselves are coarse. The reason for this is that tolerances for capacitors are $\pm 20\%$ or worse; therefore, a closer spacing of capacitance values is almost meaningless. Larger capacitors come in better increments, and a 3900 $\mu$f capacitor can be purchased.

Using the preceding circuit values, a digital simulation of this system reveals the filtering characteristics of the circuit. Since we are interested only in the steady-state response, we can start the sytem with the initial condition equal to the mean (average) steady-state value of the output voltage. In this case, the mean output is equal to the mean input times the static gain:

$$e_{0\,mean} = \frac{1}{\left(1 + \dfrac{R_s}{R_L}\right)} e_{s\,mean} \tag{10.12}$$

The mean input of a rectified sine wave of amplitude $A$ is

$$e_{s\,mean} = A\frac{2}{\pi} \tag{10.13}$$

Thus, the expected steady-state value of the output voltage is

$$e_{0\,mean} = \frac{2}{\pi}\frac{A}{\left(1 + \dfrac{R_s}{R_L}\right)} = \frac{2}{\pi}\frac{12\text{ volt}}{\left(1 + \dfrac{8}{48}\right)} = 6.55\text{ volts} \tag{10.14}$$

We should start the simulation with this inital condition to avoid dynamic transients.

The dynamic simulation of this system is shown in Figure 10.3. Notice that the voltage charges faster than it decays, due to the difference in supply resistance (8 ohms) and the load resistance (48 ohms). Notice further that the ripple is 6.2%, which is very close to our allowable value.

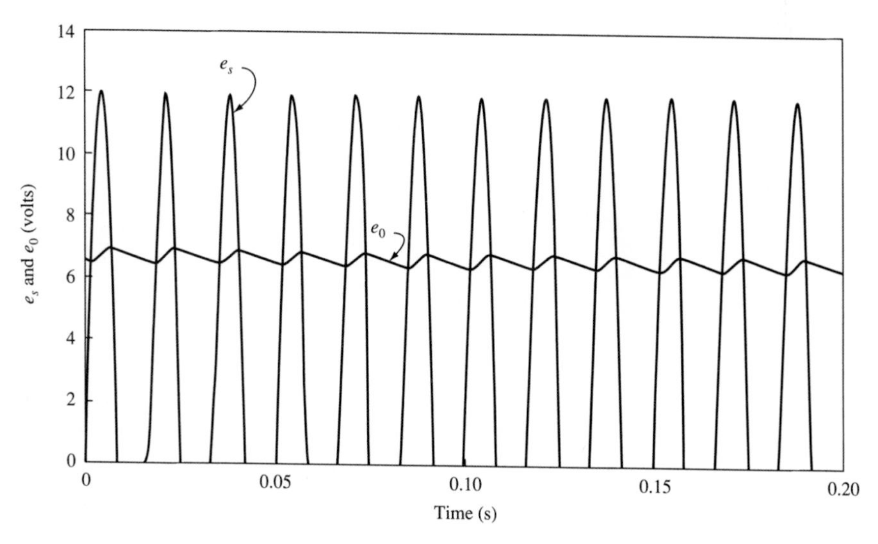


**Figure 10.3** Transient response of voltage supply filter.

## 10.3 AUTOMOTIVE BRAKE SYSTEM

Let us now consider the brake system typical of a race car. The system has two master cylinders, with a bias bar to adjust the bias of braking effort in the front and rear. The goal of an optimum system is to distribute the braking efforts between the front and the rear in order to compensate for variations in the road surface, wearing of the tires, and shifts in the distribution of weight from the rear to the front during braking.

This example considers weight transfer due to the dynamic operation of a system and illustrates the practical modeling of a complete system. It is included as an example of how to generate criteria that are sufficient to size and select the components of an actual engineering system.

The overall configuration of the brake assembly is shown in Figure 10.4. The ball of the driver's foot actuates a foot pedal that is connected to the master cylin-

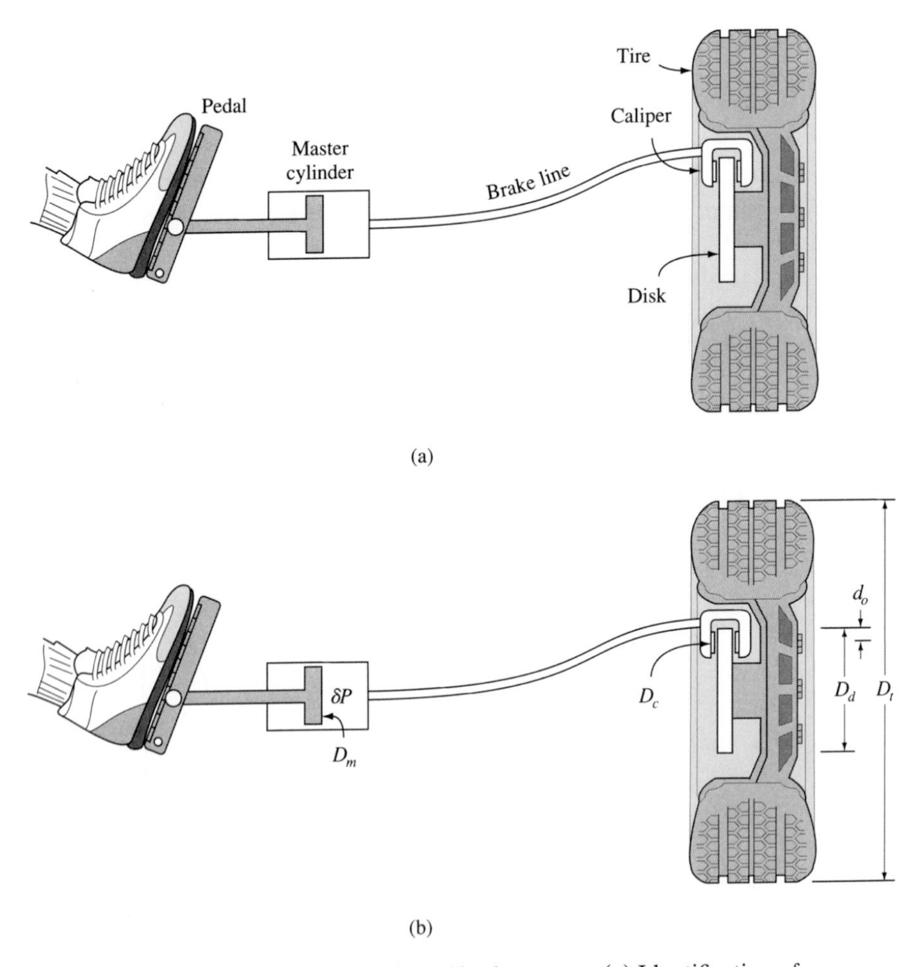


**Figure 10.4** Overall configuration of brake system. (a) Identification of brake system components. (b) Variables and dimensions.

ders through a bias bar. The bias bar ratios the foot force between the front and rear master cylinders to achieve the necessary front and rear bias, depending upon the static weight distribution of the car and the weight transfer during deceleration. The pressure generated by the master cylinder is fed to the brake calipers, which actuates a piston that forces the brake pads against the brake disk. The friction between the pads and the disk causes a torque in the wheel, which produces the adhesion force in the tire that stops the car.

Since this example is rather lengthy and involved, the nomenclature used in analyzing the system is presented next.

## NOMENCLATURE FOR BRAKE SYSTEM ANALYSIS

$a$ acceleration of vehicle

$F_{foot}$ force applied to pedal by driver's foot

$F_{bias}$ force transmitted to bias bar

$L_p$ length of pedal lever to ball of driver's foot

$L_b$ length of pedal lever to bias bar

$G_m$ mechanical advantage of pedal and intermediate lever ratios

$\delta x_{foot}$ motion of foot

$\delta x_{bias}$ motion of bias bar

$Y_f, Y_r$ length of bias bar to front and rear master cylinders

$Y_b$ total length of bias bar $= Y_f + Y_r$

$X_{bf}, X_{br}$ bias of forces to front and rear master cylinders

$F_{mf}, F_{mr}$ force to front and rear master cylinders

$\delta x_{mf}, \delta x_{mr}$ displacement of front and rear master cylinders when brakes are applied

$D_{mf}, D_{mr}$ diameter of front and rear master cylinder pistons

$A_{mf}, A_{mr}$ area of front and rear master cylinder pistons $= \pi D_{mf}^2/4$ and $\pi D_{mr}^2/4$

$\delta P_f, \delta P_r$ pressure generated in front and rear master cylinders

$D_{cf}, D_{cr}$ diameter of front and rear brake caliper pistons

$A_{cf}, A_{cr}$ area of front and rear brake caliper pistons $= \pi D_{cf}^2/4$ and $\pi D_{cr}^2/4$

$\delta x_{cf}, \delta x_{cr}$ displacement of front and rear caliper pistons

$\delta V_{cf}, \delta V_{cr}$ volume of fluid displaced in front and rear calipers as piston strokes

$\delta V_{mf}, \delta V_{mr}$ volume of fluid displaced in front and rear master cylinders as piston strokes

$\delta V_{\beta f}, \delta V_{\beta r}$ volume of fluid displaced in front and rear master cylinders due to compressibility

$L_{cf}, L_{cr}$ equivalent length of front and rear caliper cylinders

$L_{mf}, L_{mr}$ equivalent length of front and rear master cylinders
$V_f, V_r$ total volume of fluid in front and rear brake systems (with no pressure)
$\beta$ bulk modulus of brake fluid
$\mu_b$ coefficient of friction of brake pad on disk
$d_{of}, d_{or}$ offset distance to the center of front and rear calipers
$D_{df}, D_{dr}$ diameter of front and rear brake disks
$n_f, n_r$ number of disks in the front and rear
$D_{tf}, D_{tr}$ diameter of front and rear tires
$T_f, T_r$ front and rear torque generated by one brake disk
$F_{df}, F_{dr}$ front and rear tire braking force generated by one brake disk
$F_f, F_r$ front and rear total braking force from both tires
$W$ total weight of vehicle and driver
$h$ height from ground to the center of gravity of vehicle and driver
$Z_f, Z_r$ distances from front and rear tires to the horizontal location of the center of gravity
$B$ wheelbase of vehicle
$X_{wf}, X_{wr}$ static weight distribution on front and rear tires
$W_f, W_r$ weight on front and rear tires (including static and weight transfer)
$G$ deceleration of vehicle in $g$'s $= -a/g$
$\mu$ coefficient of adhesion between tire and pavement

### 10.3.1 Foot Pedal Geometry and Forces

The foot pedal assembly shown in Figure 10.5 has an input force from the ball of the foot that is amplified by the pedal lever ratio to produce a force to the bias bar as illustrated in the figure. From torque balances about the main pedal pivot, we know that the force transmitted to the bias bar is

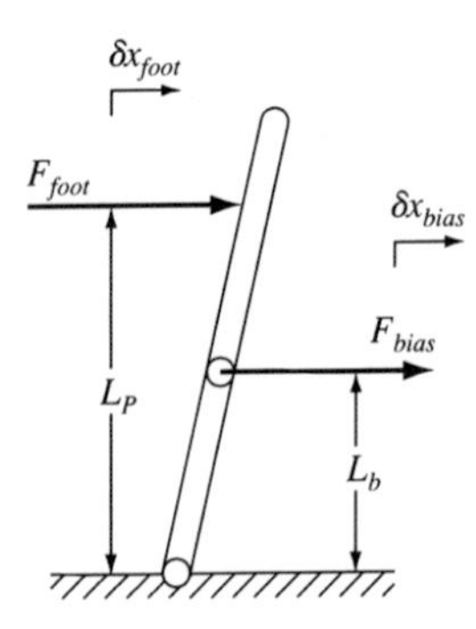


**Figure 10.5** Foot pedal lever to bias bar.

$$F_{bias} = G_m F_{foot} \tag{10.15}$$

where $G_m$ is the mechanical advantage of the lever

$$G_m = \frac{L_p}{L_b} \tag{10.16}$$

For small angular motions of the pedal, the kinematics of the motion between the foot and the bias bar can be stated as

$$\delta x_{bias} = \frac{\delta x_{foot}}{G_m} \tag{10.17}$$

The force from the bias bar is distributed to the master cylinders according to the adjustment of the bar, as illustrated in Figure 10.6.

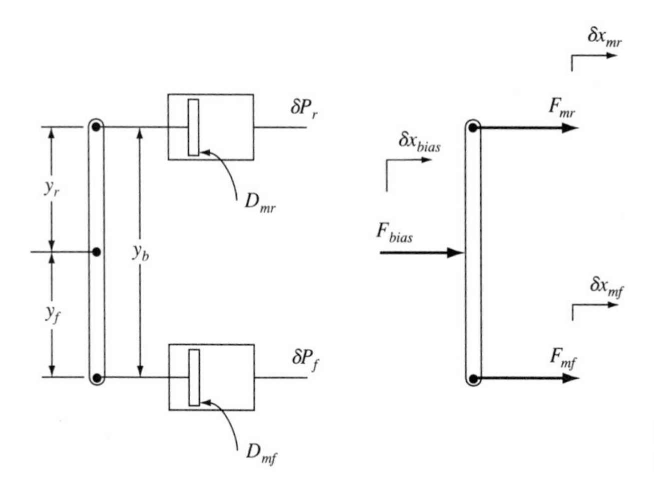


**Figure 10.6** Bias bar force distribution.

A force balance on the bias bar reveals the relation between the force on the bar and the forces transmitted to the master cylinders:

$$F_{bias} = F_{mf} + F_{mr} \tag{10.18}$$

Using torque balances about the pivot points at the ends of the bar, we can determine the forces to the master cylinders. It is desirable that the bias bar setting $X_{bf}$, be at 50% for the nominal maximum braking situation. We have

$$F_{mf} = X_{bf}\, F_{bias} \qquad F_{mr} = X_{br}\, F_{bias} \tag{10.19a,b}$$

where

$$X_{bf} = \frac{Y_r}{Y_b} \qquad X_{br} = \frac{Y_f}{Y_b} \tag{10.20a,b}$$

and

$$Y_b = Y_f + Y_r \qquad \text{or} \qquad 1 = X_{bf} + X_{br} \tag{10.21a,b}$$

The kinematics of the deflections of the bias bar can be stated as follows (for small angular motions):

$$\delta x_{bias} = X_{bf}\,\delta x_{mf} + X_{br}\,\delta x_{mr} \quad \text{and} \quad \delta x_{foot} = G_m[X_{bf}\,\delta x_{mf} + X_{br}\,\delta x_{mr}] \tag{10.22a,b}$$

### 10.3.2 Brake Hydraulics and Pad Friction

The forces to the master cylinders produce static pressures that depend upon the area of the piston (neglecting any small friction due to the motion of the piston itself):

$$\delta P_f = \frac{F_{mf}}{A_{mf}} \qquad \delta P_r = \frac{F_{mr}}{A_{mr}} \tag{10.23a,b}$$

These pressures cause the pistons in the brake calipers to apply a force to the disks by friction of the brake pads. This force is the pressure times the caliper piston area (whether the caliper has one or two pistons). The center of the force is offset a distance from the outer edge of the brake disk. The coefficient of friction of the brake pad material on the disk causes a force to be transmitted to the disk (force = $\mu_b \times$ normal force). Since there are two pads (one on each side), the force is doubled. These forces cause a torque on the wheel by the action of the frictional forces at a radius of half of the disk diameter minus the offset distance of the center of the caliper piston. Thus, the torque on the individual wheel is

$$T_{df} = C_f\left(\frac{D_{df}}{2} - d_{of}\right)\delta P_f \qquad T_{dr} = C_r\left(\frac{D_{dr}}{2} - d_{or}\right)\delta P_r \tag{10.24a,b}$$

The caliper coefficient $C$ is twice the area of the piston (since there are two pads, each causing frictional forces on the disk) times the coefficient of friction of the pad $\mu_b$:

$$C_f = 2\,A_{cf}\,\mu_b \qquad C_r = 2\,A_{cr}\,\mu_b \tag{10.25a,b}$$

The brake calipers deflect a small amount $\delta x_c$ when actuated and therefore displace a small volume $\delta V_c$ of fluid in the front and rear brakes:

$$\delta V_{cf} = n_f A_{cf}\,\delta x_{cf} \qquad \delta V_{cr} = n_r A_{cr}\,\delta x_{cr} \tag{10.26a,b}$$

The volume displaced in the master cylinder, $\delta V_m$, is equal to the volume displaced in the caliper plus the volume of fluid required due to the compressibility of the fluid. The compressibility of the volume of fluid is related to the bulk modulus of the fluid, $\beta$, and the pressure in the system. We obtain

$$\delta V_{mf} = \delta V_{cf} + \frac{V_f}{\beta}\,\delta P_f \qquad \delta V_{mr} = \delta V_{cr} + \frac{V_r}{\beta}\,\delta P_r \tag{10.27a,b}$$

The total volume of fluid in each system is given by

$$V_f = A_{mf} L_{mf} + n_f \, A_{cf} \, L_{cf} \qquad V_r = A_{mr} \, L_{mr} + n_r \, A_{cr} \, L_{cr} \quad (10.28a,b)$$

Note that the volume of fluid in the interconnecting lines and the mechanical compliance of the lines themselves are neglected in the analysis.

The displaced volume of the master cylinders allows us to calculate the strokes of the cylinders:

$$\delta x_{mf} = \frac{\delta V_{mf}}{A_{mf}} = \frac{n_f \, A_{cf} \, L_{cf} + \left(\dfrac{A_{mf} L_{mf} + n_f A_{cf} L_{cf}}{\beta}\right)\delta P_f}{A_{mf}} \quad (10.29a)$$

$$\delta x_{mr} = \frac{\delta V_{mr}}{A_{mr}} = \frac{n_r \, A_{cr} \, L_{cr} + \left(\dfrac{A_{mr} \, L_{mr} + n_r \, A_{cr} \, L_{cr}}{\beta}\right)\delta P_r}{A_{mr}} \quad (10.29b)$$

These strokes can now be used to determine whether the master cylinder has enough stroke for the brakes and to calculate how much the foot moves:

$$\delta x_{foot} = G_m X_{bf} \left\{ \frac{n_f A_{cf} L_{cf} + \dfrac{V_f}{\beta}\delta P_f}{A_{mf}} \right\} + G_m X_{br} \left\{ \frac{n_r \, A_{cr} \, L_{cr} + \dfrac{V_r}{\beta}\delta P_r}{A_{mr}} \right\} \quad (10.30)$$

### 10.3.3 Braking Forces on the Car

The torques generated on the wheels due to the brakes cause forces on the tire at the point of contact with the road. These forces are held through the tire adhesion and are the torque divided by the radius of the tire:

$$F_{df} = \frac{T_{df}}{D_{tf}/2} \qquad F_{dr} = \frac{T_{dr}}{D_{tr}/2} \quad (10.31a,b)$$

The total braking forces are the forces on each disk times the number of disks in the front and rear:

$$F_f = n_f \, F_{df} \qquad F_r = n_r \, F_{dr} \quad (10.32a,b)$$

Thus, the combined expressions for the braking forces generated in the front and the rear can be stated as

$$F_f = \frac{n_f C_f (D_{df}/2 - d_{of}) \, X_{bf}}{(D_{tf}/2) \; A_{mf}} G_m \, F_{foot} \quad (10.33a)$$

$$F_r = \frac{n_r \, C_r (D_{dr}/2 - d_{or}) \, (1 - X_{bf})}{(D_{tr}/2) \;\; A_{mr}} G_m \, F_{foot} \quad (10.33b)$$

Notice that this configuration of two master cylinders and a bias bar produces pressures that increase linearly with the force of the foot and that the front and rear pressures also have a linear relationship, namely,

$$\delta P_r = \frac{A_{mf}(1 - X_{bf})}{A_{mr}\, X_{bf}} \delta P_f \tag{10.34}$$

as shown in Figure 10.7a.

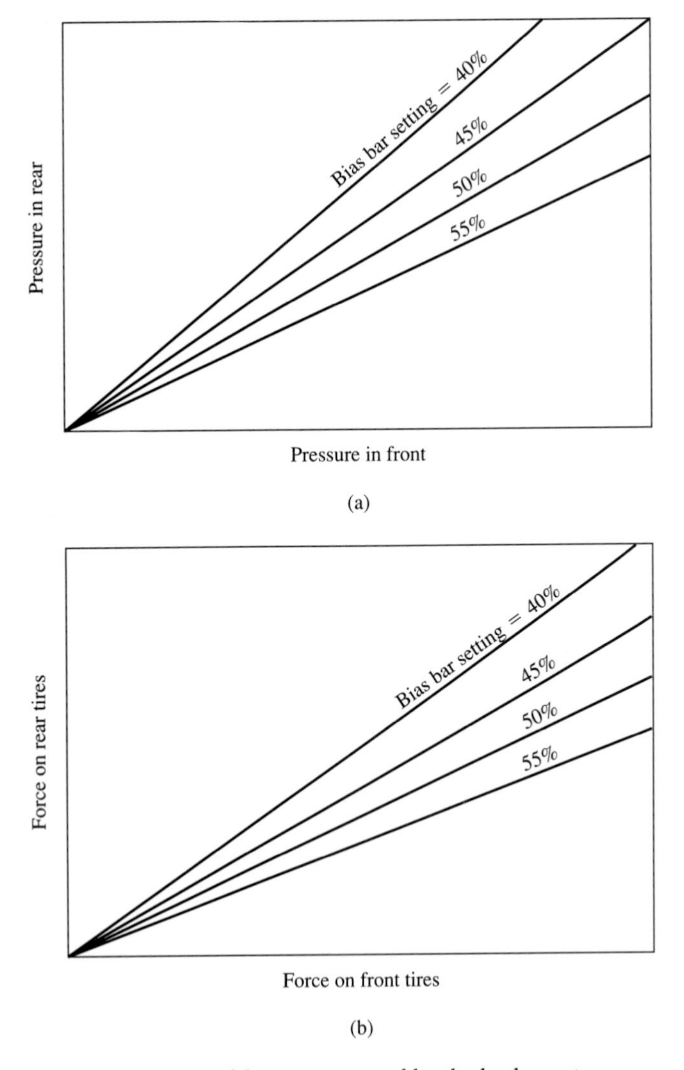


**Figure 10.7** Pressures and forces generated by the brake system. (a) Front and rear pressures. (b) Stopping forces (front and rear).

The total braking forces on the front and rear also have a linear relationship, namely,

$$F_r = \frac{n_r C_r (D_{dr}/2 - d_{or}) D_{tf} A_{mf}(1 - X_{bf})}{n_f C_f (D_{df}/2 - d_{of}) D_{tr} A_{mr} X_{bf}} F_f \tag{10.35}$$

This relationship is shown in Figure 10.7b, and the slope of the graph can be set with the bias bar.

### 10.3.4 Weight Transfer with Deceleration

To determine what the car needs, we show, in Figure 10.8a, a car with its wheelbase and the location of its center of gravity illustrated. Figure 10.8b illustrates the forces acting on the car. The force of gravity, $Mg$, and the deceleration force, $Ma$, act at the center of gravity of the car; the reaction forces at the wheels illustrate the force or weight on the front and rear wheels, as well as the adhesion forces of the tire with the road.

First, note that the weight of the car is the mass times the pull of gravity $g$:

$$W = M g \tag{10.36}$$

Second, observe that the force due to deceleration $a$ can be expressed as a function of the weight and the normalized acceleration, or the acceleration divided by the pull of gravity:

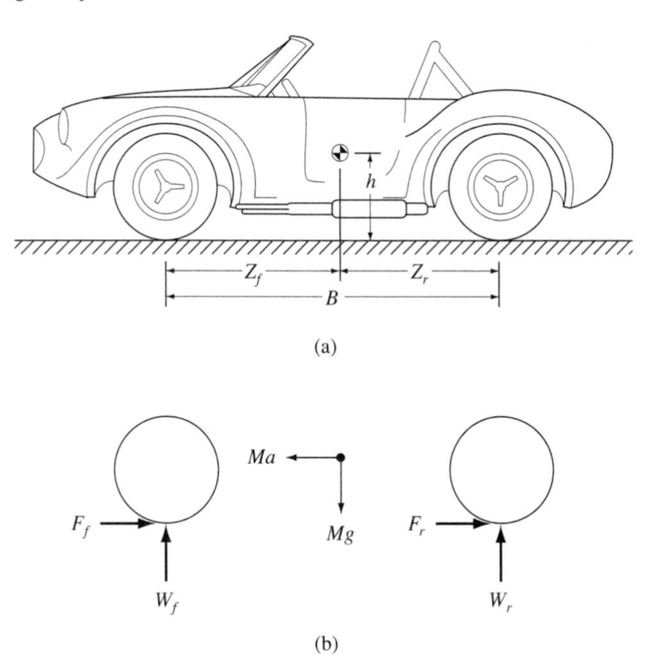


**Figure 10.8** Car dimensions and forces. (a) Geometry of the car. (b) Forces acting on the car.

$$M\,a = M\,g\,\frac{a}{g} = W\frac{a}{g} = W\,G \tag{10.37}$$

The ratio $a/g$ is termed the $g$'s of deceleration:

$$G = \frac{a}{g} \tag{10.38}$$

A force balance in the direction of gravity relates the instantaneous weight on each set of wheels to the weight of the car:

$$W = W_f + W_r \tag{10.39}$$

A force balance in the direction of motion relates the deceleration to the tire adhesion forces:

$$WG = F_f + F_r \quad \text{or} \quad \frac{F_f}{W} + \frac{F_r}{W} = G \tag{10.40}$$

From this result, we see that the sum of the forces from the tire adhesion is related to the deceleration $g$'s. Thus, lines of constant deceleration can be shown in the graph of Figure 10.9.

A static torque balance at the point of contact of the front and rear tires with the road defines the static weight distribution on the front and rear tires:

$$W_f = \frac{Z_r}{B}W \qquad W_r = \frac{Z_f}{B}W \tag{10.41a,b}$$

Thus, the static weight distributions on the front and rear wheels are given by

$$X_{wf} = \frac{Z_r}{B} \qquad X_{wr} = \frac{Z_f}{B} \tag{10.42a,b}$$

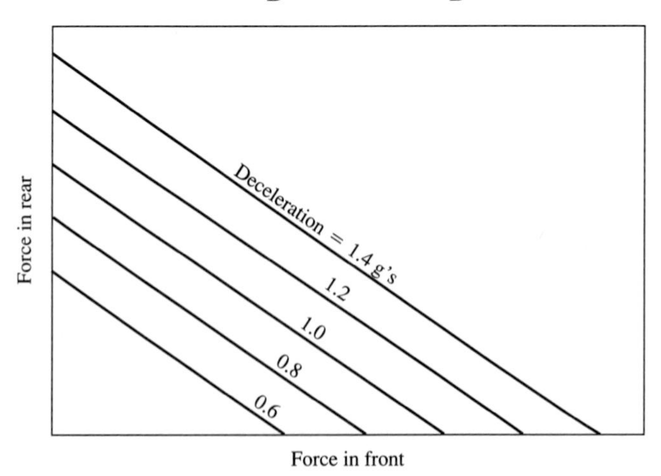


**Figure 10.9** Lines of constant deceleration.

where

$$Z_f + Z_r = B \qquad \text{and thus,} \qquad X_{wf} + X_{wr} = 1 \tag{10.43a,b}$$

A torque balance at the points of contact of the front and rear tires, considering deceleration, yields the weight transfer equations

$$BW_f = Z_r W + hWG \qquad BW_r = Z_f W - hWG \tag{10.44a,b}$$

or

$$\frac{W_f}{W} = X_{wf} + \frac{h}{B}G \qquad \frac{W_r}{W} = X_{wr} - \frac{h}{B}G \tag{10.45a,b}$$

### 10.3.5 Tire Adhesion

Tire adhesion is somewhat like friction, except in friction we normally have relative motion or slip between the two surfaces, whereas with tires we want there to be adhesion without the sliding motion. In this case, we define slip as the velocity of the tire relative to the road as a percentage of the car's speed. If we apply the brakes to the maximum and cause a total tire lockup, then we have 100% slip; if we apply the brakes until the tires are slipping slightly (and probably just starting to squeal), we will have the maximum braking effect, as shown in Figure 10.10. In the figure, the braking adhesion is the ratio of the tire adhesion force to the total normal force, or weight on the tire, at that instant (static weight plus weight transfer). This ratio of forces is also equal to the normalized deceleration, or *g*'s.

**Figure 10.10** Tire adhesion characteristics and maximum $\mu$.

Notice from Figure 10.10 that as the braking effort is increased, there is more deceleration, until a peak is observed (at about 5% slip in this case). Applying more braking effort will cause more wheel slip and a decrease in braking adhesion. Thus, the optimum deceleration occurs with a limited amount of slip, and further braking effort from the driver's foot will have less deceleration and, possibly, including 100% slip, or tire lockup. Similar to the coefficient of friction, the coefficient of adhesion of a tire $\mu$ is the maximum value shown in the figure.

At this juncture, we point out the equivalence of the force ratio of the deceleration force to the normal force (or weight on the tire), the coefficient of adhesion, and the normalized deceleration (or $g$'s).

### 10.3.6 Optimum Brake Distribution

From the previous discussion, we can see that if the braking forces are less than $\mu$ times the weight on the tire, then we have adhesion. That is,

$$\text{if } F_f < \mu W_f \quad \text{and} \quad F_r < \mu W_r \text{ then adhesion} \tag{10.46a,b}$$

The maximum braking effectiveness, or the optimum braking force, is achieved when the braking forces are exactly equal to $\mu$ times the weight on the tire:

$$F_f^* = \mu W_f \qquad F_r^* = \mu W_r \tag{10.47a,b}$$

Further, in optimal braking, the braking forces in the front and rear achieve the optimum values at the same time. Thus, the optimum brake distribution can be stated as

$$\frac{F_f^*}{W} = \mu \frac{W_f}{W} = \mu\left[X_{wf} + \frac{h}{B}G\right] \qquad \frac{F_r^*}{W} = \mu \frac{W_r}{W} = \mu\left[X_{wr} - \frac{h}{B}G\right] \tag{10.48a,b}$$

Recalling that the sum of the weight distribution is equal to 1.0, noting that the sum of the previous two equations is thus equal to $\mu$, and then remembering that the sums of the front and rear braking forces are each equal to $W\,G$, we can see that the maximum deceleration of the car is equal to $\mu$ if the brake bias is set to the optimum. In other words,

$$\text{since } \frac{F_f}{W} + \frac{F_r}{W} = G \quad \text{and} \quad \frac{F_f^*}{W} + \frac{F_r^*}{W} = \mu \quad \text{then } G_{max} = \mu \tag{10.49}$$

Substituting $G_{max} = \mu$ into the brake distribution equations yields the optimum relation of the brake distribution:

$$\frac{F_f^*}{W} = \mu\left[X_{wf} + \frac{h}{B}\mu\right] \qquad \frac{F_r^*}{W} = \mu\left[X_{wr} - \frac{h}{B}\mu\right] \tag{10.50a,b}$$

The relationship of these optimum braking forces as a function of $\mu$ is shown in Figure 10.11. Notice that this is now a nonlinear relationship such that the optimum ratio of the front to rear forces is a function of the $\mu$ of the road. (Recall Figure 10.9.)

In order to better interpret the graph of Figure 10.11, Figure 10.12 illustrates that, for a given $\mu$ of the road, if the brake bias bar is set too high or too low, then the brakes

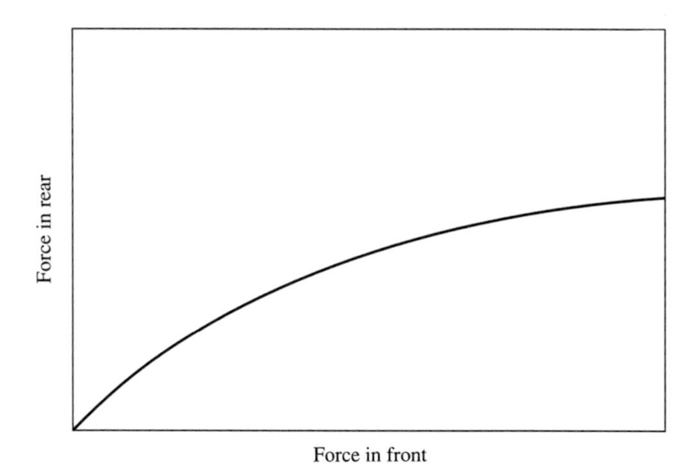


**Figure 10.11** Optimum brake distribution.

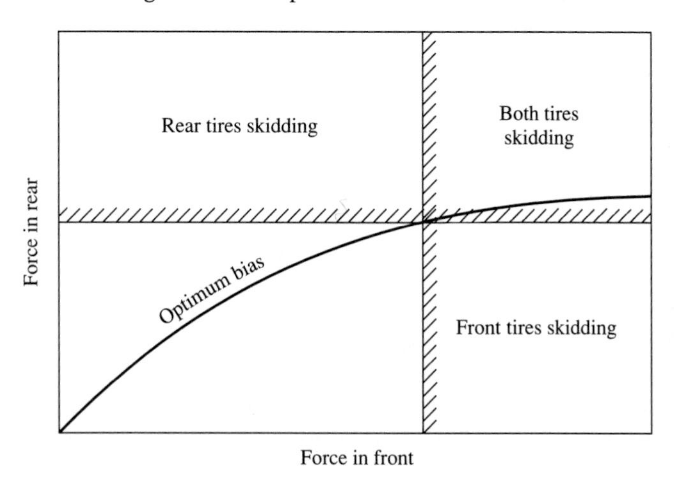


**Figure 10.12** Illustration of braking forces exceeding the optimal value, causing skidding.

will apply more pressure to the front or rear tires than is optimal. Therefore, any tire braking forces in excess of the optimal value will result in tire slip or tire lockup.

In terms of pressure, the optimum braking performance can be stated as

$$\delta P_f^* = \frac{D_{tf} W \mu \left[ X_{wf} + \frac{h}{B}\mu \right]}{2\, n_f C_f \left( \frac{D_{df}}{2} - d_{of} \right)} \qquad \delta P_r^* = \frac{D_{tr} W \mu \left[ X_{wr} - \frac{h}{B}\mu \right]}{2\, n_r C_r \left( \frac{D_{dr}}{2} - d_{or} \right)} \qquad (10.51a,b)$$

In order to achieve this optimum brake distribution, the bias bar must be set to pass through the optimum braking point for a given $\mu$ of the road surface, as is illustrated in Figure 10.13.

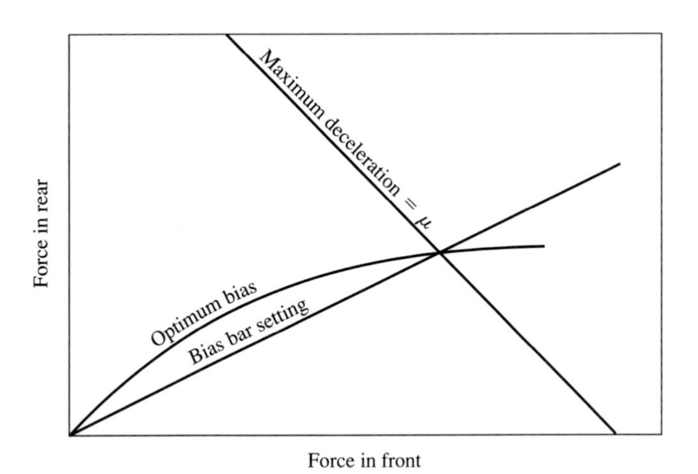


**Figure 10.13** Optimum braking forces and bias bar adjustment for a given $\mu$.

This optimum bias bar setting can be expressed as

$$x_{bf} = \cfrac{1}{1 + \cfrac{n_f\, C_f\,(D_{df}/2 - d_{of})\, D_{tr}\, A_{mr}\left(X_{wr} - \frac{h}{B}\mu\right)}{n_r\, C_r\,(D_{dr}/2 - d_{or})\, D_{tf}\, A_{mf}\left(X_{wf} + \frac{h}{B}\mu\right)}} \tag{10.52}$$

### 10.3.7 Engineering Sizing for a Specific Application

In order to illustrate the sizing and selection of brake components and parameters, we will consider a small race car known as Formula SAE® and used in intercollegiate engineering competitions. The size and specifications of a typical car are as follows:

$$W = 650 \text{ lbf (including driver)}$$

$$B = 75 \text{ inch}$$

$$h = 12 \text{ inch}$$

$$X_{wf} = 0.45 \qquad \text{thus} \quad X_{wr} = 0.55$$

$$D_{tf} = 20 \text{ inches} \qquad \text{and} \quad D_{tr} = 20 \text{ inches}$$

In this application, two brake disks are used in the front and only one brake disk in the rear, so

$$n_f = 2 \qquad n_r = 1$$

The slick race tires can hold 1.4 $g$'s in braking on a typical road surface. The typical brake components are rated at 1000 psi; therefore, we will use about 600 psi to include a safety factor. A typical driver would expect to push the brake pedal with a foot force of about 100 lbf to get 1.4 *g's*. Hence,

$$\mu = 1.4\, g's$$

$$P_{max} = 600 \text{ psi}$$

$$F_{footmax} = 100 \text{ lbf}$$

It is desired that the bias bar be set to 50% at maximum braking, to allow minor adjustments for variations in $\mu$ on different road surfaces or as the tire wears.

The front brake disks must fit inside the wheels. We want to use the largest diameter of disk for better braking and for heat dissipation. Thus, the maximum possible diameter of the disk is 10.5 inches, but we will try to use a 10.0 inch disk for more clearance. We can use a disk of 9 to 11 inches in the rear. The calipers have an offset of 0.875 inch to the center of the caliper piston. Mathematically,

$$D_{df} = 10 \text{ in}$$

$$D_{dr} = 9 \text{ to } 11 \text{ in}$$

$$d_{of} = d_{or} = 0.875 \text{ in}$$

The first selection is the coefficients of the calipers, from the tire forces required for deceleration at 1.4 *g's*:

$$F^*_{f\,max} = \mu\left[X_{wf} + \frac{h}{B}\mu\right]W = 1.4\left[0.45 + \frac{12}{75}1.4\right]650 \text{ lbf} = 613 \text{ lbf} \qquad (10.53a)$$

$$F^*_{r\,max} = \mu\left[(1 - X_{wf}) - \frac{h}{B}\mu\right]W = 1.4\left[0.55 - \frac{12}{75}1.4\right]650 \text{ lbf} = 297 \text{ lbf} \qquad (10.53b)$$

The torque that can be generated by the brake system at the maximum selected pressure is

$$T^*_{f\,max} = n_f C_f (D_{df}/2 - d_{of}) P_{f\,max} = \frac{D_{tf}}{2} F^*_{f\,max} \qquad (10.54a)$$

$$T^*_{r\,max} = n_r C_r (D_{dr}/2 - d_{or}) P_{r\,max} = \frac{D_{tr}}{2} F^*_{r\,max} \qquad (10.54b)$$

Thus, the coefficients of the calipers are

$$C_f = \frac{\dfrac{D_{tf}}{2} F^*_{f\,max}}{n_f (D_{df}/2 - d_{of})\, P_{f\,max}} = \frac{\dfrac{20}{2} \text{ in } 613 \text{ lbf}}{2\,(10/2 - 0.875) \text{ in } 600 \text{ psi}} = 1.24 \,\frac{\text{lbf}}{\text{psi}} \qquad (10.55a)$$

$$C_r = \frac{\dfrac{D_{tr}}{2} F^*_{r\,max}}{n_r (D_{dr}/2 - d_{or}) P_{r\,max}} = \frac{\dfrac{20}{2} \text{ in } 297 \text{ lbf}}{1(10/2 - 0.875) \text{ in } 600 \text{ psi}} = 1.20 \,\frac{\text{lbf}}{\text{psi}} \qquad (10.55b)$$

A survey of available brake calipers reveals that a piston with a diameter of 1.75 inches is popular. This results in a brake caliper coefficient of 1.44 lbf/psi, assuming a $\mu_b$ of 0.30. If we use 1.75 inch calipers in the front and the rear, the maximum pressures would be 517 psi and 500 psi, respectively.

Thus, the final selections for our parameters are as follows:

$$D_{df} = 10 \text{ in} \qquad D_{dr} = 10 \text{ in}$$

$$D_{cf} = 1.75 \text{ in} \qquad D_{cr} = 1.75 \text{ in}$$

$$A_{cf} = 2.405 \text{ in}^2 \qquad A_{cr} = 2.405 \text{ in}^2$$

$$C_{cf} = 1.44 \text{ lbf/psi} \qquad C_{cr} = 1.44 \text{ lbf/psi}$$

$$\delta P_f = 517 \text{ psi} \qquad \delta P_r = 500 \text{ psi}$$

We still have several parameters to select that are multiplied together. We need some rationale, however, for selecting them individually rather than selecting their product. We can select an approximate value for the area of the master cylinder if we know the strokes of the caliper and master cylinder pistons. However, the equation for the area of the master cylinder is nonlinear; still, we can make some simplifying approximations to get a preliminary answer. First, we will neglect the compressibility effects and solve for the area from the deflection equations. We must allow for the compressibility by making the area larger; and further, the actual stroke of the master cylinder must be less than the maximum allowable stroke. Thus, the area must actually be larger than that calculated in what follows.

Typical brake calipers have a pad deflection of about 0.015 inch per pad, for a total displacement of 0.030 inch. A typical maximum stroke of the master cylinder is 0.500 inch, but we will use about half of the maximum stroke to accommodate the possibility of air in the brakes and the increased stroke required if air is present. We can use these numbers for an approximate calculation of the area. For the front, we have

$$A_{mf} > \frac{n_f A_{cf}\, \delta x_{cf}}{\delta x_{mf}} = \frac{2 \times 2.405 \text{ in}^2\, 0.030 \text{ in}}{0.250 \text{ in}} = 0.577 \text{ in}^2 \qquad (10.56a)$$

This area results in a front master cylinder piston diameter of

$$D_{mf} > 0.857 \text{ in}$$

For the rear, we have

$$A_{mr} > \frac{n_r A_{cr}\, \delta x_{cr}}{\delta x_{mr}} = \frac{1 \times 2.405 \text{ in}^2\, 0.030 \text{ in}}{0.250 \text{ in}} = 0.289 \text{ in}^2 \qquad (10.56b)$$

This results in a rear master cylinder piston diameter of

$$D_{mr} > 0.606 \text{ in}$$

The preceding analysis gives us approximate values for the master cylinders. We will now go back and consider compressibility effects. Typical values for the effective lengths of the calipers and master cylinders are as follows:

$$L_{cf} = 1.0 \text{ in} \qquad L_{cr} = 1.0 \text{ in}$$

$$L_{mf} = 1.0 \text{ in} \qquad L_{mr} = 1.0 \text{ in}$$

With a bulk modulus of about 200,000 psi, the volume displaced by the master cylinders that is required to compensate for the compressibility is

$$\delta V_{\beta f} = \left[\frac{A_{mf} L_{mf} + n_f A_{cf} L_{cf}}{\beta}\right] \delta P_f$$

$$= \left[\frac{0.577 \text{ in}^2 \, 1.0 \text{ in} + 2 \times 2.405 \text{ in}^2 \, 1.0 \text{ in}}{200{,}000 \text{ psi}}\right] 517 \text{ psi} = 0.0139 \text{ in}^3 \qquad (10.57a)$$

$$\delta V_{\beta r} = \left[\frac{A_{mr} L_{mr} + n_r A_{cr} L_{cr}}{\beta}\right] \delta P_r$$

$$= \left[\frac{0.289 \text{ in}^2 \, 1.0 \text{ in} + 1 \times 2.405 \text{ in}^2 \, 1.0 \text{ in}}{200{,}000 \text{ psi}}\right] 500 \text{ psi} = 0.0067 \text{ in}^3 \qquad (10.57b)$$

This incremental volume required for compressibility must be compared to the displaced volume required of the calipers to find the significance on the stroke of the master cylinders:

$$\delta V_{mf} = \delta V_{cf} + \delta V_{\beta f} = 2 \times 2.405 \text{ in}^2 0.030 \text{ in} + 0.0139 \text{ in}^3 = 0.1582 \text{ in}^3 \qquad (10.58a)$$

$$\delta V_{mr} = \delta V_{cr} + \delta V_{\beta r} = 1 \times 2.405 \text{ in}^2 0.030 \text{ in} + 0.0067 \text{ in}^3 = 0.0789 \text{ in}^3 \qquad (10.58b)$$

Thus, the compressibility represents a displacement volume increase of 10% and 9% for the front and rear master cylinders, respectively.

Now if we go back and reevaluate the master cylinder area requirements, taking into account the compressibility, we can find updated values for the areas of the master cylinders:

$$A_{mf} > \frac{\delta V_{mf}}{\delta x_{mf}} = \frac{0.1582 \text{ in}^3}{0.250 \text{ in}} = 0.633 \text{ in}^2 \qquad (10.59a)$$

$$A_{mr} > \frac{\delta V_{mr}}{\delta x_{mr}} = \frac{0.0789 \text{ in}^3}{0.250 \text{ in}} = 0.316 \text{ in}^2 \qquad (10.59b)$$

These areas result in master cylinder diameters of 0.898 inch and 0.634 inch in the front and rear, respectively. Since we want the bias bar to be set in the middle (or about 50%) for the nominal case, the forces on the front and rear master cylinders will be about the same. Thus, the ratio of the master cylinder areas must be proportional to the pressure ratios:

$$\frac{A_{mf}}{A_{mr}} = \frac{X_{bf}}{(1 - X_{bf})} \frac{\delta P_r}{\delta P_f} \tag{10.60}$$

Since the front and rear pressures are about the same, and the bias bar is about 0.50, the master cylinders should be the same size. A review of available master cylinder sizes reveals that diameters of 0.625, 0.700, 0.750, and 0.875 inch are available. Considering the trade-off between area and stroke, we make the following final selections:

$$D_{mf} = 0.750 \text{ in} \quad \text{with} \quad A_{mf} = 0.442 \text{ in}^2 \quad \text{and} \quad \delta X_{mf} = 0.358 \text{ in}$$

$$D_{mr} = 0.750 \text{ in} \quad \text{with} \quad A_{mr} = 0.442 \text{ in}^2 \quad \text{and} \quad \delta X_{mr} = 0.179 \text{ in}$$

Notice that the front master cylinder strokes slightly more than half the stroke and the rear master cylinder strokes less than half the stroke.

With these areas selected, we can now consider the forces required on the master cylinder push rods, namely,

$$F_{mf} = A_{mf} P_f = 0.442 \text{ in}^2 \, 517 \text{ psi} = 229 \text{ lbf} \tag{10.61a}$$

$$F_{mr} = A_{mr} P_r = 0.442 \text{ in}^2 \, 500 \text{ psi} = 221 \text{ lbf} \tag{10.61b}$$

The force on the bias bar can be calculated as

$$F_{bias} = F_{mf} + F_{mr} = 229 \text{ lbf} + 221 \text{ lbf} = 450 \text{ lbf} \tag{10.62}$$

The driver force at the pedal is about 100 lbf; hence, the overall pedal lever ratio is

$$G_m = \frac{F_{bias}}{F_{foot}} = \frac{450 \text{ lbf}}{100 \text{ lbf}} = 4.50 \tag{10.63}$$

With these selected values, the actual optimum brake bias setting is 50.8%, and the pressure performance illustrating optimum pressures, pressures with various bias bar settings, and various foot efforts is shown in Figure 10.14.

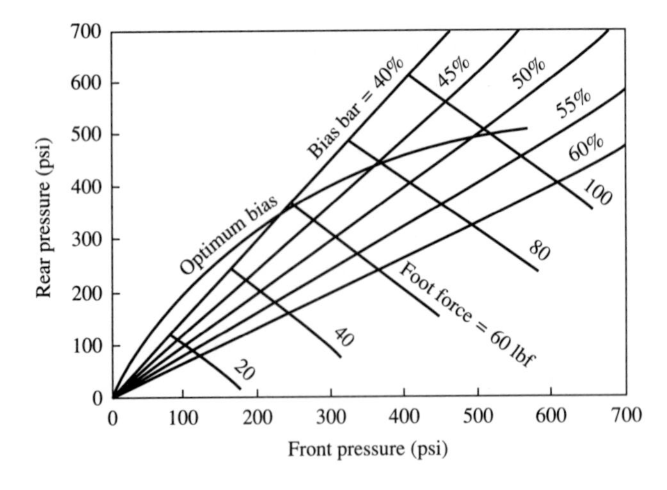


**Figure 10.14** Pressure performance of brake system.

The example has illustrated the use of engineering analysis and the rationale for sizing all of the required component values in this automotive brake system.

## 10.4 DC MOTOR SPEED SERVO CONTROL SYSTEM

As our next example of sizing and selecting components, consider the DC motor speed servo control system of Chapter 7 that is represented by Figure 10.15. (See Section 7.2.1 for the modeling of this closed-loop control system.)

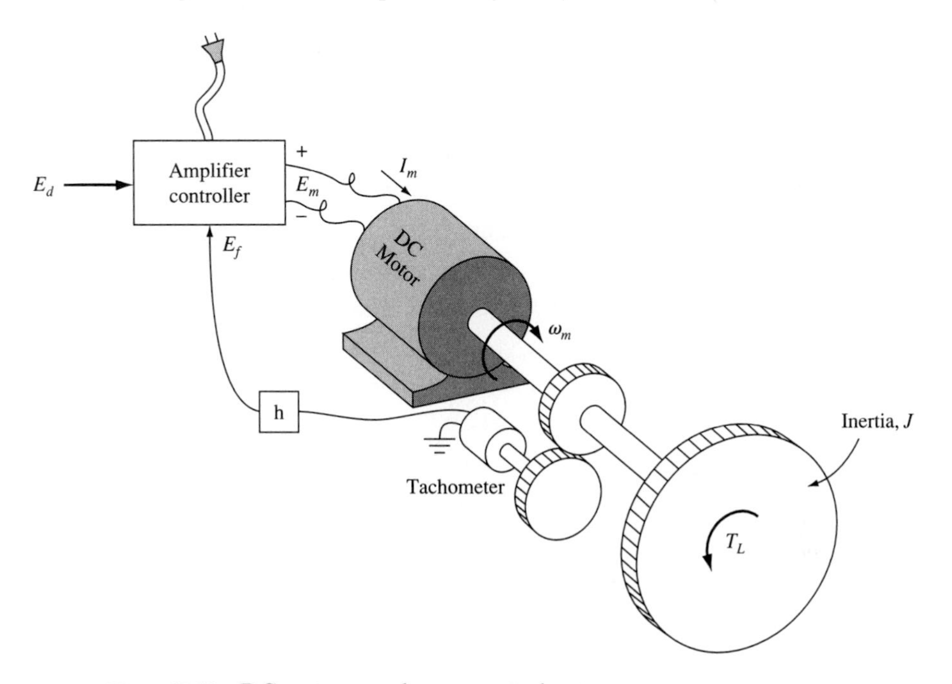


**Figure 10.15** DC motor speed servo control system.

The transfer function for the system, from Chapter 7, is

$$\omega_m = \frac{\dfrac{1/h}{\left(1 + \left(\dfrac{1 + \dfrac{B}{K^2/R_o}}{\dfrac{Gh}{K}}\right)\right)} E_d - \dfrac{R_o/K^2}{\left(1 + \dfrac{B}{K^2/R_o} + \dfrac{Gh}{K}\right)} T_L}{\dfrac{\dfrac{J}{K^2/R_o}}{\left(1 + \dfrac{B}{K^2/R_o} + \dfrac{Gh}{K}\right)} D + 1} \tag{10.64}$$

What we know about the system is the inertia $J$ and the friction $B$ in all of the bearings. The performance requirements dictate that the system should have 1000 rpm for a 10 volt input; thus, we know the desired closed-loop gain $G_s^*$. We want the system to have a time constant of $\tau = 0.100$ second. In addition, we want to use a 48 volt power amplifier to drive the motor. These are the only initial specifications that we have, but we must generate further conditions to determine the remaining unknown coefficients ($G$, $R_o$, $K$, and $h$). We use the following nomenclature in our analysis.

### NOMENCLATURE FOR DC MOTOR SPEED CONTROL

- $E_d$ voltage input command
- $\omega_m$ motor speed
- $T_L$ disturbance torque applied to load
- $G$ amplifier gain
- $R_o$ amplifier output impedance
- $K$ motor constant
- $J$ mass moment of inertia of disk
- $M$ mass of disk
- $r$ radius of disk
- $B$ viscous friction of load
- $h$ gain of feedback sensor
- $E_f$ voltage output from feedback sensor
- $G_L$ loop gain
- $Z_r$ impedance ratio
- $\tau$ desired time constant of the closed-loop system
- $\tau_{natural}$ natural time constant of the load
- $P_\tau$ time constant parameter
- $e$ error signal
- $e^*$ maximum error signal allowable before saturation occurs
- $E_m^*$ supply voltage used to drive the motor
- $E_d^*$ maximum voltage used on the input command

### 10.4.1 Equations And Criteria For Sizing And Selection Of Components

In order to expedite our analysis and sizing, we will restate the transfer function in terms of the loop gain $G_L$ and the impedance ratio $Z_r$:

$$\omega_m = \frac{\dfrac{1/h}{\left(1 + \left(\dfrac{1 + Z_r}{G_L}\right)\right)} E_d - \dfrac{Z_r/B}{(1 + Z_r + G_L)} T_L}{\dfrac{\dfrac{J}{B} Z_r}{(1 + Z_r + G_L)} D + 1} \quad (10.65)$$

The loop gain (the product of all components from the amplifier input $e$ back through the feedback voltage $E_f$) is

$$G_L = \frac{G\,h}{K} \quad (10.66)$$

Notice that the loop gain is dimensionless (or rather, has units of volt/volt).

The impedance ratio $Z_r$ is the ratio of the mechanical impedance $B$ to the output impedance of the amplifier as reflected through the motor, $K^2/R_o$:

$$Z_r = \frac{B}{K^2/R_o} \quad (10.67)$$

Notice that the reflected amplifier impedance has the units of mechanical impedance, even though the reflected impedance is determined by an electrical impedance and the constant of the electromagnetic motor.

The maximum power transfer from the motor to the load occurs when the output impedance of the motor matches (is equal to) the input impedance of the load, $B$. In steady-state operation, this impedance ratio $Z_r$ should be equal to 1.0 for maximum power transfer. However, considering that the system operates dynamically at its maximum frequency ($\omega_{max} = 1/\tau$), the impedance ratio should take into account the impedance of the inertia, as well as the impedance of the friction. A dynamic impedance match will make the motor output impedance $K^2/R_o$ equal to the dynamic load impedance, $J/\tau + B$. The impedance ratio can then be expressed as

$$Z_r = \frac{1}{\dfrac{J}{B\,\tau} + 1} \text{ (for dynamic operation)} \quad (10.68)$$

Thus, the impedance ratio should fall somewhere between the dynamic case and the static case, depending upon whether we want to emphasize static operation or dynamic operation:

$$\frac{1}{\dfrac{J}{B\,\tau} + 1} \leq Z_r \leq 1.0 \quad (10.69)$$

Before we start selecting coefficients, we should look at the system and determine which performance criteria should be used to select each coefficient. In most systems, the static gain is determined chiefly by the feedback gain; the output impedance of the amplifier affects the dynamic response, and the loop gain affects the effectiveness of the servo system.

The first coefficient that we can select is based upon the static gain. The parameter that has the most effect upon static gain is the feedback gain $h$, through the equation

$$G_s = \frac{1/h}{1 + \left(\frac{1 + Z_r}{G_L}\right)} \tag{10.70}$$

Thus, the feedback gain should be

$$h = \frac{\frac{1}{G_s}}{1 + \left(\frac{1 + Z_r}{G_L}\right)} \tag{10.71}$$

The dynamic response is determined by the system time constant

$$\tau = \frac{\frac{J}{B} Z_r}{(1 + Z_r + G_L)} \tag{10.72}$$

The natural time constant, $\tau_{natural}$, of the open-loop system is $J/B$, which usually will be very large (indicating a slow system response). The closed-loop time constant will be very small (indicating a fast system response) compared to that of the open-loop system. We can express our desired closed-loop time constant as a function of the open-loop time constant by defining the time-constant parameter

$$P_\tau = \frac{\tau}{\tau_{natural}} = \frac{Z_r}{(1 + Z_r + G_L)} \tag{10.73}$$

Since $J$, $B$, and $\tau$ are all known, $P_\tau$ is known, and the impedance ratio can be found as a function of the loop gain:

$$Z_r = \frac{P_\tau(1 + G_L)}{(1 - P_\tau)} \tag{10.74}$$

This impedance ratio should be selected to be in the range specified in Eq. (10.69).

Amplifiers have a maximum output value that is limited by the supply voltage. If the input caused the output to reach this maximum value, then further increases in the input will not cause a proportionate increase in the output, and the output will remain fixed as the input increases. This nonlinear effect is termed **saturation** and will limit the rate at which the servo system responds. Because the response will no longer be predicted by the linear performance factors, it is usually desirable to avoid saturation effects in a system.

The servo system we are examining will become saturated if the input signal to the amplifier $e$ is larger than the critical value $e^*$, which is determined by the maximum output voltage $E_m^*$ and the amplifier gain $G$. That is,

$$e \leq e^* = \frac{E_m^*}{G} \tag{10.75}$$

The steady-state error signal, found from the system transfer function (with $T_L = 0$), is (after reduction using Eq. (10.65))

$$e = E_d^* - h\,\omega_m = \frac{E_d^*}{1 + \dfrac{G_L}{1 + Z_r}} \tag{10.76}$$

Using the preceding two equations, we can solve for the maximum allowable amplifier gain to avoid saturation:

$$G \leq \frac{E_m^*}{E_d^*}\left(1 + \frac{G_L}{1 + Z_r}\right) \tag{10.77}$$

We find the motor constant $K$ from the loop gain and the selected values for $h$ and $G$:

$$K = \frac{G\,h}{G_L} \tag{10.78}$$

Finally, we can calculate the required amplifier output impedance $R_o$ from the definition of the impedance ratio:

$$R_o = \frac{Z_r K^2}{B} \tag{10.79}$$

We can now begin selecting the values of our components. We select the loop gain $G_L$ to cause the impedance ratio to be between our specified dynamic and static limits. Then, knowing $Z_r$, we can calculate $h$, $G$, $K$, and $R_o$ as illustrated in the next subsection.

### 10.4.2 Selection of Numeric Values

The system under consideration has a uniform 10 inch diameter disk that weighs 2 pounds. Thus, the inertia is

$$J = \frac{Mr^2}{2} = \frac{2\text{ lbf } 5^2\text{ in}^2\text{ s}^2\text{ ft}}{2 \times 32.18\text{ ft } 12\text{ in}} = 0.06474\text{ in lbf s}^2 \tag{10.80}$$

The motor and bearings have 0.333 in lbf of friction at 1000 rpm. Hence, the friction coefficient can be calculated as

$$B = \frac{0.333\text{ in lbf}}{1000\,\dfrac{\text{rev}}{60\text{ s}}\,\dfrac{2\pi\text{ rad}}{\text{rev}}} = 0.003183\text{ in lbf s} \tag{10.81}$$

The natural time constant, or the time constant of the open-loop system, is

$$\tau_{natural} = \frac{J}{B} = \frac{0.06474 \text{ in lbf s}^2}{0.003183 \text{ in lbf s}} = 20.34 \text{ s} \tag{10.82}$$

The desired time constant of the system, $\tau$, is 0.100 second. Thus, we can calculate the time-constant parameter as

$$P_\tau = \frac{\tau}{\tau_{natural}} = \frac{0.100 \text{ s}}{20.34 \text{ s}} = 0.004917 \tag{10.83}$$

From this parameter, we can calculate the impedance ratio. Using a loop gain of $G_L = 0.5$, we obtain

$$Z_r = \frac{P_\tau(1 + G_L)}{1 - P_\tau} = \frac{0.004917(1 + 0.5)}{1 - 0.004917} = 0.007412 \tag{10.84}$$

We must compare this value of the impedance ratio to the values specified previously. For a dynamic impedance match, we have

$$Z_r = \frac{1}{\dfrac{J}{B\,\tau} + 1} = \frac{1}{\dfrac{20.34 \text{ s}}{0.10 \text{ s}} + 1} = 0.004892 \tag{10.85}$$

The impedance ratio required for a dynamic impedance match is 0.00489; for a static impedance match, it is 1.0. Thus, our impedance ratio of 0.007412 is 1.5 times the dynamic (or minimum) value and is much less than the value required for a static match.

The feedback gain can be calculated from the desired static gain $G_s^*$ and the foregoing parameters:

$$h = \frac{\dfrac{1}{G_s}}{1 + \left(\dfrac{1 + Z_r}{G_L}\right)} = \frac{\dfrac{\text{volt}}{100 \text{ rpm}}}{1 + \left(\dfrac{1 + 0.007412}{0.5}\right)} = 0.003317 \frac{\text{volt}}{\text{rpm}} \tag{10.86}$$

Hence, the feedback sensor would produce 3.317 volts at 1000 rpm.

The amplifier gain $G$ should be as high as possible without causing saturation; we therefore select the maximum gain:

$$G = \frac{E_m^*}{E_d^*}\left(1 + \frac{G_L}{1 + Z_r}\right) = \frac{48 \text{ volt}}{10 \text{ volt}}\left(1 + \frac{0.5}{1 + 0.007412}\right) = 7.182 \frac{\text{volt}}{\text{volt}} \tag{10.87}$$

The motor constant $K$ can now be calculated:

$$K = \frac{G\,h}{G_L} = \frac{7.182 \dfrac{\text{volt}}{\text{volt}}\; 0.00332 \dfrac{\text{volt}}{\text{rpm}}}{0.5} = 0.04765 \frac{\text{volt}}{\text{rpm}} \tag{10.88}$$

This motor would produce 1007 rpm at 48 volts.

The amplifier output impedance $R_o$ can be calculated as follows (note that some rather involved conversions from ft lbf/s to watt or volt amp are required):

$$R_o = \frac{Z_r K^2}{B} = \frac{0.007412\ 0.04765^2 \dfrac{\text{volt}^2\ 60^2\ \text{s}^2\ \text{rev}^2}{\text{rev}^2\ (2\pi)^2\ \text{rad}^2}}{0.003183 \dfrac{\text{in lbf s ft}}{12\ \text{in}} \dfrac{1.356\ \text{volt amp}}{\dfrac{\text{ft lbf}}{\text{s}}}} = 4.267 \frac{\text{volt}}{\text{amp}} \quad (10.89)$$

To summarize, we have calculated the following values for the parameters as our best selections:

Feedback gain: $h = 0.003317 \dfrac{\text{volt}}{\text{rpm}}$ (3.317 volts @ 1000 rpm)

Amplifier gain: $G = 7.182 \dfrac{\text{volt}}{\text{volt}}$ (48 volt output for 6.68 volt input)

Motor constant: $K = 0.04765 \dfrac{\text{volt}}{\text{rpm}}$ (1007 rpm @ 48 volts)

Output impedance: $R_o = 4.267 \dfrac{\text{volt}}{\text{amp}}$ (11.25 amps from 48 volt amplifier)

Now we are faced with the challenge of finding these components from suppliers. The sensor gain and the amplifier gain can be trimmed from existing components very easily; however, the motor constant can be selected only in discrete values from existing motors. Thus, we probably cannot find a motor with a constant exactly equal to 0.04765 volt/rpm. However, a close available substitute is a motor with a constant of 0.04 volt/rpm (1200 rpm @ 48 volts). In a similar manner, an amplifier with this output impedance is also somewhat difficult to find. A substitute with an output impedance of 4.0 ohms is available and will be used instead.

With these new values selected, we can adjust the other parameters. We will maintain the static gain and time constant as specified. However, the loop gain must change to accommodate the stock components. Since $B$, $R_o$, and $K$ are all known, we have

$$Z_r = \frac{B\,R_o}{K^2} = \frac{0.003183\ \text{in lbf s}\ 4.0 \dfrac{\text{volt}}{\text{amp}} \dfrac{1.356\ \text{volt amp s}}{\text{ft lbf}}}{0.04^2 \left(\dfrac{\text{volt}\ 60\ \text{s}}{2\ \pi}\right)^2 12 \dfrac{\text{in}}{\text{ft}}} = 0.00986 \quad (10.90)$$

This value, 2.0 times the minimum, is still within our range.

The new loop gain can be calculated from the time-constant parameter equation:

$$G_L = \frac{(1 - P_\tau)}{P_\tau} Z_r - 1 = \frac{(1 - 0.00492)}{0.00492} 0.00986 - 1 = 0.996 \quad (10.91)$$

The new feedback gain is

$$h = \frac{\dfrac{1}{G_s}}{1 + \left(\dfrac{1 + Z_r}{G_L}\right)} = \frac{\dfrac{\text{volt}}{100\ \text{rpm}}}{1 + \left(\dfrac{1 + 0.00986}{0.996}\right)} = 0.004965\ \frac{\text{volt}}{\text{rpm}} \tag{10.92}$$

The new amplifier gain, based upon the loop gain and selected components, is

$$G = \frac{G_L K}{h} = \frac{0.996\ 0.04\ \dfrac{\text{volt}}{\text{rpm}}}{0.004965\ \dfrac{\text{volt}}{\text{rpm}}} = 8.024\ \frac{\text{volt}}{\text{volt}} \tag{10.93}$$

The gain required to avoid saturation is 9.533 volt/volt; thus, $G$ is less than the saturation gain.

The final values selected for the parameters of our DC motor speed servo, using components available from stock, are as follows:

$$K = 0.040\ \frac{\text{volt}}{\text{rpm}} \qquad \text{(1200 rpm @ 48 volts)}$$

$$R_o = 4.0\ \frac{\text{volt}}{\text{amp}} \qquad \text{(12 amps from 48 volt amplifier)}$$

$$h = 0.004965\ \frac{\text{volt}}{\text{rpm}} \qquad \text{(4.965 volts @ 1000 rpm)}$$

$$G = 8.024\ \frac{\text{volt}}{\text{volt}} \qquad \text{(48 volt output for 5.98 volt input)}$$

The transfer function for the system with these selected components is

$$\omega_m = \frac{\dfrac{\dfrac{1}{h}}{\left[1 + \left(\dfrac{1 + Z_r}{G_L}\right)\right]} E_d - \dfrac{\dfrac{Z_r}{B}}{(1 + Z_r + G_L)} T_L}{\dfrac{\dfrac{J}{B} Z_r}{(1 + Z_r + G_L)} D + 1} \tag{10.94a}$$

$$\omega_m = \frac{\dfrac{\dfrac{\text{rpm}}{0.004965\ \text{volt}}}{\left[1 + \left(\dfrac{1 + 0.00986}{0.996}\right)\right]} E_d - \dfrac{\dfrac{0.00986\ \text{rev}\ 60\ \text{s}}{0.003183\ \text{in lbf s}\ 2\ \pi\ \text{min}}}{(1 + 0.00986 + 0.996)} T_L}{\dfrac{20.34\ \text{s}\ 0.00986}{(1 + 0.00986 + 0.996)} D + 1} \tag{10.94b}$$

$$\omega_m = \frac{100 \dfrac{\text{rpm}}{\text{volt}} E_d - 14.7 \dfrac{\text{rpm}}{\text{in lbf}} T_L}{(0.10 \text{ s}) D + 1} \tag{10.94c}$$

These components will result in a servo system with 1000 rpm output for 10 volts input, will have a time constant of 0.100 second, and will degrade 14.7 rpm per in lbf of torque disturbance.

## 10.5 PNEUMATIC POSITION SERVO SYSTEM

Pneumatic position servo systems are used in numerous applications because of their ability to position loads with high dynamic response and to augment the force required to move the loads. Pneumatic systems are also very reliable. The power brake system on most cars is actually a pneumatic position servo that actuates the brake master cylinder with a stroke identical to that produced by the foot pedal, but with a force greater than that applied by the foot.

For our sizing example, we consider the pneumatic servo modeled in Chapter 7. We intend to use this servo to actuate a 25 lbm mass and to position the mass within a range of motion of $\pm 2.5$ inches. The servo must be fast enough to track a specified ramp input and must be stable; otherwise, no further specifications are initially stated here. However, we need more specific criteria that will be discussed as the design progresses.

We refer again to Section 7.3.2 for a review of the modeling equations and the final differential equations. Figure 10.16 illustrates this system.

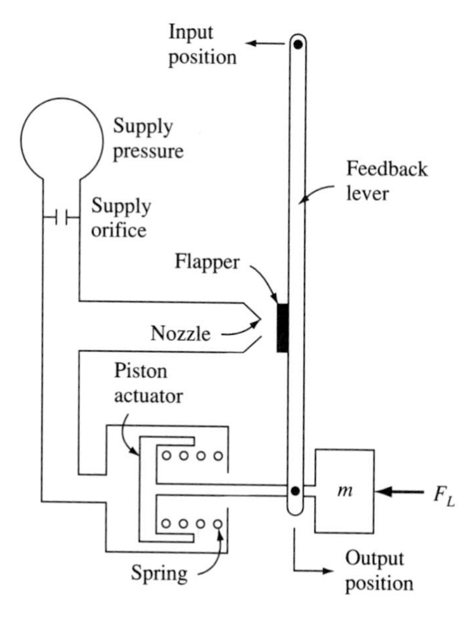


**Figure 10.16** Pneumatic position servo control system.

The transfer function for the linearized model of the system, from Chapter 7, is

$$\delta z = \frac{G_s\,\delta u + \delta z_{ss} - \dfrac{\left(R_o \dfrac{V^*}{\beta^*} D + 1\right)}{k_s} F_L}{R_o \dfrac{V^* M}{\beta^* k_s} D^3 + \dfrac{M}{k_s} D^2 + \dfrac{R_o A^2}{k_s}\left(1 + \dfrac{V^*/\beta^*}{A^2/k}\right) D + 1} \tag{10.95}$$

We use the following nomenclature.

## NOMENCLATURE FOR PNEUMATIC POSITION SERVO

$a$ input lever length, $a' = a/(a + b)$
$b$ feedback lever length, $b' = b/(a + b)$
$\delta P_s$ supply pressure driving valve
$\delta P_o^*$ null output pressure of valve
$M$ mass of load
$\beta^*$ mean bulk modulus of fluid in actuator
$\delta z_{max}$ maximum value of actuator motion from null
*stroke* stroke of actuator
$z_{null}$ null position of actuator
$\bar{z}_{null}$ ratio of null position to stroke
$z_p$ spring preload compression
$\bar{z}_p$ ratio of spring preload to actuator stroke
$F_{Lmax@0}$ maximum load force capability at null position
$F_{Lmax@\delta z}$ maximum load force capability at full stroke
$\tau^*$ maximum time for ramped step input
$a_1, a_2, a_3$ coefficients of characteristic equation
$y_0$ null value of valve input
$\delta y_{sat}$ maximum stroke of flapper valve
$d_n$ diameter of nozzle
$d_s$ diameter of supply orifice
$k$ stiffness of spring
$k_s$ closed-loop stiffness of servo system
$A$ area of actuator
$d_p$ diameter of actuator
$R_o$ output impedance of valve
$Q_o$ output flow from valve
$Q_s^*$ null flow rate to valve from pneumatic supply
$G_p$ valve pressure gain

$V_0$ dead volume in actuator
$\delta u_{max}$ maximum input command
$G_s$ closed-loop gain

A review of the nonlinear and the linear models for this system reveals that there are nine coefficients or parameters that must be determined: $\delta P_s$, $M$, $V_0$, $z_p$, $z_{null}$, $k$, $A$, $a'$, and $y_0$. From $y_0$, we can determine $d_s$ and $Q_s^*$ for the nonlinear model or $d_s$, $Q_s^*$, $G_p$, and $R_o$ for the linear model. The linear model is used to select the coefficients, and then the linear and nonlinear models are tested to evaluate the response.

### 10.5.1 Static Operation Criteria

The mass is given as an inertial load and is thus fixed. Further, we will be operating the servo from a supply pressure of 15 psig, which is the only source available. Thus,

$$M = 25 \text{ lbm}$$

$$\delta P_s = 15 \text{ psig}$$

Attendant upon the selection of the supply pressure is the determination of the null output pressure from the valve, as well as the nominal bulk modulus for the linearized model:

$$\delta P_o^* = 0.671\ \delta P_s = 10.07 \text{ psi} \tag{10.96}$$

$$\beta^* = 1.4\,(\delta P_o^* + 14.69) = 34.66 \text{ psi} \tag{10.97}$$

Hence, two of our parameters are specified, and we have seven parameters remaining. One of these, $V_0$, is parasitic and can be specified in general terms as a function of the size of the actuator. The dead volume is the volume of fluid in the actuator and fittings when the actuator is fully retracted. Normally, it is negligible, but we cannot put it into the model as zero, since it directly affects one of the system coefficients and, consequently, the dynamics of the system. The dead volume is the piston area times a fraction of the diameter. A reasonable expectation is that the dead distance is about one-tenth of a diameter:

$$V_0 = A(0.1 d_p) \tag{10.98}$$

Thus, the nominal volume in the actuator for the linearized model is

$$V^* = A(0.1 d_p + z_{null}) \tag{10.99}$$

This application requires symmetric operation of $\pm 2.5$ inches from a null point at the center of stroke of the actuator. Thus, an actuator with a stroke of 5 inches is required. From that criterion, we can find the optimum value of the null position of the actuator and the maximum stroke:

$$stroke = 5 \text{ in}$$

$$z_{null} = \frac{stroke}{2} \tag{10.100}$$

$$\delta z_{max} = \frac{stroke}{2} \tag{10.101}$$

It is convenient to define the following normalized variables:

$$\bar{z}_{null} = \frac{z_{null}}{stroke} = 0.5 \tag{10.102}$$

$$\delta \bar{z}_{max} = \frac{\delta z_{\max}}{stroke} = 0.5 \tag{10.103}$$

To have the servo null at the center position of the actuator, we must have the $\delta z_{ss}$ term from Chapter 7 be zero. This gives the optimum relation of $A/k$ after some rearrangement of terms:

$$\left(\frac{A}{k}\right)^{*} = \frac{z_p + z_{null}}{\delta P_o^*} \tag{10.104}$$

In order to determine the maximum force output at the null position ($\delta z = 0$), at steady state ($Q_o = 0$), and with the maximum output pressure ($\delta P_o = \delta P_s$), we can use the dynamic force balance equation from Chapter 7 in the steady state:

$$F_{Lmax@0} = A\delta P_s - k(z_p + z_{null}) = k\left[\left(\frac{A}{k}\right)^{*} \delta P_s - (z_p + z_{null})\right] \tag{10.105}$$

Solving this equation for $k$ and substituting the expressions for $(A/k)^*$, and $\delta P_o^*/\delta P_s$ into the result yields the following design guide for the stiffness of the spring:

$$k = \frac{F_{Lmax@0}}{0.49\,(z_p + z_{null})} \tag{10.106}$$

Since a spring opposes the actuator's increasing stroke, we must ensure that the actuator will both take a full stroke and have some excess force capability at that full stroke. Again, from the static force balance, we can see that at full stroke,

$$F_{Lmax@\delta z} = A\delta P_s - k(z_p + z_{null} + \delta z_{max})$$

$$= k\left[\left(\frac{A}{k}\right)^{*} \delta P_s - (z_p + z_{null} + \delta z_{max})\right] \tag{10.107}$$

Once more, solving for $k$ and making the appropriate substitutions from the preceding equations, we find that

$$F_{Lmax@\delta z} = k(z_p + z_{null})\left[\frac{1}{\delta P_o^*/\delta P_s} - 1 - \frac{\delta z_{max}}{(z_p + z_{null})}\right] \tag{10.108}$$

Now, by taking the ratio of the force available at maximum stroke to the force available at null, we obtain

$$\bar{F} = \frac{F_{L\,max@\delta z}}{F_{L\,max@0}} = 1 - \frac{\delta\bar{z}_{max}}{0.49\,(\bar{z}_p + \bar{z}_{null})} \tag{10.109}$$

From this result, we can see that the force available at the end of the stroke is a function of the preload compression we put on the spring. With the values already stated for $\bar{z}_{null} = 0.5$, and $\delta\bar{z}_{\max} = 0.5$, the full-stroke force is zero with $\bar{z}_p = 0.52$. The full-stroke force is thus 32% of the null force with $\bar{z}_p = 1.00$. Hence, the full-stroke force increases with the preload compression. This occurs because the spring must get softer to meet the criteria with more preload compression; therefore, the spring force doesn't subtract as much force from the pressure force.

Given the difficulties of installing a spring that has been compressed to a small percentage of its initial length, it is in turn difficult to use a preload compression of the spring that is more than the length of the actuator. Therefore, a good compromise might be to use $\bar{z}_p = 1.00$. We have

$$\bar{z}_p = 1.00 \qquad \text{thus,}\ z_p = 5.0\ \text{in} \tag{10.110}$$

$$\bar{z}_{null} = 0.50 \qquad \text{thus,}\ z_{null} = 2.5\ \text{in} \tag{10.111}$$

$$\delta\bar{z}_{max} = 0.50 \quad \text{thus,}\ \delta z_{max} = 2.5\ \text{in} \tag{10.112}$$

With these values selected, we can now solve for the optimum spring rate. We get

$$k = \frac{F_{Lmax@0}}{0.49\,(z_p + z_{null})} = \frac{100\ \text{lbf}}{0.49\,(5.0 + 2.5)\ \text{in}} = 27.2\ \text{lbf/in} \tag{10.113}$$

Now using the optimum $A/k$ ratio, we can find the area of the actuator:

$$A = k\left(\frac{A}{k}\right)^* = k\left[\frac{z_p + z_{null}}{0.671\ \delta P_s}\right] = 27.2\ \frac{\text{lbf}}{\text{in}}\left[\frac{(5.0 + 2.5)\ \text{in}}{10.07\ \text{psi}}\right] = 20.27\ \text{in}^2 \tag{10.114}$$

This area results in a piston diameter of 5.08 inches.

Up to now, we have selected nominal values for seven of our nine parameters. The last two values (for $a'$ and $y_0$) come from dynamic response criteria.

### 10.5.2 Dynamic Response Criteria

Next, we need a dynamic specification that could be in the form of a transient response, a frequency response, or a specification of the roots of the characteristic equation of the system. We also need the system to be stable, even if we have to compromise other specifications. The dynamic specification for this system is that it will be able to track a ramp input without saturating the valve. The ramp will be a full positive input $\delta u_{\max}$, in $\tau^*$ seconds. During this ramp input, we want the output to track the input command closely enough that the valve doesn't saturate (i.e., $|\ \delta y\ |$

must be less than $y_0$). For this example, we want $\tau^*$ to be 200 ms, which is about how fast a human operator can move the input command with a smooth motion.

Ramp input specifications define the value of the normalized coefficient, $a_1$ of the $D$ term in the characteristic equation. (See Appendix E.) If the input is a ramp specified by

$$\delta u(t) = u_0 t \quad \text{where} \quad u_0 = \frac{\delta u_{max}}{\tau^*}, \tag{10.115}$$

then the output response of the system in the steady state is

$$\delta z(t) = G_s u_0 (t - a_1) \tag{10.116}$$

Thus, the error signal is

$$\delta y = a' \, G_s u_0(t - a_1) - (1 - a') \, u_0 t \tag{10.117}$$

Substituting

$$G_s = \frac{\dfrac{(1 - a')}{a'}}{\left(1 + \dfrac{k}{G_p A a'}\right)}$$

into the preceding equation and rearranging yields the following expression for the maximum error:

$$\left|\delta y_{max}\right| = \frac{u_0(1 - a')}{\left(1 + \dfrac{k}{G_p A a'}\right)} \left[a_1 + \frac{k}{G_p A a'} \tau^*\right] \tag{10.118}$$

The first normalized coefficient of the system is

$$a_1 = \frac{R_o A^2}{k_s} \left[1 + \frac{V^*/\beta^*}{A^2/k}\right] \tag{10.119}$$

Thus, for $|\delta y_{max}| < y_0$, we must have

$$\frac{u_0 \,(1 - a')}{\left(1 + \dfrac{k}{G_p A a'}\right)} \left[a_1 + \frac{k}{G_p A a'} \tau^*\right] < y_0 \tag{10.120}$$

This ramp input tracking gives us a requirement on the settling time of the system. The settling time must be less than $\tau^*$; however, since we are dealing with a third-order system, it is difficult to determine the settling time without knowing all of the system coefficients. As an approximation and a guideline, we will use the first normalized coefficient $a_1$ as the time constant of the system, so that

$$4 \, a_1 < \tau^* \tag{10.121}$$

Thus, for our pneumatic system, we have

$$\frac{R_o A^2}{k_s}\left[1 + \frac{V^*/\beta^*}{A^2/k}\right] < \frac{\tau^*}{4} \tag{10.122}$$

The feedback gain $a'$ can greatly affect the stability of a third-order system. The null position of the flapper-nozzle valve, $y_0$, determines the nozzle diameter $d_n$, the upstream orifice diameter $d_s$, the maximum supply flow $Q_s^*$, the pressure gain $G_p$, and the output impedance $R_o$.

The previous inequalities give us guidelines on how to select the final two parameters. Since the inequalities are not linear in the two parameters that we have to vary, we must solve them by iteration. This is best done in a spreadsheet.

If we look at possible values of $a'$ and $y_0$ that meet the foregoing two inequality requirements, we must simultaneously check for the stability of the system. Since the system is linear, we can calculate the polynomial coefficients of the characteristic equation and solve for the roots using MATLAB.

In either case, we find that the system is unstable for large (or normal) values of feedback $a'$ and that we must decrease the feedback gain to almost unreasonable values (we were hoping to use $a' = 0.5$) and use a large value of $y_0$ (which requires a large nozzle and upstream orifice, and consequently, a large flow rate from the supply) before we can stabilize the system.

A good compromise design is to use

$$a' = 0.040$$

and

$$y_0 = 0.0265 \text{ in}$$

With these selections, the design is complete, and the following is a summary of the preliminary values of the components and the system performance parameters:

$$\delta P_s = 15 \text{ psi} \qquad \delta P_o^* = 10.07 \text{ psi}$$

$$M = 25 \text{ lbm} \qquad \beta^* = 34.7 \text{ psi}$$

$$\delta z_{max} = \pm 2.50 \text{ in} \qquad stroke = 5 \text{ in}$$

$$\bar{z}_{null} = 0.50 \qquad z_{null} = 2.50 \text{ in}$$

$$\bar{z}_p = 1.00 \qquad z_p = 5.00 \text{ in}$$

$$F_{L\,max@0} = 100 \text{ lbf} \qquad F_{L\,max@\delta z} = 32 \text{ lbf}$$

$$\tau^*/4 = 50 \text{ ms} \quad > \quad a_1 = 32.7 \text{ ms}$$

$$y_0 = 0.0265 \text{ in} \quad > \quad \delta y_{sat} = 0.0263 \text{ in}$$

$$d_n = 0.212 \text{ in} \qquad d_s = 0.170 \text{ in}$$

$$k = 27.21 \text{ lbf/in} \qquad k_s = 229.6 \text{ lbf/in}$$

$$A = 20.27 \text{ in}^2 \qquad d_p = 5.080 \text{ in}$$

$$R_o = 0.01638 \text{ lbf s/in}^5 \qquad Q_s^* = 705 \text{ in}^3\text{/s} = 24.5 \text{ SCFM}$$

$$G_p = 250 \text{ psi/in} \qquad V_0 = 10.3 \text{ in}^3$$

$$\delta u_{max} = 0.104 \text{ in} \qquad G_s = 21.16 \text{ in/in}$$

With these values, we can find the coefficients and the roots of the characteristic equation $a_3D^3 + a_2D^2 + a_1D + 1 = 0$:

$$a_3 = 8.1264 \times 10^{-6}\,\text{s}^3 \qquad \text{roots: } -31.36, \text{ and } -1.67 \pm j62.62 \text{ rad/s}$$

$$a_2 = 2.8194 \times 10^{-4}\,\text{s}^2 \qquad (\omega = 62.6 \text{ with } \zeta = 0.027)$$

$$a_1 = 3.2736 \times 10^{-2}\,\text{s}$$

Of course, we cannot buy a cylinder with a diameter of 5.080 inches, nor should we build one with that diameter. We could have a spring custom built with a rate of 27.21 lbf/in, but that would be costly. Therefore, we can use existing components that are available from manufacturers as follows:

$$k = 30.0 \text{ lbf/in} \qquad d_p = 5.000 \text{ in}$$

Stock springs with a free length very close to the correct value for the preload compression to be equal to the stroke are also readily available.

With these changes, the servo will not exactly center itself at midstroke anymore. The small offset in the steady state would be

$$\delta z_{ss} = \frac{k(z_p + z_{null}) - A\,\delta P_0^*}{k_s} = \frac{30\,\frac{\text{lbf}}{\text{in}}(5.0 + 2.5)\text{in} - 19.63 \text{ in}^2\, 10.07\,\frac{\text{lbf}}{\text{in}^2}}{222.4 \text{ lbf/in}}$$

$$= 0.123 \text{ in} \tag{10.123}$$

However, if we readjust the spring preload compression slightly by the use of shims or machined washers, then we can make the steady-state offset term be equal to zero if we select the following value for the spring compression preload:

$$z_p = \frac{A}{k}\,\delta P_0^* - z_{null} = \frac{19.635 \text{ in}^2}{30 \text{ lbf/in}} 10.067 \text{ psi} - 2.5 \text{ in} = 4.089 \text{ in} \tag{10.124}$$

Thus, we can use the different-sized spring and actuator if we use a spring compression of 4.089 inch instead of 5.000 inch and still have the servo null itself at the midpoint of the stroke.

We will have to build the flapper-nozzle valve, since one of the derived size is not commonly available on the market. If we readjust the nozzle diameter to achieve the desired performance with the stock spring and actuator, then the system parameters shift slightly as follows:

$$\tau^*/4 = 50 \text{ ms} \quad > \quad a_1 = 31.0 \text{ ms}$$

$$y_0 = 0.0270 \text{ in} \quad > \quad \delta y_{sat} = 0.0269 \text{ in}$$

$$d_n = 0.216 \text{ in} \qquad d_s = 0.173 \text{ in}$$

$$k = 30 \text{ lbf/in} \qquad k_s = 222 \text{ lbf/in}$$

$$A = 19.63 \text{ in}^2 \qquad d_p = 5.000 \text{ in}$$

$$R_o = 0.01578 \text{ lbf s/in}^5 \qquad Q_s^* = 732 \text{ in}^3\text{/s} = 25.4 \text{ SCFM}$$

$$G_p = 245 \text{ psi/in} \qquad V_0 = 9.82 \text{ in}^3$$

$$\delta u_{max} = 0.104 \text{ in} \qquad G_s = 20.76 \text{ in/in}$$

With these new values, we can find the coefficients and the roots of the characteristic equation:

$$a_3 = 7.8068 \times 10^{-6} \text{ s}^3 \qquad \text{roots: } -33.38, \text{ and } -1.95 \pm j61.91 \text{ rad/s}$$

$$a_2 = 2.9107 \times 10^{-4} \text{ s}^2 \qquad (\omega = 62.6 \text{ with } \zeta = 0.027)$$

$$a_1 = 3.0972 \times 10^{-2} \text{ s}$$

This completes the sizing of the system. The resulting transfer function, using the selected components, is

$$\delta z = \frac{G_s\,\delta u + \delta z_{ss} - \dfrac{\left(R_o \dfrac{V^*}{\beta^*} D + 1\right)}{k_s} F_L}{R_o \dfrac{V^* M}{\beta^* k_s} D^3 + \dfrac{M}{k_s} D^2 + \dfrac{R_o A^2}{k_s}\left(1 + \dfrac{V^*/\beta^*}{A^2/k}\right) D + 1} \tag{10.125}$$

$$\delta z = \frac{20.76 \dfrac{\text{in}}{\text{in}}\,\delta u - \dfrac{[(0.00446 \text{ s})\,D + 1]}{222} \dfrac{\text{in}}{\text{lbf}} F_L}{(7.807 \times 10^{-6} \text{ s}^3)\,D^3 + (291 \times 10^{-6} \text{ s}^2)\,D^2 + (30.97 \times 10^{-3} \text{ s})\,D + 1} \tag{10.126}$$

The nonlinear state-space model with the selected components is

$$\delta \dot{z} = v \tag{10.127}$$

$$\dot{v} = \frac{A}{M}\,\delta P_o - \frac{k}{M}(z_p + z_{null} + \delta z) - \frac{1}{M} F_L \tag{10.128}$$

$$\delta \dot{P}_o = \frac{nP_{atm} Q_s^* \left(1 + \dfrac{\delta P_o}{P_{atm}}\right)\left\{\sqrt{1 - \dfrac{\delta P_o}{\delta P_s}} - \alpha^*\left(1 + a' \dfrac{\delta z}{y_0} - (1 - a')\dfrac{\delta u}{y_0}\right)\sqrt{\dfrac{\delta P_o}{\delta P_s}}\right\} - nP_{atm} A\left(1 + \dfrac{\delta P_o}{P_{atm}}\right) v}{V_0\left[1 + \dfrac{A\,z_{null}}{V_0}\left(1 + \dfrac{\delta z}{z_{null}}\right)\right]} \tag{10.129}$$

$$\delta \dot{z} = v \tag{10.130}$$

$$\dot{v} = 303.3 \frac{\text{in/s}^2}{\text{psi}} \delta P_o - 463.4 \frac{\text{in/s}^2}{\text{in}} (6.589 \text{ in} + \delta z) - 15.45 \frac{\text{in/s}^2}{\text{lbf}} F_L \tag{10.131}$$

$$\delta \dot{P}_o = \frac{1533 \frac{\text{psi}}{\text{s}} \left(1 + \frac{\delta P_o}{14.7}\right) \left\{ \sqrt{1 - \frac{\delta P_o}{14.7}} - 0.7 (1 + 1.48\, \delta z - 35.6\, \delta u) \sqrt{\frac{\delta P_o}{14.7}} \right\} - 41.1 \frac{\text{psi}}{\text{in}} \left(1 + \frac{\delta P_o}{14.7}\right) v}{\left[1 + 5.0 \left(1 + \frac{\delta z}{2.5}\right)\right]} \tag{10.132}$$

### 10.5.3 Dynamic Response Simulation

In the previous subsection, we used a linearized approximation to size the pneumatic position servo system, a very nonlinear and marginally stable system. It is therefore very important to check the nonlinear system response with the values obtained, as well as to rely on the information about stabilitiy gained from the roots of the linear system.

The response of the system is tested by analyzing the nonlinear and the linear models for their response to a ramp input from zero to one-half of the maximum input in half of the maximum time, $\tau^*/2 = 0.100$ second. The results are shown in Figure 10.17.

The figure indicates that the linear and nonlinear system responses agree very well. The nonlinear system model is slightly slower than the linear system model. In conducting these simulations, it is important to keep in mind the fact that the initial conditions for the nonlinear state space model are as follows:

$$\delta z(0) = 0$$

$$v(0) = 0$$

$$\delta P_o(0) = \delta P_o^* = 0.671\, \delta P_s = 10.07 \text{ psi}$$

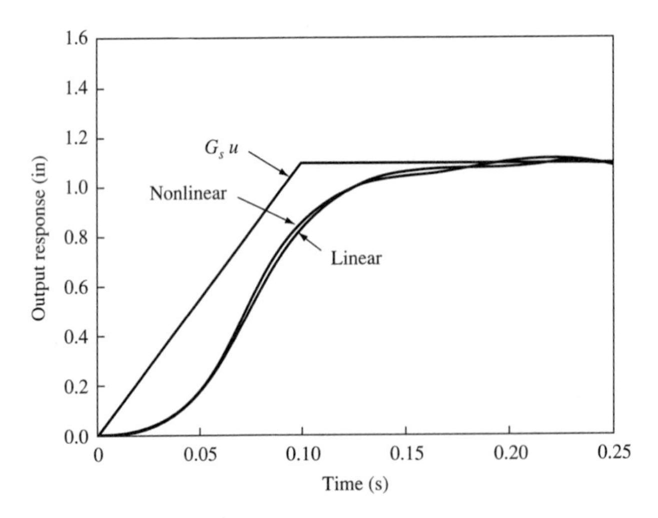


**Figure 10.17** Transient response of pneumatic servo system.

If the wrong initial conditions are used for this system (e.g., $\delta P_o = 0$), then the system response will look very unstable, and the system will take a long time to settle. In other words, even if the system is starting from rest, you cannot always assume that the initial conditions on all variables are zero!

Further, it must also be kept in mind that, in order to digitally simulate terms such as $\sqrt{\delta P_o}$, one must use $\sqrt{|\delta P_o|}$ sign ($\delta P_o$) (the square root of the absolute value of $\delta P_o$ times the sign of $\delta P_o$) so that the computer will not take the square root of a negative number (if the pressure ever goes negative).

## 10.6 SUMMARY

In this chapter, we have stressed some of the considerations that come into play in selecting the components of a system, as well as some further considerations pertaining to what components are available from suppliers. These concepts are generally not presented in college textbooks, but do bring a necessary aspect of engineering to light. In sizing components, a knowledge of the static behavior, range of operation (or maximum values), dynamic response requirements, and performance requirements of the system aid in specifying the performance requirements of every component that we must select. It is interesting to note that the analyses we have presented in regard to the sizing and selection of components are longer and more involved than the actual modeling of the system.

## REFERENCES

10.1 Rath, H., and Roggenbuck, G. "Deceleration Conscious Apportioning Valves for Passenger Cars and Light Trucks," SAE paper #790395, 1979.

## PROBLEMS

**10.1** A 10 pound instrument is to be mounted in a moving vehicle in such a manner that the instrument is isolated from the vibrations and motion of the frame of the vehicle. It is proposed that this be done with a mechanical system, as shown in Figure P10.1a. In this system, a soft spring with a lot of preload is used to isolate the mass. Linear viscous damping is provided by placing oil inside the cylinder and allowing the oil to pass through a laminar flow viscous resistance.

In the specifications for the system, it is mentioned that frequencies above 0.5 Hz should be attenuated as much as possible. The expected input disturbances can be characterized by a smooth bump represented by $x_f = (x_f^*/2)(1 - \cos\omega_f t)$ for one cycle, as shown in Figure P10.1b. The bumps are expected to be 0.5 foot in amplitude. Bump frequencies of $\omega_f = 0.1$, 0.5, and 2.5 Hz are expected. In no case should the output response be greater in amplitude than the input disturbance. You may use any of the SAE oils (from 10 though 70). The length of the fluid resistance passage is 2 inches.

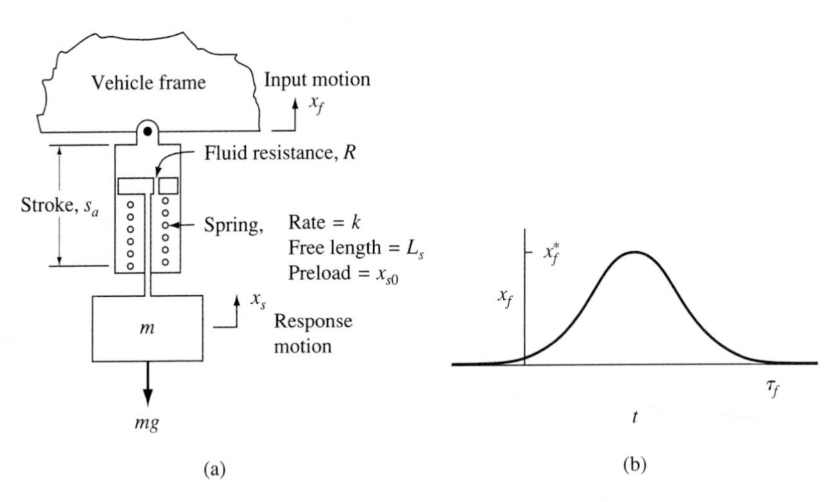


**Figure P10.1** Vibration isolation system. (a) Configuration (b) Bump input.

Additionally the stroke of the actuator must be large enough so that the piston does not hit either end of the cylinder. At steady state, the piston should sit at midstroke.

Care must be taken in selecting the damping. Notice that, since the damping is transmitted through the input motion, a large force is transmitted to the mass during a transient if the damping is high. In other words, the conventional wisdom of selecting damping of 0.707 or higher for good attenuation might not work here.

Based upon the stated performance goals and other criteria, *select* values for the actuator diameter and stroke; the spring stiffness, preload, and free length; and the diameter of the fluid resistor passage, as well as the viscosity of the oil to be used in the cylinder. *Perform* time simulations and a frequency response analysis of the system to illustrate that the final designed system meets the expectations.

**10.2** A chemical manufacturer wants to build a handheld liquid dispenser so that its customers can mix and dispense their cleaning products. The "squirt bottle" containing the chemical must have enough pressure to spray the liquid a considerable distance, but must dispense a large quantity of liquid per actuation. This is the only specification criterion given to you from the marketing department (other than that the dispenser must look good and be colored brown!). A conceptual sketch is provided in Figure P10.2a, with dimensions shown in Figure P10.2b.

As a good engineer, you start your own testing to develop more criteria. The first thing that you do is test for what pressure is necessary in the dispenser. The distance that liquid can be dispensed from a nozzle and the velocity of the fluid at the exit of the nozzle are determined by pressure only; theoretically, the area of the orifice does not affect the distance or velocity. As you plot distance squirted versus pressure, you find that distance increases with pressure up to about 20 psi; then the distance does not increase substantially with pressure. Therefore, you conclude that an actual nozzle pressure of 20 psi during delivery is a good goal.

The next thing that you do is measure the forces available from the hand of the user. You find that most people can develop a maximum static force of 15 pounds between their fingers and the palm of their hand without discomfort. However, it is very impor-

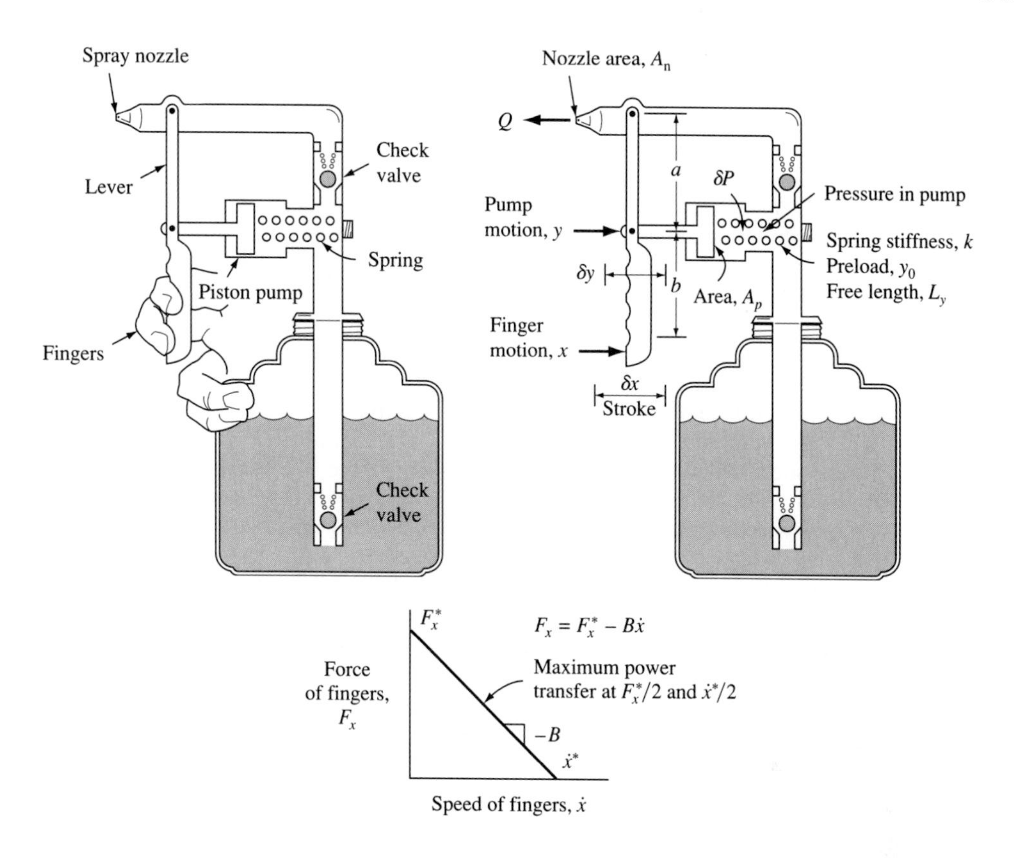


**Figure P10.2** Liquid dispenser bottle. (a) Configuration. (b) Details of liquid dispenser bottle. (c) Force-speed characteristics of a human hand.

tant to realize that this force capability decreases with speed of actuation. This is analogous to the voltage of a battery dropping as more current is drawn from the battery (until the point at which the battery voltage is zero, at a short circuit). Therefore, you test the maximum speed that a person can move his or her fingers without any force or load. You find this maximum no-load speed to be 50 in/s. You also test a few points in between the conditions of maximum force (with no motion) and maximum speed (with no force) by using a long, soft spring with a lot of preload (so that the force is constant during the stroke distance). The result is given in Figure P10.2c. The slope of this graph represents the mechanical output impedance ($B = F_x^*/\dot{x}^* = 0.30$ lbf s/in) of the human hand. Notice that most people like to stroke their fingers about 1.5 inches for this application.

A very significant concept in the operation of this bottle (or any other application in any discipline) in which power must be transferred from the hand to the pump is that the best combination of pressure and flow rate will result when the hand is operated at maximum power. Maximum power transfer (from the hand to the pump lever) will occur when the hand is being operated at the maximum power output point. Notice that, since power is the product of force and speed, at the stalled condition (maximum force with zero speed) and at the no-load condition (maximum speed with zero force), the power from the hand is zero. A little mathematics should convince you that the maximum power from the hand occurs at the half-force, half-speed point marked on the graph. (This is always true for a linear system.) As you test people with squirt bottle pumps that are sized to allow this maximum power operation point, you find that people prefer that maximum power operation much more than one which is so stiff, that it doesn't seem to move as you try to squirt it, or one that offers little or no resistance as you try to squirt it. Therefore, you conclude that your design should be sized such that, in operation, the pressure and the flow require the maximum-power force-speed operation point from your hand. Accordingly, you must select the lever ratio, piston area, and nozzle diameter in an optimum manner to achieve this maximum power operation from the hand. In so doing, you will be assured that the maximum volume per actuation will be dispensed at the proper pressure.

Next, you must develop the mathematical model for the system, starting with the force-speed characteristics of the hand given in Figure P10.2c and then obtaining equations to relate (1) the forces and deflections on the lever at the hand and at the pump, (2) the pressure and flow from the force and velocity of the pump, and (3) the pressure and flow through the orifice. Initially, you should assume that both friction and inertia are small. In your model, you should find a natural grouping of terms, $A_p a'$; this is the primary gain of the system and both represents the volume of liquid dispensed per stroke of the fingers and determines the pressure that you can get from the force at the hand.

Now you find that you have many more parameters and coefficients to pick than you have criteria. So you need to develop some more criteria. First, observe that the check valves require about 2 psi of "cracking pressure." This means that the pressure differential must be 2 psi before the spring-loaded ball can be lifted off its seat in the direction of forward flow. However, once the ball is lifted, the flow area is so large that there is very little pressure drop due to flow as the flow increases. Therefore, a 2 psi pressure drop is required to open the check valve, but that pressure is basically constant with flow. This has the effect of causing a small pressure drop in the flow going through the top check valve as the liquid goes to the nozzle, but it does determine how much spring force is required to generate the pressure (vacuum) in the piston chamber during the recharge stroke of the pump to open the bottom check valve. (Notice that in this recharge flow condition, $F_x^*$ and $B$ are zero.) You probably should select a spring that has a preload $y_0$ of double the pump stroke $\delta y$ and then select a stiffness $k$ (along with the pump area $A_p$) that will create the required vacuum of 2 psi. However, be sure to provide a large enough spring force to provide the necessary suction and to overcome friction or increased fluid resistance so as to move the piston back quickly after its actuation.

The next thing to do is realize that the pump can be placed only within a narrow range of 1.0 inch $< a <$ 2.5 inches relative to the nozzle and pivot point of the lever. Further, the fingers cannot be lower than the pump by more than 1 inch without interfering with the bottle. (Thus, $b <$ 1.0 inch.) Note that the pump shaft can withstand some misalignment as it is stroked over its distance. Due to the nature of the cup seal, however, it cannot withstand a misalignment resulting from the lever handle moving more than $\pm 12$ degrees (assuming that the pump shaft is perpendicular to the lever at midstroke).

The nozzle can be modeled as an orifice whose geometry is such that the discharge coefficient is estimated to be 0.90. The nozzle must be sized so that the operating pressure is 20 psi when the hand is at its maximum power point ($F_x^*/2$ and $\dot{x}^*/2$) during the stroke. This means that the maximum pressure that can be generated (with no flow rate) is about 40 psi.

Based upon the stated criteria and the secondary criteria generated, *select* all of the parameters for this system, including $a$, $b$, $k$, $y_0$, the free length of the spring, $A_p$, and $A_n$. *What* is the volume dispensed per stroke? Use your equations to show that the design is optimum. *Test* a spray bottle at home, and determine the volume dispensed per actuation. *How* does it compare to your analysis? *How* does it feel: too stiff? too soft?

**10.3** The hydraulic servo position control system of Section 7.3.1 is used to position a load weighing 100 pounds, as shown in Figure P10.3. The static gain should be 1.0, and the stroke of the actuator should be $\pm 1.5$ inches. The system should be as fast as possible and as stiff as possible; however, it must have a frequency response of at least 20 Hz with no more than 45 degrees of phase lag. The stiffness must be large enough so that a 250 pound load disturbance does not deflect the output more than 0.025 inch. The damping ratio should be between 0.4 and 0.9.

The system supply pressure is 1000 psi. There are three candidate valves that might be used for this application:

| Valve | $G_p$(psi/in) | $R_o$(lbf/in$^5$) |
|---|---|---|
| 1 | 80,000 | 520 |
| 2 | 74,000 | 350 |
| 3 | 67,000 | 260 |

For a given valve, the only parameter to select is the area of the piston actuator, and we have enough criteria for that. Usually, this is not the case; we have to determine further criteria to select all of the parameters. In this design, however, the preceding specifications are actually enough to select the coefficients in the differential equation. Note that the natural frequency and stiffness are both affected by the actuator area in the same direction (i.e., a larger area makes the natural frequency and the stiffness increase, as is desirable). However, increasing the area increases the damping ratio beyond reasonable limits.

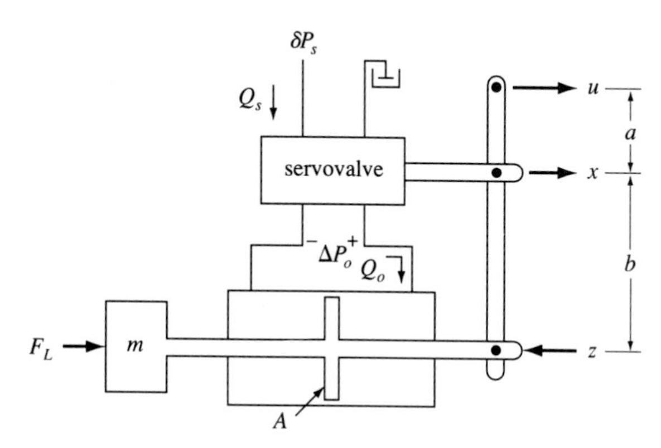


**Figure P10.3** Hydraulic position servo control system from Section 7.3.1.

The steady-state gain determines the lever ratio; however, one other criterion can be used to select the lever lengths $a$ and $b$, namely, that the angular deflection of the lever should be small. If the angular deflection of the lever is large, then there will have to be axial movement of the valve shaft and actuator rod on the lever. A small axial motion could be taken with clearance in the bearings and flexibility in the valve and actuator shafts or their mounts; however, a larger axial motion must be accommodated by slots or sliding surfaces on the lever. Given the actuator sizes and mounting stiffnesses in this application, the angular deflection should probably be less than $\pm 5$ degrees.

*Select* a valve and piston diameter for this application. Hydraulic actuators are readily available with the specified stroke in 0.25 inch increments, starting from 0.50 inch. *Determine* the spacing between the valve and actuator and the lever lengths $a$ and $b$.

*Plot* the dynamic response to a step input of 1.0 inch to show that the design is adequate. *Plot* the frequency response of your final system.

# APPENDIX A

# Units and Conversions

SI units are basically metric units (although not strictly MKS units) that have been adopted as standard units throughout the world. It is important that the reader not only be familiar with these units, but also have an intuitive feel for quantities expressed in terms of SI units, as well as quantities expressed in the British system of units of measure.

This appendix is intended to serve as an introduction to SI units, list some of the common units in various disciplines, and provide a summary of a few of the basic conversion factors between British and SI units.

**TABLE A–1** MULTIPLES AND SUBMULTIPLES

| Multiplication Factor | Prefix | Symbol |
|---|---|---|
| $10^{12}$ | tera | T |
| $10^{9}$ | giga | G |
| $10^{6}$ | mega | M |
| $10^{3}$ | kilo | k |
| $10^{2}$ | hecto | h |
| 10 | deka | da |
| $10^{-1}$ | deci | d |
| $10^{-2}$ | centi | c |
| $10^{-3}$ | milli | m |
| $10^{-6}$ | micro | $\mu$ |
| $10^{-9}$ | nano | n |
| $10^{-12}$ | pico | p |
| $10^{-15}$ | femto | f |
| $10^{-18}$ | atto | a |

In what follows, note that the unit of dynamic viscosity is the pascal-second (N s/$m^2$) and that kinematic viscosity is simply $m^2$/s and has no specific name. Furthermore, the following CGS and other units are not used in the SI system: erg, dyne, poise, stoke, gauss, oersted, maxwell, torr, kilogram-force, calorie, and micron.

**TABLE A–2** BASIC AND DERIVED SI UNITS

| Physical quantity | Name of unit | Symbol | Formula |
|---|---|---|---|
| **Basic Units** | | | |
| Length | meter | m | |
| Mass | kilogram | kg | |
| Time | second | s | |
| Electric current | ampere | A | |
| Temperature | kelvin | °K | |
| Luminous intensity | candela | cd | |
| **Derived Units** | | | |
| Area | square meter | $m^2$ | |
| Volume | cubic meter | $m^3$ | |
| Frequency | hertz | Hz | (cycle/sec) |
| Density | kilogram per cubic meter | kg/$m^3$ | |
| Velocity | meter per second | m/s | |
| Angular velocity | radian per second | rad/s | |
| Acceleration | meter per second squared | m/$s^2$ | |
| Angular acceleration | radian per second squared | rad/$s^2$ | |
| Force | newton | N | (kg m/$s^2$) |
| Pressure and stress | pascal | Pa | (N/$m^2$) |
| Kinematic viscosity | square meter per second | $m^2$/s | |
| Dynamic viscosity | pascal-second | Pa s | (N s/$m^2$) |
| Work, energy, quantity of heat | joule | J | (N m) |
| Power | watt | W | (J/s) |
| Electric charge | coulomb | C | (A s) |
| Voltage, potential difference, electromotive force | volt | V | (W/A) |
| Electric field strength | volt per meter | V/m | |
| Electric resistance | ohm | Ω | (V/A) |
| Electric capacitance | farad | f | (A s/V) |
| Magnetic flux | weber | Wb | (V s) |
| Inductance | henry | h | (V s/A) |
| Wave number | 1 per meter | $m^{-1}$ | |
| Entropy | joule per kelvin | J/°K | |
| Specific heat | joule per kilogram kelvin | J/(kg °K) | |

*(continued)*

**TABLE A–2** *(continued)*

| Physical quantity | Name of unit | Symbol | Formula |
|---|---|---|---|
| Thermal conductivity | watt per meter kelvin | W/(m °K) | |
| Radiant intensity | watt per steradian | W/sr | |
| **Supplementary units** | | | |
| Plane angle | radian | rad | |
| Solid angle | steradian | sr | |

**TABLE A–3** CONVERSION FACTORS

| Multiply British units | By | To obtain SI units |
|---|---|---|
| **Length** | | |
| angstrom | 1.000 E–10 | meter |
| foot | 3.048 E–01 | meter |
| inch | 2.540 E–02 | meter |
| micron | 1.000 E–06 | meter |
| mil | 2.540 E–05 | meter |
| **Speed** | | |
| miles/hour | 4.470 4 E–01 | meter/s |
| feet/minute | 5.08 E–03 | meter/s |
| inch/second | 2.540 E–02 | meter/s |
| **Acceleration** | | |
| foot/sec$^2$ | 3.048 E–01 | meter/s$^2$ |
| free fall (standard gravity) | 9.806 65 E+00 | meter/s$^2$ |
| inch/sec$^2$ | 2.540 E–02 | meter/s$^2$ |
| **Area** | | |
| foot$^2$ | 9.290 304 E–02 | meter$^2$ |
| inch$^2$ | 6.451 6 E–04 | meter$^2$ |
| **Volume** | | |
| fluid ounce (U.S. fluid) | 2.957 270 E–05 | meter$^3$ |
| foot$^3$ | 2.831 685 E–02 | meter$^3$ |
| gallon (U.K. liquid) | 4.546 087 E–03 | meter$^3$ |
| gallon (U.S. dry) | 4.404 883 E–03 | meter$^3$ |
| gallon (U.S. liquid) | 3.785 412 E–03 | meter$^3$ |
| inch$^3$ | 1.638 706 E–05 | meter$^3$ |
| liter | 1.000 E–03 | meter$^3$ |
| **Mass** | | |
| lbm(pound mass, avoidrupois) | 4.535 924 E–01 | kilogram |
| slug (lbf-sec$^2$/foot) | 1.459 390 E+01 | kilogram |

*(continued)*

**TABLE A–3** *(continued)*

| Multiply British units | By | To obtain SI units |
|---|---|---|
| **Density** | | |
| gram/centimeter$^3$ | 1.000 E+03 | kilogram/meter$^3$ |
| lbm/inch$^3$ | 2.767 990 E+04 | kilogram/meter$^3$ |
| lbm/foot$^3$ | 1.601 846 E+01 | kilogram/meter$^3$ |
| slug/foot$^3$ | 5.153 788 E+02 | kilogram/meter$^3$ |
| **Force** | | |
| dyne | 1.000 E–05 | newton |
| kilogram force (kgf) | 9.806 65 E+00 | newton |
| lbf (found force avoirdupois) | 4.448 222 E+00 | newton |
| poundal | 1.382 549 E–01 | newton |
| **Pressure** | | |
| atmosphere | 1.013 25 E+05 | pascal |
| bar | 1.000 E+05 | pascal |
| dyne/centimeter$^2$ | 1.000 E–01 | pascal |
| inch of mercury (60°F) | 3.376 85 E+03 | pascal |
| inch of water (60°F) | 2.488 4 E+02 | pascal |
| kgf/centimeter$^2$ | 9.806 65 E+04 | pascal |
| lbf/foot$^2$ | 4.788 026 E+01 | pascal |
| lbf/inch$^2$ (psi) | 6.894 757 E+03 | pascal |
| millimeter of mercury (0°C) | 1.333 224 E+02 | pascal |
| torr (0°C) | 1.333 224 E+02 | pascal |
| **Temperature** | | |
| Celsius | $t_k = t_c + 273.15$ | kelvin |
| Fahrenheit | $t_k = (5/9)(t_f + 459.67)$ | kelvin |
| Fahrenheit | $t_c = (5/9)(t_f - 32)$ | Celsius |
| Rankine | $t_k = (5/9)t_R$ | kelvin |
| **Viscosity** | | |
| centistoke | 1.000 E–06 | meter$^2$/s |
| foot$^2$/sec | 9.290 304 E–02 | meter$^2$/s |
| inch$^2$/sec | 6.451 6 E–04 | meter$^2$/s |
| centipoise | 1.000 E–03 | pascal s |
| lbf sec/foot$^2$ | 4.788 026 E+01 | pascal s |
| poise | 1.000 E–01 | pascal s |
| lbf sec/inch$^2$ | 6.894 757 E+03 | pascal s |
| **Energy** | | |
| British thermal unit(mean), Btu | 1.055 87 E+03 | joule |
| calorie (mean) | 4.190 02 E+00 | joule |
| electron volt | 1.602 09 E–19 | joule |
| erg | 1.000 E–07 | joule |
| foot lbf | 1.355 82 E+00 | joule |
| joule (international of 1948) | 1.000 165 E+00 | joule (absolute) |
| watt hr | 3.600 E+03 | joule |

*(continued)*

**TABLE A–3** *(continued)*

| Multiply British units | By | To obtain SI units |
|---|---|---|
| **Energy/Area Time** | | |
| Btu(thermochemical)/(inch$^2$ sec) | 1.634 246 E+06 | watt/meter$^2$ |
| calorie (thermochemical)/(cm$^2$ minute) | 6.973 333 E+02 | watt/meter$^2$ |
| erg/(cm$^2$ sec) | 1.000 E–03 | watt/meter$^2$ |
| watt/cm$^2$ | 1.000 E+04 | watt/meter$^2$ |
| **Power** | | |
| Btu(thermochemical)/sec | 1.054 350 E+03 | watt |
| calorie(thermochemical)/sec | 4.184 E+00 | watt |
| foot lbf/sec | 1.355 818 E+00 | watt |
| horsepower (550 foot lbf/sec) | 7.456 999 E+02 | watt |

# APPENDIX B

# Properties of Mechanical System Components

## B.1 ELASTIC AND AREA PROPERTIES

In this appendix, $x$, $y$, and $z$ are centroidal axes, and with regard to uniform bodies, the terms *centroid, center of mass,* and *center of gravity* are used interchangeably.

### B.1.1 Torsion of Rods

The torsion of long, slender cylinders, as shown in Figure B.1, is governed by the equation

$$\Delta\theta = \theta_2 - \theta_1 = TL/(GI_{xx}) \tag{B.1}$$

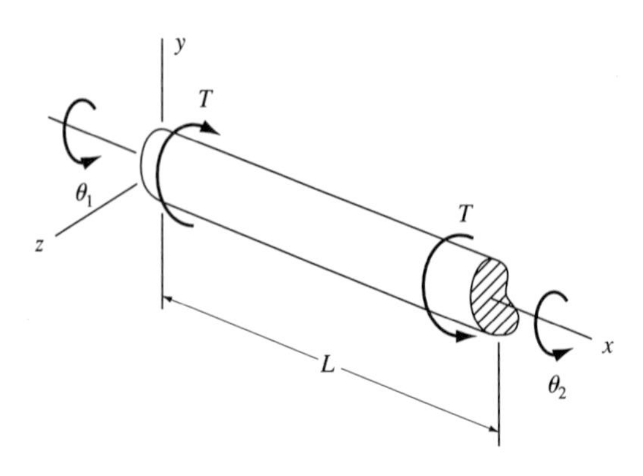


**Figure B.1** Rod in torsion.

where

$I_{xx}$ is the cross-sectional area torsional stiffness constant,
$T$ is the applied torque,
$L$ is the length, and
$G$ is the shear modulus of the material.

The cross section of the cylinder need not be circular, and expressions for $I_{xx}$ for various cross–sectional shapes are given in Table B.1.

The torsional stiffness of a shaft is

$$k = T/\Delta\theta = (GI_{xx})/L \tag{B.2}$$

**TABLE B.1** AREA PROPERTIES

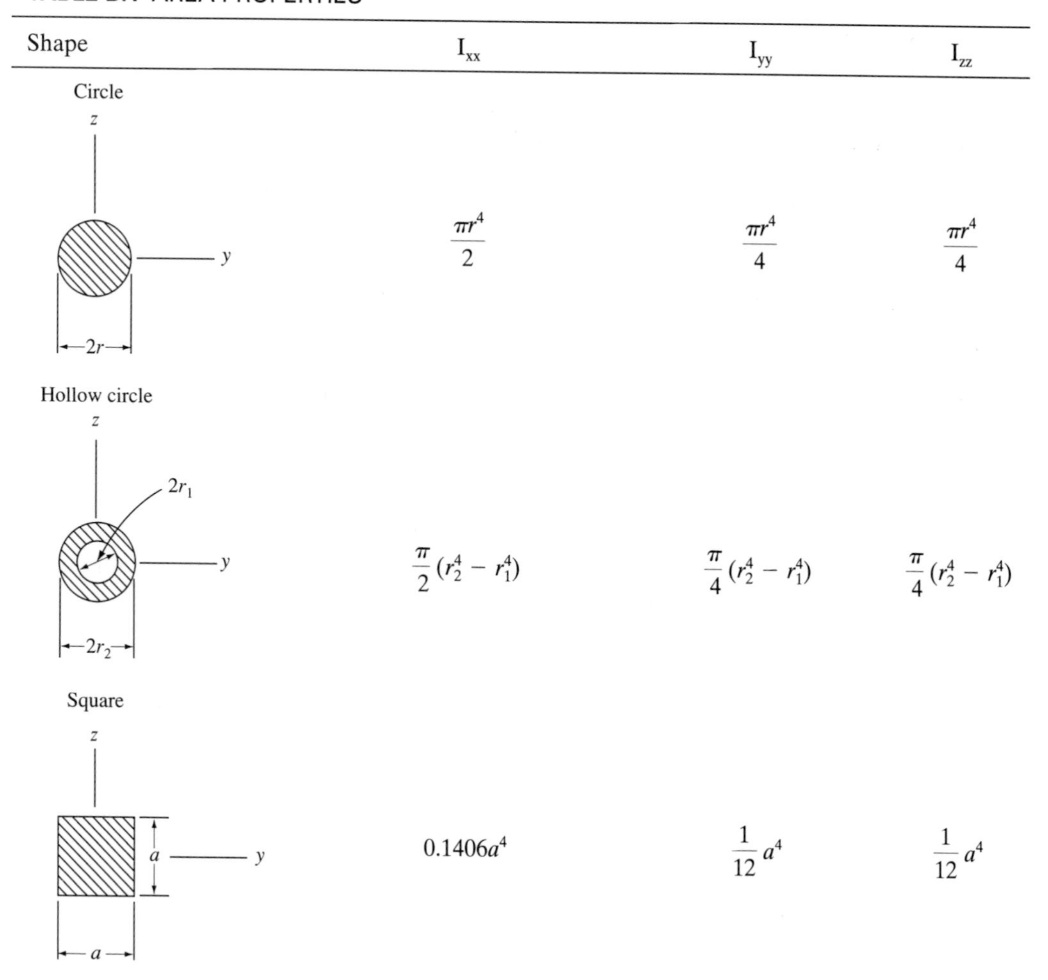

| Shape | $I_{xx}$ | $I_{yy}$ | $I_{zz}$ |
|---|---|---|---|
| Circle | $\frac{\pi r^4}{2}$ | $\frac{\pi r^4}{4}$ | $\frac{\pi r^4}{4}$ |
| Hollow circle | $\frac{\pi}{2}(r_2^4 - r_1^4)$ | $\frac{\pi}{4}(r_2^4 - r_1^4)$ | $\frac{\pi}{4}(r_2^4 - r_1^4)$ |
| Square | $0.1406a^4$ | $\frac{1}{12}a^4$ | $\frac{1}{12}a^4$ |

**TABLE B.1** AREA PROPERTIES *(continued)*

| Shape | $I_{xx}$ | $I_{yy}$ | $I_{zz}$ |
|---|---|---|---|
| Rectangle 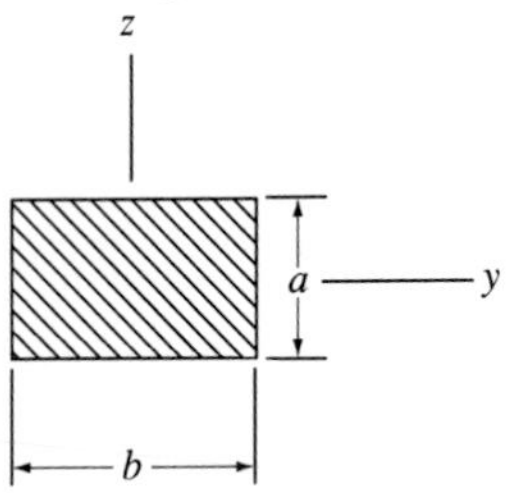  | $ab^3\left[\frac{1}{3} - \frac{3.36}{16}\frac{b}{a}\left(1 - \frac{b^4}{12a^4}\right)\right]$ | $\frac{1}{12}ba^3$ | $\frac{1}{12}ab^3$ |
| Equilateral triangle 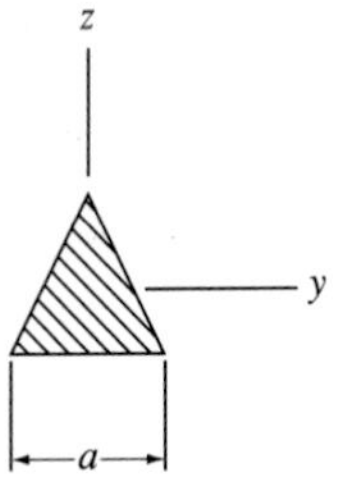  | $\frac{\sqrt{3}}{80}a^4$ | $\frac{\sqrt{3}}{96}a^4$ | $\frac{\sqrt{3}}{96}a^4$ |

## B.1.2 Bending of Beams

Bending theory governs the behavior of long, slender uniform beams in flexure, where $I_{yy}$ and $I_{zz}$ are the cross-sectional area bending stiffness constants, which are area moments of inertia about principal axes. Expressions for $I_{yy}$ and $I_{zz}$ for various cross–sectional shapes are given in Table B.1. Some results from bending deformation analysis for various loadings and boundary conditions are given in Table B.2; in the table, $E$ is the elastic modulus of the material, $I$ is the principal flexural inertia of the cross section of the beam, and $L$ is the length of the beam.

Beam equivalent transverse deformation or rotational deformation spring stiffnesses are given by the following equations:

$$\text{Transverse:} \quad k = P/\delta \tag{B.3}$$

$$\text{Rotational:} \quad k = M/\theta \tag{B.4}$$

**TABLE B.2** BEAM DEFORMATIONS

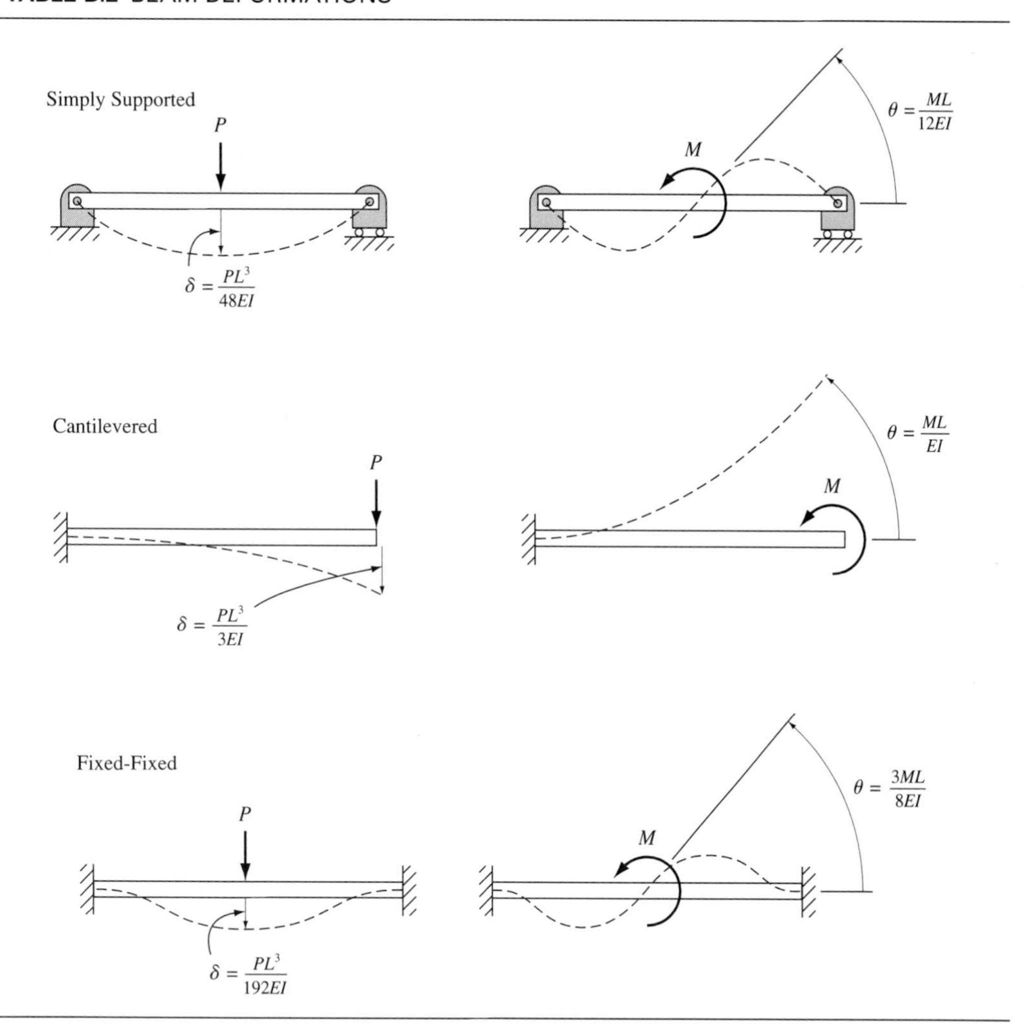


## B.2 MASS PROPERTIES

### B.2.1 Common Shapes

The mass moments of inertia with respect to principal, centroidal axes are given in Table B.3 for some common shapes.

**TABLE B.3** MASS MOMENTS OF INERTIA

| Shape | Inertias |
|---|---|
| Slender Rod | 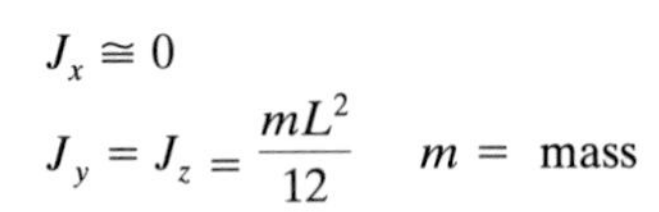 $J_x \cong 0$ <br> $J_y = J_z = \dfrac{mL^2}{12}$ $m$ = mass |
| Thin disk 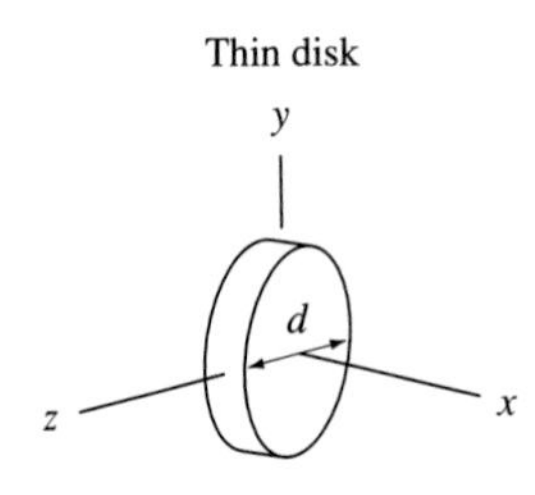 | $J_x = \dfrac{md^2}{8}$ <br> $J_y = J_z = \dfrac{md^2}{16}$ |
| Hollow disk 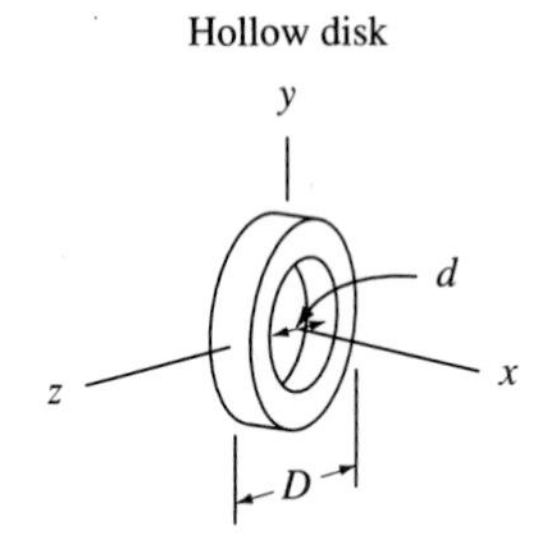 | $J_x = m\,\dfrac{D^2 + d^2}{8}$ <br> $J_y = J_z = m\,\dfrac{D^2 + d^2}{16}$ |
| Cylinder 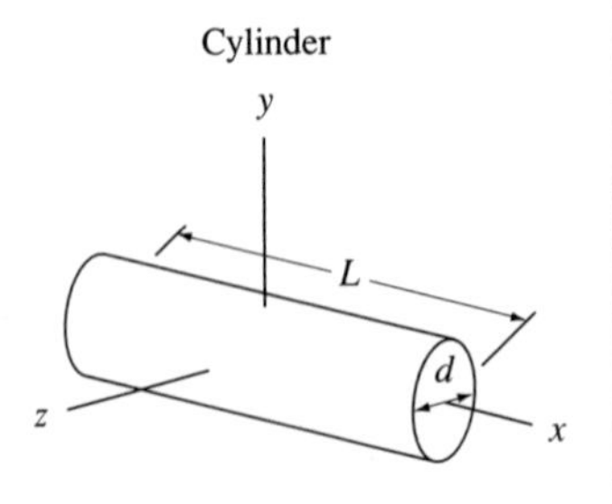 | $J_x = \dfrac{md^2}{8}$ <br> $J_y = J_z = \dfrac{m}{48}(3d^2 + 4L^2)$ |

*Continued*

**TABLE B.3** MASS MOMENTS OF INERTIA *(continued)*

| Shape | Inertias |
|---|---|
| Hollow cylinder 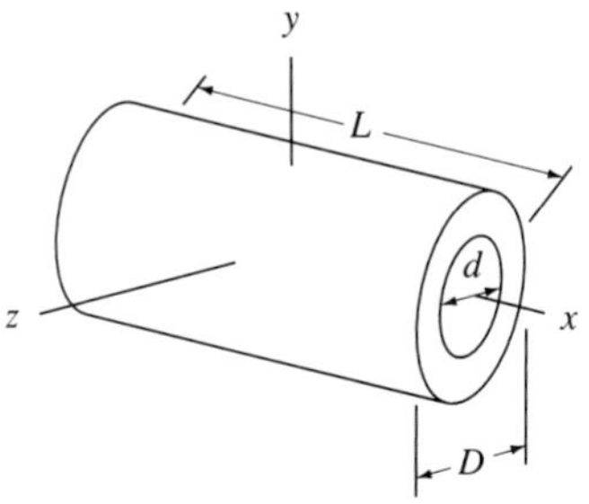 | $J_x = m\frac{D^2 + d^2}{8}$<br>$J_y = J_z = \frac{m}{4}\left(\frac{D^2 + d^2}{4} + \frac{L^3}{3}\right)$ |
| Block 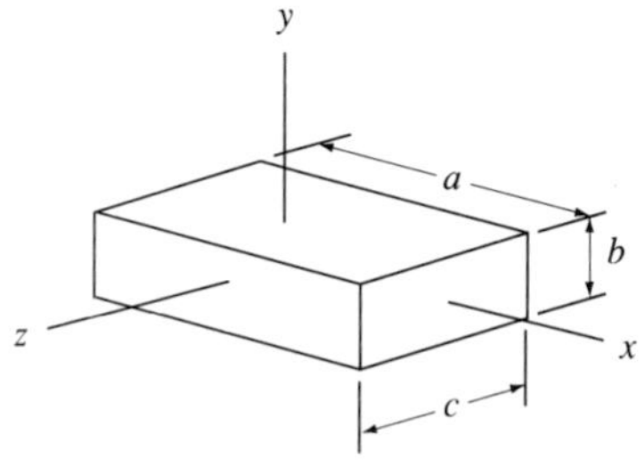 | $J_x = \frac{m}{12}(b^2 + c^2)$<br>$J_y = \frac{m}{12}(a^2 + c^2)$<br>$J_z = \frac{m}{12}(a^2 + b^2)$ |
| Sphere 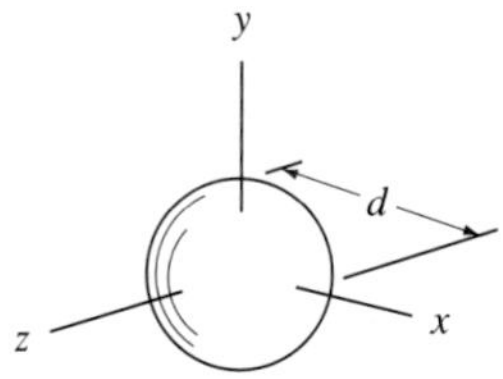 | $J_x = J_y = J_z = \frac{m}{10}d^2$ |

### B.2.2 Parallel Axis Theorem

If the mass moment of inertia $J_{cg}$ of an object of mass $m$ is known for an axis passing though the center of mass of the object, the mass moment of inertia for a parallel axis located at a distance $d$ is given by the expression

$$J = J_{cg} + md^2 \tag{B.5}$$

## B.3 MATERIAL PROPERTIES

Representative values for the properties of some common materials are given in Table B.4. Values in British system units are shown in parentheses.

**TABLE B.4** PROPERTIES OF MATERIALS

| Material | Elastic Modulus GPa (psi $10^6$) | Shear Modulus GPa (psi $10^6$) | Poisson's Ratio | Weight Density N/cm$^3$ (lbf/in$^3$) |
|---|---|---|---|---|
| Aluminum | 69 (10) | 26 (3.8) | 0.33 | .0277 (0.1) |
| Cast Iron | 103 (15) | 27.5 (4) | 0.20 | .0692 (0.25) |
| Copper | 117 (17) | 44 (6.4) | 0.36 | .0894 (0.323) |
| Steel | 207 (30) | 82.7 (12) | 0.33 | .0783 (0.283) |
| Titanium | 103 (15) | 44.8 (6.5) | 0.33 | .0451 (0.163) |
| | psi = Pa/6894.76 | | | lbf/in$^3$ = N/cm$^3$/.2768 |

## REFERENCES

B.1 Hibbeler, R. C., *Engineering Mechanics, Dynamics,* 5th ed. Macmillan Publishing Co., Inc., 1989.

B.2 Hibbeler, R.C., *Mechanics of Materials*, 2nd ed. Macmillan Publishing Co., Inc., 1994.

# APPENDIX C

# Vector and Matrix Algebra

## C.1 VECTORS

### C.1.1 Definitions

A **scalar** is a quantity that can be characterized or identified by a single number which indicates its magnitude. Temperature, age, weight, and length are all examples of scalars. A **vector** is a quantity that has both magnitude and direction; it requires two, three, or even more independent values for its complete characterization. In this text, vectors are designated by a bold lowercase letter (e.g., **f**). In handwriting, it is common to denote a vector with curly or square brackets. Force is an example of a vector, since in three dimensions, three independent quantities are required to determine the magnitude and orientation of a force. A three-dimensional vector is shown in Figure C.1.

A variety of different notations are used to represent vector quantities symbolically. The following representations are all equivalent:

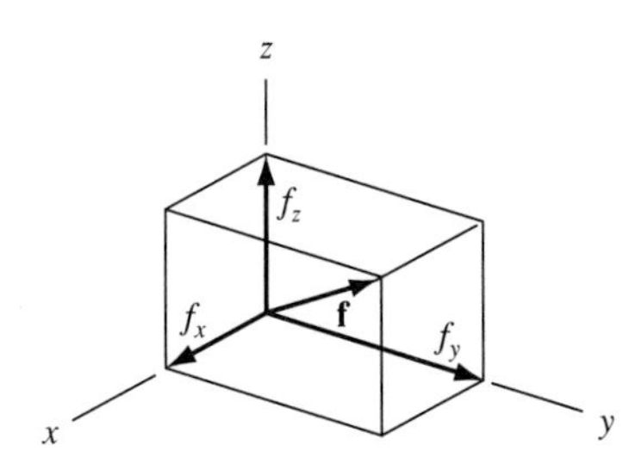


**Figure C.1** Three-dimensional vector.


359

$$\mathbf{f} = f_x\hat{\mathrm{i}} + f_y\hat{\mathrm{j}} + f_z\hat{\mathrm{k}} \qquad \text{unit vector representation}$$

$$\mathbf{f} = \begin{bmatrix} f_x \\ f_y \\ f_z \end{bmatrix} \qquad \text{column vector} \tag{C.1}$$

$$\mathbf{f}^T = (f_1, f_2, f_3) \qquad \text{row vector}$$

In the last equation, $T$ indicates **transpose** and means that the column values have been transposed into a row. Vectors must have as many elements as are necessary to characterize the quantity being represented. Thus, a vector may have $n$ components:

$$\mathbf{v} = \begin{bmatrix} v_1 \\ v_2 \\ \vdots \\ v_n \end{bmatrix} \tag{C.2}$$

### C.1.2 Operations

Two vectors are equal if and only if (iff) all of their corresponding elements are equal:

$$\mathbf{u} = \mathbf{v} \text{ iff } u_1 = v_1,\ u_2 = v_2,\ \ldots,\ u_n = v_n \tag{C.3}$$

$$\text{or } u_i = v_i,\ i = 1, 2, \ldots, n$$

Vectors with the same number of components may be **added** or **subtracted,** element by element. Thus, if

$$\mathbf{u} = \begin{bmatrix} u_1 \\ u_2 \\ u_3 \\ u_4 \end{bmatrix} = \begin{bmatrix} 1 \\ 2 \\ 3 \\ 4 \end{bmatrix} \text{ and } \mathbf{v} = \begin{bmatrix} v_1 \\ v_2 \\ v_3 \\ v_4 \end{bmatrix} = \begin{bmatrix} 2 \\ -6 \\ -3 \\ 1 \end{bmatrix} \tag{C.4}$$

then

$$\mathbf{w} = \mathbf{u} + \mathbf{v} = \begin{bmatrix} u_1 + v_1 \\ u_2 + v_2 \\ u_3 + v_3 \\ u_4 + v_4 \end{bmatrix} = \begin{bmatrix} 3 \\ -4 \\ 0 \\ 5 \end{bmatrix} \tag{C.5}$$

or

$$w_i = u_i + v_i \quad i = 1, \ldots, n$$

A **null vector** has zeros as all of its elements and is indicated by a boldface zero (**0**).

**Multiplication** of a vector by a scalar means that all elements of the vector are multiplied by the scalar quantity. Hence, if the scalar $\beta = 3$ and $\mathbf{w}$ is the vector of Eq. (C.5), then

$$\beta\mathbf{w} = 3\begin{bmatrix} w_1 \\ w_2 \\ w_3 \\ w_4 \end{bmatrix} = \begin{bmatrix} 9 \\ -12 \\ 0 \\ 15 \end{bmatrix} \tag{C.6}$$

**Differentiation of a vector** quantity means that each element of the vector is to be differentiated. Thus,

$$d\mathbf{u}/dt = \dot{\mathbf{u}} = \begin{bmatrix} \dot{u}_1 \\ \dot{u}_2 \\ \vdots \\ \dot{u}_n \end{bmatrix} \tag{C.7}$$

The **inner product,** or **scalar product,** of two vectors is defined as the scalar

$$\begin{aligned} \mathbf{u}^T\mathbf{v} &= u_1v_1 + u_2v_2 + \cdots + u_nv_n \\ &= \sum_{i=1}^{n} u_iv_i \end{aligned} \tag{C.8}$$

The scalar product of $\mathbf{u}$ and $\mathbf{v}$ as defined in Eq. (C.4) is

$$\mathbf{u}^T\mathbf{v} = (1)(2) + (2)(-6) + (3)(-3) + (4)(1) = -15 \tag{C.9}$$

Note that $\mathbf{u}^T\mathbf{v} = \mathbf{v}^T\mathbf{u}$.

Two vectors are said to be **orthogonal** if their inner product is zero. In three-dimensional Cartesian space, if two vectors are orthogonal, then they are at right angles to each other.

Let $\mathbf{d}$ represent the vector joining two points in three-dimensional space. The **distance** between the two points, or the **magnitude** of the vector, is then

$$d = \|\mathbf{d}\| = \sqrt{d_1^2 + d_2^2 + d_3^2} = \sqrt{\mathbf{d}^T\mathbf{d}} \tag{C.10}$$

For vectors with any number of elements, the square root of the inner product of the vector itself is called the **Euclidean norm**.

A **unit vector** is a vector normalized by its magnitude:

$$\hat{\mathbf{a}} = \frac{\mathbf{a}}{\|\mathbf{a}\|} = \frac{\mathbf{a}}{\sqrt{\mathbf{a}^T\mathbf{a}}} \tag{C.10a}$$

## C.2 MATRICES

### C.2.1 Definitions

A vector may be thought of as a one-dimensional array of values, since only one subscript is necessary to identify the elements of the vector. A **matrix** is a two-dimensional, rectangular array of elements:

$$\mathbf{A} = \begin{bmatrix} a_{11} & a_{12} & \cdots & \cdots & \cdots & a_{1m} \\ a_{21} & a_{22} & \cdots & \cdots & \cdots & a_{2m} \\ \cdots & \cdots & \cdots & \cdots & \cdots & \cdots \\ \cdots & \cdots & \cdots & a_{ij} & \cdots & \cdots \\ \cdots & \cdots & \cdots & \cdots & \cdots & \cdots \\ a_{n1} & a_{n2} & \cdots & \cdots & \cdots & a_{nm} \end{bmatrix} \tag{C.11}$$

The array of Eq. (C.11) has $n$ rows and $m$ columns and is said to be of order $n \times m$. To specify any particular element $a_{ij}$ of the matrix, the subscripts $i$ and $j$ must be designated. A matrix is denoted by a bold uppercase letter (e.g., **A**). In handwritten work, square brackets are often used to denote matrices, as, for example, in [A].

A **square matrix** is a matrix with the same number of rows and columns, i.e., $n = m$.

The **principal diagonal** of a square matrix is composed of the elements $a_{ii}$ and is shown by the dashed line in the following equation:

$$\mathbf{A} = \begin{bmatrix} 2 & -6 & 4 \\ 1 & 10 & -3 \\ 8 & 7 & -5 \end{bmatrix} \tag{C.12}$$

### C.2.2 Operations

Two matrices are equal iff $a_{ij} = b_{ij}$ for all $i = 1, \ldots, n$ and all $j = 1, \ldots, m$.

The **multiplication of a matrix by a scalar** is

$$\beta\mathbf{A} = \begin{bmatrix} \beta a_{11} & \beta a_{12} & \cdots \\ \beta a_{21} & \beta a_{22} & \cdots \\ \cdots & \cdots & \cdots \end{bmatrix} \tag{C.13}$$

Two matrices with the same number of rows and the same number of columns may be **added** or **subtracted**. Thus,

$$\mathbf{C} = \mathbf{A} + \mathbf{B} \qquad c_{ij} = a_{ij} + b_{ij}, \;\; i = 1, \ldots, n, \;\; j = 1, \ldots, m \tag{C.14}$$

**Multiplication of matrices** is defined as follows: The matrix **A** of order $m \times n$ and the matrix **B** of order $n \times p$ can be multiplied together to form a matrix **C** of order $m \times p$. The element $c_{ij}$ is formed by taking the inner product of the $i$th row of **A** with the $j$th column of **B**. Thus;

$$c_{ij} = \sum_{k=1}^{n} a_{ik} b_{kj} \quad i = 1, \ldots, m, \; j = 1, \ldots, p \tag{C.15}$$

To be compatible for multiplication, **A** must have the same number of columns as **B** has rows. Then **C** will have the same number of rows as **A** and the same number of columns as **B**. In the equation $\mathbf{C} = \mathbf{AB}$, **A** is the **premultiplier** and **B** is the **postmultiplier**.

**Example C.1**

If

$$\underset{2 \times 3}{\mathbf{A}} = \begin{bmatrix} 1 & 0 & 2 \\ 1 & 1 & 3 \end{bmatrix} \quad \underset{3 \times 2}{\mathbf{B}} = \begin{bmatrix} 2 & 1 \\ -2 & 0 \\ 4 & 3 \end{bmatrix} \tag{C.16}$$

then

$$\mathbf{C} = \mathbf{AB} = \underset{2 \times 3}{\begin{bmatrix} 1 & 0 & 2 \\ 1 & 1 & 3 \end{bmatrix}} \underset{3 \times 2}{\begin{bmatrix} 2 & 1 \\ -2 & 0 \\ 4 & 3 \end{bmatrix}} = \underset{2 \times 2}{\begin{bmatrix} 10 & 7 \\ 12 & 10 \end{bmatrix}} \tag{C.17}$$

Multiplication of a column vector by a matrix gives a column vector. That is, we may regard a column vector as an $n \times 1$ matrix. Thus,

$$\mathbf{Ax} = \mathbf{y} \tag{C.18}$$

If **A** has the numerical values indicated in Eq. (C.16) and **x** has elements (1,2,3), then

$$\mathbf{y} = \begin{bmatrix} 1 & 0 & 2 \\ 1 & 1 & 3 \end{bmatrix} \begin{bmatrix} 1 \\ 2 \\ 3 \end{bmatrix} = \begin{bmatrix} (1)(1) + (0)(2) + (2)(3) \\ (1)(1) + (1)(2) + (3)(3) \end{bmatrix} = \begin{bmatrix} 7 \\ 12 \end{bmatrix} \tag{C.19}$$

The **transpose** of a matrix is obtained by interchanging its rows and columns. For the $2 \times 3$ matrix of Eq. (C.16), the transpose is the $3 \times 2$ matrix

$$\mathbf{A}^T = \begin{bmatrix} a_{11} & a_{21} \\ a_{12} & a_{22} \\ a_{13} & a_{23} \end{bmatrix} = \begin{bmatrix} 1 & 1 \\ 0 & 1 \\ 2 & 3 \end{bmatrix} \tag{C.20}$$

The transpose of a matrix product is equal to the product of the transposed matrices, but in reverse order:

$$[\mathbf{ABC}]^T = \mathbf{C}^T \mathbf{B}^T \mathbf{A}^T \tag{C.21}$$

### C.2.3 Properties

It is important to note that matrix multiplication is **not commutative**. That is, in general,

$$\mathbf{AB} \neq \mathbf{BA} \tag{C.22}$$

Matrix multiplication is, however, **associative** and left (and right) **distributive** over addition:

$$[\mathbf{AB}]\mathbf{C} = \mathbf{A}[\mathbf{BC}] \tag{C.23a}$$

$$\mathbf{A}[\mathbf{B} + \mathbf{C}] = \mathbf{AB} + \mathbf{AC} \tag{C.23b}$$

### C.2.4 Special Types of Matrices

All elements of the **null matrix** are zero. Only the principal or main diagonal elements are different from zero in the $n \times n$ **diagonal matrix**:

$$\mathbf{A} = \begin{bmatrix} a_{11} & 0 & 0 & \dots & 0 & 0 \\ 0 & a_{22} & 0 & \dots & 0 & 0 \\ \dots & \dots & \dots & \dots & \dots & \dots \\ \dots & \dots & \dots & a_{ij} & \dots & \dots \\ \dots & \dots & \dots & \dots & \dots & \dots \\ 0 & 0 & 0 & \dots & 0 & a_{nm} \end{bmatrix} \tag{C.24}$$

The **identity matrix** or **unity matrix** is a diagonal matrix whose elements are all ones:

$$\mathbf{I} = \begin{bmatrix} 1 & 0 & 0 \\ 0 & 1 & 0 \\ 0 & 0 & 1 \end{bmatrix} \tag{C.25}$$

In matrix algebra, multiplication by **I** has the same effect as multiplication by unity in scalar algebra. A **symmetric matrix** is a square matrix that is symmetrical about its main diagonal, i.e., $a_{ij} = a_{ji}$. The following matrix is an example:

$$\mathbf{A} = \begin{bmatrix} 1 & -2 & 3 & 7 \\ -2 & 8 & 4 & -5 \\ 3 & 4 & 16 & 2 \\ 7 & -5 & 2 & 1 \end{bmatrix} \tag{C.26}$$

In a **lower triangular matrix,** all elements above the main diagonal are zero:

$$\mathbf{A} = \begin{bmatrix} 21 & 0 & 0 & 0 \\ 7 & 1 & 0 & 0 \\ -9 & 0 & 2 & 0 \\ 3 & 5 & -2 & 11 \end{bmatrix} \tag{C.27}$$

Analogously, in an **upper triangular matrix,** all elements below the main diagonal are zero. A **band-structured matrix** is a matrix whose nonzero terms are clustered in a band about the main diagonal:

$$\mathbf{A} = \begin{bmatrix} 11 & -3 & 7 & 0 & 0 \\ 2 & 14 & 10 & 5 & 0 \\ -1 & 4 & 21 & 3 & 1 \\ 0 & 7 & -8 & 1 & -4 \\ 0 & 0 & 2 & 9 & 25 \end{bmatrix} \tag{C.28}$$

It is sometimes convenient to **partition** matrices into smaller groups of elements, or submatrices. This may be done as indicated in the following equations, where $\mathbf{A}_{11}$, etc., denotes a **submatrix**.

$$\mathbf{A} = \left[\begin{array}{cc:c} a_{11} & a_{12} & a_{13} \\ a_{21} & a_{22} & a_{23} \\ \hdashline a_{31} & a_{32} & a_{33} \end{array}\right] = \begin{bmatrix} A_{11} & A_{12} \\ A_{12} & A_{22} \end{bmatrix} \tag{C.29a}$$

where

$$\mathbf{A}_{11} = \begin{bmatrix} a_{11} & a_{12} \\ a_{21} & a_{22} \end{bmatrix} \tag{C.29b}$$

$$\mathbf{A}_{12} = \begin{bmatrix} a_{13} \\ a_{23} \end{bmatrix} \tag{C.29c}$$

$$\mathbf{A}_{21} = [a_{31}\ a_{32}] \tag{C.29d}$$

$$\mathbf{A}_{22} = [a_{33}] \tag{C.29e}$$

Two partitioned matrices can be multiplied together if all submatrices which are to be multiplied together have compatible numbers of rows and columns.

## C.3 DETERMINANTS

### C.3.1 Definitions

The **determinant** of a square matrix is a single expression formed in a definite way using all of the elements of the matrix. The determinant gives a single quantity that

is a measure of a characteristic of the matrix. If all of the elements of the matrix are constants, the determinant is a single numerical value.
The determinant of a $2 \times 2$ matrix is

$$\det \mathbf{A} = |\mathbf{A}| = \begin{vmatrix} a_{11} & a_{12} \\ a_{21} & a_{22} \end{vmatrix} = a_{11}a_{22} - a_{21}a_{12} \tag{C.30}$$

The determinant of a $3 \times 3$ matrix is

$$\det \mathbf{A} = |\mathbf{A}| = \begin{vmatrix} a_{11} & a_{12} & a_{13} \\ a_{21} & a_{22} & a_{23} \\ a_{31} & a_{32} & a_{33} \end{vmatrix} \tag{C.31a}$$

or

$$\det \mathbf{A} = a_{11}a_{22}a_{33} + a_{12}a_{23}a_{31} + a_{13}a_{21}a_{32} - a_{13}a_{22}a_{31} - a_{11}a_{23}a_{32} - a_{12}a_{21}a_{33} \tag{C.31b}$$

As examples,

$$\text{if } \mathbf{A} = \begin{bmatrix} 1 & 2 \\ 3 & 4 \end{bmatrix}, \quad \det \mathbf{A} = 4 - 6 = -2$$

and

$$\text{if } \mathbf{A} = \begin{bmatrix} 1-s & 2 \\ 3 & 4-s \end{bmatrix}, \quad \det \mathbf{A} = (1-s)(4-s) - 6 = s^2 - 5s - 2$$

where $s$ is a variable. Note that Eqs. (C.30) and (C.31) apply only to $2 \times 2$ and $3 \times 3$ matrices, respectively. For $4 \times 4$ and higher order matrices, we must use different methods, which require the introduction of some additional definitions.

The **minor** of the element $a_{ij}$ of a matrix is the determinant of the array formed by removing the $i$th row and $j$th column of the matrix. Thus, there are as many minors as there are elements of the matrix. So if $\mathbf{A}$ is $3 \times 3$, it has nine minors. We will denote minors by $M_{ij}$. Accordingly, let

$$\mathbf{A} = \begin{bmatrix} a_{11} & a_{12} & a_{13} \\ a_{21} & a_{22} & a_{23} \\ a_{31} & a_{32} & a_{33} \end{bmatrix} \tag{C.32a}$$

Then we would have

$$M_{11} = \begin{vmatrix} a_{22} & a_{23} \\ a_{32} & a_{33} \end{vmatrix} = a_{22}a_{33} - a_{23}a_{32} \tag{C.32b}$$

and

$$M_{21} = \begin{vmatrix} a_{12} & a_{13} \\ a_{32} & a_{33} \end{vmatrix} = a_{12}a_{33} - a_{13}a_{32}, \text{ etc} \tag{C.32c}$$

The **cofactor** of element $a_{ij}$ is just the minor multiplied by a sign determined by $(-1)^{i+j}$. The cofactors of the $3 \times 3$ matrix corresponding to elements $a_{11}$ and $a_{21}$ are

$$\begin{aligned} A_{11}^c &= (-1)^2(a_{22}a_{33} - a_{23}a_{32}) = a_{22}a_{33} - a_{23}a_{32} \\ A_{21}^c &= (-1)^3(a_{12}a_{33} - a_{13}a_{32}) = -(a_{12}a_{33} - a_{13}a_{32}) \end{aligned} \tag{C.33}$$

We now are in a position to give a general method of evaluating determinants that may be applied to a matrix of any size.

### C.3.2 Evaluation by Cofactors

The **cofactor expansion method** gives the determinant as the sum of the elements of any **one row or one column,** multiplied by their cofactors. Thus,

$$\det \mathbf{A} = \sum_{i\,(\text{or } j)=1}^{n} a_{ij} A_{ij}^c \quad \text{for any } j \text{ or any } i \tag{C.34}$$

For a $3 \times 3$ matrix, this expansion, using elements of the first row, gives

$$\begin{aligned} \det \mathbf{A} = \begin{vmatrix} a_{11} & a_{12} & a_{13} \\ a_{21} & a_{22} & a_{23} \\ a_{31} & a_{32} & a_{33} \end{vmatrix} &= a_{11}(1)\begin{vmatrix} a_{22} & a_{23} \\ a_{32} & a_{33} \end{vmatrix} + a_{12}(-1)\begin{vmatrix} a_{21} & a_{23} \\ a_{31} & a_{33} \end{vmatrix} \\ &+ a_{13}(1)\begin{vmatrix} a_{21} & a_{22} \\ a_{31} & a_{32} \end{vmatrix} \\ &= a_{11}(a_{22}a_{33} - a_{23}a_{32}) - a_{12}(a_{21}a_{33} - a_{23}a_{31}) \\ &+ a_{13}(a_{21}a_{32} - a_{22}a_{31}) \end{aligned} \tag{C.35}$$

Expanding this expression shows that it is equal to Eq. (C.31).

### C.3.3 Properties

Following are several useful properties of determinants:

- **(a)** If two rows or two columns of a matrix are interchanged, the sign of the determinant of the matrix is changed.
- **(b)** Det $\mathbf{A}^T$ = Det $\mathbf{A}$.
- **(c)** If one row or one column of a matrix is multiplied by a scalar $\beta$, the determinant of the matrix is multiplied by $\beta$.
- **(d)** If all of the elements of an $n \times n$ matrix are multiplied by a scalar $\beta$, the determinant of the matrix is multiplied by $\beta^n$. Mathematically, Det $\beta\mathbf{A} = \beta^n$ Det $\mathbf{A}$.

**(e)** The determinant of a matrix is zero when:
one or more rows or columns of the matrix are zero,
two or more rows or columns of the matrix are equal,
two or more rows or columns of the matrix are proportional.

A matrix is said to be **singular** if its determinant is zero.

## C.4 MATRIX INVERSION

### C.4.1 Definitions

A single scalar equation with unknown $x$ and the solution of that equation are

$$Ax = \beta, \quad A^{-1}Ax = A^{-1}\beta, \quad x = \beta/A, \quad \text{if } A \neq 0 \tag{C.36}$$

where it is clear what is meant by the reciprocal of the scalar $A$. For a matrix $\mathbf{A}$ and vectors $\mathbf{x}$ and $\mathbf{b}$,

$$\mathbf{Ax} = \mathbf{b}, \quad \mathbf{A}^{-1}\mathbf{Ax} = \mathbf{A}^{-1}\mathbf{b}, \quad \mathbf{x} = \mathbf{A}^{-1}\mathbf{b}, \quad \text{if } \mathbf{A}^{-1} \text{ exists} \tag{C.37}$$

The **reciprocal matrix** or **inverse matrix** is defined as

$$\mathbf{A}^{-1}\mathbf{A} = \mathbf{A}\mathbf{A}^{-1} = \mathbf{I}, \text{ where } \mathbf{I} \text{ is the identity matrix} \tag{C.38}$$

There are a variety of ways to determine the inverse of a square matrix. We discuss here the **cofactor method** because of its fundamental importance. The **cofactor matrix** is formed by replacing each element of the matrix by its cofactor. The **adjoint matrix** is then formed by taking the transpose of the cofactor matrix:

$$\mathbf{A}^a = [\mathbf{A}^c]^T \tag{C.39}$$

Now, the product of a matrix with its adjoint matrix gives a diagonal matrix, each element of which is the determinant of the matrix $\mathbf{A}$:

$$\mathbf{A}\mathbf{A}^a = |\mathbf{A}|\mathbf{I} = \begin{bmatrix} |\mathbf{A}| & 0 & 0 & 0 \\ 0 & |\mathbf{A}| & 0 & 0 \\ 0 & 0 & |\mathbf{A}| & 0 \\ 0 & 0 & 0 & \cdots \end{bmatrix} \tag{C.40}$$

If Eq. (C.40) is divided by Det $\mathbf{A}$, the identity matrix is obtained, and we may define the inverse as the adjoint matrix divided by matrix determinant:

$$\mathbf{A}^{-1} = \frac{1}{|\mathbf{A}|}\mathbf{A}^a = \frac{1}{|\mathbf{A}|}[\mathbf{A}^c]^T \tag{C.41}$$

Note that if the determinant is zero, the matrix is said to be singular, and its inverse does not exist.

**Example C.2**

Find the inverse of the 2 × 2 matrix **A.**

**Solution**

$$\mathbf{A} = \begin{bmatrix} a_{11} & a_{12} \\ a_{21} & a_{22} \end{bmatrix}, \qquad \text{Det } \mathbf{A} = a_{11}a_{22} - a_{12}a_{21}$$

$$a_{11}^c = a_{22}, \quad a_{12}^c = -a_{21}, \quad a_{21}^c = -a_{12}, \quad a_{22}^c = a_{11}$$

$$\mathbf{A}^c = \begin{bmatrix} a_{22} & -a_{21} \\ -a_{12} & a_{11} \end{bmatrix}, \qquad \mathbf{A}^a = \begin{vmatrix} a_{22} & -a_{12} \\ -a_{21} & a_{11} \end{vmatrix}$$

$$\mathbf{A}^{-1} = \frac{1}{a_{11}a_{22} - a_{12}a_{21}} \begin{bmatrix} a_{22} & -a_{12} \\ -a_{21} & a_{11} \end{bmatrix}$$

$$\text{If } \mathbf{A} = \begin{bmatrix} 1 & 2 \\ 3 & 4 \end{bmatrix} \text{ then } \mathbf{A}^{-1} = \frac{1}{-2}\begin{bmatrix} 4 & -2 \\ -3 & 1 \end{bmatrix}$$

**Example C.3**

Find the inverse of the matrix

$$\mathbf{A} = \begin{bmatrix} 1 & 3 & -2 \\ 2 & 4 & -1 \\ 8 & -3 & 2 \end{bmatrix}$$

**Solution**

$$\det \mathbf{A} = 1(8 - 3) - 3(4 + 8) - 2(-6 - 32) = 45$$

$$\mathbf{A}^c = \begin{bmatrix} 5 & -12 & -38 \\ 0 & 18 & 27 \\ 5 & -3 & -2 \end{bmatrix}$$

$$\mathbf{A}^{-1} = \frac{1}{45}\begin{bmatrix} 5 & 0 & 5 \\ -12 & 18 & -3 \\ -38 & 27 & -2 \end{bmatrix}$$

Use of the cofactor method for inverting matrices of large order can quickly become numerically unwieldy because of the large number of determinants that must be calculated. In this circumstance, methods based on Gaussian elimination should be used. (See Appendix D.)

## C.4.2 Properties

Following are some important properties of the inverse matrix:

**(a)** The inverse of the inverse is the matrix itself, i.e., $[\mathbf{A}^{-1}]^{-1} = \mathbf{A}$.

**(b)** The inverse of the diagonal matrix **D** is a diagonal matrix, the elements of which are reciprocals of the diagonal entries of **D**.

**(c)** The inverse of a product is the product of the inverses, but in reverse order. That is,

$$[\mathbf{AB}]^{-1} = \mathbf{B}^{-1}\mathbf{A}^{-1}$$

**(d)** If **A** is symmetric, its inverse is symmetric.

**(e)** The inverse of a matrix multiplied by a scalar is the inverse of the matrix multiplied by the reciprocal of the scalar:

$$[\beta\mathbf{A}]^{-1} = \frac{1}{\beta}\mathbf{A}^{-1}$$

**(f)** The inverse of a transpose of a matrix is the transpose of the inverse matrix; i.e.,

$$[\mathbf{A}^T]^{-1} = [\mathbf{A}^{-1}]^T$$

**(g)** An orthogonal matrix is one with its inverse equal to its transpose.

$$\mathbf{A}^{-1} = \mathbf{A}^T$$

## PROBLEMS

C.1 *Determine* $\mathbf{a} + 3\mathbf{b}$ if

$$\mathbf{a}^T = [2, -6, 1, 9, 2, 9, -1]$$

$$\mathbf{b}^T = [1, -6, 5, -2, 3, 7, 2]$$

C.2 *Find* the magnitude of the vector $\mathbf{c} = \mathbf{a} + \mathbf{b}$ if

$$\mathbf{a}^T = [2, 6, 1, 9]$$

$$\mathbf{b}^T = [-2, -7, -3, 5]$$

C.3 *Evaluate* the expression $\mathbf{e} = [\mathbf{a}^T\mathbf{b}][\mathbf{c} + \mathbf{d}]$ if

$$\mathbf{a}^T = [1, -1, 0, 3] \quad \mathbf{b}^T = [0, 2, 4, -1]$$

$$\mathbf{c}^T = [5, 4, 2, 6, 1] \quad \mathbf{d}^T = [9, -2, -9, 7, 3]$$

C.4 *Compute* the inner product $\mathbf{c}^T\mathbf{d}$ if

$$\mathbf{c}^T = [8, 12, 1, 2]$$

$$\mathbf{d}^T = [-1, 3, 7, 11]$$

C.5 *Find* the scalar product $\mathbf{v}^T\mathbf{u}$ if

$$\mathbf{v}^T = [1, -2, 0.5, 6, -3, -1]$$

$$\mathbf{u}^T = [0.8, 2, -1, -0.5, 0.2, 0.9]$$

C.6 *Find* the inner product between the vectors **u** and **v** if

$$\mathbf{u}^T = [2, -3, 6, -7]$$

$$\mathbf{v}^T = [-0.1, 0.1, 0.5, -0.2]$$

C.7 *Calculate* $\mathbf{a}^T\mathbf{b}$ if

$$\mathbf{a}^T = [2, -6, 1, 9, 2, -9, 1]$$

$$\mathbf{b}^T = [-2, 7, -3, 2, 5, -6, 1]$$

C.8 *Determine* the vector with unit magnitude if

$$\mathbf{b} = \mathbf{a}/\sqrt{\mathbf{a}^T\mathbf{a}}$$

$$\mathbf{a}^T = [9, 7, 0, 2, 5]$$

C.9 *Determine* $\|\mathbf{c}\|$, $\|\mathbf{d}\|$, and $\mathbf{c} - 2\mathbf{d}$ if

$$\mathbf{c}^T = [2, -4, 7, 0]$$

$$\mathbf{d}^T = [-4, 2, -8, 3]$$

C.10 *Evaluate* the expression $\mathbf{f} = \|\mathbf{a}\|\ \mathbf{u}^T\mathbf{v}\mathbf{e}$ if

$$\mathbf{a}^T = [2, -4, -2, 9, 0]$$

$$\mathbf{u}^T = [8, -2, 1, -1]$$

$$\mathbf{v}^T = [3, 2, 9, 1]$$

$$\mathbf{e}^T = [6, -6, 9, 3, -8]$$

C.11 *What* must be the magnitude of element $a_4$ if vectors **a** and **b** are to be orthogonal, given that

$$\mathbf{a}^T = [1, 2, -1, a_4, 2]$$

$$\mathbf{b}^T = [2, -1, 4, 2, -6]$$

C.12 *Find* the magnitude or norm of $\mathbf{u}^T = [2, -3, 6, 7]$.

C.13 If $\mathbf{x}^T = [\cos 3t, e^{-2t}, 3.5t^2 + t]$, *differentiate* the elements of $\boldsymbol{x}$ to obtain

$$\dot{\mathbf{x}} = d\mathbf{x}/dt$$

C.14 *Find* $\mathbf{c}^T\mathbf{d}$ and $\mathbf{c}\mathbf{d}^T$ if

$$\mathbf{c}^T = [1, 2, 3]$$

$$\mathbf{d}^T = [6, -2, 1]$$

C.15 *Find* the matrix product **AB** if

$$\mathbf{A} = \begin{bmatrix} 1 & -2 \\ 5 & 3 \\ -4 & 6 \end{bmatrix} \quad \mathbf{B} = \begin{bmatrix} 9 & -6 & 3 & -2 \\ -1 & 8 & 7 & 4 \end{bmatrix}$$

C.16 *Find* $\mathbf{A}^T\mathbf{b}$ and $\mathbf{b}^T\mathbf{A}$ if

$$\mathbf{A} = \begin{bmatrix} 3 & 1 \\ -1 & 2 \\ 0 & -4 \end{bmatrix} \quad \mathbf{b} = \begin{bmatrix} 2 \\ 1 \\ 3 \end{bmatrix}$$

C.17 *Calculate* the result of **A** premultiplied by **B** if

$$\mathbf{A} = \begin{bmatrix} 1 & 3 \\ 7 & 2 \end{bmatrix} \quad \mathbf{B} = \begin{bmatrix} 2 & 0 \\ 8 & 5 \end{bmatrix}$$

C.18 *Determine* the product $\mathbf{d} = \mathbf{Be}$ if

$$\mathbf{B} = \begin{bmatrix} 4 & 4 & 2 & 9 \\ 4 & 7 & 8 & 2 \end{bmatrix}$$

$$\mathbf{e}^T = [3, 2, 3, 8]$$

C.19 *Determine* the matrix product **AB** if

$$\mathbf{A} = \begin{bmatrix} 2 & 7 & 5 \\ 6 & 3 & 6 \\ 1 & 2 & 1 \end{bmatrix} \quad \mathbf{B} = \begin{bmatrix} 2 & 6 & 1 & 9 \\ 2 & 9 & 1 & 0 \\ 4 & 5 & 5 & 2 \end{bmatrix}$$

C.20 *Determine* the matrix obtained if **C** is postmultiplied by **D,** given that

$$\mathbf{C} = \begin{bmatrix} 4 & 8 & 0 \\ 6 & 4 & 9 \\ 5 & 1 & 5 \end{bmatrix} \quad \mathbf{D} = \begin{bmatrix} 7 & 5 & 1 \\ 2 & 3 & 3 \\ 6 & 2 & 5 \end{bmatrix}$$

C.21 *Find* $[\mathbf{A} + \mathbf{B}]\mathbf{C}$ if

$$\mathbf{A} = \begin{bmatrix} 1 & 2 & 3 \\ 3 & 1 & 4 \end{bmatrix} \quad \mathbf{B} = \begin{bmatrix} 0 & 2 & 1 \\ -1 & 3 & 0 \end{bmatrix}$$

$$\mathbf{C}^T = [1, 0, -1]$$

C.22 *Compute* $[\mathbf{AB}]^T$ and show that it is equal to $\mathbf{B}^T\mathbf{A}^T$ if

$$\mathbf{A} = \begin{bmatrix} 8 & 3 & 7 \\ 1 & 6 & 2 \end{bmatrix} \quad \mathbf{B} = \begin{bmatrix} 4 & 5 \\ 1 & 2 \\ 3 & 1 \end{bmatrix}$$

C.23 *Find* $\mathbf{AB}$ and $\mathbf{B}^T\mathbf{A}^T$, and *show* that $[\mathbf{AB}]^T = \mathbf{B}^T\mathbf{A}^T$, if

$$\mathbf{A} = \begin{bmatrix} 2 & -1 \\ 0 & 2 \\ 1 & 3 \end{bmatrix} \quad \mathbf{B} = \begin{bmatrix} 4 & 1 & 5 \\ 2 & -3 & 1 \end{bmatrix}$$

C.24 *Determine* $[\mathbf{ABC}]^T$, and *show* that it is equal to $\mathbf{C}^T\mathbf{B}^T\mathbf{A}^T$, if

$$\mathbf{A} = \begin{bmatrix} 3 & 8 \\ 8 & 4 \\ 1 & 2 \end{bmatrix} \quad \mathbf{C} = \begin{bmatrix} 1 & 4 & 6 & 0 \\ 2 & 3 & 7 & 1 \end{bmatrix}$$

$$\mathbf{B} = \begin{bmatrix} 1 & 7 \\ 9 & 3 \end{bmatrix}$$

C.25 *Determine* $\mathbf{A} = \mathbf{B}[\mathbf{c} + \mathbf{d}]$ if

$$\mathbf{B} = \begin{bmatrix} 1 & 4 & 6 \\ 2 & 1 & 3 \end{bmatrix}, \quad \mathbf{c} = \begin{bmatrix} 1 \\ 0 \\ 3 \end{bmatrix}, \quad \mathbf{d} = \begin{bmatrix} 8 \\ 7 \\ 1 \end{bmatrix}$$

C.26 *Evaluate* the expression $\mathbf{A} = \mathbf{B}[\mathbf{C} + \mathbf{D}]$ if

$$\mathbf{B} = \begin{bmatrix} 3 & 2 & 5 \\ 4 & 2 & 7 \\ 5 & 6 & 3 \end{bmatrix}$$

$$\mathbf{C} = \begin{bmatrix} 7 & 8 \\ 3 & 1 \\ 5 & 7 \end{bmatrix} \quad \mathbf{D} = \begin{bmatrix} 1 & 6 \\ 5 & 1 \\ 4 & 2 \end{bmatrix}$$

C.27 *Calculate* the triple product

$$\beta = \mathbf{v}^T\mathbf{B}\mathbf{v}$$

if

$$\mathbf{v}^T = [3, 6, 1]$$

$$\mathbf{B} = \begin{bmatrix} 6 & 3 & 0 \\ 7 & 1 & 9 \\ 2 & 0 & 2 \end{bmatrix}$$

C.28 *Determine* the matrix vector product

$$\beta = \mathbf{u}^T\mathbf{A}\mathbf{u}$$

if

$$\mathbf{u}^T = [8, 3, 2, 9] \quad \mathbf{A} = \begin{bmatrix} 2 & 7 & 5 & 0 \\ 6 & 1 & 9 & 1 \\ 1 & 5 & 7 & 4 \\ 8 & 1 & 2 & 5 \end{bmatrix}$$

C.29 *Show* that $|\mathbf{A}| = |\mathbf{A}^T|$ if

$$\mathbf{A} = \begin{bmatrix} a_{11} & a_{12} \\ a_{21} & a_{22} \end{bmatrix}$$

C.30 *Find* $|\mathbf{B}|$ if

$$\mathbf{B} = \begin{bmatrix} 6 & 3 & 0 \\ 7 & 11 & 9 \\ 2 & 0 & 2 \end{bmatrix}$$

C.31 *Find* the determinant of **A** if

$$\mathbf{A} = \begin{bmatrix} 1 & 3 & 3 \\ 1 & 3 & 4 \\ 1 & 4 & 3 \end{bmatrix}$$

C.32 *Compute* the determinant of **B** if

$$\mathbf{B} = \begin{bmatrix} 2 & 1 & 5 \\ 1 & 1 & -3 \\ 3 & 6 & -2 \end{bmatrix}$$

C.33 *Compute* the determinant of **B** if

$$\mathbf{B} = \begin{bmatrix} 3 & -1 & 2 \\ 1 & 4 & 3 \\ 2 & -1 & 1 \end{bmatrix}$$

C.34 *Find* the determinant of **A** if

$$\mathbf{A} = \begin{bmatrix} 1 & 2 & 3 \\ 4 & 5 & 6 \\ 7 & 8 & 9 \end{bmatrix}$$

C.35 *Evaluate* the determinant

$$\mathbf{A} = \begin{bmatrix} 1 & 0 & 0 & 2 \\ 3 & 4 & 0 & 0 \\ 7 & 2 & 0 & 1 \\ 1 & 3 & 1 & 0 \end{bmatrix}$$

C.36 *Show* that the determinant of

$$\mathbf{G} = \begin{bmatrix} g_{11} & 0 & 0 & 0 \\ g_{21} & g_{22} & 0 & 0 \\ g_{31} & g_{32} & g_{33} & 0 \\ g_{41} & g_{42} & g_{43} & g_{44} \end{bmatrix}$$

is $g_{11}\, g_{22}\, g_{33}\, g_{44}$.

C.37 *Find* the determinant of

$$\mathbf{D} = \begin{bmatrix} \cos\theta & -\sin\theta & 0 \\ \sin\theta & \cos\theta & 0 \\ 0 & 0 & 1 \end{bmatrix}$$

C.38 *Compute* the determinant of $\mathbf{A} = 2\mathbf{BC}$ if

$$\mathbf{B} = \begin{bmatrix} 1 & 3 & 4 \\ 2 & 1 & 0 \end{bmatrix} \quad \mathbf{C} = \begin{bmatrix} 1 & -1 \\ -3 & 0 \\ 1 & 2 \end{bmatrix}$$

C.39 *Find* the value of $a_{13}$ if Det $\mathbf{A} = 0$ for

$$\mathbf{A} = \begin{bmatrix} 4 & -2 & a_{13} \\ 3 & 1 & 0 \\ 5 & 0 & 2 \end{bmatrix}$$

C.40 *Find* the value of the determinant of

$$\mathbf{A} = \mathbf{u}\mathbf{v}^T$$

if

$$\mathbf{u}^T = [6, -2, u_3]$$

and

$$\mathbf{v}^T = [1, 2, 3]$$

C.41 *Find* the inverse of the $3 \times 3$ matrix

$$\mathbf{A} = \begin{bmatrix} 6 & 3 & 0 \\ 7 & 1 & 9 \\ 2 & 0 & 2 \end{bmatrix}$$

C.42 *Compute* the inverse of

$$\mathbf{A} = \begin{bmatrix} 1 & 3 & 3 \\ 1 & 3 & 4 \\ 1 & 4 & 3 \end{bmatrix}$$

Check your result by finding $\mathbf{A}\mathbf{A}^{-1}$.

C.43 *Determine* the inverse of

$$\mathbf{A} = \begin{bmatrix} 2 & 3 & 1 \\ 1 & 2 & 3 \\ 3 & 1 & 2 \end{bmatrix}$$

C.44 *Find* $\mathbf{A}^{-1}$ if

$$\mathbf{A} = \begin{bmatrix} 2 & 2 & 0 \\ 0 & 3 & 1 \\ 1 & 0 & 1 \end{bmatrix}$$

C.45 *Find* the inverse of

$$\mathbf{A} = \begin{bmatrix} 0 & 2 & 2 \\ 1 & 0 & 3 \\ 1 & 1 & 0 \end{bmatrix}$$

C.46 *Find* the inverse of the matrix

$$\mathbf{A} = \begin{bmatrix} 1 & -2 & 4 \\ 0 & 1 & 3 \\ 0 & 0 & 1 \end{bmatrix}$$

*Check* the result by calculating $\mathbf{A}^{-1}\mathbf{A}$.

C.47 *Prove* that $[\mathbf{AB}]^{-1} = \mathbf{B}^{-1}\mathbf{A}^{-1}$.

C.48 *Find* the inverse of the symmetric matrix

$$\mathbf{B} = \begin{bmatrix} 1 & 2 & 3 \\ 2 & 4 & 5 \\ 3 & 5 & 6 \end{bmatrix}$$

C.49 *Which* of the following matrices possess inverses?

$$\mathbf{A} = \begin{bmatrix} 1 & 2 & 3 \\ 4 & 5 & 6 \\ 7 & 8 & 9 \end{bmatrix} \quad \mathbf{B} = \begin{bmatrix} 3 & 2 & 1 \\ 1 & 4 & 5 \end{bmatrix}$$

$$\mathbf{C} = \begin{bmatrix} \cos\theta & \sin\theta \\ -\sin\theta & \cos\theta \end{bmatrix}$$

C.50 *Calculate* $[\mathbf{AB}]^{-1}$ and *show* that it is equal to $\mathbf{B}^{-1}\mathbf{A}^{-1}$ if

$$\mathbf{A} = \begin{bmatrix} 1 & 2 \\ 3 & 4 \end{bmatrix} \quad \mathbf{B} = \begin{bmatrix} 6 & 1 \\ 2 & 1 \end{bmatrix}$$

C.51 *What* values must $a_{12}$ and $a_{21}$ have if

$$\mathbf{A} = \begin{bmatrix} 0.5 & a_{12} \\ a_{21} & 0.5 \end{bmatrix}$$

is to be an orthogonal matrix?

C.52 *Show* that

$$\mathbf{C} = \begin{bmatrix} \cos\theta & -\sin\theta \\ \sin\theta & \cos\theta \end{bmatrix}$$

is an orthogonal matrix.

C.53 *Compute* the inverse of the matrix

$$\mathbf{A} = \begin{bmatrix} 2 & 1 & 0 & 0 \\ 1 & 2 & 1 & 0 \\ 0 & 1 & 2 & 1 \\ 0 & 0 & 1 & 2 \end{bmatrix}$$

# APPENDIX D

# Systems of Algebraic Equations

## D.1 INTRODUCTION

The modeling and simulation of physical systems require the derivation and solution of various sets of simultaneous algebraic and/or differential equations. Since the equations are linear in most cases, matrix-vector methods are very helpful in finding the solution.

## D.2 NONHOMOGENEOUS ALGEBRAIC EQUATIONS

The linear algebraic matrix equation

$$\mathbf{Ax} = \mathbf{b} \tag{D.1}$$

is designated as **homogeneous** when the vector $\mathbf{b}$ is zero and as **nonhomogeneous** when $\mathbf{b}$ has one or more nonzero elements. As is often the case, the coefficient matrix $\mathbf{A}$ is known and the solution vector $\mathbf{x}$ is to be found.

### D.2.1 Matrix Inversion

The solution to Eq. (D.1) can be obtained by finding the inverse of the coefficient matrix. As in Eq. (C.37), the solution is given by the product $\mathbf{A}^{-1}\mathbf{b}$ if the inverse matrix exists. That is,

$$\mathbf{x} = \mathbf{A}^{-1}\mathbf{b} \tag{D.2}$$

provided that **A** is nonsingular.

**Example D.1**

Find the solution vector **x** if

$$\begin{bmatrix} 1 & 2 \\ 3 & 4 \end{bmatrix} \begin{bmatrix} x_1 \\ x_2 \end{bmatrix} = \begin{bmatrix} 8 \\ -10 \end{bmatrix}$$

$$\text{where } \mathbf{A} = \begin{bmatrix} 1 & 2 \\ 3 & 4 \end{bmatrix} \text{ and } \mathbf{b} = \begin{bmatrix} 8 \\ -10 \end{bmatrix}$$

**Solution** From Example C.2,

$$\mathbf{A}^{-1} = -\frac{1}{2}\begin{bmatrix} 4 & -2 \\ -3 & 1 \end{bmatrix}$$

Thus,

$$\mathbf{x} = \begin{bmatrix} x_1 \\ x_2 \end{bmatrix} = -\frac{1}{2}\begin{bmatrix} 4 & -2 \\ -3 & 1 \end{bmatrix}\begin{bmatrix} 8 \\ -10 \end{bmatrix} = \begin{bmatrix} -26 \\ 17 \end{bmatrix}$$

$$x_1 = -26, \ x_2 = 17$$

### D.2.2 Cramer's Rule

Each element of the solution vector can be determined independently of the others using Cramer's rule (Gabriel Cramer, Swiss mathematician, 1704–52). According to this technique, the value of the element $x_i$ is found by calculating the ratio of two determinants. The denominator is the determinant of the coefficient matrix. The numerator is the determinant of the matrix obtained by replacing the $i$th column with the constant vector **b**. The ratio of these two determinants is the desired solution. We illustrate the method in the following example.

**Example D.2**

Using the data of Example D.1, find only $x_2$ by using Cramer's rule.

**Solution**

$$|\mathbf{A}| = -2, \ x_2 = \frac{\begin{vmatrix} 1 & 8 \\ 3 & -10 \end{vmatrix}}{-2} = \frac{-10 - 24}{-2}$$

Hence,

$$x_2 = 17$$

### D.2.3 Gaussian Elimination

A system of linear algebraic equations may also be solved by eliminating variables among the various equations in the set until only one equation in one unknown remains. Once this single equation has been solved, the result can be substituted into a previous equation with only two unknowns—the one just found and another unknown. This equation can now be solved for an additional unknown and the process repeated until all the unknowns have been found. For small numbers of equations, this elimination process can be done by inspection, as is illustrated in the next example.

**Example D.3**

Find **x** by the Gaussian elimination process just described:

$$\begin{bmatrix} 2 & -1 & 0 \\ 3 & 2 & -4 \\ 1 & 4 & -2 \end{bmatrix} \begin{bmatrix} x_1 \\ x_2 \\ x_3 \end{bmatrix} = \begin{bmatrix} 0 \\ -5 \\ 3 \end{bmatrix} \tag{a}$$

**Solution** From the first equation, $2x_1 = x_2$. We then multiply the third equation by 2 and subtract it from the second:

$$3x_1 + 2x_2 - 4x_3 = -5$$

$$\underline{-2x_1 - 8x_2 + 4x_3 = -6}$$

$$x_1 - 6x_2 \qquad = -11 \tag{b}$$

Next, we substitute $x_2$ from the first expression of Eq. (a) into Eq. (b) to obtain

$$x_1 - 6(2x_1) = -11$$

$$-11x_1 = -11$$

$$x_1 = 1 \tag{c}$$

Substituting Eq. (c) into the first expression of Eq. (a) yields

$$x_2 = 2x_1 = 2 \tag{d}$$

Finally, we substitute Eqs.(c) and (d) into the second equation. The result is:

$$3(1) + 2(2) - 4x_3 = -5$$

Thus,

$$x_3 = 3$$

For a large set of equations, the preceeding procedure may be difficult to carry out, particularly if the coefficient matrix has few zeros. The Gaussian elimination method (C. F. Gauss, German mathematician and astromoner, 1777–1855) presents an organized approach to the solution of a large set of equations. The method is as follows:

**(a)** Divide the first equation by the coefficient of $x_1$. The resulting equation is then used to eliminate that coefficient in the remaining equations, following step (b) next.

**(b)** Multiply the equation obtained in (a) by the coefficient of $x_1$ in the second equation, and subtract the result from the second equation. This gives a zero coefficient for $x_1$ in the second row of the coefficient matrix.

**(c)** Repeat step (b) for the third and remaining equations. The system of equations then has the form

$$\begin{bmatrix} 1 & N & N & N & \cdots \\ 0 & N & N & N & \cdots \\ 0 & N & N & N & \cdots \\ 0 & N & N & N & \cdots \\ \cdots & \cdots & \cdots & \cdots & \cdots \end{bmatrix} \begin{bmatrix} x_1 \\ x_2 \\ x_3 \\ x_4 \\ \cdots \end{bmatrix} = \begin{bmatrix} N \\ N \\ N \\ N \\ \cdots \end{bmatrix} \tag{D.3}$$

where $N$ indicates a term that generally will not be zero.

**(d)** The process is continued by dividing the second equation by its coefficient of $x_2$ and using the resulting equation to eliminate the coefficients of $x_2$ from the remaining equations, just as in steps (a)–(c). At the end of this step, the equations take the form

$$\begin{bmatrix} 1 & N & N & N & \cdots \\ 0 & 1 & N & N & \cdots \\ 0 & 0 & N & N & \cdots \\ 0 & 0 & N & N & \cdots \\ \cdots & \cdots & \cdots & \cdots & \cdots \end{bmatrix} \begin{bmatrix} x_1 \\ x_2 \\ x_3 \\ x_4 \\ \cdots \end{bmatrix} = \begin{bmatrix} N \\ N \\ N \\ N \\ \cdots \end{bmatrix} \tag{D.4}$$

**(e)** The process is repeated until the resulting coefficient matrix is an upper triangular matrix. The last equation is then used to solve for $x_n$, the next to last to solve for $x_{n-1}$, etc., until all the unknowns have been found.

**Example D.4**

The steps in the Gaussian elimination method for a set of three equations are shown in this example. The original equations are

$$\begin{bmatrix} 1 & -2 & 3 \\ 2 & 0 & -3 \\ 1 & 1 & 1 \end{bmatrix} \begin{bmatrix} x_1 \\ x_2 \\ x_3 \end{bmatrix} = \begin{bmatrix} 2 \\ 3 \\ 6 \end{bmatrix}$$

The first step gives

$$\begin{bmatrix} 1 & -2 & 3 \\ 0 & 4 & -9 \\ 0 & 3 & -2 \end{bmatrix} \begin{bmatrix} x_1 \\ x_2 \\ x_3 \end{bmatrix} = \begin{bmatrix} 2 \\ -1 \\ 4 \end{bmatrix}$$

The second step gives

$$\begin{bmatrix} 1 & -2 & 3 \\ 0 & 1 & -2.25 \\ 0 & 0 & 4.75 \end{bmatrix} \begin{bmatrix} x_1 \\ x_2 \\ x_3 \end{bmatrix} = \begin{bmatrix} 2 \\ -0.25 \\ 4.75 \end{bmatrix}$$

From the third equation, we obtain $4.75\, x_3 = 4.75$, so $x_3 = 1$. From the second equation, we get $x_2 = -.25 + 2.25x_3 = 2$. Finally, from the first equation, $x_1 = 2 + 2x_2 - 3x_3 = 3$.

Note that at each stage in the process the diagonal term of the coefficient matrix is used as a divisor. Thus, it is generally more accurate to rearrange the equations, if necessary, so that at the first step the equation with the largest coefficient of $x_1$ is used as the first equation. Then, on the second step, the equation with the largest coefficient of $x_2$ is put in the second row, etc. This rearranging of the equations, as necessary, at each step is called **pivoting**. If no rearrangement can be found that avoids the development of one or more zeros on the diagonal of the matrix, then the matrix is singular. This follows from the fact that the determinant of a triangular matrix is just the product of the diagonal terms.

## D.3 HOMOGENEOUS EQUATIONS

The matrix equation

$$\mathbf{Ax} = \mathbf{0} \tag{D.5}$$

represents a set of homogeneous algebraic equations. If the solution is written by Cramer's rule, the numerator is zero in each case, since $\mathbf{b} = \mathbf{0}$ in Eq. (D.5). Thus,

$$x_1 = \frac{0}{|\mathbf{A}|}, \; x_2 = \frac{0}{|\mathbf{A}|}, \text{ etc., or } \mathbf{x} = \frac{1}{|\mathbf{A}|}\mathbf{0} \tag{D.6}$$

Equation (D.6) may be rewritten by multiplying through by $|\mathbf{A}|$ to obtain

$$|\mathbf{A}|\,\mathbf{x} = \mathbf{0} \tag{D.7}$$

This equation has two possible solutions. The **trivial solution** occurs when $\mathbf{x} = \mathbf{0}$. For a **nontrivial solution,** with $\mathbf{x} \neq \mathbf{0}$, it is necessary for $|\mathbf{A}| = 0$. Thus, $\mathbf{A}$ must be singular in order that a nontrivial solution exist for a linear set of homogeneous algebraic equations.

**Example D.5**

Determine the solution vector for the following set of equations:

$$\begin{bmatrix} 2 & 6 \\ 1 & 2 \end{bmatrix} \begin{bmatrix} x_1 \\ x_2 \end{bmatrix} = \begin{bmatrix} 0 \\ 0 \end{bmatrix}$$

**Solution**

$$\det \mathbf{A} = 4 - 6 = -2$$

Since $\det \mathbf{A} \neq \mathbf{0}$, the only solution possible is $\mathbf{x} = \mathbf{0}$, or $x_1 = x_2 = 0$

**Example D.6**

Determine the solution vector for the set of equations

$$\begin{bmatrix} 3 & 6 \\ 1 & 2 \end{bmatrix} \begin{bmatrix} x_1 \\ x_2 \end{bmatrix} = \begin{bmatrix} 0 \\ 0 \end{bmatrix}$$

**Solution** In this case, $\det \mathbf{A} = 0$ because the two equations are not independent. From either equation,

$$x_1 = -2\,x_2$$

and the solutions to the equations are given by

$$\begin{bmatrix} x_1 \\ x_2 \end{bmatrix} = \beta \begin{bmatrix} -2 \\ 1 \end{bmatrix}$$

where $\beta$ is *any* arbitrary scalar multiplier. Thus, the nontrivial solution to a homogeneous set of equations is not unique (more on this later in this appendix).

## D.4 LINEARITY, SINGULARITY, AND RANK

To characterize the solutions to linear algebraic equations more clearly, it is necessary to introduce additional concepts.

To satisfy linearity requirements, a function $f(x)$ must possess a number of characteristics. First, evaluation of the function with a scalar multiple $\alpha$ of the independent variable $x$ must result in the same multiple times the function of $x$. (In other words, a multiple of the input yields a multiple of the output.) Mathematically,

$$f(\alpha x) = \alpha f(x) \tag{D.8}$$

Second, a function using the sum of two values of the independent variable must result in the sum of the two individual functions of the independent variables (addition of inputs yields addition of outputs), or

$$f(x_1 + x_2) = f(x_1) + (x_2) \tag{D.9}$$

Hence, a linear function $f(x)$ must satisfy the general condition

$$f(\alpha_1 x_1 + \alpha_2 x_2) = \alpha_1 f(x_1) + \alpha_2 f(x_2) \tag{D.10}$$

where $x_1$ and $x_2$ are two values of the independent variable $\mathbf{x}$, and $\alpha_1$ and $\alpha_2$ are two arbitrary scalar constants. As an example of a *nonlinear* function, consider

$$f(x) = x^2 \tag{D.11}$$

This function does not satisfy the general condition of Eq. (D.10), because, in general,

$$(\alpha_1 x_1 + \alpha_2 x_2)^2 \neq \alpha_1 (x_1)^2 + \alpha_2 (x_2)^2 \tag{D.12}$$

As a further example, let us determine whether the familiar equation $y = mx + b$ is linear according to Eq. (D.10). In function notation,

$$f(x) = mx + b \tag{D.13}$$

From the first condition, Eq. (D.8), we find that

$$m(\alpha x) + b \neq \alpha(mx + b) \tag{D.14}$$

Further, the second condition, Eq. (D.9), gives

$$m(x_1 + x_2) + b \neq (mx_1 + b) + (mx_2 + b) \tag{D.15}$$

Hence, while it does describe a straight line, $y = mx + b$ is not a linear function by our definition, Eq. (D.10), because of the presence of the constant $b$. If $b = 0$, we have $y = mx$, which will satisfy Eq. (D.10). Thus, a straight line through the origin is a linear function.

A **linear combination** of several variables $\mathbf{x}$ is defined as a summation of linear functions:

$$y = \sum_{i=1}^{p} \alpha_i x_i = \alpha_1 x_1 + \alpha_2 x_2 + \ldots + \alpha_p x_p \tag{D.16}$$

A **linear differential equation** is a linear combination of the successive derivatives of a variable. That is,

$$u = \sum_{i=0}^{p} \alpha_i D^i x_i \tag{D.17}$$

where $D$ is the differential operator $d/dt$.

A matrix is said to be **singular** if its determinant is zero. If one or more of the rows (or columns) of a matrix can be expressed as a linear combination of the remaining rows (or columns), as in Eq. (D.16), where the $x$ and $y$ are now regarded as vectors, then the matrix is singular, and the rows (or columns) are said to be **linearly dependent**.

The **rank** of a matrix is defined to be the size of the largest square matrix with a nonzero determinant that can be formed using the elements of the matrix. We will denote the rank of the matrix **A** by $r(\mathbf{A})$.

**Example D.7**

Find the rank of the matrix

$$A = \begin{bmatrix} 1 & 2 & 3 \\ 3 & 6 & 9 \end{bmatrix}$$

**Solution** The determinants of all possible $2 \times 2$ submatrices of **A** are zero:

$$\begin{vmatrix} 1 & 2 \\ 3 & 6 \end{vmatrix} = 0, \quad \begin{vmatrix} 1 & 3 \\ 3 & 9 \end{vmatrix} = 0, \quad \begin{vmatrix} 2 & 3 \\ 6 & 9 \end{vmatrix} = 0$$

Thus, the rank of **A** is 1, since the largest matrix with nonzero determinant is $1 \times 1$.

The rank of a matrix is equal to the number of independent rows or columns in the matrix. Note that in Example D.7 the second row is three times the first, and the second and third columns are two and three times the first column, respectively. The rank of 1 indicates that **A** has only one independent row and one independent column.

The definition of the rank of a matrix can be used to determine the character of the solutions of a linear set of algebraic equations. Let $\mathbf{A}^b$ denote the coefficient matrix **A** with an added column composed of the vector **b**. Hence, **A** is $n \times n$, and $\mathbf{A}^b$ is $n \times (n + 1)$. A comparison of the rank of **A** with the rank of $\mathbf{A}^b$ leads to the following three possibilities:

**(a)** If $r(\mathbf{A}) = r(\mathbf{A}^b) = n$, a unique solution to the equation exists. (In the case of homogeneous equations, this is the trivial solution, as in Example D.5).

**(b)** If $r(\mathbf{A}) = r(\mathbf{A}^b) = p < n$, there is no unique solution; of the $n$ unknowns, $n - p$ may be chosen arbitrarily, as in Example D.6.

**(c)** If $r(\mathbf{A}) < r(\mathbf{A}^b)$, the equations are inconsistent and no solution exists. Several examples with two equations are given next to illustrate the various cases possible.

**Example D.8**

Determine the character of the solution of the system

$$\begin{bmatrix} 2 & 1 \\ 3 & 2 \end{bmatrix} \begin{bmatrix} x_1 \\ x_2 \end{bmatrix} = \begin{bmatrix} 2 \\ 3 \end{bmatrix}$$

**Solution** Let

$$\mathbf{A} = \begin{bmatrix} 2 & 1 \\ 3 & 2 \end{bmatrix}$$

Then

$$\mathbf{A}^b = \begin{bmatrix} 2 & 1 & 2 \\ 3 & 2 & 3 \end{bmatrix}$$

Here, $r(\mathbf{A}) = r(\mathbf{A}^b) = 2$ with the unique solution $x_1 = 1$, $x_2 = 0$. This system of two equations may be interpreted geometrically, and the solution is found to be the intersection of the two lines described by the equations. See Figure D.1.

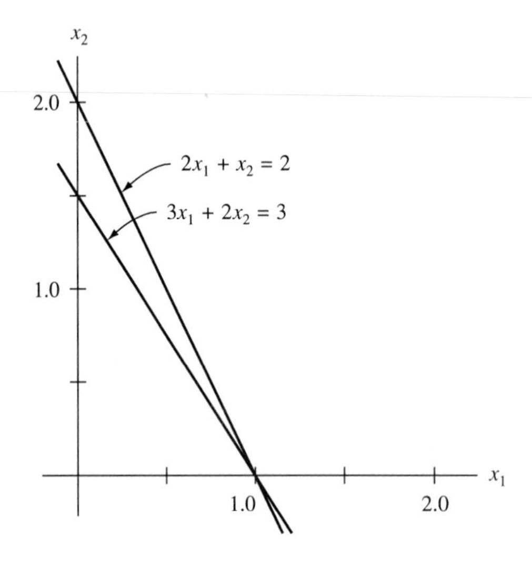


**Figure D.1** Solution for Example D.8

**Example D.9**

Characterize the solution to the sytem

$$\begin{bmatrix} 2 & 1 \\ 6 & 3 \end{bmatrix}\begin{bmatrix} x_1 \\ x_2 \end{bmatrix} = \begin{bmatrix} 2 \\ 6 \end{bmatrix}$$

**Solution** Here, $r(\mathbf{A}) = r(\mathbf{A}^b) = 1 < 2$, and there is no unique solution. However, one variable may be determined in terms of the remaining one. Thus, we find $2x_1 = 2 - x_2$ from either equation. If we choose $x_2 = 1$, then $x_1 = 1/2$. Geometrically, the two equations lie on top of one another, since one is just a multiple of the other.

**Example D.10**

An example of a matrix equation with no solution is

$$\begin{bmatrix} 1 & 2 \\ 3 & 6 \end{bmatrix}\begin{bmatrix} x_1 \\ x_2 \end{bmatrix} = \begin{bmatrix} 14 \\ 6 \end{bmatrix}$$

Here, $r(\mathbf{A}) = 1$, $r(\mathbf{A}^b) = 2$, and there is no solution. A plot shows that the two equations are parallel and thus do not intersect.

**Example D.11**

For the homogeneous set

$$\begin{bmatrix} 1 & 3 \\ 2 & 2 \end{bmatrix} \begin{bmatrix} x_1 \\ x_2 \end{bmatrix} = \begin{bmatrix} 0 \\ 0 \end{bmatrix}$$

$r(\mathbf{A}) = r(\mathbf{A}^b) = 2$, and the unique solution is $x_1 = x_2 = 0$. In this case, the equations intersect at the origin.

**Example D.12**

The homogeneous set

$$\begin{bmatrix} 2 & -4 \\ -6 & 12 \end{bmatrix} \begin{bmatrix} x_1 \\ x_2 \end{bmatrix} = \begin{bmatrix} 0 \\ 0 \end{bmatrix}$$

has a nontrivial solution. Here, $r(\mathbf{A}) = r(\mathbf{A}^b) = 1 < 2$, and again, $x_1$ may be determined only in terms of $x_2$ or vice versa, as $x_1 = 2x_2$. If $x_2 = 1$, then $x_1 = 2$, or $\mathbf{x}^T = (2, 1)$. Any multiple of this solution is also a solution in the homogeneous case. Thus,

$$\mathbf{x} = \beta \begin{bmatrix} 2 \\ 1 \end{bmatrix}$$

**Example D.13**

Find the solution to the $3 \times 3$ system

$$\begin{bmatrix} 7 & 8 & 9 \\ 1 & 2 & 3 \\ 4 & 5 & 6 \end{bmatrix} \begin{bmatrix} x_1 \\ x_2 \\ x_3 \end{bmatrix} = \begin{bmatrix} 0 \\ 0 \\ 0 \end{bmatrix}$$

**Solution** For this system, $r(\mathbf{A}) = r(\mathbf{A}^b) = 2 < 3$. Therefore, we may solve for two of the unknowns in terms of the third. That is, $3 - 2 = 1$ unknown may be chosen arbitrarily.

Using the second two equations, we get

$$1x_1 + 2x_2 = -3x_3$$

$$4x_1 + 5x_2 = -6x_3$$

Let $x_3 = 1$. Then

$$\begin{bmatrix} 1 & 2 \\ 4 & 5 \end{bmatrix} \begin{bmatrix} x_1 \\ x_2 \end{bmatrix} = \begin{bmatrix} -3 \\ -6 \end{bmatrix}$$

Solving this system, we obtain

$$\begin{bmatrix} x_1 \\ x_2 \end{bmatrix} = \begin{bmatrix} 1 \\ -2 \end{bmatrix}$$

The solution to the original problem, then, is

$$\begin{bmatrix} x_1 \\ x_2 \\ x_3 \end{bmatrix} = \beta \begin{bmatrix} 1 \\ -2 \\ 1 \end{bmatrix}$$

Since the system is homogeneous, $\beta$ is any multiplier.

## D.5 MATRIX EIGENVALUE PROBLEMS

### D.5.1 Eigenvalues and Eigenvectors

Often in the analysis of linear systems, a solution to the following equation is desired.

$$\mathbf{A}\,\mathbf{x} = \lambda\mathbf{x} \quad \text{or} \quad (\mathbf{A} - \lambda\mathbf{I})\,\mathbf{x} = 0 \tag{D.18}$$

Here, the matrix **A** is known, and the vector **x** and the scalar $\lambda$ are unknowns.

The scalar $\lambda$ is known as the **eigenvalue** of the matrix **A,** and **x** is the associated **eigenvector**. From the previous discussion of homogeneous equations, the system of Eq. (D.18) has a nontrivial solution only if the coefficient matrix $[\mathbf{A} - \lambda\mathbf{I}]$ is singular; hence, for a nontrivial solution,

$$\det[\mathbf{A} - \lambda\mathbf{I}] = 0 \tag{D.19}$$

If the elements of Eq. (D.19) are written out in full, we obtain

$$\begin{vmatrix} (a_{11} - \lambda) & a_{12} & a_{13} & \cdots \\ a_{21} & (a_{22} - \lambda) & a_{23} & \cdots \\ a_{31} & a_{32} & (a_{33} - \lambda) & \cdots \\ \cdots & \cdots & \cdots & \cdots \end{vmatrix} = 0 \tag{D.20}$$

When expanded, Eq. (D.20) gives an $n$th-order polynomial in $\lambda$. For example, if **A** is $2 \times 2$, we have

$$(a_{11} - \lambda)(a_{22} - \lambda) - a_{12}a_{21} = 0 \tag{D.21}$$

$$\lambda^2 - (a_{11} + a_{22})\lambda + (a_{11}a_{22} - a_{12}a_{21}) = 0$$

The polynomial in $\lambda$ is called the **characteristic equation** of the matrix. In general, then, there are $n$ possible values of $\lambda$ for which nontrivial solutions to Eq. (D.18) can be found. To each of these eigenvalues $\lambda$, there corresponds an eigenvector **x**. However, since $r(\mathbf{A}) = r(\mathbf{A}^b) = p < n$, the eigenvector is not unique. We may thus solve for $p$ elements of the unknown vector in terms of $n - p$ elements.

**Example D.14**

Find the eigenvalues and eigenvectors of

$$\mathbf{A} = \begin{bmatrix} 5 & -2 \\ -2 & 2 \end{bmatrix}$$

**Solution** Expanding the determinant gives $(5 - \lambda)(2 - \lambda) - 4 = \lambda^2 - 7\lambda + 6 = 0$. By factoring or through the use of the quadratic equation, we find two solutions:

$$\lambda_1 = 6$$

$$\lambda_2 = 1$$

To find the eigenvectors, we substitute the eigenvalues, one by one, into the equations and solve for $\mathbf{x}$. The Eq. (D.18) form gives

$$\begin{bmatrix} (5 - \lambda) & -2 \\ -2 & (2 - \lambda) \end{bmatrix} \begin{bmatrix} x_1 \\ x_2 \end{bmatrix} = \begin{bmatrix} 0 \\ 0 \end{bmatrix}$$

From the first equation, with $\lambda = \lambda_1 = 6$, we get

$$(5 - 6)x_1 - 2x_2 = 0 \quad \text{or} \quad -x_1 = 2x_2$$

Thus, if $x_2 = 1$, then $x_1 = -2$, and

$$\begin{bmatrix} x_1 \\ x_2 \end{bmatrix}_1 = \beta \begin{bmatrix} -2 \\ 1 \end{bmatrix}$$

where $\beta$ is any multiplier. Similarly, for $\lambda = \lambda_2 = 1$,

$$\begin{bmatrix} x_1 \\ x_2 \end{bmatrix}_2 = \beta \begin{bmatrix} 1 \\ 2 \end{bmatrix}$$

**Example D.15**

Find the eigensolution to the system

$$\begin{bmatrix} -2 & -13 \\ 1 & 4 \end{bmatrix} \begin{bmatrix} x_1 \\ x_2 \end{bmatrix} = \lambda \begin{bmatrix} x_1 \\ x_2 \end{bmatrix}$$

**Solution** The characteristic equation is

$$(-2 - \lambda)(4 - \lambda) + 13 = \lambda^2 - 2\lambda + 5 = 0$$

with roots

$$\lambda_{1,2} = 1 \pm 2j \quad \text{where } j = \sqrt{-1}$$

Substituting the eigenvalues gives the eigenvectors

$$\begin{bmatrix} x_1 \\ x_2 \end{bmatrix}_1 = \beta \begin{bmatrix} 13 \\ -3 - 2j \end{bmatrix} \quad \text{and} \quad \begin{bmatrix} x_1 \\ x_2 \end{bmatrix}_2 = \beta \begin{bmatrix} 13 \\ -3 + 2j \end{bmatrix}$$

**Example D.16**

Determine the eigenvector of the matrix

$$\mathbf{A} = \begin{bmatrix} 2 & -2 & 2 \\ 1 & 1 & 1 \\ 1 & 3 & -1 \end{bmatrix}$$

corresponding to the eigenvalue $\lambda = -2$.

**Solution** Since $r(\mathbf{A} - \lambda\mathbf{I}) = r[(\mathbf{A} - \lambda\mathbf{I})^b] = 2 < 3$, we can determine two unknowns in terms of the third arbitrarily chosen one. Substituting $\lambda = -2$ into the first two equations yields

$$[2 - (-2)]x_1 - 2x_2 = -2x_3$$

$$x_1 + [1 - (-2)]x_2 = -x_3$$

or

$$\begin{bmatrix} 4 & -2 \\ 1 & 3 \end{bmatrix} \begin{bmatrix} x_1 \\ x_2 \end{bmatrix} = -x_3 \begin{bmatrix} 2 \\ 1 \end{bmatrix}$$

Using the inversion method, we obtain

$$\begin{bmatrix} x_1 \\ x_2 \end{bmatrix} = x_3 \begin{bmatrix} -\frac{4}{7} \\ -\frac{1}{7} \end{bmatrix}$$

and the complete eigenvector is

$$\begin{bmatrix} x_1 \\ x_2 \\ x_3 \end{bmatrix} = \beta \begin{bmatrix} -\frac{4}{7} \\ -\frac{1}{7} \\ 1 \end{bmatrix} = \frac{\beta}{-7} \begin{bmatrix} 4 \\ 1 \\ -7 \end{bmatrix}$$

### D.5.2 Reduction to Diagonal Form

An $n \times n$ matrix $\mathbf{A}$ can be transformed into a diagonal matrix if $\mathbf{A}$ has $n$ linearly independent eigenvectors. Let $\mathbf{M}$ be an $n \times n$ matrix, each column of which is an eigenvector of $\mathbf{A}$; then it can be shown that

$$\mathbf{M}^{-1}\mathbf{A}\,\mathbf{M} = \mathbf{L} \tag{D.22}$$

where

$$\mathbf{L} = \begin{bmatrix} \lambda_1 & 0 & 0 & \cdots \\ 0 & \lambda_2 & 0 & \cdots \\ 0 & 0 & \lambda_3 & \cdots \\ \cdots & \cdots & \cdots & \cdots \end{bmatrix} \quad \text{(D.23)}$$

is a diagonal matrix whose diagonal entries are the eigenvalues of **A**.

**Example D.17**

We demonstrate the property of Eq. (D.22) using the matrix of Example (D.14). Selecting $\beta = 1$ we have

$$\mathbf{M} = \begin{bmatrix} -2 & 1 \\ 1 & 2 \end{bmatrix} \quad \text{and} \quad \mathbf{M}^{-1} = -\frac{1}{5}\begin{bmatrix} 2 & -1 \\ -1 & -2 \end{bmatrix}$$

Then

$$\mathbf{M}^{-1}\mathbf{A}\,\mathbf{M} = -\frac{1}{5}\begin{bmatrix} 2 & -1 \\ -1 & -2 \end{bmatrix}\begin{bmatrix} 5 & -2 \\ -2 & 2 \end{bmatrix}\begin{bmatrix} -2 & 1 \\ 1 & 2 \end{bmatrix} = \begin{bmatrix} 6 & 0 \\ 0 & 1 \end{bmatrix} = \mathbf{L}$$

Reduction to diagonal form using the eigenvectors of a matrix is a very useful property and forms the basis for an important solution method used in dynamic systems analysis and design.

## D.6 SYSTEMS OF NONLINEAR ALGEBRAIC EQUATIONS

In some cases, it may be necessary to include nonlinear algebraic relationships in the development of simulation models. The force developed in a nonlinear spring, for example, might be represented by an equation of the form $f = kx^2$. (See Section 3.3.1.) If there are two or more variables involved in the problem, then the number of equations that must be considered increases accordingly. For example we might wish to determine the values of $x_1$ and $x_2$ that satisfy the equations

$$x_1 + x_1 x_2 = x_2^2 + 7$$

and (D.24)

$$x_1^3 + x_2 - x_2^3 = -9$$

The correct approach to obtaining a solution is not obvious.

Previously, we wrote systems of linear equations in the form

$$\mathbf{A}\mathbf{x} = \mathbf{b} \quad \text{(D.25)}$$

Suppose the vector **b** is moved to the left side of the equation:

$$\mathbf{A}\mathbf{x} - \mathbf{b} = \mathbf{0} \quad \text{(D.26)}$$

Then we can consider finding roots **x** of the system that satisfy the system of equations described by

$$\mathbf{f}(\mathbf{x}) = 0 \tag{D.27}$$

It is convenient to write systems of nonlinear equations in this form. For Eq. (D.24), we would write

$$\begin{aligned} f_1 &= x_1 + x_1x_2 - x_2^2 - 7 = 0 \\ f_2 &= x_1^3 + x_2 - x_2^3 + 9 = 0 \end{aligned} \tag{D.28}$$

Newton's method is a good approach for finding a value $x$ that satisfies a single nonlinear equation $f(x) = 0$. We illustrate the technique in Figure D.2.

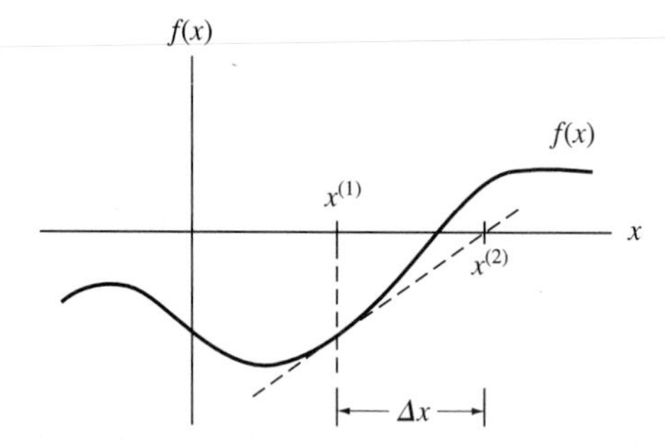


**Figure D.2** Newton's method.

Suppose we know the approximate location of the root of $f(x) = 0$. Call this solution estimate $x^{(1)}$. If we can calculate the slope of the function, we can get closer to the solution by projecting the slope. That is, notice from Figure D.2 that

$$\frac{f(x^{(1)})}{\Delta x} \text{ is proportional to } \frac{df(x^{(1)})}{dx}$$

$$\text{Thus, } \Delta x = \frac{-f(x^{(1)})}{\dfrac{df(x^{(1)})}{dx}} \tag{D.29}$$

(The negative sign is required because the function $f(x_1)$, as shown, would be a negative number.) Based upon this geometric development, we can set up an iteration of the form

$$x^{(j+1)} = x^{(j)} - \frac{f(x^{(j)})}{\dfrac{df(x^{(j)})}{dx}} \tag{D.30}$$

A graphical representation of this process is shown in Figure D.3.

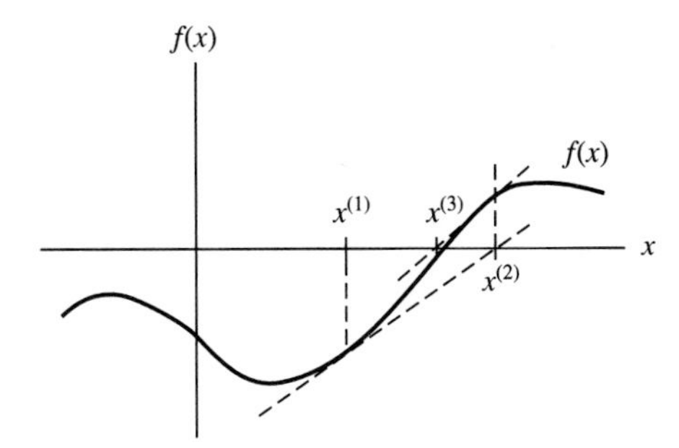


**Figure D.3** Newton's method showing two iterations.

Newton's method is easy to use, the convergence is rapid, and results are accurate, *provided* that the initial estimate of the solution is sufficiently close to the desired root *and* there are no unusual circumstances (such as the function being tangent to the $x$-axis; see Figure D.4b.)

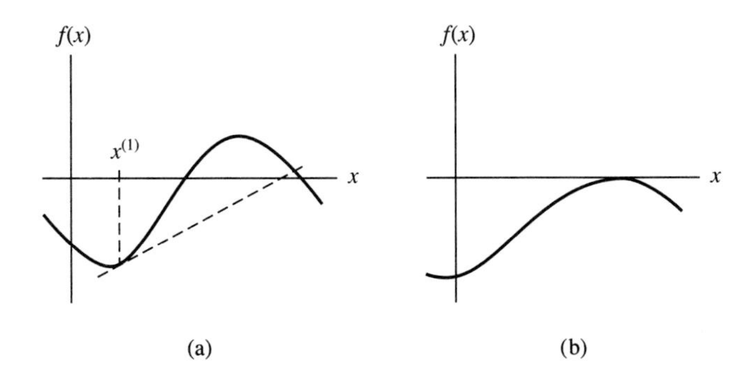


**Figure D.4** When Newton's method may fail. (a) Projection misses root. (b) Curve is tangent to $x$ axis.

The method of Newton can also be developed from a consideration of the Taylor series of $f(x)$ about some point:

$$f(x + \Delta x) = f(x) + \Delta x \frac{df}{dx} + \frac{1}{2}\Delta x^2 \frac{d^2 f}{dx^2} + \ldots \tag{D.31}$$

If we retain only the linear terms, then

$$f(x + \Delta x) = f(x) + \Delta x \frac{df}{dx} \tag{D.32}$$

We also suppose that $x + \Delta x$ satisfies the given equation, i.e., that $f(x + \Delta x) = 0$. Then

$$\Delta x = \frac{-f(x)}{\dfrac{df}{dx}} = -\left(\frac{df}{dx}\right)^{-1} f(x) \tag{D.33}$$

Since $f(x + \Delta x)$ may not be exactly zero, we must define the iteration shown in Eq. (D.30) in order to get closer to the solution.

Newton's method can easily be extended to multiple functions involving several unknowns. As a start, consider two equations in $x_1$ and $x_2$. We want to find the values of $x_1$ and $x_2$ that satisfy the two functions

$$f_1(x_1,x_2) = 0$$

and

$$f_2(x_1,x_2) = 0 \tag{D.34}$$

Extrapolating the preceding development for a single nonlinear equation, we expand these two functions in a Taylor series:

$$f_1(x_1 + \Delta x_1, x_2 + \Delta x_2) = f_1(x_1, x_2) + \Delta x_1 \frac{\partial f_1}{\partial x_1} + \Delta x_2 \frac{\partial f_1}{\partial x_2} + \dots \tag{D.35a}$$

$$f_2(x_1 + \Delta x_1, x_2 + \Delta x_2) = f_2(x_1, x_2) + \Delta x_1 \frac{\partial f_2}{\partial x_1} + \Delta x_2 \frac{\partial f_2}{\partial x_2} + \dots \tag{D.35b}$$

Now we assume, as before, that $f_1(x_1 + \Delta x_1, x_2 + \Delta x_2) = 0$ and $f_2(x_1 + \Delta x_1, x_2 + \Delta x_2) = 0$, and solve for $\Delta x_1$ and $\Delta x_2$.

$$\begin{bmatrix} f_1(x_1 + \Delta x_1, x_2 + \Delta x_2) - f_1(x_1, x_2) \\ f_2(x_1 + \Delta x_1, x_2 + \Delta x_2) - f_2(x_1, x_2) \end{bmatrix} = \begin{bmatrix} \frac{\partial f_1}{\partial x_1} & \frac{\partial f_1}{\partial x_2} \\ \frac{\partial f_2}{\partial x_1} & \frac{\partial f_2}{\partial x_2} \end{bmatrix} \begin{bmatrix} \Delta x_1 \\ \Delta x_2 \end{bmatrix} = [\mathbf{J}] \begin{bmatrix} \Delta x_1 \\ \Delta x_2 \end{bmatrix} \tag{D.36}$$

Thus,

$$\begin{bmatrix} \Delta x_1 \\ \Delta x_2 \end{bmatrix} = - \begin{bmatrix} \frac{\partial f_1}{\partial x_1} & \frac{\partial f_1}{\partial x_2} \\ \frac{\partial f_2}{\partial x_1} & \frac{\partial f_2}{\partial x_2} \end{bmatrix}^{-1} \begin{bmatrix} f_1(x_1, x_2) \\ f_2(x_1, x_2) \end{bmatrix} \tag{D.37}$$

To obtain the changes in the variables ($\Delta x_1$ and $\Delta x_2$) we solve this set of linear algebraic equations. Of course, these are only estimates of the roots, and the functions may not be zero at $x_i + \Delta x_i, i = 1, 2$. Accordingly, we define the following iterative equation to find successively better estimates of the solution:

$$[\mathbf{x}]^{(j+1)} = [\mathbf{x}]^{(j)} - [\mathbf{J}]^{-1}[\boldsymbol{f}([\mathbf{x}]^{(j)})] \tag{D.38}$$

Here, $\mathbf{J}$ is the matrix of partial derivatives evaluated at $[\mathbf{x}]^j$. This iteration may be used for any number of equations by extending the number of functions and variables appropriately.

**Example D.18**

Use Newton's method to find a solution to the following system of equations:

$$f_1(x_1, x_2) = 3x_1 - x_1x_2 + 2x_2^2 - 11 = 0$$

$$f_2(x_1, x_2) = 2x_1^2 + x_2^2 - 22 = 0$$

**Solution** First, we find the matrix $\mathbf{J}$:

$$\frac{\partial f_1}{\partial x_1} = 3 - x_2, \quad \frac{\partial f_1}{\partial x_2} = -x_1 + 4x_2$$

$$\frac{\partial f_2}{\partial x_1} = 4x_1, \quad \frac{\partial f_2}{\partial x_2} = 2x_2$$

Next, we set up a calculation procedure to implement Eq. (D.38) and select starting values of $x_1^{(1)} = 1.5, x_2^{(1)} = 2.5$. The following results are obtained:

| $x^{(1)}$ | $x^{(2)}$ |
|---|---|
| 1.5000 | 2.5000 |
| 3.7036 | 2.1057 |
| 3.0747 | 1.9748 |
| 3.0008 | 2.0005 |
| 3.0000 | 2.0000 |
| 3.0000 | 2.0000 |

Notice that the process converges very quickly, and you can verify for yourself that the values satisfy the given equations.

Suppose however, we had selected $x_1^{(1)} = -1.5$ and $x_2^{(1)} = -2.5$ as the starting values. Then the following values would be obtained from Newton's method:

| $x^{(1)}$ | $x^{(2)}$ |
|---|---|
| −1.5000 | −2.5000 |
| −2.2882 | −3.8041 |
| −2.1707 | −3.5587 |
| −2.1680 | −3.5496 |
| −2.1680 | −3.5496 |
| −2.1680 | −3.5496 |

Again, you can verify that these values satisfy the given functions.

Thus, we observe that nonlinear systems of equations may have multiple solutions, the solutions may be easy or difficult to find, and an iterative process such as

Newton's method can converge very quickly *if* you start sufficiently close to the solution you want to find *and* the functions are well behaved.

## PROBLEMS

D.1 *Find* the solution vector **x** if

$$\begin{bmatrix} 2 & -3 & 5 \\ 2 & 4 & 1 \\ -1 & 3 & -2 \end{bmatrix} \begin{bmatrix} x_1 \\ x_2 \\ x_3 \end{bmatrix} = \begin{bmatrix} 21 \\ -11 \\ -16 \end{bmatrix}$$

D.2 *Determine* a solution to

$$\begin{bmatrix} 1 & 2 & -3 \\ 3 & -2 & 4 \\ 6 & 1 & -5 \end{bmatrix} \begin{bmatrix} x_1 \\ x_2 \\ x_3 \end{bmatrix} = \begin{bmatrix} 4 \\ -1 \\ 11 \end{bmatrix}$$

D.3 *Determine* the solution vector **x** if

$$x_1 + x_2 + x_3 = 2$$

$$x_1 - x_2 + x_3 = -2$$

$$2x_1 - x_2 + x_3 = -1$$

D.4 *Solve* for $x_1, x_2, x_3$:

$$\begin{bmatrix} 2 & 4 & -2 \\ 1 & 3 & -4 \\ -1 & 2 & 3 \end{bmatrix} \begin{bmatrix} x_1 \\ x_2 \\ x_3 \end{bmatrix} = \begin{bmatrix} 14 \\ 16 \\ 1 \end{bmatrix}$$

D.5 Use Cramer's rule to *find* $x_3$ for the following system of equations:

$$2x_1 + 3x_2 + x_3 = 9$$

$$x_1 + 2x_2 + 3x_3 = 6$$

$$3x_1 + x_2 + 2x_3 = 8$$

D.6 *Determine* the solution for **x** in problem D.5 by matrix inversion.

D.7 *Solve* for $x_1$ and $x_2$ by using Cramer's rule:

$$\begin{bmatrix} 1 & 2 & 3 \\ -4 & 2 & -1 \\ 3 & 2 & 5 \end{bmatrix} \begin{bmatrix} x_1 \\ x_2 \\ x_3 \end{bmatrix} = \begin{bmatrix} 1 \\ 1 \\ 1 \end{bmatrix}$$

D.8 Use Cramer's rule to *find* the solution for $\mathbf{x}$ if

$$2x_1 - 3x_2 = 6$$

$$x_1 - 2x_2 = 3$$

D.9 Use Gaussian elimination to *solve* problem D.7.

D.10 Use matrix inversion to *solve* problem D.8.

D.11 *Can* a unique solution be found to the following linear system? If yes, *find* it.

$$\begin{bmatrix} 2 & 1 & 5 \\ 1 & 1 & -3 \\ 3 & 6 & -2 \end{bmatrix} \begin{bmatrix} x_1 \\ x_2 \\ x_3 \end{bmatrix} = \begin{bmatrix} 3 \\ 1 \\ 0 \end{bmatrix}$$

D.12 *Find* the solution to the following system of equations:

$$x_1 + 2x_2 + 9x_3 = 0$$

$$2x_1 + 2x_3 = 0$$

$$3x_1 - 2x_2 + 5x_3 = 0$$

D.13 *Determine* the solution to problem D.1 by Gaussian elimination.

D.14 *Find* a solution to

$$\begin{bmatrix} 7 & 8 & 9 \\ 1 & 2 & 3 \\ 4 & 5 & 6 \end{bmatrix} \begin{bmatrix} x_1 \\ x_2 \\ x_3 \end{bmatrix} = \begin{bmatrix} 0 \\ 0 \\ 0 \end{bmatrix}$$

D.15 *Find* the solution to

$$\begin{bmatrix} 1 & 2 & -3 \\ 3 & -2 & 4 \\ 6 & 1 & -5 \end{bmatrix} \begin{bmatrix} x_1 \\ x_2 \\ x_3 \end{bmatrix} = \begin{bmatrix} 0 \\ 0 \\ 0 \end{bmatrix}$$

D.16 *Find* the rank of $\mathbf{E}$, the rank of $\mathbf{F}$, and the rank of $\mathbf{E}\,\mathbf{F}$ if

$$\mathbf{E} = \begin{bmatrix} 1 & 5 & 4 \\ 6 & 1 & 2 \end{bmatrix}, \quad \mathbf{F} = \begin{bmatrix} 7 & 8 \\ 3 & 1 \\ 5 & 7 \end{bmatrix}$$

D.17 *Find* a solution vector $\mathbf{x}$ for the following system of equations:

$$-x_1 + 2x_2 + 4x_3 = -3$$

$$4x_1 - 8x_2 - 16x_3 = 12$$

$$-2x_1 + 4x_2 + 8x_3 = -6$$

D.18 *Determine* the rank of

$$\mathbf{A} = \begin{bmatrix} 2 & 1 & -3 & 5 \\ 0 & 0 & 1 & 2 \\ 0 & 0 & 8 & 0 \\ 0 & 0 & 0 & 3 \end{bmatrix}$$

D.19 *Compute* the rank of the following matrices:

$$\mathbf{A} = \begin{bmatrix} 1 & 2 \\ 3 & 4 \end{bmatrix} \quad \mathbf{B} = \begin{bmatrix} 4 & -2 & 1 \\ 3 & 1 & 0 \end{bmatrix}$$

D.20 *Compute* $r(\mathbf{D})$ if

$$\mathbf{D} = \begin{bmatrix} 1 & 2 & 3 \\ 7 & 8 & 9 \\ 4 & 5 & 6 \end{bmatrix}$$

D.21 *Compute* $r(\mathbf{A})$ and $r(\mathbf{A}^b)$ for the matrices of problem D.2.

D.22 *Compute* $r(\mathbf{A})$ and $r(\mathbf{A}^b)$ for the system of equations of problem D.3.

D.23 *Compute* $r(\mathbf{A})$ and $r(\mathbf{A}^b)$ for the matrices of problem D.4.

D.24 *Find* the eigenvalues and eigenvectors of

$$\mathbf{A} = \begin{bmatrix} 5 & -2 \\ -2 & 3 \end{bmatrix}$$

D.25 *Find* the eigenvalues and eigenvectors of

$$\mathbf{A} = \begin{bmatrix} 3 & 0 & -8 \\ 2 & 1 & -4 \\ 0 & 0 & -1 \end{bmatrix}$$

D.26 *Find* the eigenvalues and eigenvectors of

$$\mathbf{A} = \begin{bmatrix} 2 & -3 \\ -5 & 4 \end{bmatrix}$$

D.27 *Find* the eigenvalues and eigenvectors of

$$\mathbf{A} = \begin{bmatrix} 2 & -6 \\ 1 & 2 \end{bmatrix}$$

D.28 *Determine* the eigenvalues and eigenvectors of

$$\mathbf{B} = \begin{bmatrix} 1 & -2 & 4 \\ 0 & 3 & 3 \\ 0 & 0 & 2 \end{bmatrix}$$

D.29 *Find* the eigenvalues of

$$\mathbf{C} = \begin{bmatrix} 0 & 2 & 2 \\ 1 & 0 & 3 \\ 1 & 1 & 0 \end{bmatrix}$$

D.30 *Determine* the eigenvalues and eigenvectors of

$$\mathbf{B} = \begin{bmatrix} 2 & 7 \\ 3 & 2 \end{bmatrix}$$

D.31 *Find* the eigenvalues of

$$\mathbf{A} = \begin{bmatrix} 2 & -2 & 2 \\ 1 & 1 & 1 \\ 1 & 3 & -1 \end{bmatrix}$$

D.32 *Determine* the eigenvalues and eigenvectors of

$$\mathbf{D} = \begin{bmatrix} 1 & 2 \\ 3 & 4 \end{bmatrix}$$

D.33 *Compute* the eigenvalues and eigenvectors of

$$\mathbf{E} = \begin{bmatrix} 2 & -7 \\ 3 & 2 \end{bmatrix}$$

D.34 *Find* the eigenvalues and eigenvectors of

$$\mathbf{F} = \begin{bmatrix} 4 & 2 \\ -6 & 1 \end{bmatrix}$$

D.35 *Determine* the eigenvalues and eigenvectors of

$$\mathbf{G} = \begin{bmatrix} 1 & -1 & 0 \\ -1 & 2 & -1 \\ 0 & -1 & 1 \end{bmatrix}$$

# APPENDIX E

# Solutions of Differential Equations by Classical Methods

## E.1 INTRODUCTION

The behavior of a large and very important class of physical systems can be adequately and accurately represented through the use of models that utilize linear ordinary differential equations. Since closed-form expressions for the solutions to the governing equations for these models can be easily obtained, knowledge of the resulting behavior characteristics constitutes an important part of our understanding of the performance of the system.

In this appendix, classical methods for solving these systems are reviewed, and several representative cases are treated. In the following two sections, we develop solutions of first- and second-order system differential equations. Since the former is the simpler of the two, it is examined first, in the next section.

## E.2 FIRST-ORDER SYSTEMS

The general form of first-order systems is

$$a_1\dot{x} + a_0 x = b_0\, u(t) \tag{E.1a}$$

or, dividing through by $a_0$,

$$\frac{a_1}{a_0}\dot{x} + x = \frac{b_0}{a_0} u(t) \tag{E.1b}$$

In this representation, $x$ is the dependent or response variable, $t$ is the independent variable, and $u(t)$ is the input function. Note that, as before in this text, we use $\dot{x}$ to mean $dx/dt$, and we also use the $D$-operator notation: $D = d/dt$ such that $Dx = dx/dt$.

We replace $a_1/a_0$ with $\tau$, $b_0/a_0$ with $G$, and obtain the characterization of the response of linear first-order systems:

$$\tau\dot{x} + x = G\,u(t), \text{ or } \tau Dx + x = Gu(t) \tag{E.1c}$$

### E.2.1 Initial-Condition Response

Consider what happens to a first-order system if it is given a nonzero initial condition and no input ($u(t) = 0$):

$$\tau\dot{x} + x = 0 \tag{E.2}$$

The initial condition (when time is just slightly before $t = 0$, designated by $0^-$) is known to be $x_0$, so that

$$x(0^-) = x_0 \tag{E.3}$$

Rearranging the governing equation, we have

$$\dot{x} = -\frac{1}{\tau}x \tag{E.4}$$

Therefore, the solution $x(t)$ must be some function of time such that its derivative is proportional to the function itself. It is known that an exponential function satisfies this condition; therefore, the homogeneous solution is of the form

$$x(t) = A\,e^{\lambda t} \tag{E.5}$$

where $A$ and $\lambda$ are constants that are to be found. Substituting this function into the differential equation yields

$$\lambda Ae^{\lambda t} = -\frac{1}{\tau}Ae^{\lambda t} \tag{E.6}$$

from which we find the characteristic equation for the first-order system, which has one root, namely,

$$\lambda = -\frac{1}{\tau} \tag{E.7}$$

Hence, the solution for the free response is of the form

$$x(t) = A\,e^{-t/\tau} \tag{E.8}$$

Applying the initial condition $x(0^-) = x_0$, we obtain $A$:

$$x(0^-) = x_0 = A, \quad \text{or} \quad A = x_0 \tag{E.9}$$

The complete solution is then

$$x(t) = x_0 e^{-t/\tau} \tag{E.10}$$

A plot of the response is shown in Figure E.1. The quantity $\tau$ in this equation is called the **time constant.** It has units of time and is a measure of the speed of the response of a first-order system, as dictated by its physical parameters. Systems with small time constants respond quickly to external inputs or to initial conditions; those with large time constants are sluggish. In any case, first-order systems will be very near to their final response after a time equal to four time constants has elapsed. Figure E.1 shows this for the response to an initial condition $x_0$. Note that the response is normalized by the magnitude of the initial condition, and time is normalized by the system time constant.

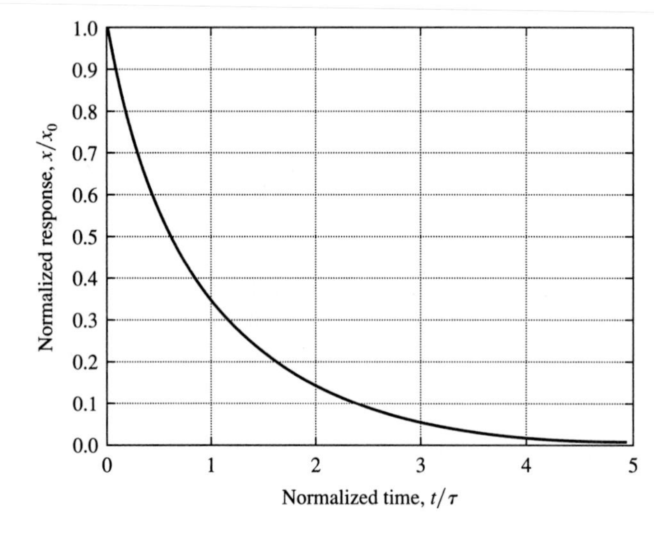


**Figure E.1** First-order system response.

### E.2.2 Step Input Response

Let us now examine the response of the preceding first-order system to the step input shown in Figure E.2. The governing equation for $t \geq 0$ is

$$\tau \dot{x} + x = Gu \tag{E.11}$$

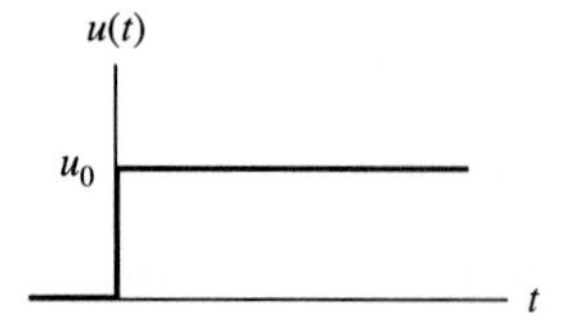


**Figure E.2** Step input.

with initial condition

$$x(0^-) = x_0 \tag{E.12}$$

We construct the solution of this equation as the sum of the solution of the homogeneous equation corresponding to $u_0 = 0$ and a particular solution when $u_0 \neq 0$:

$$x(t) = x_h(t) + x_p(t) \tag{E.13}$$

We found the homogeneous solution previously:

$$x_h(t) = A\,e^{-t/\tau} \tag{E.14}$$

The characteristics of the right-hand side of the governing equation, Eq. (E.13), dictate the form of the particular solution. Since the excitation in this case is a constant, a particular solution consisting of a constant (say, $B$) will satisfy the differential equation. Let

$$x_p(t) = B \tag{E.15}$$

Then

$$\dot{x}_p = 0 \tag{E.16}$$

Substitution of this particular solution into the differential equation yields

$$B = Gu_0 \tag{E.17}$$

Therefore, the form of the complete solution to the differential equation is

$$x(t) = Ae^{-t/\tau} + Gu_0 \tag{E.18}$$

To find the unknown constant $A$, *apply the initial condition to the complete solution*. Note that the initial conditions must always be applied to the complete solution after it has been found: Since $x(0) = x_0$, then $x_0 = A + Gu_0$, or $A = x_0 - Gu_0$, and therefore,

$$x(t) = x_0 e^{-t/\tau} + Gu_0(1 - e^{-t/\tau}) \tag{E.19}$$

This equation represents the response of the first-order system to a step input. A plot of the equation is shown in Figure E.3 for the initial condition $x_0 = 0$. Note that $x(t)$ begins at zero and, as time increases, approaches $Gu_0$, the value that results after all dynamic effects have died away. The quantity $Gu_0$ can also be thought of as the displacement which would occur if the input were applied statically, i.e., so slowly that all derivative effects are negligible. Observe from the figure that the time constant can be used to measure the response time of the system. That is, the position $x(t)$ is 98% of the system response from the initial value to the final value after a time equal to four time constants has elapsed. The response $x(t)$ is also shown in Table E.1 at equally spaced times given as multiples of the time constant $\tau$.

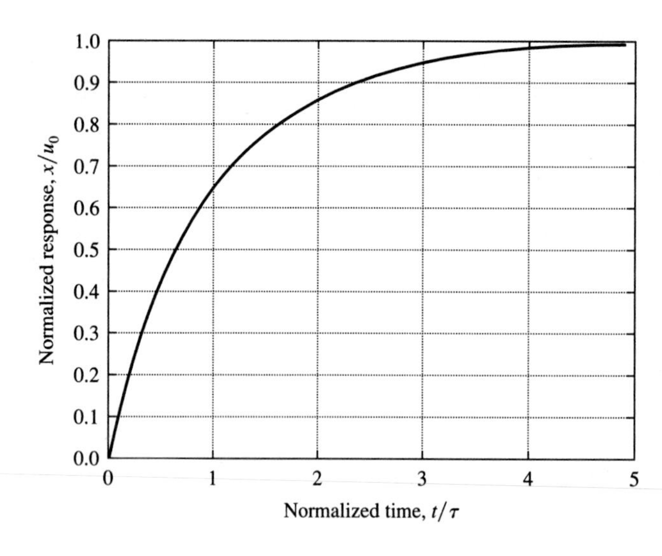


**Figure E.3** First-order system response to step input.

**TABLE E.1** RESPONSE OF FIRST-ORDER SYSTEM TO A STEP INPUT

| $t$ | $x(t)/Gu_0$ |
|---|---|
| 0 | 0.000 |
| $1\tau$ | 0.632 |
| $2\tau$ | 0.865 |
| $3\tau$ | 0.950 |
| $4\tau$ | 0.982 |
| $5\tau$ | 0.993 |
| $6\tau$ | 0.998 |

### E.2.3 Ramp Input Response

Next, consider what happens to the first-order system if an input whose magnitude increases linearly with time is applied. This type of excitation is called a ramp and is shown in Figure E.4. The magnitude of the slope of the input signal is $u_0$.

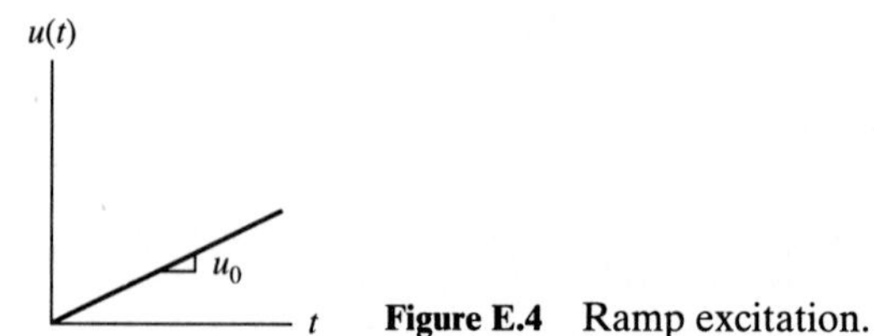


**Figure E.4** Ramp excitation.

The expression for the equation of motion of the system is

$$\tau\dot{x} + x = Gu_0 t \tag{E.20}$$

As before, the homogeneous solution is given by Eq. (E.5), and the particular solution should be selected to have the same form as the excitation. Since the input is a linear function of time, we will try a similar function. Let $Q$ and $R$ be constants that we will find by substitution:

$$x_p(t) = Q + Rt \tag{E.21}$$

Substituting into the governing equation gives

$$\tau R + (Q + Rt) = Gu_0 t \tag{E.22}$$

The constant terms and the coefficients of the time functions in this equation are independent. That is, for the equation to be satisfied, the constant terms must be related by

$$\tau R + Q = 0 \tag{E.23}$$

and the coefficients multiplying $t$ must independently be related by

$$R = Gu_0 \tag{E.24}$$

Solving these equations gives the constants $R$ and $Q$:

$$R = Gu_0 \tag{E.25}$$

$$Q = -\tau R = -Gu_0\tau \tag{E.26}$$

Thus, the particular solution is

$$x_p(t) = -Gu_0\tau + Gu_0 t = Gu_0(t - \tau) \tag{E.27}$$

Adding the homogeneous solution and the particular solution gives the following form for the complete solution:

$$x(t) = Ae^{-t/\tau} + Gu_0(t - \tau) \tag{E.28}$$

The constant $A$ is found using the initial conditions on the system, i.e., $x(0) = x_0$; thus,

$$A = x_0 + Gu_0\tau \tag{E.29}$$

Therefore, the complete system response is

$$x(t) = x_0 e^{-t/\tau} + Gu_0[t - \tau(1 - e^{-t/\tau})] \tag{E.30}$$

### E.2.4 Sinusoidal Input Response

Useful information concerning the behavior of a system can also be obtained by determining its response to a harmonic excitation. Consider again the first-order system, but now let the input be a cosine function of time with amplitude $u_0$ and frequency $\omega$:

$$\tau\dot{x} + x = Gu_0 \cos \omega t \tag{E.31}$$

Again, we develop the complete solution as the sum of the homogeneous plus the particular solution. As before,

$$x_h(t) = Ae^{-t/\tau} \tag{E.32}$$

For the particular solution, we look to the form of the input. Since the system is being excited at frequency $\omega$, we would expect a harmonic response at the same frequency. To account for the possibility of the response lagging behind or getting ahead of the excitation, we will include a phase angle $\phi$. We try a particular solution of the form

$$x_p(t) = B\cos(\omega t + \phi) \tag{E.33}$$

Typical excitation and response functions of time for this type of system are shown in Figure E.5.

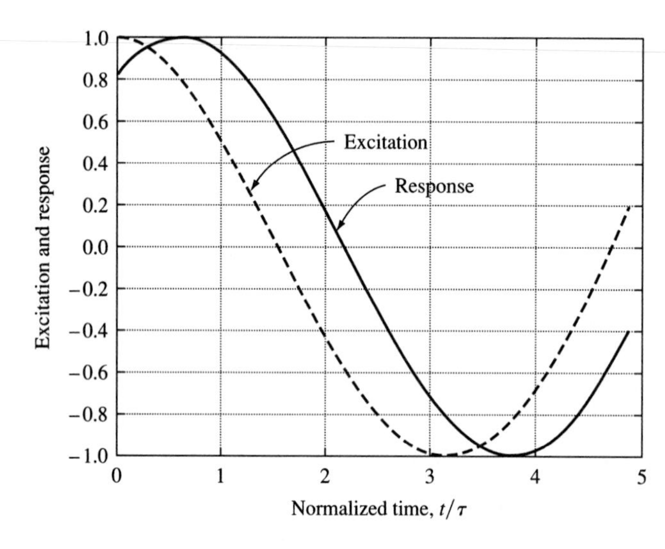


**Figure E.5** Typical harmonic excitation and response functions.

To obtain the steady-state amplitude $B$ and phase angle $\phi$, we substitute the particular solution into the governing differential equation:

$$-\omega\tau B\sin(\omega t + \phi) + B\cos(\omega t + \phi) = Gu_0\cos\omega t \tag{E.34}$$

If the sum-of-angles identities are used to expand the left-hand side we obtain

$$-\omega\tau B[\sin\omega t\cos\phi + \cos\omega t\sin\phi] + B[\cos\omega t\cos\phi - \sin\omega t\sin\phi] = Gu_0\cos\omega t \tag{E.35}$$

Now, the $\sin\omega t$ and $\cos\omega t$ functions are independent. Hence, the coefficients of $\sin\omega t$ on each side of this equation must be equal, and, independently, the coefficients of $\cos\omega t$ on each side must be equal. That is, the following two equations must hold:

$$-\omega\tau B\cos\phi - B\sin\phi = 0 \tag{E.36}$$

$$-\omega\tau B\sin\phi + B\cos\phi = Gu_0 \tag{E.37}$$

The constant $B$ cannot be zero, for if it were, no harmonic response would occur. Thus, we can divide Eq. (E.36) by $B$ to obtain the following result, after some algebraic manipulation:

$$\frac{\sin\phi}{\cos\phi} = \tan\phi = -\omega\tau \qquad \text{(E.38)}$$

Then, from the triangle of Figure E.6, $\sin\phi$ and $\cos\phi$ can be written as

$$\sin\phi = \frac{-\omega\tau}{\sqrt{1+\omega^2\tau^2}} \qquad \text{(E.39)}$$

and

$$\cos\phi = \frac{1}{\sqrt{1+\omega^2\tau^2}} \qquad \text{(E.40)}$$

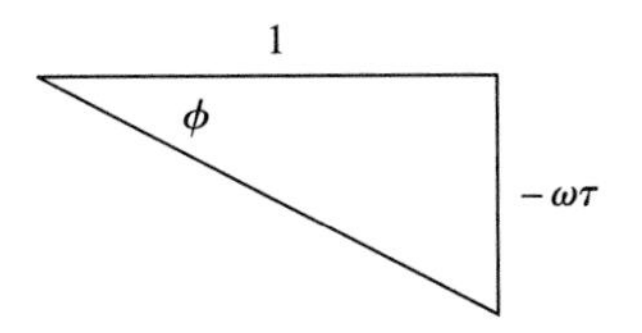


**Figure E.6** Phase angle components.

Equation (E.37) now becomes

$$B\left(\frac{\omega^2\tau^2}{\sqrt{1+\omega^2\tau^2}} + \frac{1}{\sqrt{1+\omega^2\tau^2}}\right) = Gu_0 \qquad \text{(E.41)}$$

from which we obtain

$$B = \frac{Gu_0}{\sqrt{1+\omega^2\tau^2}} \qquad \text{(E.42)}$$

Thus, the particular solution is

$$x_p(t) = \frac{Gu_0\cos(\omega t + \phi)}{\sqrt{1+\omega^2\tau^2}} \qquad \text{(E.43)}$$

where, from Eq. (E.38),

$$\phi = \tan^{-1}(-\omega\tau) \qquad \text{(E.44)}$$

The form for the solution is the sum of the homogeneous and particular solutions:

$$x(t) = Ae^{-t/\tau} + B\cos(\omega t + \phi) \qquad \text{(E.45)}$$

homogeneous + particular
(transient) (steady state)

The first term of this equation (the homogenous solution) is called the **transient response** portion, since it dies away with time and is essentially zero after an elapsed time of about four time constants. We refer to the second term (the particular solution) as the **steady-state response,** since it represents the system behavior after the starting transients have died away and steady operational conditions exist. Applying the initial condition to the complete solution, we find the following:

$$A = x_0 - \frac{Gu_0}{[1 + \omega^2\tau^2]} \tag{E.46}$$

$$x(t) = x_0 e^{-t/\tau} + Gu_0\left(\frac{\cos(\omega t + \phi)}{\sqrt{1 + \omega^2\tau^2}} - \frac{e^{-t/\tau}}{[1 + \omega^2\tau^2]}\right) \tag{E.47}$$

It is now of interest to examine the steady-state response to input signals of various frequencies. That is, what are the steady-state amplitude $|X_{ss}| = B$ and phase angle $\phi$ as the excitation frequency $\omega$ changes? (We wait until the transient portion of the response has died away in each case.) The necessary results can be written in terms of the system time constant, $\tau$ as

$$|X_{ss}| = B = \frac{Gu_0}{\sqrt{1 + \omega^2\tau^2}} \tag{E.48}$$

and

$$\phi = \tan^{-1}(-\omega\tau) \tag{E.49}$$

These results are plotted in Figure E.7. In this dimensionless plot, the response amplitude is normalized by the input amplitude $Gu_0$, and the excitation frequency is normalized by multiplying by the time constant $\tau$. Notice that when $\omega = 0$, the response is the static value, and the phase between the input and resulting response is

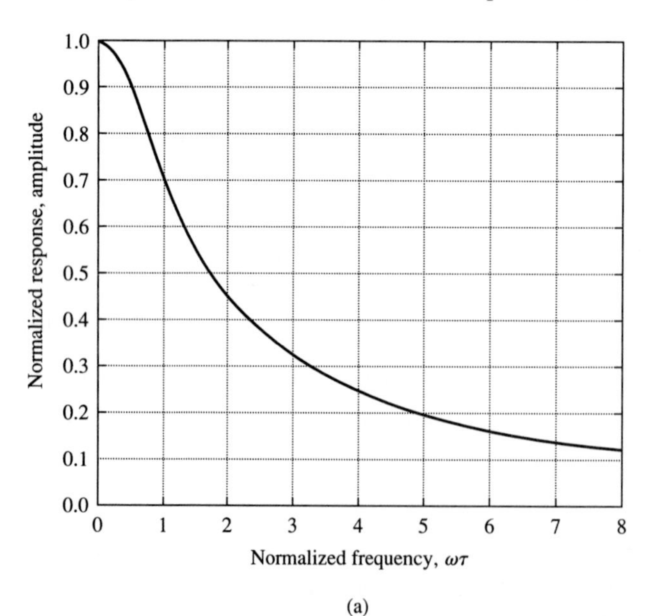


**Figure E.7** First-order system frequency response. (a) Amplitude.

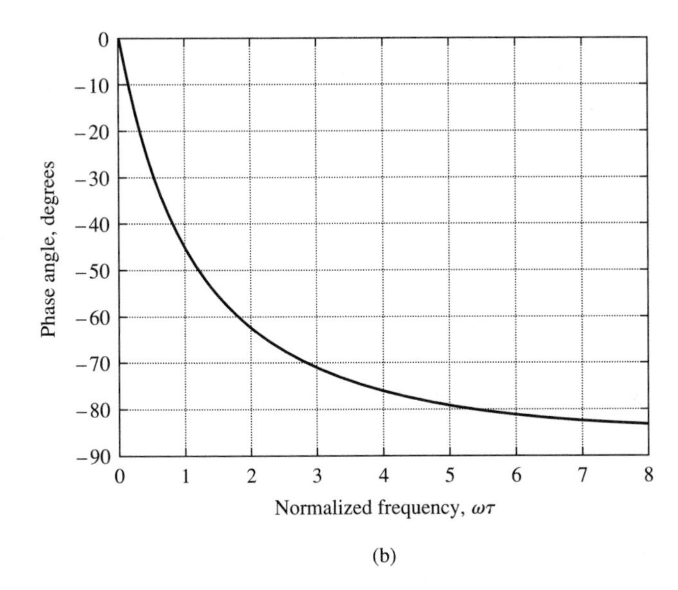


**Figure E.7** First-order system frequency response. (b) Phase.

zero. By "static value," we again mean the response that would occur if the amplitude of the input were applied so slowly that no dynamics would result.

As the input frequency $\omega$ increases, the response amplitude decreases, and the phase angle becomes negative. As $\omega$ increases indefinitely, $X_{ss}$ approaches zero, and the phase angle $\phi$ approaches 90 degrees. The angle $\phi$ represents the phase between the input $u_0$ and the output $x$. If $\omega = 0$, then $\phi = 0$, meaning that at any point in its cycle when the input is acting in a positive sense, the output response is in the positive sense also.

In many applications, it is more useful to plot the response amplitude on a base-10 logarithmic scale. In particular, it is helpful to use the definition

$$\text{dB} = 20 \log_{10}(B/Gu_0) \tag{E.50}$$

where dB is an abbreviation for **decibel**. The result of converting the magnitude to decibels and plotting is shown in Figure E.8. Note that the response curve is tangent to two straight lines when the results are displayed in this way.

Plots such as those shown in Figures E.7 and E.8 give information concerning the frequency response characteristics of the system being studied and form an important and essential tool for the design and analysis of dynamic systems.

## E.3 SECOND-ORDER SYSTEMS

Consider a second-order system with an arbitrary input function $u(t)$, as given by the equation

$$a_2\ddot{x} + a_1\dot{x} + a_0x = b_0u(t) \tag{E.51a}$$

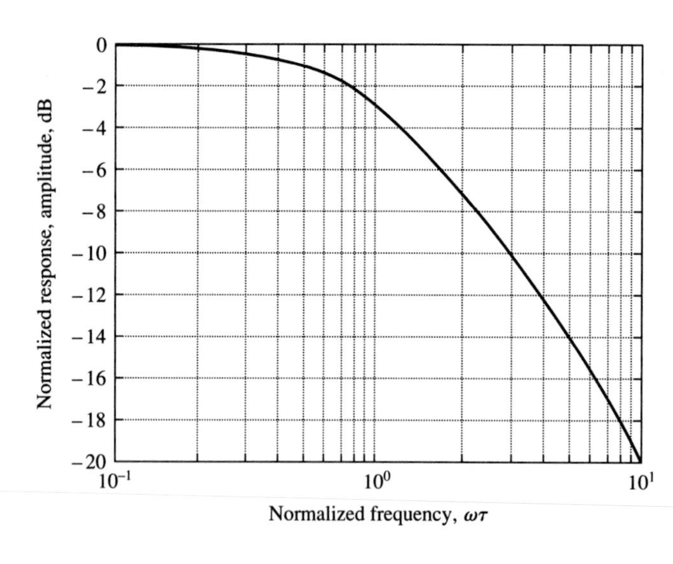


**Figure E.8** Decibel (dB) plot of first-order system frequency response.

In the study of dynamic systems, it is convenient to normalize this expression using $a_0$ and to choose the symbols for the constants so that the equation takes the form

$$\frac{1}{\omega_n^2}\ddot{x} + \frac{2\zeta}{\omega_n}\dot{x} + x = Gu(t) \tag{E.51b}$$

or, using the $D$-operator,

$$\frac{1}{\omega_n^2}D^2x + \frac{2\zeta}{\omega_n}Dx + x = Gu(t) \tag{E.51c}$$

Equation (E.51c) is representative of a general linear second-order system. The constant $\omega_n$ is called the **natural frequency** of the system, and $\zeta$ is the **damping ratio**. Their significance in the system response solutions will be seen later in our development. Solutions for a free response, a response to a step input, a response to a ramp input, and a response to a harmonic input signal are developed in the remainder of this appendix.

### E.3.1 Free Response

The equation of free motion of a second-order system is

$$\frac{1}{\omega_n^2}\ddot{x} + \frac{2\zeta}{\omega_n}\dot{x} + x = 0 \tag{E.52}$$

with initial conditions

$$x(0) = x_0 \tag{E.53}$$

and

$$\dot{x}(0) = \dot{x}_0 \tag{E.54}$$

A general exponential will be assumed for the solution function:

$$x(t) = Ce^{\lambda t} \tag{E.55}$$

Here, $C$ and $\lambda$ are constants to be determined from the differential equation. Substituting the assumed solution in the differential equation yields

$$C\frac{\lambda^2}{\omega_n^2}e^{\lambda t} + C\frac{2\zeta}{\omega_n}\lambda e^{\lambda t} + Ce^{\lambda t} = 0 \tag{E.56}$$

If $C = 0$, no motion occurs; likewise, $e^{\lambda t}$ is not zero for finite values of $t$. Therefore, we can divide by those terms to obtain the **characteristic equation** of the system:

$$\frac{1}{\omega_n^2}\lambda^2 + \frac{2\zeta}{\omega_n}\lambda + 1 = 0 \tag{E.57}$$

This equation is quadratic in $\lambda$, and its roots will determine the form of the solution. The roots are

$$\lambda_{1,2} = \frac{\omega_n^2}{2}\left(-\frac{2\zeta}{\omega_n} \pm \sqrt{\frac{4\zeta^2}{\omega_n^2} - \frac{4}{\omega_n^2}}\right) \tag{E.58a}$$

or

$$\lambda_{1,2} = \omega_n\left(-\zeta \pm \sqrt{\zeta^2 - 1}\right) \tag{E.58b}$$

We will assume that $\omega_n$ and $\zeta$ are positive constants and distinguish between three possible outcomes of Eq. (E.58b), depending upon the value of the quantity inside the square root.

**Underdamped Case, $\zeta < 1$** If the quantity $\zeta^2 - 1$ is negative, the roots of Eq. (E.58b) are complex and can be written as

$$\lambda_{1,2} = \omega_n\left(-\zeta \pm j\sqrt{1 - \zeta^2}\right) \tag{E.59}$$

where the complex operator is $j = \sqrt{-1}$. In this instance, the system is said to be **underdamped,** the free response will be oscillatory, and it is customary to define the **damped natural frequency** by the equation

$$\omega_d = \omega_n\sqrt{1 - \zeta^2} \tag{E.60}$$

Thus, the roots can be expressed as

$$\lambda_{1,2} = -\zeta\omega_n \pm j\omega_d \tag{E.61}$$

The solution may now be written as

$$x(t) = C_1 e^{(-\zeta\omega_n + j\omega_d)t} + C_2 e^{(-\zeta\omega_n - j\omega_d)t} \tag{E.62a}$$

or

$$x(t) = e^{-\zeta\omega_n t}(C_1 e^{j\omega_d t} + C_2 e^{-j\omega_d t}) \tag{E.62b}$$

Either of these equations may be represented in terms of harmonic functions through the use of Euler's identity $e^{j\psi} = \cos\psi + j\sin\psi$. The harmonic functions are cyclic in time and dictate an oscillatory free response. Thus,

$$x(t) = e^{-\zeta\omega_n t}(A \sin \omega_d t + B \cos \omega_d t) \tag{E.63}$$

where $A$ and $B$ are real constants.

**Overdamped Case, $\zeta > 1$** If the quantity, $\zeta^2 - 1$, is positive, then the roots $\lambda_1$ and $\lambda_2$ are real, and the solution is of the form

$$x(t) = C_1 e^{\lambda_1 t} + C_2 e^{\lambda_2 t} \tag{E.64}$$

where $C_1$and $C_2$ are constants. In this case the solution is of an exponential character with the displacement decreasing with time, since both exponents are negative. This exponential type of response is called **overdamped** because, given an initial value, the system will return to its original state without oscillation.

**Critical Damping, $\zeta = 1$** The boundary between an oscillatory and a nonoscillatory free response occurs when $\zeta^2 - 1 = 0$. Thus, $\zeta = 1$ is called the **critical damping ratio.** Like the overdampened case, the response with critical damping is exponential and decreasing. It is unlikely that you would ever actually build a physical system with a damping ratio that was exactly equal to 1. We will not show the solution for the critically damped case, since it can be determined quite closely by using the underdamped solution with $\zeta$ set very nearly equal to 1 – say, $\zeta = 0.9999$ – or by using the overdamped solution with $\zeta$ equal to, say, 1.0001.

The constants in the equations for the free response of the second-order system can now be evaluated in terms of the initial conditions. Suppose the initial values are $x_0$ and $\dot{x}_0$. We evaluate the constants $A$ and $B$ for the **underdamped** case:

$$x(0) = x_0 = B \tag{E.65}$$

$$\dot{x}(0) = \dot{x}_0 = -\zeta\omega_n B + \omega_d A \tag{E.66}$$

Solving for $A$ and $B$ and substituting into the response solution yields, for the underdamped free response solution,

$$x(t) = x_0 e^{-\zeta\omega_n t}\left\{\cos \omega_d t + \frac{\zeta}{\sqrt{1-\zeta^2}} \sin \omega_d t\right\} + \frac{\dot{x}_0}{\omega_d} e^{-\zeta\omega_n t} \sin \omega_d t \tag{E.67}$$

This free-response function is shown in Figure E.9. Note that it is an oscillatory function with an exponential envelope. Note also that the oscillation occurs at the damped natural frequency, since the argument of each of the three harmonic functions is $\omega_d t$.

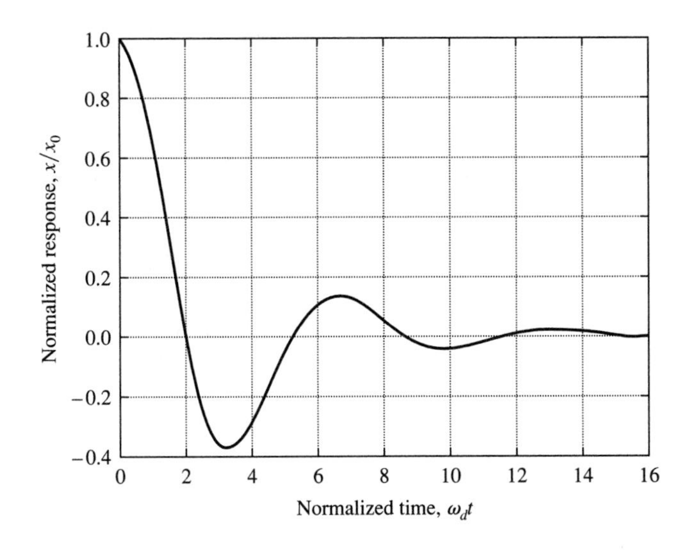


**Figure E.9** Free response of underdamped second-order system.

Using equation (E.64) for the **overdamped** case, we can find the constants $C_1$ and $C_2$:

$$x(0) = x_0 = C_1 + C_2 \tag{E.68}$$

$$\dot{x}(0) = \dot{x}_0 = \lambda_1 C_1 + \lambda_2 C_2 \tag{E.69}$$

Solving for the constants gives

$$x(t) = \frac{x_0}{\lambda_1 - \lambda_2}\{-\lambda_2 e^{\lambda_1 t} + \lambda_1 e^{\lambda_2 t}\} + \frac{\dot{x}_0}{\lambda_1 - \lambda_2}\{e^{\lambda_1 t} - e^{\lambda_2 t}\} \tag{E.70}$$

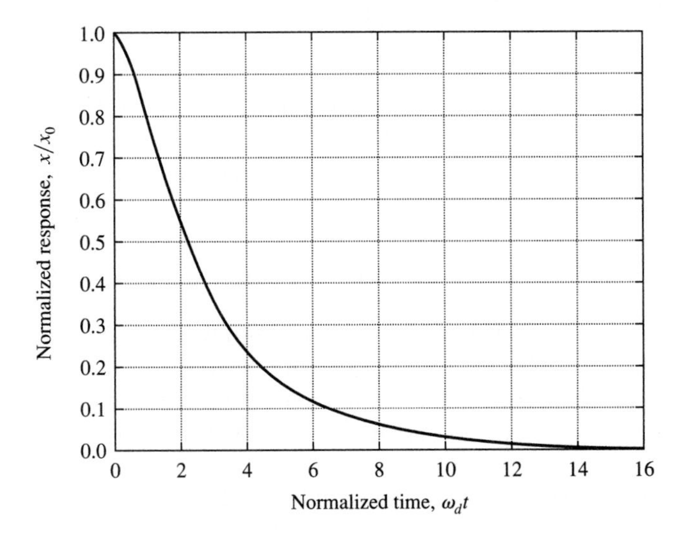


**Figure E.10** Free response of overdamped second-order system.

where

$$\lambda_1 = -\omega_n(\zeta - \sqrt{\zeta^2 - 1}) \tag{E.71}$$

$$\lambda_2 = -\omega_n(\zeta + \sqrt{\zeta^2 - 1}) \tag{E.72}$$

$$\lambda_1 - \lambda_2 = 2\omega_n\sqrt{\zeta^2 - 1} \tag{E.73}$$

An examination of the terms in this response equation shows that the free response is a nonoscillatory decay from the initial condition, as shown in Figure E.10.

### E.3.2 Response to a Step Input

If a step input $u(t) = u_0$ is applied to the second-order system, the equation

$$\frac{1}{\omega_n^2}\ddot{x} + \frac{2\zeta}{\omega_n}\dot{x} + x = Gu(t) = Gu_0 \tag{E.74}$$

with $x(0) = x_0$ and $\dot{x}(0) = \dot{x}_0$ must be solved. The homogeneous solution is Eq. (E.63) or Eq. (E.64), depending upon the amount of the damping. In either case, the particular solution is the constant $Gu_0$, which we see satisfies Eq. (E.74). Thus,

$$x(t) = x_h(t) + Gu_0 \tag{E.75}$$

Consider the **underdamped** case first. The form of the complete solution is

$$x(t) = e^{-\zeta\omega_n t}\{A \sin \omega_d t + B \cos \omega_d t\} + Gu_0 \tag{E.76}$$

Applying the initial conditions to this equation, we find the following complete solution to the step input:

$$x(t) = x_0 e^{-\zeta\omega_n t}\left\{\cos \omega_d t + \frac{\zeta}{\sqrt{1 - \zeta^2}} \sin \omega_d t\right\} + \frac{\dot{x}_0}{\omega_d} e^{-\zeta\omega_n t} \sin \omega_d t$$

$$+ Gu_0\left\{1 - e^{-\zeta\omega_n t}\left[\cos \omega_d t + \frac{\zeta}{\sqrt{1 - \zeta^2}} \sin \omega_d t\right]\right\} \tag{E.77}$$

This equation is applicable for $0 \le \zeta < 1$. The case of zero damping can be found by substituting $\zeta = 0$:

$$x(t) = x_0 \cos \omega_n t + \frac{\dot{x}_0}{\omega_n} \sin \omega_n t + Gu_0\{1 - \cos \omega_n t\} \tag{E.78}$$

This response is shown in Figure E.11 for the case in which the initial values of $x$ and $\dot{x}$ are zero. The static response would be $Gu_0$. Because of the dynamics of the system, we observe that the response first overshoots the static value, but then converges to it with increasing time as the system damping causes the transients to die away.

Similar results for the **overdamped** case are shown in Figure E.12, but because of the amount of damping present, no oscillation or overshoot occurs.

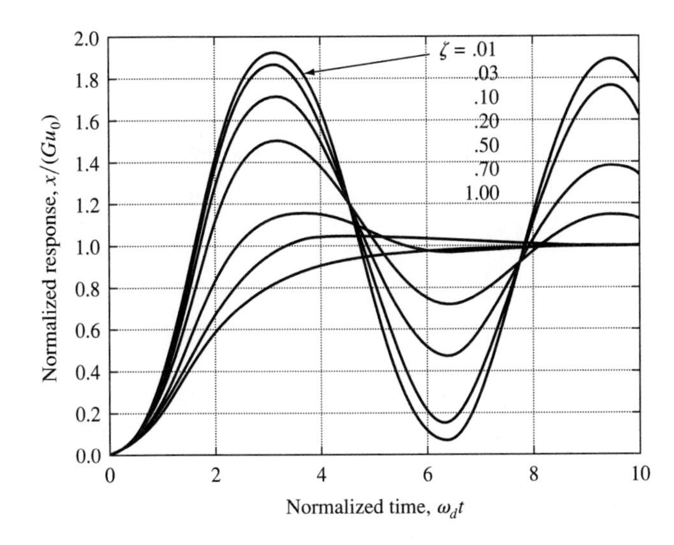


**Figure E.11** Response of underdamped second-order system to a step input.

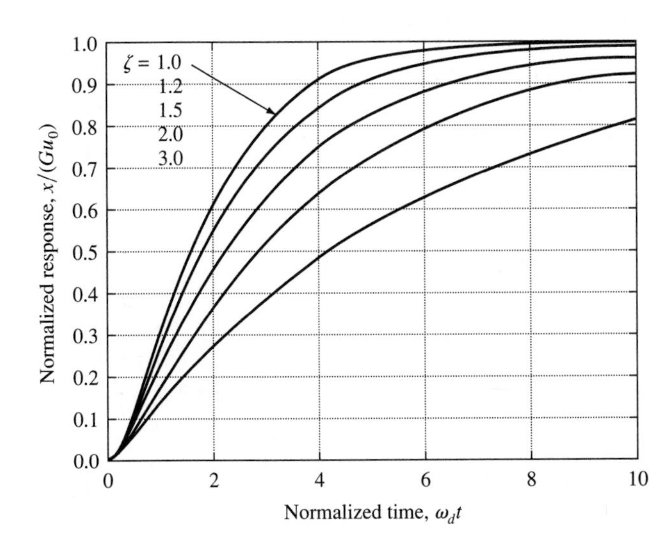


**Figure E.12** Response of overdamped second-order system to a step input.

### E.3.3 Response to a Ramp Input

In the case of a ramp input, the expression for the equation of motion is

$$\frac{1}{\omega_n^2}D^2x + \frac{2\zeta}{\omega_n}Dx + x = Gu(t) = Gu_0 t \tag{E.79}$$

Again, we try a particular solution of the form

$$x_p(t) = Q + Rt \tag{E.80}$$

Substituting this into the equation of motion gives

$$\frac{2\zeta}{\omega_n} R + (Q + Rt) = Gu_0 t \tag{E.81}$$

Equating like coefficients of the constant and time terms gives the values for $R$ and $Q$:

$$R = Gu_0 \tag{E.82}$$

$$Q = -\frac{2\zeta}{\omega_n} Gu_0 \tag{E.83}$$

Thus, the particular solution is

$$x_p(t) = -\frac{2\zeta}{\omega_n} Gu_0 + Gu_0 t = Gu_0\left(t - \frac{2\zeta}{\omega_n}\right) \tag{E.84}$$

For the **underdamped** case, the complete solution is

$$x(t) = e^{-\zeta\omega_n t}(A \sin \omega_d t + B \cos \omega_d t) + Gu_0\left(t - \frac{2\zeta}{\omega_n}\right) \tag{E.85}$$

The constants $A$ and $B$ may be found by using the initial conditions:

$$A = x_0 \frac{\zeta}{\sqrt{1 - \zeta^2}} + \dot{x}_0 \frac{1}{\omega_d} - Gu_0 \frac{1 - 2\zeta^2}{\omega_d} \tag{E.86}$$

$$B = x_0 + Gu_0 \frac{2\zeta}{\omega_n} \tag{E.87}$$

Thus, the complete solution is

$$x(t) = x_0 e^{-\zeta\omega_n t}\left[\cos \omega_d t + \frac{\zeta}{\sqrt{1 - \zeta^2}} \sin \omega_d t\right] + \frac{\dot{x}_0}{\omega_d} e^{-\zeta\omega_n t} \sin \omega_d t$$

$$+ Gu_0\left[\left\{t - \frac{2\zeta}{\omega_n}\right\} + e^{-\zeta\omega_n t}\left\{\frac{2\zeta}{\omega_n} \cos \omega_d t - \frac{1 - 2\zeta^2}{\omega_d} \sin \omega_d t\right\}\right] \tag{E.88}$$

### E.3.4 Response to a Harmonic Excitation

To complete our discussion of second-order systems, we determine the characteristics of the response of a damped second-order system when a harmonic excitation of amplitude $u_0$ and frequency $\omega$ is the input. The governing equation is

$$\frac{1}{\omega_n^2} \ddot{x} + \frac{2\zeta}{\omega_n} \dot{x} + x = Gu_0 \cos \omega t \tag{E.89}$$

The homogeneous solution depends upon the amount of damping, as discussed in Section E.3.1. Because damping is present, the particular solution must include the possibility of an arbitrary phase relationship between the input and the output and is given by

$$x_p(t) = C\cos(\omega t + \phi) \tag{E.90}$$

Substitution of this particular solution into the differential equation gives

$$-\frac{\omega^2}{\omega_n^2}C\cos(\omega t + \phi) - 2\zeta\frac{\omega}{\omega_n}C\sin(\omega t + \phi) + C\cos(\omega t + \phi) = Gu_0\cos\omega t \tag{E.91}$$

This equation can be reduced to a form from which $C$ and $\phi$ can be found by expanding $\cos(\omega t + \phi)$ and $\sin(\omega t + \phi)$ using sum-of-angle trigonometric identities and equating like coefficients of $\cos\omega t$ and $\sin\omega t$ independently. The expansion results in

$$\begin{aligned}&-\frac{\omega^2}{\omega_n^2}C\{\cos\omega t\cos\phi - \sin\omega t\sin\phi\}\\&-2\zeta\frac{\omega}{\omega_n}C\{\sin\omega t\cos\phi + \cos\omega t\sin\phi\}\\&+C\{\cos\omega t\cos\phi - \sin\omega t\sin\phi\} = Gu_0\cos\omega t\end{aligned} \tag{E.92}$$

Equating the terms multiplying $\cos\omega t$ yields

$$-\frac{\omega^2}{\omega_n^2}C\cos\phi - 2\zeta\frac{\omega}{\omega_n}C\sin\phi + C\cos\phi = Gu_0 \tag{E.93}$$

and equating those multiplying $\sin\omega t$ gives

$$+\frac{\omega^2}{\omega_n^2}C\sin\phi - 2\zeta\frac{\omega}{\omega_n}C\cos\phi - C\sin\phi = 0 \tag{E.94}$$

From this equation, we can find the phase angle and its trigonometric relations. For simplification, let $\overline{\omega} = \omega/\omega_n$. Then

$$\frac{\sin\phi}{\cos\phi} = \tan\phi = \frac{-2\zeta\frac{\omega}{\omega_n}}{1 - \frac{\omega^2}{\omega_n^2}} = \frac{-2\zeta\overline{\omega}}{1 - \overline{\omega}^2} \tag{E.95}$$

From Figure E.13,

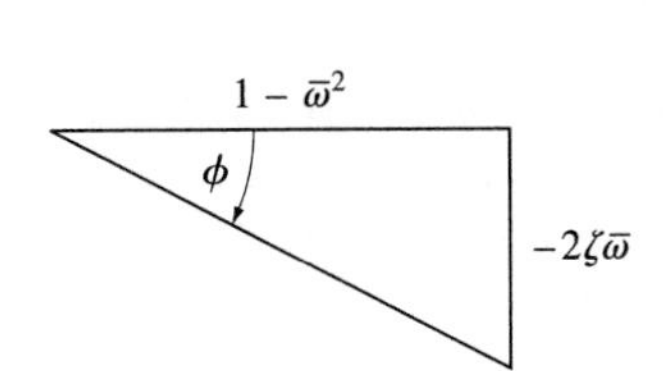


**Figure E.13** Phase angle components for a second-order system.

$$\sin\phi = \frac{-2\zeta\overline{\omega}}{\sqrt{(1 - \overline{\omega}^2)^2 + (2\zeta\overline{\omega})^2}} \tag{E.96}$$

and

$$\cos\phi = \frac{1 - \overline{\omega}^2}{\sqrt{(1 - \overline{\omega}^2)^2 + (2\zeta\overline{\omega})^2}} \tag{E.97}$$

These results, when substituted into the differential equation (E.89), yield

$$C = \frac{Gu_0}{\sqrt{(1 - \overline{\omega}^2)^2 + (2\zeta\overline{\omega})^2}} \tag{E.98}$$

The complete solution for the underdamped case is

$$x(t) = e^{-\zeta\omega_n t}\{A \sin \omega_d t + B \cos \omega_d t\} + \frac{Gu_0 \cos(\omega t + \phi)}{\sqrt{(1 - \overline{\omega}^2)^2 + (2\zeta\overline{\omega})^2}} \tag{E.99}$$

transient + steady state

Where $\phi$ is found from Eq. (E.95) to be

$$\phi = \tan^{-1}\left\{-2\zeta\frac{\overline{\omega}}{(1 - \overline{\omega}^2)}\right\} \tag{E.100}$$

Note that the transient solution dies away with increasing time, since it is multiplied by $e^{-\zeta\omega_n t}$. However, it is important to remember that in lightly damped systems, the transient portion might persist for long periods if $\zeta\omega_n$ is a small number. Utilizing the initial conditions for the problem, we solve for the constants $A$ and $B$, giving the response solution

$$\begin{aligned} x(t) = {} & x_0 e^{-\zeta\omega_n t}\left[\cos \omega_d t + \frac{\zeta}{\sqrt{1 - \zeta^2}} \sin \omega_d t\right] + \frac{\dot{x}_0}{\omega_d} e^{-\zeta\omega_n t} \sin \omega_d t \\ & + Gu_0\left[\frac{\cos(\omega t + \phi)}{\sqrt{(1 - \overline{\omega}^2)^2 + (2\zeta\overline{\omega})^2}}\right. \\ & \left. + e^{-\zeta\omega_n t}\left(\frac{(\overline{\omega}^2 - 1)\cos \omega_d t + \left[\left\{1 - \frac{2\zeta}{\sqrt{1 - \zeta^2}}\right\}\overline{\omega}^2 - 1\right]\sin \omega_d t}{(1 - \overline{\omega}^2)^2 + (2\zeta\overline{\omega})^2}\right)\right] \end{aligned} \tag{E.101}$$

where $\overline{\omega} = \omega/\omega_n$ is the normalized frequency.

The magnitude of the steady-state response to harmonic excitation can be examined by plotting Eq. (E.98) as a function of the excitation frequency $\omega$ normalized with respect to $\omega_n$. Figure E.14 shows the response amplitude $C$ as a function of $\omega/\omega_n$ with the amount of damping, $\zeta$, as a parameter. Many interesting properties of the damped second-order system are displayed in this figure. For small values of $\zeta$, the output response at **resonance** ($\omega = \omega_n$) is very large. This causes a dynamic amplification, which means that the dynamic response is much larger than if the input were applied statically. Note, however, that when $\zeta$ is 0.707 or greater, no amplification occurs. Also, we see from this equation that when $\omega = \omega_n$, the magnitude of the response is inversely proportional to the amount of damping present. In addition, as the damping increases, the response amplitude decreases. These results are based upon an input of constant amplitude; if the input is of some other form, the analysis must be modified accordingly.

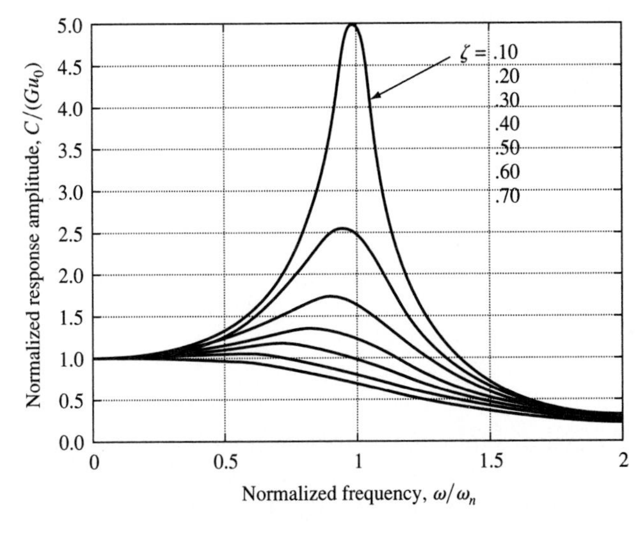


**Figure E.14** Frequency response of a second-order system for various damping.

The variation in the phase of response with frequency is displayed in Figure E.15 with $\zeta$ as a parameter. Here it is apparent that for nonzero values of damping, the steady-state phase is a slowly varying continuous function of the excitation frequency.

If the amplitude function of Eq. (E.98) is plotted using logarithmic scales on both axes, the behavior at high frequency becomes much clearer. This is shown in Figure E.16.

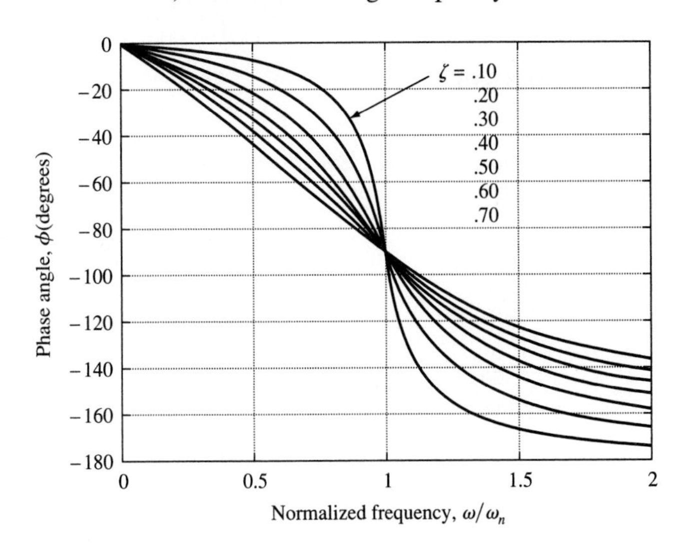


**Figure E.15** Phase response of a second-order system for various amounts of damping.

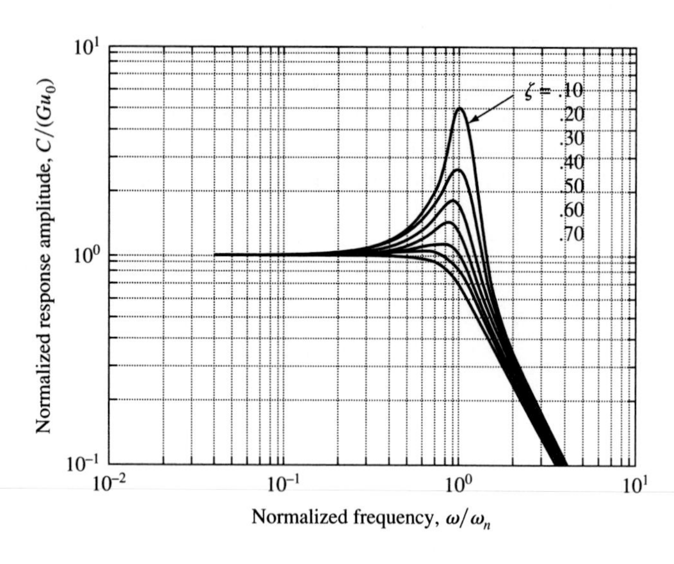


**Figure E.16** Log-log plot of frequency response amplitude.

## E.4 SUMMARY

In this appendix, we have used classical differential equation analysis methods to develop the dynamic response for a variety of modeling and input situations. Familiarity with the results provides the engineer with a knowledge base for understanding many common dynamic systems and is a prerequisite for further system analysis and simulation studies.

Table E.2 gives the conversions for coefficients, roots, and dynamic system parameters for second-order systems. Tables E.3 and E.4 summarize the responses of first- and second-order systems to a variety of inputs.

## REFERENCE

E.1 Goode, Stephen W. *An Introduction to Differential Equations and Linear Algebra.* Prentice Hall, Englewood Cliffs, NJ, 1991.

**TABLE E.2** ROOT CHARACTERISTICS OF SECOND-ORDER SYSTEMS

| | Polynomial Form | Factored Form | Dynamic Form |
|---|---|---|---|
| **Equation** | $a_2\lambda^2 + a_1\lambda + a_0 = 0$<br>$\lambda^2 + \frac{a_1}{a_2}\lambda + \frac{a_0}{a_2} = 0$ | $(\lambda + p_1)(\lambda + p_2) = 0$<br>$(\lambda + A)^2 + B^2 = 0$ (if complex)<br>$\lambda_{1,2} = A \pm jB \quad j = \sqrt{-1}$ | $\lambda^2 + 2\zeta\omega_n\lambda + \omega_n^2 = 0$<br>$\left(\frac{\lambda}{\omega_n}\right)^2 + 2\zeta\left(\frac{\lambda}{\omega_n}\right) + 1 = 0$ |
| **Roots** | $\lambda_{1,2} = -\frac{a_1/a_2}{2} \pm \sqrt{\left(\frac{a_1/a_2}{2}\right)^2 - \frac{a_0}{a_2}}$<br>$\lambda_{1,2} = -\frac{a_1/a_2}{2}\left[1 \pm \sqrt{1 - \frac{4a_0/a_2}{(a_1/a_2)^2}}\right]$ | $\lambda_{1,2} = -p_1, -p_2$<br>$p_{1,2} = -A \pm jB \quad$ (if complex) | $\lambda_{1,2} = -\zeta\omega_n \pm \sqrt{(\zeta\omega_n)^2 - \omega_n^2}$<br>$\lambda_{1,2} = -\omega_n\left[\zeta \pm \sqrt{\zeta^2 - 1}\right]$ |
| **Conversions** | $\frac{a_0}{a_2} = p_1 p_2 \qquad \frac{a_1}{a_2} = p_1 + p_2$<br>$\frac{a_0}{a_2} = A^2 + B^2 \qquad \frac{a_1}{a_2} = 2A$<br>$\frac{a_0}{a_2} = \omega_n^2 \qquad \frac{a_1}{a_2} = 2\zeta\omega_n$ | $p_1 = \frac{a_1/a_2}{2}\left[1 + \sqrt{1 - \frac{a_0/a_2}{\left(\frac{a_1/a_2}{2}\right)}}\right]$<br>$p_2 = \frac{a_1/a_2}{2}\left[1 - \sqrt{1 - \frac{a_0/a_2}{\left(\frac{a_1/a_2}{2}\right)}}\right]$<br>$p_1 = \omega_n\left[\zeta + \sqrt{\zeta^2 - 1}\right]$<br>$p_2 = \omega_n\left[\zeta - \sqrt{\zeta^2 - 1}\right]$<br>$A = \zeta\omega_n;\ B = \omega_n\sqrt{1 - \zeta^2} = \omega_d$ | $\omega_n = \sqrt{\frac{a_0}{a_2}}$<br>$\zeta = \frac{a_1/a_2}{2\sqrt{a_0/a_2}} = \frac{a_1}{2\sqrt{a_0 a_2}}$<br>$\omega_n = \sqrt{p_1 p_2}$<br>$\zeta = \frac{(p_1 + p_2)}{2\sqrt{p_1 p_2}}$<br>$\omega_n = \sqrt{A^2 + B^2}$<br>$\zeta = \frac{A}{\sqrt{A^2 + B^2}} = \frac{A}{\omega_n}$ |

**TABLE E.3** RESPONSE OF FIRST-ORDER SYSTEMS, $[\tau D + 1]x = Gu(t)$

| | | |
|---|---|---|
| **Initial Condition**<br><br>$u(t) = 0$ | $[\tau D + 1]x = 0$<br><br>$x(t) = x_0 e^{-t/\tau}$ | with $x(0) = x_0$ |
| **Step Input**<br><br>$u(t) = u_0$ | $[\tau D + 1]x = Gu_0$<br><br>$x(t) = x_0 e^{-t/\tau} + Gu_0(1 - e^{-t/\tau})$ | with $x(0) = x_0$ |
| **Ramp Input**<br><br>$u(t) = u_0 t$ | $[\tau D + 1]x = Gu_0 t$<br><br>$x(t) = x_0 e^{-t/\tau} + Gu_0[t - \tau(1 - e^{-t/\tau})]$ | with $x(0) = x_0$ |
| **Harmonic Input**<br><br>$u(t) = u_0 \cos \omega t$ | $[\tau D + 1]x = Gu_0 \cos \omega t$<br><br>$x(t) = x_0 e^{-t/\tau} + Gu_0\left(\frac{\cos(\omega t + \phi)}{\sqrt{1 + \omega^2\tau^2}} - \frac{e^{-t/\tau}}{[1 + \omega^2\tau^2]}\right)$<br><br>where $\phi = \tan^{-1}(-\omega\tau)$ | with $x(0) = x_0$ |

**TABLE E.4** RESPONSE OF SECOND-ORDER SYSTEMS $\left[\frac{D^2}{\omega_n^2} + 2\zeta\frac{D}{\omega_n} + 1\right]x = Gu(t)$

| | Underdamped Case, $0 \le \zeta < 1$ where $\omega_d = \omega_n\sqrt{1-\zeta^2}$ |
|---|---|
| **Initial Condition** | $\frac{1}{\omega_n^2}D^2x + \frac{2\zeta}{\omega_n}Dx + x = 0$ with $x(0) = x_0$ and $Dx(0) = \dot{x}_0$ |
| $u(t) = 0$ | $x(t) = x_0e^{-\zeta\omega_n t}\left\{\cos\omega_d t + \frac{\zeta}{\sqrt{1-\zeta^2}}\sin\omega_d t\right\} + \frac{\dot{x}_0}{\omega_d}e^{-\zeta\omega_n t}\sin\omega_d t$ |
| **Step Input** | $\frac{1}{\omega_n^2}D^2x + \frac{2\zeta}{\omega_n}Dx + x = Gu_0$ with $x(0) = x_0$ and $Dx(0) = \dot{x}_0$ |
| $u(t) = u_0$ | $x(t) = x_0e^{-\zeta\omega_n t}\left\{\cos\omega_d t + \frac{\zeta}{\sqrt{1-\zeta^2}}\sin\omega_d t\right\} + \frac{\dot{x}_0}{\omega_d}e^{-\zeta\omega_n t}\sin\omega_d t$ |
| | $+ Gu_0\left[1 - e^{-\zeta\omega_n t}\left\{\cos\omega_d t + \frac{\zeta}{\sqrt{1-\zeta^2}}\sin\omega_d t\right\}\right]$ |
| **Ramp Input** | $\frac{1}{\omega_n^2}D^2x + \frac{2\zeta}{\omega_n}Dx + x = Gu_0t$ with $x(0) = x_0$ and $Dx(0) = \dot{x}_0$ |
| $u(t) = u_0t$ | $x(t) = x_0e^{-\zeta\omega_n t}\left\{\cos\omega_d t + \frac{\zeta}{\sqrt{1-\zeta^2}}\sin\omega_d t\right\} + \frac{\dot{x}_0}{\omega_d}e^{-\zeta\omega_n t}\sin\omega_d t$ |
| | $+ Gu_0\left[\left\{t - \frac{2\zeta}{\omega_n}\right\} + \frac{2\zeta}{\omega_n}e^{-\zeta\omega_n t}\left\{\cos\omega_d t - \frac{1-2\zeta^2}{2\zeta\sqrt{1-\zeta^2}}\sin\omega_d t\right\}\right]$ |
| **Harmonic Input** | $\frac{1}{\omega_n^2}D^2x + \frac{2\zeta}{\omega_n}Dx + x = Gu_0\cos\omega t$ with $x(0) = x_0$ and $Dx(0) = \dot{x}_0$ |
| $u(t) = u_0\cos\omega t$ | $x(t) = x_0e^{-\zeta\omega_n t}\left\{\cos\omega_d t + \frac{\zeta}{\sqrt{1-\zeta^2}}\sin\omega_d t\right\} + \frac{\dot{x}_0}{\omega_d}e^{-\zeta\omega_n t}\sin\omega_d t$ |
| $+ Gu_0\left[\frac{\cos(\omega t + \phi)}{\sqrt{(1-\overline{\omega}^2)^2 + (2\zeta\overline{\omega})^2}} + e^{-\zeta\omega_n t}\left(\frac{(\overline{\omega}^2 - 1)\cos\omega_d t + \left[\left\{1 - \frac{2\zeta}{\sqrt{1-\zeta^2}}\right\}\overline{\omega}^2 - 1\right]\sin\omega_d t}{(1-\overline{\omega}^2)^2 + (2\zeta\overline{\omega})^2}\right)\right]$ | |
| | where $\overline{\omega} = \omega/\omega_n$ and $\phi = \tan^{-1}[-2\zeta\overline{\omega}/(1-\overline{\omega}^2)]$ |

*Continued*

**TABLE E.4** *(continued)*

| | Overdamped Case, $\zeta > 1$ $\lambda_1 = \omega_n(\zeta - \sqrt{\zeta^2 - 1})$; $\lambda_2 = \omega_n(\zeta + \sqrt{\zeta^2 - 1})$ |
|---|---|
| **Initial Condition**<br><br>$u(t) = 0$ | $\frac{1}{\omega_n^2} D^2x + \frac{2\zeta}{\omega_n} Dx + x = 0$ with $x(0) = x_0$ and $Dx(0) = \dot{x}_0$<br><br>$x = x_0\left[\frac{\lambda_2 e^{-\lambda_1 t} - \lambda_1 e^{-\lambda_2 t}}{\lambda_2 - \lambda_1}\right] + \frac{\dot{x}_0}{\omega_n}\left[\frac{e^{-\lambda_1 t} - e^{-\lambda_2 t}}{(\lambda_2 - \lambda_1)/\omega_n}\right]$ |
| **Step Input**<br><br>$u(t) = u_0$ | $\frac{1}{\omega_n^2} D^2x + \frac{2\zeta}{\omega_n} Dx + x = G u_0$ with $x(0) = x_0$ and $Dx(0) = \dot{x}_0$<br><br>$x = x_0\left[\frac{\lambda_2 e^{-\lambda_1 t} - \lambda_1 e^{-\lambda_2 t}}{\lambda_2 - \lambda_1}\right] + \frac{\dot{x}_0}{\omega_n}\left[\frac{e^{-\lambda_1 t} - e^{-\lambda_2 t}}{(\lambda_2 - \lambda_1)/\omega_n}\right]$<br><br>$+ Gu_0\left[1 - \left(\frac{\lambda_2 e^{-\lambda_1 t} - \lambda_1 e^{-\lambda_2 t}}{\lambda_2 - \lambda_1}\right)\right]$ |
| **Ramp Input**<br><br>$u(t) = u_0 t$ | $\frac{1}{\omega_n^2} D^2x + \frac{2\zeta}{\omega_n} Dx + x = Gu_0 t$ with $x(0) = x_0$ and $Dx(0) = \dot{x}_0$<br><br>$x = x_0\left[\frac{\lambda_2 e^{-\lambda_1 t} - \lambda_1 e^{-\lambda_2 t}}{\lambda_2 - \lambda_1}\right] + \frac{\dot{x}_0}{\omega_n}\left[\frac{e^{-\lambda_1 t} - e^{-\lambda_2 t}}{(\lambda_2 - \lambda_1)/\omega_n}\right]$<br><br>$+ Gu_0\left[\left(t - \frac{2\zeta}{\omega_n}\right) - \frac{2\zeta}{\omega_n}\left(\frac{\lambda_2 e^{-\lambda_1 t} - \lambda_1 e^{-\lambda_2 t}}{\lambda_2 - \lambda_1}\right) - \left(\frac{e^{-\lambda_1 t} - e^{-\lambda_2 t}}{\lambda_2 - \lambda_1}\right)\right]$ |
| **Harmonic Input**<br><br>$u(t) = u_0 \cos \omega t$ | $\frac{1}{\omega_n^2} D^2x + \frac{2\zeta}{\omega_n} Dx + x = G u_0 \cos \omega t$ with $x(0) = x_0$ and $Dx(0) = \dot{x}_0$<br><br>$x = x_0\left[\frac{\lambda_2 e^{-\lambda_1 t} - \lambda_1 e^{-\lambda_2 t}}{\lambda_2 - \lambda_1}\right] + \frac{\dot{x}_0}{\omega_n}\left[\frac{e^{-\lambda_1 t} - e^{-\lambda_2 t}}{(\lambda_2 - \lambda_1)/\omega_n}\right]$<br><br>$+ \frac{Gu_0}{[(1 - \overline{\omega}^2)^2 + (2\zeta\overline{\omega})^2]}\left[\begin{array}{l} 2\zeta\overline{\omega}\sin\omega t + (1 - \overline{\omega}^2)\cos\omega t \\ - (1 - \overline{\omega}^2)\left(\frac{\lambda_2 e^{-\lambda_1 t} - \lambda_1 e^{-\lambda_2 t}}{\lambda_2 - \lambda_1}\right) - 2\zeta\overline{\omega}^2\left(\frac{e^{-\lambda_1 t} - e^{-\lambda_2 t}}{(\lambda_2 - \lambda_1)/\omega_n}\right) \end{array}\right]$<br><br>where $\overline{\omega} = \omega/\omega_n$ |

## PROBLEMS

**E.1** *Rearrange* the following first-order differential equations to the form $(\tau D + 1)z = Gu$:

**a.** $8\dfrac{dx}{dt} + 4x = 4u$ with $x(0) = -1$

$$u(t) = 1 \quad \text{for } t \geq 0$$

**b.** $0.01\dot{x} + x = 0$ with $x(0) = 1$

**c.** $0.05\dot{z} + 10z = u$ with $z(0) = 1.0$, $u(t) = 1.0$

*What* are the time constants and steady-state gains for each? *Derive* the analytic response for each. *Sketch* the time response.

**E.2** *Derive* the analytic response to a ramp input of the first-order system

$$[\tau D + 1]z = Gu \quad \text{where } u(t) = At \ (t \geq 0)$$

$$u(t) = 0 \quad (t < 0)$$

$$\text{with } z(0) = z_0$$

**E.3** *Derive* the analytic response of a first-order system with a sinusoidal input, i.e.,

$$[\tau D + 1]z = Gu \quad \text{where } u(t) = A \sin \omega t \ (t \geq 0)$$

$$u(t) = 0 \ (t < 0)$$

$$\text{with } z(0) = z_0$$

**E.4** *Find* the roots of the following characteristic equations. *What* are their natural frequencies and damping ratios?

**a.** $0.01D^2 + 0.02\,D + 1 = 0$
**b.** $D^2 + 20\,D + 25 = 0$
**c.** $10\,D^2 + 600\,D + 1000 = 0$
**d.** $100\,D^2 + 400\,D + 1 = 0$

**E.5** *Find* and *sketch* the time response solution to the following second-order systems:

**a.** $[D^2 + 2D + 100]x = 0$ with $x(0) = 10$, $\dot{x}(0) = 0$

**b.** $[D^2z + Dz + z] = 5u$ with $z(0) = 0$, $\dot{z}(0) = 0$, $u(t) = u_0 \quad t \geq 0$

**c.** $0.01\,D^2z + 0.4Dz + z = u$ with $z(0) = 0$, $\dot{z}(0) = 0$, $u(t) = u_0$

# APPENDIX F

# Laplace Transform Solution of Differential Equations

## F.1 INTRODUCTION

The Laplace transform is an integral transform that has applications in several areas of mathematics and can be used in a variety of ways. In this appendix, we will be concerned with the use of the Laplace transform technique to solve differential equations [Refs. F.1, F.2]. In that regard, we are not as interested in the mathematics of the transform itself, but we are interested in the solution of differential equations that have been transformed and manipulated in such a manner that we can use the Laplace transform pairs (Laplace transform and inverse Laplace transform) which have been tabulated for a variety of functions. Accordingly, we will normally not have to calculate a Laplace transform; rather, we will just use algebra to solve differential equations. Therefore, we concentrate here on using tables of Laplace transforms that have already been generated; however, we will still need to go through some of the preliminaries so that the Laplace transform technique can be fully understood.

## F.2 THE LAPLACE TRANSFORM

The **Laplace transform** is an integral transform involving the time domain function $f(t)$ and a complex variable $s$. The time domain function is multiplied by $e^{-st}$ and is then integrated from $t = -\infty$ to $t = +\infty$. The result is termed the Laplace transform $F(s)$ and is a function of the complex variable $s$, since the variable $t$ disappears upon integration. Mathematically,

$$F(s) = \mathcal{L}_2[f(t)] = \int_{-\infty}^{+\infty} f(t)e^{-st}\,dt \tag{F.1}$$

This equation represents a transformation from the time domain to the complex plane. Notice that we have used a lowercase $f(t)$ to designate the time domain function and an uppercase $F(s)$ to designate the Laplace transform. Notice further that in order to carry out the transformation, we must know the time function from its inception to the present. Since $f(t)$ could represent the transform of the response of a dynamic system, we know that it is impractical to think that we could know the response from $-\infty$ to the present time ($t = 0$). However, we really need to know only the value of $f(t)$ at the instant of the present time, $f(0)$, and then carry out the integration from $t = 0$ to $t = \infty$. Therefore, we will just use the single-sided Laplace transform

$$F(s) = \mathcal{L}[f(t)] = \int_{0^-}^{\infty} f(t)e^{-st}\,dt \tag{F.2}$$

and we will not really need the two-sided transform given by Eq. (F.1).

**Example F.1 Step Input**

As an example, let us take the Laplace transform of a step input—that is, a constant $u_o$. We have

$$f(t) = u_0 \quad \text{for } t \geq 0 \tag{F.3}$$

$$F(s) = \int_0^{+\infty} u_o e^{-st}\,dt = u_0\int_0^{+\infty} e^{-st}\,dt = u_o \left.\frac{e^{-st}}{-s}\right|_0^{+\infty} \tag{F.4a}$$

$$F(s) = \frac{u_0}{-s}(0 - 1) = \frac{u_o}{s} \tag{F.4b}$$

**Example F.2 Exponential Input**

As a second example, consider the negative exponential function

$$f(t) = Ae^{-rt} \tag{F.5}$$

Transforming, we obtain

$$F(s) = \int_0^{+\infty} Ae^{-rt}e^{-st}\,dt = A\int_0^{+\infty} e^{-(s+r)t}\,dt = \left.\frac{Ae^{-(s+r)t}}{-(s+r)}\right|_0^{+\infty} \tag{F.6a}$$

$$F(s) = \frac{A(0 - 1)}{-(s + r)} = \frac{A}{(s + r)} \tag{F.6b}$$

**Example F.3 Sinusoidal Input**

Finally, consider a sine function of frequency $\omega$, i.e.,

$$f(t) = A\sin(\omega t) \tag{F.7}$$

Applying the Laplace transform and using tables of integration or integrating by parts twice results in

$$F(s) = \int_0^{+\infty} A\sin(\omega t)e^{-st}\,dt = \frac{A\omega}{(s^2 + \omega^2)} \tag{F.8}$$

## F.3 MATHEMATICAL PROPERTIES OF THE LAPLACE TRANSFORM

Several very important properties of the Laplace transform are necessary in the solution of differential equations. Of particular note are the linearity properties:

$$\mathcal{L}[a\, f(t)] = a\, \mathcal{L}[f(t)] = a\, F(s) \tag{F.9}$$

$$\mathcal{L}[f_1(t) + f_2(t)] = \mathcal{L}[f_1(t)] + \mathcal{L}[f_2(t)] = F_1(s) + F_2(s) \tag{F.10}$$

These properties can be clearly understood by considering the nature of integration. The proof is left as an exercise for the reader.

Another critical property for the solution of differential equations by the Laplace transform technique is the Laplace transform of the derivative of a function:

$$\mathcal{L}\left[\frac{df(t)}{dt}\right] = s\, F(s) - f(0^-) \tag{F.11}$$

The proof of this property can be found by integrating the right-hand side of Eq. (F.1) by parts and rearranging the result. This, too, is left as an exercise for the student.

The Laplace transform of the second derivative of a function is

$$\mathcal{L}\left[\frac{d^2 f(t)}{dt^2}\right] = s^2\, F(s) - s f(0^-) - \dot{f}(0^-) \tag{F.12}$$

which, together with the Laplace transform of the third derivative of a function, is shown in Table F.1. This table also lists several important mathematical properties of the Laplace transform. You should pay careful attention to the initial-value theorem and the final-value theorem, which can be used to find the initial value of a function or the steady-state value of a system, respectively.

## F.4 THE INVERSE LAPLACE TRANSFORM

The inverse Laplace transform is an integral transformation from the complex plane $s$ back to the time domain. However, since the inverse Laplace transform of a Laplace transform results in the original time domain function, we really need not ever take the inverse Laplace transform; we can simply look it up in the tables. The following definition is stated merely for reference ($c$ is a real constant that is greater than the real parts of all singular points of $F(s)$):

$$f(t) = \mathcal{L}^{-1} F(s) = \frac{1}{2\pi j} \int_{c-j\infty}^{c+j\infty} F(s) e^{st}\, ds \tag{F.13}$$

The time function and the Laplace transform form a **transform pair,** since one can transform a function from the time domain to the complex plane (with the Laplace transform) or from the complex plane to the time domain (with the inverse Laplace transform) and obtain the same results in either direction. There-

**TABLE F.1** MATHEMATICAL PROPERTIES OF THE LAPLACE TRANSFORM

| Property | Time domain function, $f(t)$ | Laplace transform, $F(s)$ |
|---|---|---|
| Laplace Transform (Definition) | $f(t)$ | $F(s) = \int_{0^-}^{\infty} e^{-st} f(t)\, dt$ |
| Inverse Laplace Transform (Definition) | $f(t) = \frac{1}{2\pi j}\int_{c-j\infty}^{c+j\infty} F(s)e^{st}\, ds$ | $F(s)$ |
| Linearity (Multiplication by a Constant) | $a\, f(t)$ | $a\, F(s)$ |
| Linearity (Addition of Functions) | $f_1(t) + f_2(t)$ | $F_1(s) + F_2(s)$ |
| Time Scaling | $f\left(\frac{t}{a}\right)$ | $a\, F\,(a\, s)$ |
| Initial Value | $f(0^+)$ | $\lim_{s\to\infty} s\, F(s)$ |
| Final Value[1] | $f(\infty)$ | $\lim_{s\to 0} s\, F(s)$ |
| Derivatives | $\frac{df}{dt}$ with $f(0^-)$ | $sF(s) - f(0^-)$ |
| | $\frac{d^2f}{dt^2}$ with $f(0^-)$, $\dot{f}(0^-)$ | $s^2 F(s) - sf(0^-) - \dot{f}(0^-)$ |
| | $\frac{d^3f}{dt^3}$ with $f(0^-)$, $\dot{f}(0^-)$, $\ddot{f}(0^-)$ | $s^3F(s) - s^2f(0^-) - s\dot{f}(0^-) - \ddot{f}(0^-)$ |
| Integral | $\int_0^t f(\lambda)\, d\lambda$ | $\frac{1}{s} F(s)$ |

[1] All poles of $F(s)$ (other than $s = 0$) must have negative real parts; otherwise, the computed number for the final value will be incorrect.

fore, a table of transform pairs gives both the Laplace transform and the inverse Laplace transform.

Table F.2 lists several frequently used Laplace transform pairs that arise in solving for the response of dynamic systems. Note that there is no limit to the number of Laplace transform pairs that could be tabulated; however, we will see that we need only a few terms that occur in our analysis, and these terms will be summed together to obtain the response of very complicated systems.

**TABLE F.2** LAPLACE TRANSFORM PAIRS

| | Time domain function, $f(t)$ | Laplace transform, $F(s)$ |
|---|---|---|
| **Time Functions** | | |
| 1 | $\delta(t)$ = impulse function, $\int_{0^-}^{0^+} \delta(t)\,dt = 1$ | $1$ |
| 2 | $u(t)$ = step function, or constant = 1 | $\frac{1}{s}$ |
| 3 | ramp function, $t$ | $\frac{1}{s^2}$ |
| 4 | $\frac{t^{n-1}}{(n-1)!} \quad n = 1, 2, 3, \ldots.$ | $\frac{1}{s^n}$ |
| **Real Roots** | | Root = $-r$ |
| 5 | $e^{-rt}$ | $\frac{1}{(s+r)}$ |
| 6 | $\frac{1-e^{-rt}}{r}$ | $\frac{1}{s(s+r)}$ |
| 7 | $\frac{[rt-(1-e^{-rt})]}{r^2}$ | $\frac{1}{s^2(s+r)}$ |
| 8 | $t\,e^{-rt}$ | $\frac{1}{(s+r)^2}$ |
| 9 | $\frac{e^{-r_1 t}-e^{-r_2 t}}{(r_2-r_1)}$ | $\frac{1}{(s+r_1)(s+r_2)} \quad r_1 \neq r_2$ |
| 10 | $\frac{1}{r_1 r_2}\left[1-\left(\frac{r_2 e^{-r_1 t}-r_1 e^{-r_2 t}}{r_2-r_1}\right)\right]$ | $\frac{1}{s(s+r_1)(s+r_2)} \quad r_1 \neq r_2$ |
| 11 | $\frac{r_1 e^{-r_1 t}-r_2 e^{-r_2 t}}{(r_1-r_2)}$ | $\frac{s}{(s+r_1)(s+r_2)} \quad r_1 \neq r_2$ |
| 12 | $\frac{(r_1-q)e^{-r_1 t}-(r_2-q)e^{-r_2 t}}{(r_1-r_2)}$ | $\frac{(s+q)}{(s+r_1)(s+r_2)} \quad r_1 \neq r_2$ |
| 13 | $\frac{e^{-r_1 t}}{(r_2-r_1)(r_3-r_1)}+\frac{e^{-r_2 t}}{(r_1-r_2)(r_3-r_2)}+\frac{e^{-r_3 t}}{(r_1-r_3)(r_2-r_3)}$ | $\frac{1}{(s+r_1)(s+r_2)(s+r_3)} \quad r_1 \neq r_2 \neq r_3$ |
| 14 | $\frac{(q-r_1)e^{-r_1 t}}{(r_2-r_1)(r_3-r_1)}+\frac{(q-r_2)e^{-r_2 t}}{(r_1-r_2)(r_3-r_2)}+\frac{(q-r_3)e^{-r_3 t}}{(r_1-r_3)(r_2-r_3)}$ | $\frac{(s+q)}{(s+r_1)(s+r_2)(s+r_3)} \quad r_1 \neq r_2 \neq r_3$ |

*Continued*

**TABLE F.2** *continued*

| | ComplexRoots | $0 \le \zeta < 1$, roots $= -r \pm j\omega$, $\omega_n = \sqrt{r^2 + \omega^2}$, $\zeta = r/\omega_n$ |
|---|---|---|
| 15 | $\dfrac{\sin \omega t}{\omega}$ | $\dfrac{1}{s^2 + \omega^2}$ |
| 16 | $\cos \omega t$ | $\dfrac{s}{s^2 + \omega^2}$ |
| 17 | $\dfrac{1 - \cos \omega t}{\omega^2}$ | $\dfrac{1}{s(s^2 + \omega^2)}$ |
| 18 | $\dfrac{\omega t - \sin \omega t}{\omega^3}$ | $\dfrac{1}{s(s^2 + \omega^2)}$ |
| 19 | $e^{-rt} \dfrac{\sin \omega t}{\omega}$ | $\dfrac{1}{(s + r)^2 + \omega^2}$ |
| 20 | $e^{-rt}\left[\cos \omega t - \dfrac{r}{\omega} \sin \omega t\right]$ | $\dfrac{s}{(s + r)^2 + \omega^2}$ |
| 21 | $\dfrac{1 - e^{-rt}\left[\cos \omega t - \dfrac{r}{\omega} \sin \omega t\right]}{r^2 + \omega^2}$ | $\dfrac{1}{s[(s + r)^2 + \omega^2]}$ |
| 22 | $e^{-rt} \cos \omega t$ | $\dfrac{(s + r)}{(s + r)^2 + \omega^2}$ |
| 23 | $e^{-rt}\left[\cos \omega t - \dfrac{(r - q)}{\omega} \sin \omega t\right]$ | $\dfrac{(s + q)}{(s + r)^2 + \omega^2}$ |
| 24 | $\dfrac{e^{-rt}}{2\omega^3} e^{-rt}[\sin \omega t - \omega t \cos \omega t]$ | $\dfrac{1}{[(s + r)^2 + \omega^2]^2}$ |
| 25 | $\dfrac{e^{-rt}}{2\omega^2}\left[rt \cos \omega t + \left(\omega t - \dfrac{r}{\omega}\right) \sin \omega t\right]$ | $\dfrac{s}{[(s + r)^2 + \omega^2]^2}$ |
| 26 | $\dfrac{e^{-rt}}{2\omega^2}\left[(r - q)t \cos \omega t + \left\{\omega t - \dfrac{(r - q)}{\omega}\right\} \sin \omega t\right]$ | $\dfrac{(s + q)}{[(s + r)^2 + \omega^2]^2}$ |
| 27 | $\dfrac{e^{-rt}}{\omega}\left\lvert\dfrac{N(s = -r + j\omega)}{D(s = -r + j\omega)}\right\rvert \sin(\omega t + \phi)$<br>where $\phi =$ angle $N(s)/D(s)\rvert_{s=-r+j\omega}$<br>(gives portion of inverse associated with complex pair $r \pm j\omega$) | $\dfrac{N(s)}{D(s)[(s + r)^2 + \omega^2]}$ |
| 28 | $\dfrac{e^{-\zeta\omega_n t}}{\omega_n\sqrt{1 - \zeta^2}} \sin(\omega_n\sqrt{1 - \zeta^2}\, t)$ | $\dfrac{1}{s^2 + 2\zeta\omega_n s + \omega_n^2}$ |
| 29 | $\dfrac{e^{-\zeta\omega_n t}}{\sqrt{1 - \zeta^2}} \sin(\omega_n\sqrt{1 - \zeta^2}\, t + \phi)$<br>where $\phi = \cos^{-1}\zeta$ | $\dfrac{s}{s^2 + 2\zeta\omega_n s + \omega_n^2}$ |

*Continued*

**TABLE F.2** *continued*

| | | |
|---|---|---|
| 30 | $\frac{1}{\omega_n^2}\left[1 - \frac{e^{-\zeta\omega_n t}}{\sqrt{1-\zeta^2}}\sin(\omega_n\sqrt{1-\zeta^2}\,t + \phi)\right]$ where $\phi = \cos^{-1}\zeta$ | $\frac{1}{s(s^2 + 2\zeta\omega_n s + \omega_n^2)}$ |
| 31 | $\frac{e^{-\zeta\omega_n t}}{\sqrt{1-\zeta^2}}\cos(\omega_n\sqrt{1-\zeta^2}\,t)$ | $\frac{(s + 2\zeta\omega_n)}{s^2 + 2\zeta\omega_n s + \omega_n^2}$ |
| 32 | $\sqrt{\left(\frac{q}{\omega_n}\right)^2 - 2\zeta\frac{q}{\omega_n} + 1}\;\frac{e^{-\zeta\omega_n t}}{\sqrt{1-\zeta^2}}\sin(\omega_n\sqrt{1-\zeta^2}\,t + \phi)$ where $\phi = \tan^{-1}\frac{\sqrt{1-\zeta^2}}{\left(\frac{q}{\omega_n} - \zeta\right)}$ if $\frac{q}{\omega_n} - \zeta > 0$ then $0 \le \phi \le \frac{\pi}{2}$; if $\frac{q}{\omega_n} - \zeta < 0$ then $\frac{\pi}{2} \le \phi \le \pi$ | $\frac{(s + q)}{s^2 + 2\zeta\omega_n s + \omega_n^2}$ |

Recall that the trigonometric functions can be transformed as follows ($\phi$ = positive acute angle):

$$A\cos\omega t + B\sin\omega t = C\sin(\omega t + \phi) \quad \text{where} \quad C = \sqrt{A^2 + B^2} \text{ and } \phi = \tan^{-1}(A/B)$$

## F.5 SOLUTION OF DIFFERENTIAL EQUATIONS BY LAPLACE TRANSFORM TECHNIQUES

The techniques for using Laplace transforms for solving differential equations are best shown by a few simple examples and then generalized.

### F.5.1 Step Input to a First-Order System

Let us begin by considering a first-order system with a step input, namely,

$$(\tau D + 1)z(t) = Gu(t) \tag{F.14}$$

$$\text{with } z(0^-) = z_0$$

$$\text{and with } u(t) = u_0 \quad \text{for } t \ge 0 \tag{F.15}$$

Since Eq. (F.14) is an identity in the time domain, the Laplace transforms of each side of that equation will be equal. Taking the Laplace transform of both sides and using the linearity properties of the Laplace operator, we obtain the following result:

$$\mathcal{L}[(\tau D + 1)z(t)] = \tau\mathcal{L}[Dz(t)] + \mathcal{L}[z(t)] = G\mathcal{L}[u(t)] \tag{F.16}$$

$$\tau[sZ(s) - z(0^-)] + Z(s) = GU(s) \tag{F.17}$$

$$[\tau s + 1]Z(s) = GU(s) + \tau z(0^-) \tag{F.18}$$

Finally, dividing both sides of Eq. (F.18) by $\tau s + 1$ gives $Z(s)$, as an explicit function of $s$ and known system constants (notice that lowercase $z$ refers to the time domain values of the response $z(t)$):

$$Z(s) = \frac{GU(s) + \tau z_0}{[\tau s + 1]} \tag{F.19}$$

Since our input is a step input of value $u_0$, and the initial condition is $z_0$, this equation can be rearranged and stated as follows (recall that $U(s) = u_0/s$ for a step input):

$$Z(s) = \frac{Gu_0}{s[\tau s + 1]} + \frac{\tau z_0}{[\tau s + 1]} \tag{F.20}$$

One final rearrangement puts this function in the form shown in Table F.2:

$$Z(s) = \frac{Gu_0/\tau}{s[s + 1/\tau]} + \frac{z_0}{[s + 1/\tau]} \tag{F.21}$$

To this point, we have just been performing trivial algebraic operations on the transform of the differential equation; however, the resulting expression is now in the form wherein we can take the inverse Laplace transform (by using the table) and find the time domain response of the system.

Recalling the linearity properties, and noticing that the first term in Eq. (F.21) is from transform pair number 6 and the second term is from transform pair number 5, we can simply write the time domain response of the system by using the time domain functions from numbers 6 and 5 as follows:

$$z(t) = \frac{Gu_0}{\tau}[\tau(1 - e^{-t/\tau})] + z_0 e^{-t/\tau} \tag{F.22}$$

A minor algebraic alteration yields the final time domain response,

$$z(t) = z_0 e^{-t/\tau} + Gu_0(1 - e^{-t/\tau}) \tag{F.23}$$

### F.5.2 Ramp Input to a First-Order System

The solution to a ramp input to a first-order differential equation can be found from the generalized response equation of a first-order system given in Eq. (F.19). Let us suppose that the initial condition is zero ($z(0) = 0$) and that the input is the ramp function

$$u(t) = u_0 t \quad \text{for} \quad t \geq 0 \tag{F.24}$$

Then

$$U(s) = \frac{u_0}{s^2} \tag{F.25}$$

Substituting these results into the general form given by Eq. (F.19) results in

$$Z(s) = \frac{Gu_0}{s^2(\tau s + 1)} = \frac{Gu_0/\tau}{s^2(s + 1/\tau)} \tag{F.26}$$

Now, from transform pair (number 7 in Table F.2), we can see that the time domain response is

$$z(t) = Gu_0\tau[t/\tau - (1 - e^{-t/\tau})] \tag{F.27}$$

This is the same result that we obtained in Appendix E.

## F.6 GENERAL APPROACH TO THE SOLUTION OF DIFFERENTIAL EQUATIONS

Plainly it is possible to transform a differential equation using the Laplace transform, solve explicitly for the transformed response variable, rearrange the resulting equation to a form that is in Table F.2, and then simply write the time domain response (using linearity properties). This technique is very good, especially when the system is of low order and all of the transform pairs are listed in the table. However, when we have higher order differential equations and more complicated inputs, we will have transfer functions in $s$ that will not be listed in the table. This presents no barrier, because it is very easy to express the transfer functions in partial fraction expansion format and solve the system equations, as discussed in the next section.

## F.7 PARTIAL FRACTION EXPANSION

When we take the Laplace transform of a differential equation and solve for the transform of the response variable, the result is a ratio of two polynomials in $s$. The proper form for expressing this ratio of polynomials is to have numerator and denominator polynomials with positive exponents of $s$. (We can always rearrange to avoid negative exponents.) This is the **polynomic form** of the transfer function and can be expressed as

$$F(s) = \frac{N(s)}{D(s)} = \frac{[b_k s^k + \cdots + b_1 s + b_0]}{[a_n s^n + \cdots + a_1 s + a_0]} \tag{F.28}$$

Since the numerator polynomial $N$(s) and the denominator polynomial $D$(s) both have roots, the function could be expressed in terms of the root factors. The roots of the denominator polynomial are called the **poles** of $F(s)$, and the roots of the numerator polynomial are called the **zeros** of $F(s)$. In dynamic systems, the roots usually have negative real parts; therefore, it is customary to express their factors in terms of positive constants $p_i$ and $z_j$. There are $n$ poles and $k$ zeros, and the $p$'s and $z$'s are related to the poles and zeros as follows:

$$-p_i = i\text{th pole (root of denominator)} \tag{F.29}$$

$$-z_j = j\text{th zero (root of numerator)} \tag{F.30}$$

Using this notation, we can express the transfer function in **factored form** as

$$F(s) = \frac{K(s + z_1)(s + z_2)\dots(s + z_k)}{(s + p_1)(s + p_2)\dots(s + p_n)} \tag{F.31}$$

Since the roots are unaffected by multiplying the polynomial by a constant, the conversion from a polynomial to the factored form could be missing the constant term required for equality. In some cases, we don't really need to know the value of the constant term; however, if you do need it, it can be found by using the product of the $p$'s and $z$'s, i.e.,

$$K = \frac{b_0}{a_0}\frac{\prod_{i=1}^{n} p_i}{\prod_{j=1}^{k} z_j} \tag{F.32}$$

If a transfer function is too complicated to be listed in a table, then it can be written in partial fraction form. Whereas the factored form of a transfer function has the product of pole terms $(s + p_i)$ in the denominator, the **partial fraction form** for the expression of a transfer function is the summation of the individual pole terms in the denominator with appropriate coefficients in the numerator of each term.

For example, consider the transfer function

$$F(s) = \frac{1}{(s + p_1)(s + p_2)} \tag{F.33}$$

This function can be written in partial fraction form as

$$F(s) = \frac{c_1}{(s + p_1)} + \frac{c_2}{(s + p_2)} \tag{F.34}$$

It is clear that the right-hand sides of Eqs. (F.33) and (F.34) could be made equivalent with the proper choice of $c_1$ and $c_2$. This example is for real roots; the expansions for complex roots and repeated roots have slightly different forms. The challenge in partial fraction expansion is to know the correct form for the expansion and how to find the coefficients $c_i$. We deal with this next.

### F.7.1 Partial Fractions with Distinct Real Roots

If all of the roots of a transfer function are real (as opposed to having complex parts) and distinct (i.e., the roots are not repeated, or, put another way, no two roots have the same value), then the factored form of the transfer function can be written as

$$F(s) = \frac{N(s)}{(s + p_1)(s + p_2)\dots(s + p_n)} \tag{F.35}$$

The corresponding proper form for the partial fraction expansion is

$$F(s) = \frac{c_1}{(s + p_1)} + \frac{c_2}{(s + p_2)} + \cdots + \frac{c_n}{(s + p_n)} \tag{F.36}$$

The numerator coefficients $c_i$ are called the **residues** of the poles and are given by

$$c_i = \left[(s + p_i)\frac{N(s)}{D(s)}\right]_{s=-p_i} \tag{F.37}$$

The practical application of this result is to take the transfer function expressed in factored form, remove the term corresponding to the $i$th pole, and evaluate the remainder with $s = -p_i$.

Equation (F.37) can be clearly visualized by the following third-order example in which we solve for $c_2$:

$$F(s) = \frac{K(s + z_1)}{(s + p_1)(s + p_2)(s + p_3)} = \frac{c_1}{(s + p_1)} + \frac{c_2}{(s + p_2)} + \frac{c_3}{(s + p_3)} \tag{F.38a}$$

$$(s + p_2)F(s) = \frac{(s + p_2)\,K(s + z_1)}{(s + p_1)(s + p_2)(s + p_3)} =$$

$$\frac{c_1(s + p_2)}{(s + p_1)} + \frac{c_2(s + p_2)}{(s + p_2)} + \frac{c_3(s + p_2)}{(s + p_3)} \tag{F.38b}$$

Now, with $s = -p_2$, all but the second term on the right side of Eq. (F.38b) are equal to zero, and we can solve for $c_2$ directly, obtaining

$$c_2 = \left[\frac{K(s + z_1)}{(s + p_1)(s + p_3)}\right]_{s=-p_2} = \frac{K(-p_2 + z_1)}{(-p_2 + p_1)(-p_2 + p_3)} \tag{F.39}$$

**Example F.4 Distinct Real Roots**

As an example of the use of the preceding technique, consider the third-order system

$$F(s) = \frac{100(s + 6)}{(s + 1)(s + 3)(s + 10)} \tag{F.40}$$

The proper form for the partial fraction expansion is

$$F(s) = \frac{c_1}{(s + 1)} + \frac{c_2}{(s + 3)} + \frac{c_3}{(s + 10)} \tag{F.41}$$

The coefficients can be found by Eq.( F.37):

$$c_1 = \left[\frac{100\,(s + 6)}{(s + 3)(s + 10)}\right]_{s=-1} = \frac{100(-1 + 6)}{(-1 + 3)(-1 + 10)} = 27.778 \tag{F.42}$$

$$c_2 = \left[\frac{100\,(s + 6)}{(s + 1)(s + 10)}\right]_{s=-3} = \frac{100(-3 + 6)}{(-3 + 1)(-3 + 10)} = -21.429 \tag{F.43}$$

$$c_3 = \left[\frac{100\,(s + 6)}{(s + 1)(s + 3)}\right]_{s=-10} = \frac{100(-10 + 6)}{(-10 + 1)(-10 + 3)} = -6.349 \tag{F.44}$$

Therefore, the partial fraction expansion for the system is

$$F(s) = \frac{27.778}{(s+1)} - \frac{21.429}{(s+3)} - \frac{6.349}{(s+10)} \quad \text{(F.45)}$$

As a check, we can convert this result back to the factored form by using a common denominator to obtain

$$F(s) = \frac{c_1(s+3)(s+10) + c_2(s+1)(s+10) + c_3(s+1)(s+3)}{(s+1)(s+3)(s+10)} \quad \text{(F.46)}$$

By carrying out this algebra in the numerator, we see that with the values given for the $c$'s, the numerator does indeed equal $100(s+6)$.

### F.7.2 Partial Fractions with Repeated Real Roots

If the roots of a transfer function are real, but two or more roots have the exact same value, then we must alter the partial fraction expansion form to accommodate the repeated roots. We will examine the case where all of the roots are repeated, and we will then consider the case where some of the roots are distinct and some are repeated.

If all of the roots are repeated, then the factored transfer function can be written as

$$F(s) = \frac{N(s)}{(s+p_1)^m} \quad \text{(F.47)}$$

The correct partial fraction expansion form is

$$F(s) = \frac{c_1}{(s+p_1)} + \frac{c_2}{(s+p_1)^2} + \cdots + \frac{c_m}{(s+p_1)^m} \quad \text{(F.48)}$$

It can be shown that the coefficients are as follows:

$$c_m = \left[(s+p_1)^m \frac{N(s)}{D(s)}\right]_{s=-p_1} \quad \text{(F.49)}$$

$$c_{m-1} = \frac{d}{ds}\left[(s+p_1)^m \frac{N(s)}{D(s)}\right]_{s=-p_1} \quad \text{(F.50)}$$

$$c_{m-2} = \frac{1}{2!}\frac{d^2}{ds^2}\left[(s+p_1)^m \frac{N(s)}{D(s)}\right]_{s=-p_1} \quad \text{(F.51)}$$

$$c_{m-3} = \frac{1}{3!}\frac{d^3}{ds^3}\left[(s+p_1)^m \frac{N(s)}{D(s)}\right]_{s=-p_1} \quad \text{(F.52)}$$

**Example F.5 Repeated Real Roots**

As an example of repeated real roots, consider the transfer function

$$F(s) = \frac{10(s+2)}{s^3} \quad \text{(F.53)}$$

The proper form for the partial fraction expansion is

$$F(s) = \frac{c_1}{s} + \frac{c_2}{s^2} + \frac{c_3}{s^3} \quad \text{(F.54)}$$

The coefficients are:

$$c_3 = \left[\frac{s^3 10(s+2)}{s^3}\right]_{s=0} = 20 \quad (F.55)$$

$$c_2 = \frac{d}{ds}\left[\frac{s^3 10(s+2)}{s^3}\right]_{s=0} = 10 \quad (F.56)$$

$$c_1 = \frac{1}{2!}\frac{d^2}{ds^2}\left[\frac{s^3 10(s+2)}{s^3}\right]_{s=0} = 0 \quad (F.57)$$

Therefore, the partial fraction expansion for the transfer function is

$$F(s) = \frac{10}{s^2} + \frac{20}{s^3} \quad (F.58)$$

As a check, you can obtain a common denominator and confirm that this answer is correct.

### F.7.3 Partial Fractions with Combined Distinct and Repeated Real Roots

The case in which there are some distinct roots and some repeated roots can be handled by combining the two techniques just presented. If we suppose that we have $n$ roots and that $m$ of these roots are repeated, then the factored transfer function can be expressed as

$$F(s) = \frac{N(s)}{(s+p_1)^m(s+p_{m+1})\cdots(s+p_n)} \quad (F.59)$$

The proper partial fraction expansion form is

$$F(s) = \frac{c_1}{(s+p_1)} + \frac{c_2}{(s+p_1)^2} + \cdots + \frac{c_m}{(s+p_1)^m} + \frac{c_{m+1}}{(s+p_{m+1})} + \cdots + \frac{c_n}{(s+p_n)} \quad (F.60)$$

The coefficients are then found by a combination of the techniques used for distinct roots and repeated roots.

**Example F.6 Distinct and Repeated Real Roots**

Consider the third-order system

$$F(s) = \frac{4(s+3)}{s^2(s+1)} \quad (F.61)$$

This system has two repeated roots at the origin ($s = 0$) and one distinct root at $s = -1$. The partial fraction expansion form is

$$F(s) = \frac{c_1}{s} + \frac{c_2}{s^2} + \frac{c_3}{(s+1)} \quad (F.62)$$

The coefficients are:

$$c_2 = \left[\frac{4(s+3)}{(s+1)}\right]_{s=0} = 12 \quad (F.63)$$

$$c_1 = \frac{d}{ds}\left[\frac{4(s+3)}{(s+1)}\right]_{s=0} = \left[\frac{(s+1)4 - 4(s+3)1}{(s+1)^2}\right]_{s=0} = -8 \tag{F.64}$$

$$c_3 = \left[\frac{4(s+3)}{s^2}\right]_{s=-1} = 8 \tag{F.65}$$

Thus, the partial fraction expansion for the transfer function is

$$F(s) = \frac{-8}{s} + \frac{12}{s^2} + \frac{8}{(s+1)} \tag{F.66}$$

The reader should verify this result by finding a common denominator.

### F.7.4 Partial Fractions with Distinct Complex Roots

When the roots are complex, they appear in complex pairs with one root the conjugate of the other. Thus, when the roots are complex, we will associate them in pairs. To begin with, the factored transfer function can be expressed as

$$F(s) = \frac{N(s)}{(s+p_1)(s+p_2)(s+p_3)(s+p_4)\cdots(s+p_{n-1})(s+p_n)} \tag{F.67}$$

The partial fraction expansion for this case groups the complex pair of roots together, yielding

$$F(s) = \frac{c_1 s + c_2}{(s+p_1)(s+p_2)} + \frac{c_3 s + c_4}{(s+p_3)(s+p_4)} + \cdots + \frac{c_{n-1} s + c_n}{(s+p_{n-1})(s+p_n)} \tag{F.68}$$

It can be shown that the coefficients can be found by the equation

$$c_{i-1}s + c_i = \left[(s+p_{i-1})(s+p_i)\frac{N(s)}{D(s)}\right]_{s=-p_i} \tag{F.69}$$

Since the poles are complex, this equation will also be complex. Therefore, for two complex numbers to be equal, their real parts must be equal to each other, and their imaginary parts must be equal to each other; this will give us two equations to solve for the two coefficients $c_{i-1}$ and $c_i$. Notice that since the roots are complex conjugates, we could use either $p_i$ or $p_{i-1}$ in the evaluation of the numbers.

**Example F.7 Complex Roots**

Consider the transfer function

$$F(s) = \frac{s+10}{(s+p_1)(s+p_2)(s+p_3)(s+p_4)} \tag{F.70}$$

in which $p_1 = 2 + j3$, $p_2 = 2 - j3$, $p_3 = 1 + j$, and $p_4 = 1 - j$. The partial fraction expansion form for this case is

$$F(s) = \frac{c_1 s + c_2}{(s+p_1)(s+p_2)} + \frac{c_3 s + c_4}{(s+p_3)(s+p_4)} \tag{F.71}$$

We will treat the two terms independently and first solve for $c_1$ and $c_2$. Using the original transfer function in the general coefficient equation results in

$$c_1 s + c_2 = \left[\frac{s + 10}{(s + p_3)(s + p_4)}\right]_{s=-2+j3} =$$

$$\frac{-2 + j3 + 10}{(-2 + 3j + 1 + j)(-2 + j3 + 1 - j)} = \frac{8 + j3}{-7 - j6} \tag{F.72}$$

Substituting the value $s = -2 + j3$ into the left side of the equation and expanding the right side (through multiplication by the conjugate of the denominator) yields

$$c_1(-2 + j3) + c_2 = c_2 - 2c_1 + j3c_1 = \frac{-74 + j27}{85} \tag{F.73}$$

Since the real and the imaginary parts must both be equal, the following two equations result:

$$c_2 - 2c_1 = \frac{-74}{85} \tag{F.74}$$

$$j3c_1 = \frac{j27}{85} \tag{F.75}$$

Therefore,

$$c_1 = \frac{9}{85} \tag{F.76}$$

$$c_2 = \frac{-56}{85} \tag{F.77}$$

Using the second pair of poles yields

$$c_3 s + c_4 = \left[\frac{s + 10}{(s + p_1)(s + p_2)}\right]_{s=-p_4} =$$

$$\frac{(-1 + j + 10)}{(-1 + j + 2 + j3)(-1 + j + 2 + j3)} = \frac{9 + j}{9 + j2} \tag{F.78}$$

Substituting $s = -p_4$ into the left side of this equation and reducing the complex term on the right side yields

$$c_3(-1 + j) + c_4 = c_4 - c_3 + jc_3 = \frac{83 - j9}{85} \tag{F.79}$$

Therefore, the coefficients can be solved from the real and imaginary parts of Eq. (F.79) and are

$$c_3 = \frac{-9}{85} \tag{F.80}$$

and

$$c_4 = \frac{74}{85} \tag{F.81}$$

Notice that all of the $c$'s are real constants, and the partial fraction expansion for the transfer function of Eq. (F.70) can be stated as

$$F(s) = \frac{1}{85}\left[\frac{(9s - 56)}{(s + p_1)(s + p_2)} + \frac{(-9s + 74)}{(s + p_3)(s + p_4)}\right] \tag{F.82}$$

### F.7.5 Partial Fractions with Repeated Complex Roots

A transfer function with repeated complex roots is not a common occurrence; however, the mathematics for handling such a case is presented next. The transfer function can be stated as

$$F(s) = \frac{N(s)}{[(s + p_1)(s + p_2)]^m} \tag{F.83}$$

In this case, the roots are complex conjugates of each other and can be written $-p_1$ and $-p_2$, where the $p$'s are complex numbers.

In a manner similar to the partial fraction expansion for repeated real roots, the correct partial fraction expansion for repeated complex roots is

$$F(s) = \frac{c_1 s + c_2}{(s + p_1)(s + p_2)} + \frac{c_3 s + c_4}{[(s + p_1)(s + p_2)]^2} + \cdots + \frac{c_{2m-1} s + c_{2m}}{[(s + p_1)(s + p_2)]^m} \tag{F.84}$$

The coefficients are found in essentially the same way as are those for the repeated real roots, except that we must realize that now the roots come in pairs, or two at a time, and therefore, we must advance the derivative sequence by twos instead of ones. The general format for the coefficients is

$$c_{2m-1}s + c_{2m} = \left[\{(s + p_1)(s + p_2)\}^m \frac{N(s)}{D(s)}\right]_{s=-p_2} \tag{F.85}$$

$$c_{2m-3}s + c_{2m-2} = \frac{1}{2!}\frac{d^2}{ds^2}\left[\{(s + p_1)(s + p_2)\}^m \frac{N(s)}{D(s)}\right]_{s=-p_2} \tag{F.86}$$

$$c_{2m-5}s + c_{2m-4} = \frac{1}{4!}\frac{d^4}{ds^4}\left[\{(s + p_1)(s + p_2)\}^m \frac{N(s)}{D(s)}\right]_{s=-p_2} \tag{F.87}$$

$$[c_1 s + c_2]_{s=-p_2} = \frac{1}{(2m-2)!}\frac{d^{2m-2}}{ds^{2m-2}}\left[\{(s + p_1)(s + p_2)\}^m \frac{N(s)}{D(s)}\right]_{s=-p_2} \tag{F.88}$$

Notice that if we have only repeated complex roots, then the root product to the $m$th power exactly cancels $D(s)$, and therefore, we deal only with $N(s)$ and its derivatives; however, in the general case, there will be other terms if there are combined distinct and repeated roots.

**Example F.8 Repeated Complex Roots**

Consider the fourth-order system

$$F(s) = \frac{(s + 1)(s + 4)}{[(s + p_1)(s + p_2)]^2} \tag{F.89}$$

with two repeated complex roots. The correct partial fraction expansion is

$$F(s) = \frac{c_1 s + c_2}{(s + p_1)(s + p_2)} + \frac{c_3 s + c_4}{[(s + p_1)(s + p_2)]^2} \tag{F.90}$$

The coefficients can be found from

$$c_3 s + c_4 = [(s + 1)(s + 4)]_{s=-p_2} = [s^2 + 5s + 4]_{s=-p_2} = -11 + j3 \tag{F.91}$$

Equating the real parts and the imaginary parts separately and solving for the coefficients yields

$$c_3 = 1 \tag{F.92}$$

and

$$c_4 = -9 \tag{F.93}$$

The second term in the expansion can be solved in a similar way, resulting in

$$c_1 s + c_2 = \frac{1}{2!}\frac{d^2}{ds^2}[(s^2 + 5s + 4)]_{s=-p_2} = \frac{2}{2} = 1 + j0 \tag{F.94}$$

where the constants are

$$c_1 = 0 \tag{F-95}$$

and

$$c_2 = 1 \tag{F.96}$$

Thus, the complete transfer function can be stated in partial fraction form as

$$F(s) = \frac{1}{(s + p_1)(s + p_2)} + \frac{(s - 9)}{[(s + p_1)(s + p_2)]^2} \tag{F.97}$$

### F.7.6 Partial Fractions with Combined Distinct and Repeated Complex Roots

If the system has combined roots, some of which are distinct complex roots and others of which are repeated complex roots, then the partial fraction expansion is formed as a combination of the separate techniques used to deal with each.

**Example F.9 Distinct and Repeated Complex Roots**

Consider a sixth-order system in which two complex root pairs ($-p_1$ and $-p_2$) are repeated and one complex root pair ($-p_5$ and $-p_6$) is distinct. The transfer function for the system is

$$F(s) = \frac{(s + 1)(s + 4)}{\{(s + p_1)(s + p_2)\}^2(s + p_5)(s + p_6)} \tag{F.98}$$

The correct form for the partial fraction expansion is

$$F(s) = \frac{c_1 s + c_2}{(s + p_1)(s + p_2)} + \frac{c_3 s + c_4}{[(s + p_1)(s + p_2)]^2} + \frac{c_5 s + c_6}{(s + p_5)(s + p_6)} \tag{F.99}$$

The coefficients can be found by applying the techniques presented earlier.

### F.7.7 Partial Fractions with Combined Real and Complex Roots

The preceding techniques generally apply to special cases in which the roots of the characteristic equation are all of one kind or another; however, we can combine the techniques and the equations they apply to in any order to solve for the coefficients. The only "tricks" to partial fraction expansion are (1) to remember the form of the expansion applicable to the type of root (real or complex, distinct or repeated) and (2) to know how to solve for the coefficients used in the expansion. With these "tricks" mastered, one can apply partial fraction expansion techniques to any general transfer function.

**Example F.10 Real and Complex Roots**

Consider the fifth-order system

$$F(s) = \frac{(s + z_1)(s + z_2)(s + z_3)}{(s + p_1)^2(s + p_3)(s + p_4)(s + p_5)} \tag{F.100}$$

in which there are two repeated real roots, $-p_1$ and $-p_1$, one distinct real root, $-p_3$, one complex pair of roots, $-p_4$ and $-p_5$, and three zeros. The partial fraction expansion of this transfer function is

$$F(s) = \frac{c_1}{(s + p_1)} + \frac{c_2}{(s + p_1)^2} + \frac{c_3}{(s + p_3)} + \frac{c_4 s + c_5}{(s + p_4)(s + p_5)} \tag{F.101}$$

The coefficients can be found from the following equations:

$$c_1 = \frac{d}{ds}\left[(s + p_1)^2 \frac{N(s)}{D(s)}\right]_{s=-p_1} \tag{F.102}$$

$$c_2 = \left[(s + p_1)^2 \frac{N(s)}{D(s)}\right]_{s=-p_1} \tag{F.103}$$

$$c_3 = \left[(s + p_3) \frac{N(s)}{D(s)}\right]_{s=-p_3} \tag{F.104}$$

$$c_4 s + c_5 = \left[(s + p_4)(s + p_5) \frac{N(s)}{D(s)}\right]_{s=-p_5} \tag{F.105}$$

## F.8 SOLUTION OF DIFFERENTIAL EQUATIONS WITH PARTIAL FRACTION EXPANSION

Even when the transfer function of a system is not listed in Table F.2, an expanded transfer function composed of simplified terms can still be found using the partial fraction expansion techniques discussed in this appendix and summarized in Table F.3. Then the simplified terms can be found in Table F.2, and the time domain

**TABLE F–3** PARTIAL FRACTION EXPANSION FORMS

| | $$F(s) = \frac{N(s)}{D(s)} = \frac{K(s+z_1)(s+z_2)\cdots(s+z_k)}{(s+p_1)(s+p_2)\cdots(s+p_n)}$$ |
|---|---|
| **Distinct Real Roots** | $p_1, p_2, p_3$ are real. Roots at $-p_1, -p_2, -p_3$ |
| Transfer function: | $F(s) = \dfrac{N(s)}{(s+p_1)(s+p_2)(s+p_3)\cdots}$ |
| Partial fraction expansion format: | $F(s) = \dfrac{c_1}{(s+p_1)} + \dfrac{c_2}{(s+p_2)} + \dfrac{c_3}{(s+p_3)} + \cdots$ |
| Coefficients: | $c_i = \left[(s+p_i)\dfrac{N(s)}{D(s)}\right]_{s=-p_i}$ |
| **Repeated Real Roots** | $p_1$ is real. $m$ roots at $-p_1$ |
| Transfer function: | $F(s) = \dfrac{N(s)}{(s+p_1)^m\cdots}$ |
| Partial fraction expansion format: | $F(s) = \dfrac{c_1}{(s+p_1)} + \dfrac{c_2}{(s+p_1)^2} + \cdots + \dfrac{c_m}{(s+p_1)^m} + \cdots$ |
| Coefficients: | $c_m = \left[(s+p_1)^m\dfrac{N(s)}{D(s)}\right]_{s=-p_1}$ <br> $c_{m-1} = \dfrac{d}{ds}\left[(s+p_1)^m\dfrac{N(s)}{D(s)}\right]_{s=-p_1}$ <br> $c_{m-i} = \dfrac{1}{i!}\dfrac{d^i}{ds^i}\left[(s+p_1)^m\dfrac{N(s)}{D(s)}\right]_{s=-p_1}$ |
| **Distinct Complex Roots** | $p_1, p_2$ and $p_3, p_4$ are complex conjugates. Roots at $-p_1, -p_2$ and $-p_3, -p_4$ |
| Transfer function: | $F(s) = \dfrac{N(s)}{(s+p_1)(s+p_2)(s+p_3)(s+p_4)\cdots}$ |
| Partial fraction expansion format: | $F(s) = \dfrac{c_1 s + c_2}{(s+p_1)(s+p_2)} + \dfrac{c_3 s + c_4}{(s+p_3)(s+p_4)} + \cdots$ |
| Coefficients: | $[c_1 s + c_2]_{s=-p_2} = \left[(s+p_1)(s+p_2)\dfrac{N(s)}{D(s)}\right]_{s=-p_2}$ <br> $[c_3 s + c_4]_{s=-p_4} = \left[(s+p_3)(s+p_4)\dfrac{N(s)}{D(s)}\right]_{s=-p_4}$ |

*Continued*

**TABLE F.3** *continued*

| **Repeated Complex Roots** | $p_1, p_2$ are complex conjugates. $m$ roots at $-p_1, -p_2$ |
|---|---|
| Transfer function: | $$F(s) = \frac{N(s)}{[(s + p_1)(s + p_2)]^m \cdots}$$ |
| Partial fraction expansion format: | $$F(s) = \frac{c_1 s + c_2}{(s + p_1)(s + p_2)} + \frac{c_3 s + c_4}{\{(s + p_1)(s + p_2)\}^2} + \cdots + \frac{c_{2m-1} s + c_{2m}}{\{(s + p_1)(s + p_2)\}^m} + \cdots$$ |
| Coefficients: | $$[c_{2m-1} s + c_{2m}]_{s=-p_2} = \left[\{(s + p_1)(s + p_2)\}^m \frac{N(s)}{D(s)}\right]_{s=-p_2}$$ $$[c_{2m-3} s + c_{2m-2}]_{s=-p_2} = \frac{1}{2!}\frac{d^2}{ds^2}\left[\{(s + p_1)(s + p_2)\}^m \frac{N(s)}{D(s)}\right]_{s=-p_2}$$ $$[c_1 s + c_2]_{s=-p_2} = \frac{1}{(2m-2)!}\frac{d^{2m-2}}{ds^{2m-2}}\left[\{(s + p_1)(s + p_2)\}^m \frac{N(s)}{D(s)}\right]_{s=-p_2}$$ |

result can be stated directly. The linearity properties must be used to multiply the transfer functions by constants and to sum together the individual partial fraction terms.

The MATLAB function `residue` can be used to convert from transfer function form to residue form (and from residue form to transfer function form).

## REFERENCES


F.1 Strum, R.D., and Wand, J. R. *Laplace Transform Solution of Differential Equations—a Programmed Text.* Prentice Hall, Englewood Cliffs, NJ, 1968.

F.2 Ogata, Katsuhiko. *System Dynamics.* Prentice Hall, Englewood Cliffs, NJ, 1992.


## PROBLEMS

**F.1** *Convert* the following complex numbers to polar form:

**a.** $1 + j1$

**b.** $2 - j3$

**c.** $-1 - j1$

**F.2** *Convert* the following complex numbers in polar notation to rectangular notation:

**a.** $10 \angle 45°$

**b.** $1 \angle 90°$

**c.** $\sqrt{2} \angle -45°$

**F.3** *What* are the natural frequency and damping ratio associated with $a \pm jb$ ?

**F.4** *Perform* the following operations on the stated complex numbers and functions:

**a.** *Find* the complex conjugate of $z = 3 + j4$.

**b.** *Express* $z = 3 + j4$ in polar form.

**c.** *Express* $z = 10e^{j\frac{\pi}{6}}$ in rectangular form.

**d.** *Express* $z = -1 - j1$ in polar form.

**e.** *Express* $z = \sqrt{2} \angle -45°$ in rectangular form.

**f.** *Find* the rectangular form of the product $z = (1 + j2)(2 - j3)$.

**g.** *Express* $z = (2 + j1)/(1 - j2)$ in simple rectangular form.

**F.5** Use MATLAB to *find* the roots of the polynominal

$$s^5 + 15s^4 + 68s^3 + 214s^2 + 360s + 200 = 0$$

**F.6** Use MATLAB (function `residue`) to *find* the partial fraction expansion of the following functions:

**a.** $Z(s) = \dfrac{1}{s^2 + 4s + 1}$

**b.** $Z(s) = \dfrac{s + 20}{s^4 + 17s^3 + 85s^2 + 175s + 250}$

**F.7** *Find* the eigenvalues of the matrix

$$\mathbf{A} = \begin{bmatrix} 0 & 1 & 0 \\ -500 & -10 & .001 \\ 0 & 10^6 & 20 \end{bmatrix}$$

**F.8** Using the integral-exponential definition of the Laplace transform, *derive* the Laplace transform of the following function:

$$f(t) = Ae^{-at} \quad \text{for } t \geq 0$$
$$f(t) = 0 \quad \text{for } t < 0$$

**F.9** *Find* the Laplace transform of each of the given functions. *Express* your result in the form

$$Z(s) = \frac{N(s)}{D(s)} U(s)$$

**a.** $\dfrac{d^2z}{dt^2} + 4\dfrac{dz}{dt} + 10z = 20u \qquad$ with $z(0) = z_0$

$$\frac{dz(0)}{dt} = \dot{z}_0$$

**b.** $\dfrac{d^3z}{dt^3} + \alpha_2 \dfrac{d^2z}{dt^2} + \alpha_1 \dfrac{dz}{dt} + \alpha_0 z = \beta_1 \dfrac{du}{dt} + \beta_0 u$ with all initial conditions = 0

**F.10** Use the Laplace transform technique to *find* the time response of the following systems:

**a.** $\ddot{z} + 4\dot{z} + z = 20u \qquad$ where $u(t) = A \sin\omega t$

$$\text{with } z(0) = 0$$
$$\dot{z}(0) = 0$$

**b.** $\ddot{z} + \dot{z} + z = 5u$ where $u(t) = A \sin\omega t$

$$\text{with } z(0) = 15$$
$$\dot{z}(0) = 0$$

**c.** $Z(s) = \dfrac{(s + 20)}{(s + 10)(s + 5)(s + 1 + j2)(s + 1 - j2)} U(s)$

with all initial conditions equal to zero and with $u(t) = 5$ for $t \geq 0$

**F.11** A first-order dynamic system is subjected to a pulse input

$$\frac{dz}{dt} + 2z = 2u \quad \text{where } u(t) = 1 \quad 0 \leq t \leq 1$$

$$u(t) = 0 \quad t > 1$$

$$\text{with } z(0) = 0$$

**a.** *Find* the inverse Laplace transform of the system.

**b.** *Plot* the response over the two ranges of the input.

# APPENDIX G

# State-Space Representation of Dynamic Systems

## G.1 CONCEPT OF STATE SPACE

In analog and digital simulation, it is very convenient to perform single integrations at a time; thus, the state-space or state-variable representation of dynamic systems is advantageous. In this format, the primary or fundamental system variables (e.g., position, velocity, voltage, current, force, etc.) are selected as state variables, and a set of first-order differential equations is written in the form

$$\dot{\mathbf{x}} = \mathbf{f}(\mathbf{x}, \mathbf{u}) \tag{G.1}$$

or, written out,

$$\begin{aligned} \dot{x}_1 &= f_1(\mathbf{x}, \mathbf{u}) \\ \dot{x}_2 &= f_2(\mathbf{x}, \mathbf{u}) \\ &\vdots \\ \dot{x}_n &= f_n(\mathbf{x}, \mathbf{u}) \end{aligned} \tag{G.2}$$

where

$\mathbf{x}$ = vector of state variables

$\mathbf{u}$ = vector of inputs

$\mathbf{f}$ = set of functions of $\mathbf{x}$ and $\mathbf{u}$

Notice that the functions $\mathbf{f}$ could be either linear or nonlinear. The nonlinear case represents a very important class of problems that can be conveniently handled in state-space notation, but would be clumsy or often impossible to deal with in classical form. Hence, this case should not be overlooked; however, linear functions will be emphasized here.

The derivatives of linear functions can be written as a linear combination of the state variables and inputs. In matrix notation

$$\dot{\mathbf{x}} = \mathbf{A}\mathbf{x} + \mathbf{B}\mathbf{u} \tag{G.3}$$

where

$$\mathbf{A}, \mathbf{B} = \text{matrices of coefficients}$$

Thus, an $n$th-order differential equation or dynamic system would be represented by $n$ first-order differential equations, rather than a single $n$th-order classical differential equation. The solution, or system response, can then be obtained by a single integration of $n$ simultaneous equations rather than $n$ integrations of one equation.

The state-space format is also useful in representing complex systems with multiple inputs or nonlinearities and in modern control theory applications.

The state-variable vector $\mathbf{x}$ establishes a vector space, or state space, that fully describes the system response. There is no unique definition of the state variable vector, and in fact, a wide variety of system variables could be used to obtain a workable state-space representation. However, if the system is of order $n$, then there must be at least $n$ state variables that are independent of each other. Hence, it is possible to use more than $n$ state variables, but then some of the state variables will be dependent or linearly dependent upon each other.

Further, any observable system variable can be written as a linear combination of states, as illustrated by the matrix equation

$$\mathbf{y} = \mathbf{C}\mathbf{x} + \mathbf{E}\mathbf{u} \tag{G.4}$$

where

$$\mathbf{y} = \text{vector of outputs}$$

$$\mathbf{C}, \mathbf{E} = \text{matrices of coefficients}$$

Note that matrix notation is not applicable to nonlinear systems.

Equation (G.4) implies that if one is interested in monitoring or computing variables or responses other than those given by the state variables, any system variable can be computed with a knowledge of the state. For example, if the state variables in a mechanical spring-mass-damper system are position and velocity, then the forces in the spring or damper can be computed as a function of position and velocity.

## G.2 STATE-SPACE FORMULATION

The state of a system in the state-space format must include, first, a definition of the inputs and state variables in terms of system variables, where

$$\mathbf{u} = \text{inputs} \tag{G.5}$$
$$\mathbf{x} = \text{system variables}$$

Next, the system dynamic equations must be specified in the form

$$\dot{\mathbf{x}} = \mathbf{f}(\mathbf{x}, \mathbf{u}) \tag{G.6}$$

or, for a linear system;

$$\dot{\mathbf{x}} = \mathbf{A}\mathbf{x} + \mathbf{B}\mathbf{u} \tag{G.7}$$

In order to obtain an explicit solution, the initial conditions of the state variables must be specified. These conditions may be simple to determine from the initial configuration of the system or may be indirect functions of system variables other than state variables. In either case, the initial conditions must be known, and when they are known, they may be stated in the form

$$\mathbf{x}(0) = \begin{bmatrix} x_1(0) \\ x_2(0) \\ \vdots \\ x_n(0) \end{bmatrix} \tag{G.8}$$

The preceding equations are sufficient to obtain the response $\mathbf{x}(t)$ for a known input $\mathbf{u}(t)$. However, as those equations will only specify the state variables, any other system variable can be specified as

$$\mathbf{y} = \mathbf{g}(\mathbf{x}, \mathbf{u}) \tag{G.9}$$

for a nonlinear system and as

$$\mathbf{y} = \mathbf{C}\mathbf{x} + \mathbf{E}\mathbf{u} \tag{G.10}$$

for a linear system.

The foregoing equations and definitions will facilitate the simulation of a dynamic system on either a digital or an analog computer and will permit the response of any system variable to be printed or plotted. Notice that if the system is an $n$th-order dynamic system with $m$ inputs, $\mathbf{x}$ will be an $n$th-order vector, $\mathbf{u}$ will be an $m$th-order vector, $\mathbf{A}$ will be an $(n \times n)$ matrix, and $\mathbf{B}$ will be an $(n \times m)$ matrix. Often, $m$ is equal to 1. If $k$ output variables are to be monitored (printed or plotted), $\mathbf{y}$ will be a $k$th-order vector, $\mathbf{C}$ will be $(k \times n)$, and $\mathbf{E}$ will be $(k \times m)$.

## G.3 CONVERSION FROM CLASSICAL TO STATE VARIABLES

A classical differential equation can be converted to a state-space equation by defining the state variables, or so-called phase variables, to be used in the equation. In this regard, consider the classical $n$th-order differential equation that has been normalized with respect to its highest derivative and written in operator notation as

$$[D^n + \alpha_{n-1} D^{n-1} + \cdots + \alpha_1 D + \alpha_0]\, z(t) = \beta_0\, w(t) \tag{G.11}$$

where

$$D = \text{differential operator} = d/dt$$
$$z(t) = \text{system response (dependent variable)}$$
$$w(t) = \text{system input (independent variable)}$$
$$\alpha, \beta = \text{coefficients, with } \alpha_n = 1$$

with initial conditions

$$z(0),\ Dz(0), \cdots, D^{n-1}z(0)$$

This system has one input and can be converted to state space by defining the first state variable as the dependent variable and the successive state variables as successive derivatives of the dependent variable (hence the term "phase variables," since derivatives are 90° out of phase with each other) as follows:

| Definitions of State Variables | **x** as Functions of $z$, $Dz$, etc. |
|---|---|
| $u_1 \triangleq w(t)$ | |
| $x_1 \triangleq z(t)$ | $x_1 = z$ |
| $x_2 \triangleq \dot{x}_1$ | $x_2 = Dz$ |
| $x_3 \triangleq \dot{x}_2$ | $x_3 = D^2z$ |
| $\vdots$ | $\vdots$ |
| $x_n \triangleq \dot{x}_{n-1}$ | $x_n = D^{n-1}z$ |

(G.12)

Remember that we are seeking an equation of the form $\dot{\mathbf{x}} = \mathbf{Ax} + \mathbf{Bu}$ and that we have, in our definitions, expressions for $Dx_i$ for every state variable except $Dx_n$. Notice that by differentiating the last equation in the second column of Eq. (G.12), we obtain

$$\dot{x}_n = D^n z \tag{G.13}$$

The expression for $D^n z$ is found from the original differential equation:

$$D^n z = -x_0 z - \alpha_1 Dz - \cdots - \alpha_{n-1} D^{n-1} z + \beta_0 w \tag{G.14}$$

The $z$, $Dz$, etc., terms must be eliminated and replaced by $x_1$, $x_2$, etc., from the equations in column two of Eq. (G.12):

$$D^n z = -\alpha_0 x_1 - \alpha_1 x_2 - \cdots - \alpha_{n-1} x_n + \beta_0 u_1 \tag{G.15}$$

Finally, we obtain the following equations from our definitions in Eq. (G.12) and from Eqs. (G.13) and (G.15):

$$\dot{x}_1 = x_2$$

$$\begin{aligned}\dot{x}_2 &= x_3\\ &\vdots\\ \dot{x}_{n-1} &= x_n\\ \dot{x}_n &= -\alpha_0 x_1 - \cdots - \alpha_{n-1}x_n + \beta_0 u_1\end{aligned} \tag{G.16}$$

In matrix notation, these equations can be expressed as

$$\begin{bmatrix}\dot{x}_1\\ \dot{x}_2\\ \vdots\\ \dot{x}_{n-1}\\ \dot{x}_n\end{bmatrix} = \begin{bmatrix}0 & 1 & 0 & \dots & 0\\ 0 & 0 & 1 & \dots & 0\\ \vdots & & & & \\ 0 & 0 & 0 & \dots & 1\\ -\alpha_0 & -\alpha_1 & -\alpha_2 & \dots & -\alpha_{n-1}\end{bmatrix}\begin{bmatrix}x_1\\ x_2\\ x_3\\ \vdots\\ x_n\end{bmatrix} + \begin{bmatrix}0\\ 0\\ 0\\ 0\\ \beta_0\end{bmatrix}[u_1] \tag{G.17}$$

The equations in column two of Eq. (G.12) are valid for all values of time, including $t = 0$.

The initial conditions of $\mathbf{x}$ are functions of the initial conditions of the original system, as given by the equations in the second column of Eq. (G.12), and are expressed as

$$\begin{aligned}x_1(0) &= z(0)\\ x_2(0) &= Dz(0)\\ &\vdots\\ x_n(0) &= D^{n-1}z(0)\end{aligned} \tag{G.18}$$

**Example G.1**

Consider the third-order differential equation

$$[D^3 + 5D^2 + 3D + 7]\,z = 2w \tag{G.19}$$

with initial conditions

$$\begin{aligned}z(0) &= 10\\ Dz(0) &= 4\\ D^2z(0) &= 0\end{aligned} \tag{G.20}$$

For this system, one input and three state variables are required:

| Definitions of State Variables | $\mathbf{x}$ as Functions of $z$, $Dz$, etc |
|---|---|
| $u_1 \triangleq w$ | |
| $x_1 \triangleq z$ | $x_1 = z$ |
| $x_2 \triangleq \dot{x}_1$ | $x_2 = Dz$ |
| $x_3 \triangleq \dot{x}_2$ | $x_3 = D^2z$ |

(G.21)

Differentiating the last equation in the second column and using Eq. (G.19), we find that

$$\dot{x}_3 = D^3 z = -7z - 3Dz - 5D^2 z + 2w \tag{G.22}$$

Substituting the expressions for $z$, $Dz$, etc., from the second column into Eq. (G.21), we obtain

$$\dot{x}_3 = -7x_1 - 3x_2 - 5x_3 + 2u_1 \tag{G.23}$$

Thus, in matrix notation, the state-space representation results in

$$\dot{\mathbf{x}} = \begin{bmatrix} 0 & 1 & 0 \\ 0 & 0 & 1 \\ -7 & -3 & -5 \end{bmatrix} \mathbf{x} + \begin{bmatrix} 0 \\ 0 \\ 2 \end{bmatrix} \mathbf{u} \tag{G.24}$$

The initial conditions of **x** are given by the second column of Eq. (G.21) and Eq. (G.20):

$$\mathbf{x}(0) = \begin{bmatrix} 10 \\ 4 \\ 0 \end{bmatrix} \tag{G.25}$$

Notice that the definition of the state variables proceeds in a standard pattern and the information concerning the differential equation is included totally in the last equation.

## G.4 CONVERSION FROM CLASSICAL TO STATE VARIABLES WITH INPUT DERIVATIVES

The conversion discussed in the preceding section is quite simple, but can be complicated by the presence of derivatives of the input. Recall that the goal in defining state variables is to obtain the $\dot{\mathbf{x}} = \mathbf{Ax} + \mathbf{Bu}$ format and not to have any derivatives of the input (i.e., $\dot{\mathbf{u}}$, $\ddot{\mathbf{u}}$, etc.). The appearance of input derivatives in the final form would require numerical or electronic differentiation of the input signal, which is inaccurate and prone to problems. When the original classical form contains derivatives of the input, it takes a special definition of state variables to eliminate any input derivatives in the state-space format.

Consider a classical $n$th-order differential equation in $z$, with $k$th-order derivatives of the input $w$, that has been normalized with respect to the highest derivative of $z$ (i.e., $\alpha_n = 1.0$):

$$[D^n + \alpha_{n-1}D^{n-1} + \cdots + \alpha_1 D + \alpha_0]z(t) = [\beta_k D^k + \beta_{k-1}D^{k-1} + \cdots + \beta_1 D + \beta_0]\, w(t) \tag{G.26}$$

As before, a set of state variables can be defined using phase variables, with a constant multiple $\gamma_i$, of the input subtracted from the last $k$ state variable definitions. In the process, we will need both **x** as functions of $D^i z$, and $D^i z$ as functions of **x**; thus, it is convenient to set up a table with three columns as follows:

| Definitions of State Variables | **x** as Functions of $w, z, Dz, D^2z$, etc. | $z, Dz, D^2z$, etc. as Functions of **x** and **u** |
|---|---|---|
| $u_1 \triangleq w$ | | |
| $x_1 \triangleq z$ | $x_1 = z$ | $z = x_1$ |
| $x_2 \triangleq \dot{x}_1$ | $x_2 = Dz$ | $Dz = x_2$ |
| $\vdots$ | $\vdots$ | $\vdots$ |
| $x_{n-k+1} \triangleq \dot{x}_{n-k} - \gamma_1 u_1$ | $x_{n-k+1} = D^{n-k}z - \gamma_1 w$ | $D^{n-k}z = x_{n-k+1} + \gamma_1 u_1$ |
| $x_{n-k+2} \triangleq \dot{x}_{n-k+1} - \gamma_2 u_1$ | $x_{n-k+2} = D^{n-k+1}z - \gamma_1 Dw - \gamma_2 w$ | $D^{n-k+1}z = x_{n-k+2} + \gamma_1 Du + \gamma_2 u_1$ |
| $\vdots$ | $\vdots$ | $\vdots$ |
| $x_n \triangleq \dot{x}_{n-1} - \gamma_k u_1$ | $x_n = D^{n-1}z - \sum_{i=1}^{k} \gamma_i D^{k-i} w$ | $D^{n-1}z = x_n + \sum_{i=1}^{k} \gamma_i D^{k-i} u_1$ |

From the definition of the phase variables with a multiple of the input being subtracted from the last $k$ state variables, the equations in the first column are obtained. The equations in the second column can be obtained by performing the operations indicated in the first column and substituting the original variables $z$ and $w$. The right side of the equation in the second column is obtained by differentiating the previous state variable expression (in that column) and subtracting an appropriate $\gamma$ as dictated by the first column. The equations of the first column can be algebraically rearranged to obtain successive orders of derivatives of the original variable in terms of the state variables and state inputs, as presented in the third column.

Rearranging the definitions in the first column, we obtain equations for $\dot{x}_1$ through $\dot{x}_{n-1}$. The expression for $x_n$ is obtained by differentiating the last expression in the second column:

$$\dot{x}_n = D^n z - \sum_{i=1}^{k} \gamma_i D^{k-i+1} w \tag{G.27}$$

An expression for $D^n z$ is obtained from the original equation:

$$D^n z = -\alpha_0 z - \alpha_1 Dz - \cdots - \alpha_{n-1} D^{n-1} z + \beta_k D^k w + \cdots + \beta_1 Dw + \beta_0 w \tag{G.28}$$

The terms for $z$, $Dz$, etc., are obtained explicitly from the third column and are substituted into Eq. (G.28); then Eq. (G.28) can be substituted into Eq. (G.27). The result (after considerable effort) can be stated in general as

$$\begin{aligned}\dot{x}_n = &-\sum_{i=1}^{n} \alpha_{i-1} x_i + \left(\beta_0 - \sum_{j=1}^{k} \gamma_j \alpha_{n-k+j-1}\right) u_1 \\ &+ (\beta_k - \gamma_1) D^k u_1 + \sum_{j=1}^{k-1} \left(\beta_{k-j} - \gamma_{j+1} - \sum_{i=1}^{j} \gamma_i \alpha_{n+i-j-1}\right) D^{k-j} u_1\end{aligned} \tag{G.29}$$

The first parts of this equation involve only $\mathbf{x}$ and $u_1$, whereas the last parts involve $D^k u_1$ down through $Du_1$. As stated before, the appearance of any input derivative term must be avoided. The $\gamma$'s are selected in such a manner as to eliminate the coefficients of derivative terms. Thus, the $\gamma$'s are selected so that the coefficients of the derivative terms vanish:

$$\gamma_1 = \beta_k \tag{G.30}$$

$$\gamma_{j+1} = \beta_{k-j} - \sum_{i=1}^{j} \gamma_i \alpha_{n+i-j-1} \quad \text{for every } j = 1,2, \dots, k-1 \tag{G.31}$$

The final result can be stated as

$$\begin{aligned}
\dot{x}_1 &= x_2 \\
&\vdots \\
\dot{x}_{n-k} &= x_{n-k+1} + \gamma_1 u_1 \\
\dot{x}_{n-k+1} &= x_{n-k+2} + \gamma_2 u_1 \\
&\vdots \\
\dot{x}_n &= -\sum_{i=1}^{n} \alpha_{i-1} x_i + \left(\beta_0 - \sum_{j=1}^{k} \gamma_j \alpha_{n-k+j-1}\right) u_1
\end{aligned} \tag{G.32}$$

or, in matrix form;

$$\begin{bmatrix} \dot{x}_1 \\ \dot{x}_2 \\ \vdots \\ \dot{x}_{n-k} \\ \dot{x}_{n-k+1} \\ \vdots \\ \dot{x}_n \end{bmatrix} = \begin{bmatrix} 0 & 1 & 0 & \cdots & 0 \\ 0 & 0 & 1 & & 0 \\ & & & & \\ 0 & 0 & 0 & \ldots 1 & \ldots 0 \\ 0 & 0 & 0 & \ldots 1 & \ldots 0 \\ & & \vdots & & \\ -\alpha_0 & -\alpha_1 & -\alpha_2 & \cdots & -\alpha_{n-1} \end{bmatrix} \begin{bmatrix} x_1 \\ x_2 \\ x_3 \\ \vdots \\ \\ \\ x_n \end{bmatrix} + \begin{bmatrix} 0 \\ 0 \\ \vdots \\ \gamma_1 \\ \gamma_2 \\ \vdots \\ \beta_0 - \gamma_1 \alpha_{n-k} \quad \cdots \quad -\gamma_k \alpha_{n-1} \end{bmatrix} [u_1] \tag{G.33}$$

where the $\gamma$'s are given by

$$\begin{aligned}
\gamma_1 &= \beta_k \\
\gamma_2 &= \beta_{k-1} - \gamma_1 \alpha_{n-1} \\
\gamma_3 &= \beta_{k-2} - \gamma_1 \alpha_{n-2} - \gamma_2 \alpha_{n-1}
\end{aligned} \tag{G.34}$$

$$\gamma_4 = \beta_{k-3} - \gamma_1\alpha_{n-3} - \gamma_2\alpha_{n-2} - \gamma_3\alpha_{n-1}$$

$$\vdots$$

$$\gamma_k = \beta_1 - \gamma_1\alpha_{n-k+1} - \gamma_2\alpha_{n-k+2} - \cdots - \gamma_{k-1}\alpha_{n-1}$$

The initial conditions of $\mathbf{x}$ are found from the equations in the second column of the table on page 454 and are more complicated than before:

$$\begin{aligned}
x_1(0) &= z(0) \\
x_2(0) &= Dz(0) \\
&\vdots \\
x_{n-k+1}(0) &= D^{n-k}z(0) - \gamma_1 w(0^-) \\
&\vdots \\
x_n(0) &= D^{n-1}z(0) - \sum_{i=1}^{k}\gamma_i D^{k-i} w(0^-)
\end{aligned} \tag{G.35}$$

The initial conditions require a knowledge of the derivatives of the input at the instant before $t = 0$ (denoted $w(0^-)$). In some problems, *all* input derivatives can be considered to be zero until $t = 0$ and thus present no problem.

**Example G.2**

Consider the following fourth-order differential equation with second-order input derivatives:

$$[D^4 + 8D^3 + 7D^2 + 6D + 5]\, z = [4D^2 + 3D + 2]\, w \tag{G.36}$$

Suppose the initial conditions are

$$\begin{aligned}
z(0) &= 10 \\
Dz(0) &= 20 \\
D^2z(0) &= 30 \\
D^3z(0) &= 40
\end{aligned} \tag{G.37}$$

and the system is subjected to a step input of $w_0$.

The table for the state variable definitions and relations is as follows:

| Definitions of State Variables | $\mathbf{x}$ as Functions of $w$, $z$, $Dz$, etc. | $z$, $Dz$, etc., as Functions of $\mathbf{x}$ and $\mathbf{u}$ |
|---|---|---|
| $u_1 \triangleq w$ | | |
| $x_1 \triangleq z$ | $x_1 = z$ | $z = x_1$ |
| $x_2 \triangleq \dot{x}_1$ | $x_2 = Dz$ | $Dz = x_2$ |
| $x_3 \triangleq \dot{x}_2 - \gamma_1 u_1$ | $x_3 = D^2z - \gamma_1 w$ | $D^2z = x_3 + \gamma_1 u_1$ |
| $x_4 \triangleq \dot{x}_3 - \gamma_2 u_1$ | $x_4 = D^3z - \gamma_1 Dw - \gamma_2 w$ | $D^3z = x_4 + \gamma_1 Du_1 + \gamma_2 u_1$ |

From Eq. (G.36), we obtain

$$D^4z = -5z - 6Dz - 7D^2z - 8D^3z + 4D^2w + 3Dw + 2w \tag{G.38}$$

Substituting the expressions from the third column of the table into Eq. (G.38) yields

$$\begin{aligned} D^4z = &-5x_1 - 6x_2 - 7(x_3 - \gamma_1 u_1) + 8(x_4 + \gamma_1 Du_1 + \gamma_2 u_1) \\ &+ 4D^2u_1 + 3Du_1 + 2u_1 \end{aligned} \tag{G.39}$$

The expressions for $\dot{x}_1$, through $\dot{x}_3$ are given in the first column of the table; the expression for $\dot{x}_4$ is found by differentiating the last equation in the second column (Don't forget to differentiate the inputs!):

$$\dot{x}_4 = D^4z - \gamma_1 D^2u_1 - \gamma_2 Du_1 \tag{G.40}$$

After substituting Eq. (G.38) and rearranging, we find that

$$\begin{aligned} \dot{x}_4 = &-5x_1 - 6x_2 - 7x_3 - 8x_4 + (2 - 8\gamma_2 - 7\gamma_1)u_1 \\ &+ (4 - \gamma_1)\, D^2u_1 + (3 - \gamma_2 - 8\gamma_1)\, Du_1 \end{aligned} \tag{G.41}$$

The coefficients of $D^2u_1$ and $Du_1$ will vanish if we select

$$\gamma_1 = 4 \tag{G.42}$$

and

$$\gamma_2 = 3 - 8\gamma_1 = -29 \tag{G.43}$$

Thus, the matrix differential equations become

$$\dot{x} = \begin{bmatrix} 0 & 1 & 0 & 0 \\ 0 & 0 & 1 & 0 \\ 0 & 0 & 0 & 1 \\ -5 & -6 & -7 & -8 \end{bmatrix} \mathbf{x} + \begin{bmatrix} 0 \\ 4 \\ -29 \\ 206 \end{bmatrix} \mathbf{u} \tag{G.44}$$

with initial conditions

$$\mathbf{x}(0) \begin{bmatrix} 10 \\ 20 \\ 30 + 4w(0^-) \\ 40 + 4w(0^-) - 29w(0^-) \end{bmatrix} \tag{G.45}$$

From this formulation, the solutions for $z$, $Dz$, etc., can be calculated using

$$\begin{aligned} y_1 &\triangleq z = x_1 \\ y_2 &\triangleq Dz = x_2 \\ y_3 &\triangleq D^2z = x_3 + 4u_1 \end{aligned} \tag{G.46}$$

## G.5 CONVERSION FROM STATE-SPACE TO CLASSICAL FORMAT

In some cases, it will be required to convert a state-space format into a classical format. For a linear system, this can easily be accomplished with differential operator algebra. Thus, the state-space vector-matrix differential equation format

$$\dot{\mathbf{x}} = \mathbf{Ax} + \mathbf{Bu} \tag{G.47}$$

can be rewritten using operator notation as

$$[D\mathbf{I} - \mathbf{A}]\mathbf{x} = \mathbf{Bu} \tag{G.48}$$

For a linear system, the $[D\mathbf{I} - \mathbf{A}]$ term can be treated as a single algebraic matrix. Thus, this set of equations looks similar to a simple set of algebraic equations. We can solve for any state variable by a variety of methods, including matrix inversion, but Cramer's rule is a particularly attractive reduction technique for obtaining the classical differential equation in terms of any state variable. The initial conditions will be stated on the state variables and can be converted to the required classical initial conditions by successive eliminations in the original equations.

For example, consider the second-order system

$$\begin{bmatrix} (D - a_{11}) & -a_{12} \\ -a_{21} & (D - a_{22}) \end{bmatrix} \begin{bmatrix} x_1 \\ x_2 \end{bmatrix} = \begin{bmatrix} b_1 u_1 \\ b_2 u_1 \end{bmatrix} \tag{G.49}$$

in which the classical equation for $x_1$ is sought. By Cramer's rule,

$$x_1 = \frac{\begin{vmatrix} b_1 u_1 & -a_{12} \\ b_2 u_1 & (D - a_{22}) \end{vmatrix}}{\begin{vmatrix} (D - a_{11}) & -a_{12} \\ -a_{21} & (D - a_{22}) \end{vmatrix}} \tag{G.50}$$

or

$$x_1 = \frac{b_1 D u_1 + (b_2 a_{12} - b_1 a_{22}) u_1}{D^2 - (a_{11} + a_{22})D + (a_{11}a_{22} - a_{12}a_{21})} \tag{G.51}$$

Rearranging the latter equation to classical form yields

$$[D^2 - (a_{11} + a_{22})D + (a_{11}\, a_{22} - a_{12}\, a_{21})]x_1 = [b_1 D + (b_2 a_{12} - b_1 a_{22})]u_1 \tag{G.52}$$

The initial conditions for this system are $x_1(0)$ and $Dx_1(0)$ and can be obtained by solving Eq. (G.48) for state variables in terms of each other. We have:

$$x_1(0) = \text{known} \tag{G.53}$$

$$Dx_1(0) = a_{11}x_1(0) + a_{12}x_2(0) + b_1 u_1(0^-) \tag{G.54}$$

Equation (G.48) can be solved for $\mathbf{x}$ to yield

$$\mathbf{x} = [D\mathbf{I} - \mathbf{A}]^{-1}\,\mathbf{Bu} \tag{G.55}$$

Notice further that the outputs can be expressed as a function of the inputs $\mathbf{U}$:

$$\mathbf{Y} = \{[D\mathbf{I} - \mathbf{A}]^{-1}\mathbf{B} + \mathbf{E}\}\,\mathbf{u} \tag{G.56}$$

**Example G.3**

Consider the phase-space-defined state variables of a classical differential equation without derivatives of the input. Converting to the algebraic differential operator simultaneous equation form yields

$$\begin{bmatrix} D & -1 \\ \alpha_0 & (D + \alpha_1) \end{bmatrix} \begin{bmatrix} x_1 \\ x_2 \end{bmatrix} = \begin{bmatrix} 0 \\ \beta_0 u_1 \end{bmatrix} \tag{G.57}$$

Using Cramer's rule to solve for $x_1$, we obtain

$$x_1 = \frac{\begin{vmatrix} 0 & -1 \\ \beta_0 u_1 & (D + \alpha_1) \end{vmatrix}}{\begin{vmatrix} D & -1 \\ \alpha_0 & (D + \alpha_1) \end{vmatrix}} \tag{G.58}$$

or

$$x_1 = \frac{\beta_0 u_1}{D^2 + \alpha_1 D + \alpha_0} \tag{G.59}$$

In classical form;

$$[D^2 + \alpha_1 D + \alpha_0]x_1 = \beta_0 u_1 \tag{G.60}$$

This result is the familiar form. The following initial conditions would be given on $x_1$ and $x_2$:

$$x_1(0) = \text{known} \tag{G.61}$$

$$Dx_1(0) = x_2(0) \tag{G.62}$$

To convert $x_2(0)$ to $Dx_1(0)$, the first equation from Eq. (G.57) can be used.

## G.6 TRANSFER FUNCTIONS

The classical form of the linear differential equations discussed in this appendix can be expressed in terms of a transfer function as is typical in control theory. Using the differential operator $D = d/dt$, in place of the Laplace operator $s$, we can write

$$\frac{z}{w} = \frac{\beta_k D^k + \cdots + \beta_1 D + \beta_0}{\alpha_n D^n + \alpha_{n-1} D^{n-1} + \cdots + \alpha_1 D + \alpha_0} \tag{G.63}$$

Note that the proper form for a transfer function is the ratio of two polynomials in $D$ all with terms having positive exponents. Terms involving negative exponents (e.g., $D + 1 + (1/D)$) should be converted to the proper form.

As an example, Eq. (G.60) may be expressed as the transfer function

$$\frac{x_1}{u_1} = \frac{\beta_0}{D^2 + \alpha_1 D + \alpha_0} \tag{G.64}$$

Conversions such as those discussed in this appendix are implemented in the MATLAB functions `tf2ss` and `ss2tf`.

## PROBLEMS

**G.1** *How* many state inputs **u** and state variables **x** are required to describe the following differential equations?

**a.** $[D^4 + D^3 + D^2 + D + 1]P = P_0(t)$

**b.** $\dfrac{d^4z}{dt^4} + \dfrac{5d^2z}{dt^2} + 10z = 0$

**c.** $\ddot{z} + 40\dot{z} + 100z = 10w(t) + 50v(t)$

**d.** $\dddot{z} + \ddot{z} + \dot{z} = w(t)$

**G.2** *Define* a set of state variables, and *convert* the following classical differential equations to state-space equations including initial conditions:

**a.** $\ddot{z} + 2\dot{z} + 25z = 25w$

with $z(0) = -1.0$ and $\dot{z}(0) = 0.0$

**b.** $\dfrac{2d^2e}{dt^2} + \dfrac{de}{dt} + 15e = 4e_0(t)$

with $e(0) = 0, \dot{e}(0) = 0$

**G.3** *Define* state variables and *convert* the following equation to state-space form:

$$D^4z + 3D^3z + 4D^2z + 5Dz + 6z = 10\,v(t)$$

$$\text{with } z(0) = 20,\ Dz(0) = 0,\ D^2z(0) = 50,\ D^3z(0) = 0$$

**G.4** *Convert* the following equation to state-space form:

$$\dddot{p} + a\ddot{p} + b\dot{p} + cp = q(t)$$

$$\text{with } p(0) = 1,\ \dot{p}(0) = 2,\ \ddot{p}(0) = 0$$

**G.5** *Convert* the equation

$$\ddot{g} + 5\dot{g} + 15g = 10 \sin t$$
$$\text{with } g(0) = 0, \ \dot{g}(0) = 0$$

to state-space form. *Define* all variables.

**G.6** *Convert* the equation

$$\ddot{g} + 2\dot{g} + 20g = 10w(t) + v(t)$$

to state-space form. *Define* all variables.

**G.7** *Convert* the nonlinear differential equation

$$\frac{d^2z}{dt^2} + 5\frac{dz}{dt} + z^2 = w(t)$$
$$\text{with } z(0) = 2.0, \ \dot{z}(0) = 0$$

to state-space form. Do not use matrix notation.

**G.8** *Convert* the nonlinear equation

$$\dddot{x} + \ddot{x} + \dot{x}x + x = p(t)$$
$$\text{with } x(0) = 10, \ \dot{x}(0) = 0, \ddot{x}(0) = 100$$

to state-space form.

**G.9** *Define* state variables and *obtain* the state-space format for the following equation with input derivatives (assume that $u(t)$ is a step input):

$$\ddot{z} + 0.1\dot{z} + 0.5z = 2\dot{u} + u$$
$$\text{with } z(0) = -1, \ \dot{z}(0) = 10$$
$$\text{and } u(t) = 0.0 \text{ for } t < 0$$
$$u(t) = 1.0 \text{ for } t \geq 0$$

**G.10** *Define* and *obtain* the state-space format for the equation

$$\frac{d^2p}{dt^2} + 7\frac{dp}{dt} + 2p = 3\frac{da}{dt}$$
$$\text{with } p(0) = 10, \ \frac{dp(0)}{dt} = 0$$

**G.11** *Define* and *obtain* the state-space format for the equation

$$[D^4 + 0.5D^3 + 3D^2 + 2D + 5]z = [2D^2 + 7D + 2.5]w$$
$$\text{with all initial conditions equal to zero}$$

**G.12** *Define* and *obtain* the state-space format for the equation

$$[D^3 + .2D^2 + .3D + .8]e = [5D^2 + 7D + 9]v$$

with zero initial conditions and a continuous sinusoidal input of frequency $\omega$.

**G.13** Given

$$\dot{\mathbf{x}} = \begin{bmatrix} 0 & 1 \\ -2 & -5 \end{bmatrix}\mathbf{x} + \begin{bmatrix} 7 \\ 3 \end{bmatrix}\mathbf{u}$$

$$\mathbf{x}(0) = \begin{bmatrix} x_1(0) \\ x_2(0) \end{bmatrix}$$

*obtain* a classical differential equation, including all initial conditions:

**a.** For $x_1$ being the dependent variable.

**b.** For $x_2$ being the dependent variable.

**G.14** Given

$$\dot{\mathbf{x}} = \begin{bmatrix} 0 & 2 & 0 \\ -3 & 0 & -4 \\ 0 & 5 & -1 \end{bmatrix} \mathbf{x} + \begin{bmatrix} 0 \\ 2 \\ -3 \end{bmatrix} \mathbf{u}$$

$$x(0) = \begin{bmatrix} x_1(0) \\ x_2(0) \\ x_3(0) \end{bmatrix}$$

*obtain* a classical differential equation for $x_1$, including all initial conditions.

# APPENDIX H

# Digital Simulation Using Matlab

## H.1 INTRODUCTION

This appendix is intended to provide you with some assistance in using MATLAB™ for dynamic systems analysis and design. MATLAB is ideally suited to that task and has become one of the most widely used computational tools for the purpose.

For the preparation of the example problems in this text, we used the student edition of MATLAB running under WINDOWS™ on a PC; our comments, however, are not necessarily restricted to that environment, since MATLAB is available for a large number of computer systems and has a great degree of platform independence.

With MATLAB, you can:

1. Define **variables,** enter their numerical values, and perform calculations interactively.
2. Create **vectors** and **matrices,** and perform operations on them using the many built-in MATLAB functions.
3. **Visualize** your calculated results by plotting them directly on the computer screen and capturing a digital image or printing a copy for use in documenting your work.
4. Use the built-in MATLAB programming language to create a calculation **procedure** consisting of a number of related steps.
5. Use a text editor to write and save a **script file,** which contains a specific series of MATLAB statements that you want to use in the future. Script files are saved with the file extension `m` (i.e., as `name.m`), and are called M-files.

6. Utilize script files to define **functions** that are of use to you in future work.
7. Employ the **toolboxes** of functions developed by others and made available as script files. Of particular interest is the **Signals and Systems Toolbox,** which contains many useful functions for calculating and plotting the results of dynamic system analysis.

We have used MATLAB in calculating the results and preparing the graphics for this text, and MATLAB commands are included in the description of example problems in a number of chapters. The commands have been presented quite casually and without much of an introduction because we find the system easy to use and believe that you will, too, and will adapt to its functioning quite naturally.

## H.2 MATLAB ENVIRONMENT

The MATLAB environment provides you with an interactive work space in which to enter commands and view results. The prompt for the educational version is EDU >> . For the professional version, it is simply >>.

In the Windows PC configuration, you are offered a row menu providing *File, Edit, Options, Windows,* and *Help* options. Selecting items from among these alternatives by clicking a mouse will be familiar to users of Windows or similar systems. You can set the format used for displaying numerical values by selecting one of the items under *Options*.

A number of control-type commands are available from the command file, including the following:

| | |
|---|---|
| `cd` | Change the current working directory |
| `delete` | Delete file |
| `diary` | Save your MATLAB session as a text file |
| `dir` | List the current directory |
| `!` | Execute an operating system command |
| `quit` | Terminate MATLAB |

## H.3 VARIABLES AND STATEMENTS

Scalar, vector, and matrix quantities can be defined in MATLAB using an assignment statement as you would in a programming language. Some examples are as follows:

| | |
|---|---|
| `a = 14.7` | defines a **scalar** |
| `r = [1 2 3 4]` | defines a **row vector** |
| `c = [1.2 3.2 7.5 5.6]'` | defines a **column vector**. The `'` means "transpose." |

`b = [1 2 3; 4 5 6; 7 8 9]` defines a 3 x 3 **matrix.** Rows are separated by a semicolon. (The example is singular, by the way.)

An array element can be referenced directly, e.g., **`c`**`(2) = 3.4` or **`b`**`(3,2) = 8`.

Built in are the usual required mathematical functions such as ***sin, cos, tan, atan,*** etc. as well as the constant ***pi.*** In addition, a large number of array functions are provided to facilitate operations on matrices. These include array ***addition***, ***subtraction***, ***multiplication***, ***inversion***, ***extraction of eigenvalues,*** etc.

The MATLAB **programming language** supports **loops, branches,** and **functions**. For example, we can fill a row vector with the integers 1 to 10 by writing

```
for i = 1:10
  v(i) = i
end
```

Statements ending with a semicolon do not print the result of the calculation on the screen, so we would probably want to end each of the preceding statements with a semicolon. The command `v` would then display the vector, so we could see whether we obtained the correct results. Try it both ways and see.

Loops can also be nested, as in the following example:

```
for i = 1:n;
  for j = 1:m;
    b(i,j) = i*j;
  end
end
```

In addition, **branch control** statements may be included:

```
if j > 10
  q = 100;
else
  q = 0;
end
```

## H.4 SCRIPT FILES AND FUNCTIONS

A valid collection of MATLAB commands can be created and saved as a text file for later use. Such files are saved with the extension `m` and are referred to as **M-files.** They can be executed from the MATLAB environment by knowing their file name and the directory in which they reside.

Suppose you want to print a heading on the screen at the beginning of every calculation you perform. You can create a text file called `heading.m` that contains the commands necessary to do this. In the following file, we use the MATLAB command `disp`, which displays a text string on the screen. The file, *heading.m,* contains the following three lines:

```
disp('XYZ Corp., Inc.')
disp('J. Jones, Senior Engineer')
%This is a comment line.
```

A collection of MATLAB commands can be saved in an M-file as a named **function** and can be invoked from another procedure. For example, suppose you wish to calculate repeatedly the area of a circle of radius $r$. You might then use an M-file to define the following function and save it with the name *c_area.m*:

```
function area = c_area(r)
area = pi * r*r
```

This function can be used by invoking its name and supplying it with a numerical value for the radius. For example, the following commands would calculate the area of a circle with a radius of 2.75 units:

```
radius = 2.75
surface = c_area(radius)
```

We can also pass an array as a function argument, and if several outputs are calculated by the function, we include a function output list in the function definition as follows:

function [*output variable list*] = function_name (*input variable list*)

Functions are used in the examples in a number of chapters of the text to describe system differential equations and to provide numerical integration routines. For example, the fourth-order, fixed-step-size Runge-Kutta integration procedure is implemented in the file `rk4.m` shown in the next section. While a few advanced language options are used in this M-file, the basic ideas are the same as the simple case given before.

## H.5 FOURTH-ORDER RUNGE-KUTTA PROCEDURE

The following is an M-file that contains the fourth-order Runge-Kutta integration routine:

```
function [tout, xout] = rk4(xdotfun, t0, tfinal, x0, h,
trace)
```

```
% Simple, fixed step size, Fourth order Runge-Kutta
% numerical integration of a system of first order
% ordinary differential equations.
% Modeling and Simulation of Dynamic Systems
% Robert L. Woods & Kent L. Lawrence

% Inputs:
%   xdotfun - A string containing the name of the m-file
%             which defines the system variable
%             derivatives, xdot = f(t,x).
%   t0      - Initial value of time (independent
%             variable).
%   tfinal  - Final time.
%   x0      - Vector of initial values for the dependent
%             variables x.
%   h       - Fixed integration step size.
%   trace   - If trace is 1, intermediate values printed
%             from rk4.m. If trace is 0 or not specified,
%             values are not printed in rk4.m.
% Outputs:
%   tout    - Vector of time points at which solution is
%             tabulated.
%   xout    - Array of calculated values. Each column
%             is the time history of a dependent
%             variable.
% See Example 9.6 for a sample of use.
% Watch for use of the transpose operator, '.

% Initialize variables
  t = t0;
  x = x0;
  tout = t;
  xout = x0';

  if nargin < 6
    trace = 0;
  end

% Set aside storage
  A = zeros(1)*x;
  B = zeros(1)*x;
  C = zeros(1)*x;
  D = zeros(1)*x;

  h2 = h/2;
```

```
  if trace
    clc, t, x
  end

% Integration loop
  while(t < tfinal)

    if t + h > tfinal
      h = tfinal - t;
    end

    A = feval(xdotfun, t, x)';
    B = feval(xdotfun, t+h2, x+h2*A)';
    C = feval(xdotfun, t+h2, x+h2*B)';
    D = feval(xdotfun, t+h, x+h*C)';

    t = t + h;
    x = x + h*(A + 2*B + 2*C + D)/6;

    tout = [tout; t];
    xout = [xout; x'];

    if trace
      clc, t, x
    end

  end;
```

## H.6 PLOTTING YOUR RESULTS

MATLAB provides a rich set of functions for generating a graphic representation of your calculated results. Visualization is one of the engineer's most useful tools in the design and analysis of dynamic systems, and the MATLAB commands give you an interactive way to visualize your work. Many of the Signals and Systems Toolbox functions are M-files that create plots automatically. If you have two one-dimensional vectors, **t** and **v**, that are of the same length, you can create a plot of **v** vs. **t** simply by issuing the command

```
plot(t, v)
```

You can add asterisk symbols on the curve by typing

```
plot(t, v, '*')
```

If there are several dependent variables, each contained in a column of a matrix, you can plot all of them with the command

```
plot(t, b)
```

And you can put a grid on the plot by using

```
grid
```

If you want to plot only the second variable, (second column), you can do that with

```
plot(t, b(:2))
```

Titles and labels for figures are added using commands such as the following:

```
xlabel('Time(sec)')
ylabel('Velocity (ft/sec)')
title('Rocket trajectory')
```

You can put the current graphic display on `hold` and issue commands to add features to it. Many additional graphics features are available in MATLAB including three-dimensional and color plotting capabilities.

## H.7 SIGNALS AND SYSTEMS TOOLBOX

A number of toolboxes are available for use with MATLAB, enhancing its use for specific tasks. The Signals and Systems Toolbox is most helpful in analyzing various aspects of dynamic systems.

Next we list some of the functions that may be called directly from the MATLAB environment. For example, suppose we want to find the roots of the polynomial

$$1.2s^2 + 3s + 15.7 = 0$$

We first create a row vector **v** containing the polynomial coefficients in descending order of powers of $s$:

```
v = [1.2 3 15.7]
```

The MATLAB command `roots` calculates the roots of this polynomial for us:

```
roots(v)
```

The function `damp(v)` first finds the roots and then determines the frequency and damping of second-order components and/or the inverse of the time constant of first-order components. (See Chapter 8 for additional details.)

The Signals and Systems Toolbox functions will accept the definition of your linear dynamic system in a number of ways. Working from the classical form of the governing equation, the Laplace transform provides us with the system **transfer function** as the ratio of two polynomials in $s$, as in the following example:

$$\mathrm{TF}(s) = \frac{1.5s^2 + 40s + 1200}{2.6s^4 + 130s^3 + 2100s^2 + 4800s}$$

We define the vectors **num** and **den** containing the coefficients of the **numerator polynomial** and the **denominator polynomial,** arranged in descending order of powers of $s$:

```
num = [1.5 40 1200]
den = [2.6 130 2100 4800 0]
```

(Note that the constant term in the denominator is zero.)

To determine the **Bode** plot for this system, simply enter the command

```
bode(num, den)
```

The result will be displayed on your computer screen.

The linear system may also be written in state-space format. We can transform the equations from transfer function format to the state-space format to obtain:

$$\dot{\mathbf{x}} = \mathbf{Ax} + \mathbf{Bu}$$

$$\mathbf{y} = \mathbf{Cx} + \mathbf{Du}$$

where $\mathbf{y}$ is the vector of desired outputs.

If we supply numerical values for the arrays **A, B, C,** and **D,** we can determine the frequency response plot by entering

```
bode(A, B, C, D)
```

You can find out more about MATLAB options and the syntax of MATLAB commands by using the on-line help and the references listed at the end of this appendix.

Following is a list of selected MATLAB functions that are of interest to us:

`roots` Polynomial roots
`eig` Matrix eigenvalues and eigenvectors
`damp` Roots, frequencies, and damping
`poly` Characteristic polynomial

| | |
|---|---|
| `bode` | Bode plots |
| `rlocus` | Root locus plot |
| `residue` | Partial fraction expansion |
| `step` | Step response calculation |
| `ode23, ode45` | Variable-step Runge-Kutta integration |
| `tf2ss` | Transfer function to state-space conversion |
| `tf2zp` | Transfer function to zero-pole conversion |
| `ss2tf` | State-space to transfer function conversion |
| `ss2zp` | State space to zero-pole conversion |

## H.8 HELP!

On-line help is available within the MATLAB environment. A menu provides a **table of contents** as well as an **index**. You also can search for a particular topic. It is a good idea to keep a help file window in view as you work from the command line; that way, you can utilize the provided help instructions that are provided.

Help is also available from the command line by typing `help` *topic* or just `help`. The first provides help on the topic named; the second offers a list of topics you may peruse.

Additional imformation is available in the many reference books on MATLAB. See those listed next, for example, and check your local library and bookstores for additional sources.

## REFERENCES

H.1 *The Student Edition of MATLAB, Version 4, User's Guide*. Prentice Hall, Inc., Englewood Cliffs, NJ, 1995.

H.2 Ogata, Katsuhiko. *Solving Control Engineering Problems with MATLAB*. Prentice Hall, Inc., Englewood Cliffs, NJ, 1994.

H.3 Pratap, Rudra. *Getting Started with MATLAB*. Harcourt Brace College Publishing, Fort Worth, TX, 1996.

# APPENDIX I

# Analog Simulation

## I.1 INTRODUCTION

The **electronic analog computer** consists of a set of computing components that perform the functions of addition, scaling, integration, etc., required in representing differential equations [Refs. I.1, I.2]. The voltages of the analog computer represent the continuous solution of the differential equation programmed by the interconnection of the components (functions). The differential equation is a model of a physical (or natural) dynamic system. An analogy is made between the physical variables of the model and the voltages that represent them on the computer—thus the term "analog" computer. The analogy defines scale factors relating the units of the physical variables to volts in the computer variables. Since the solution of the physical system may be faster or slower than desired, the computer solution can be scaled in both time and magnitude.

The underlying concept of analog computation is to interconnect or wire together those components (summers, scalers, integrators, etc.) required to represent the differential equations of interest. As discussed in subsequent sections, these differential equations can be expressed in classical form (a single $n$th-order differential equation) or state-space form ($n$ first-order differential equations), although both representations will result in essentially the same wiring diagram. It is easy to perform a single integration at a time on the analog computer, so the state-space format is a more natural representation of dynamic systems for analog computation. Therefore, the setup procedure for analog computation will be developed in the state-space format; the classical approach, however, will be briefly illustrated as well.

## I.2 COMPUTING COMPONENTS

The basic computing element of the analog computer is a high-gain differential DC amplifier with feedback, called an operational amplifier or "op-amp." The op-amp has extremely high input impedance, low output impedance, and high gain. The input impedance $Z_{in}$ is on the order of $10^6$ ohms, so negligible input current is required into the amplifier. The output impedance $Z_{out}$ is on the order of 100 ohms so the loading effects upon the output voltage are also negligible. The high gain $G$ of the amp (on the order of $10^6$ volts/volt), accompanied by a sign change or inversion, is required in the computing equation derived shortly. (See Section 4.4.1 for a discussion of op-amps.)

The op-amp (see Figure I.1a) open-loop transfer equation (input-output relationship) is

$$e_o = -Ge_a \tag{I.1}$$

where $e_a$ is the amplifier input voltage, $G$ is the gain, and $e_o$ is the output voltage.

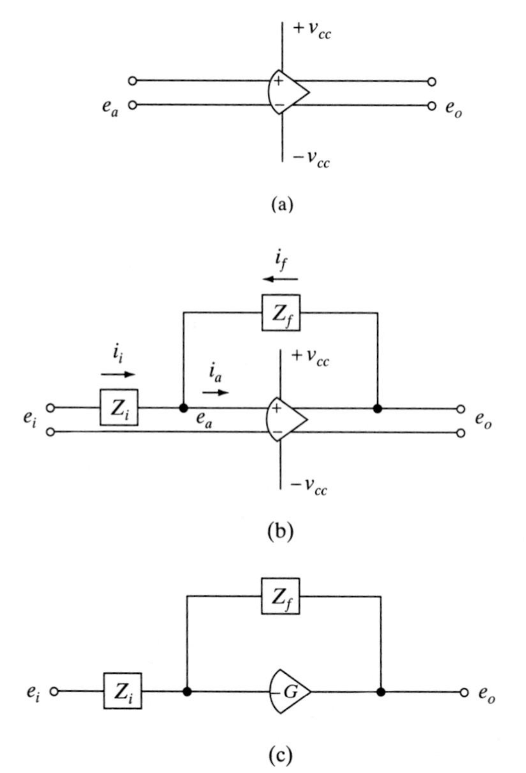


**Figure I.1** Representations of an operational amplifier circuit. (a) Isolated op-amp. (b) Op-amp with input and feedback impedances. (c) Simplified schematic of analog computer circuit.

The op-amp is always used with input and feedback impedances, as shown in Figure I.lb. Since a common ground line is always present in electronic analog computer circuits, and the op-amp is normally supplied with positive and negative voltage sources, these details can be omitted to reduce clutter in the circuit, as shown by Figure I.1c.

The circuit of Figure I.1b can be analyzed by the following component equations (neglecting the output impedance of the amplifier):

$$i_i = \frac{e_i - e_a}{Z_i} \tag{I.2}$$

$$i_f = \frac{e_o - e_a}{Z_f} \tag{I.3}$$

The node equation at the amplifier may be simplified by taking into account the fact that $I_a$ is negligibly small due to the high input impedance of the op-amp and the low input voltage to the amp:

$$i_i + i_f = i_a \approx 0 \tag{I.4}$$

Substituting the component equations into the node equation yields

$$e_o = -\frac{e_o}{G}\left(1 + \frac{Z_f}{Z_i}\right) - \left(\frac{Z_f}{Z_i}\right)e_i \tag{I.5}$$

after some algebraic rearrangement.

Since the gain $G$ is extremely large compared to $1 + (Z_f/Z_i)$, the first term can be neglected, yielding the basic computing equation

$$e_o = -\frac{Z_f}{Z_i}e_i \tag{I.6}$$

This equation indicates that the closed-loop gain between input and output is $Z_f/Z_i$, accompanied by a sign change. Different combinations of the types of circuit elements used for $Z_f$ and $Z_i$ give rise to the various computing components discussed next. (See Table I.1 for exact circuits and nomenclature.)

### I.2.1 Inverter or Sign Change

If both the feedback and the input impedances are resistive and of equal magnitude, (e.g., $Z_f = R_f$, $Z_i = R_i$, where $R_f = R_i$), then the basic computing equation reduces to a sign change or inverter:

$$e_o = -e_i \tag{I.7}$$

An **inverter** is commonly used to change the sign of a signal or variable; it is often used for subtraction in a summing junction.

### I.2.2 Scaler or Fixed-Gain Amplifier

If both impedances are resistive, but not necessarily of equal magnitude, the basic computing equation reduces to a fixed-gain relation, or scaling, between the input and output:

$$e_o = -\frac{R_f}{R_i}e_i \tag{I.8}$$

**TABLE I.1** BASIC ANALOG COMPUTER OPERATIONS.

| Operation | Computer symbol | Computer circuit |
|---|---|---|
| Amplification (with sign change)<br>$e_0 = -Ge_1$ | 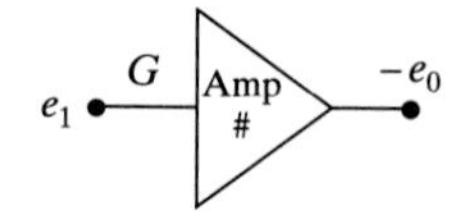 $e_1$, $G$, Amp #, $-e_0$ | 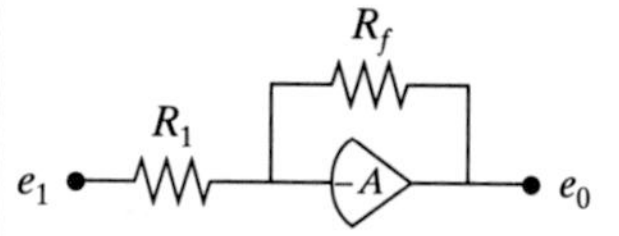 $e_1$, $R_1$, $R_f$, $A$, $e_0$<br>$G = \frac{R_f}{R_1}$ |
| Summation (only two inputs shown)<br>$e_0 = -G_1e_1 - G_2e_2$ | 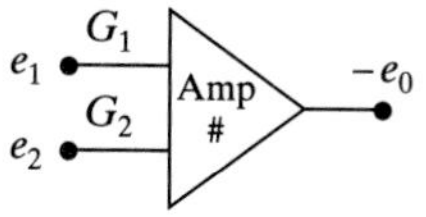 $e_1$, $G_1$, $e_2$, $G_2$, Amp #, $-e_0$ | 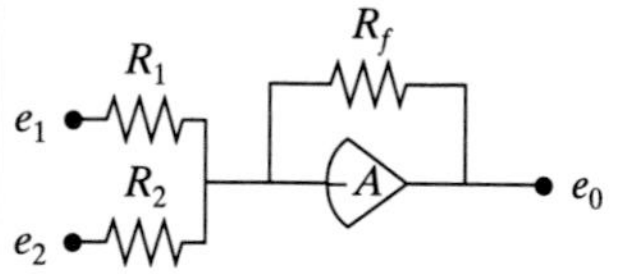 $e_1$, $R_1$, $e_2$, $R_2$, $R_f$, $A$, $e_0$<br>$G_1 = \frac{R_f}{R_1}$ $G_2 = \frac{R_f}{R_2}$ |
| Integration (shown with summation)<br>$e_o = -\int_0^t (G_1e_1 + G_2e_2)\, dt - \text{I.C.}$ | 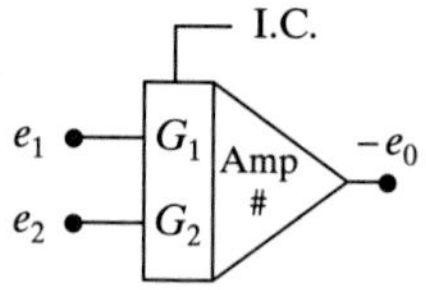 I.C., $e_1$, $G_1$, $e_2$, $G_2$, Amp #, $-e_0$ | 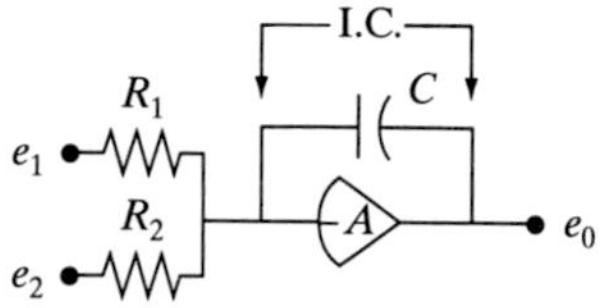 I.C., $C$, $e_1$, $R_1$, $e_2$, $R_2$, $A$, $e_0$<br>$G_1 = \frac{1}{R_1C}$ $G_2 = \frac{1}{R_2C}$ |
| Multiplication (by a constant < 1.0)<br>$e_0 = ke_1$<br>$0 < k < 1.0$ | 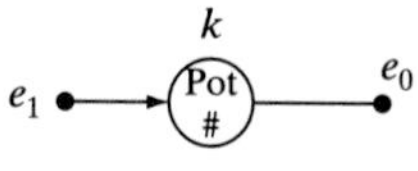 $e_1$, $k$, Pot #, $e_0$ | 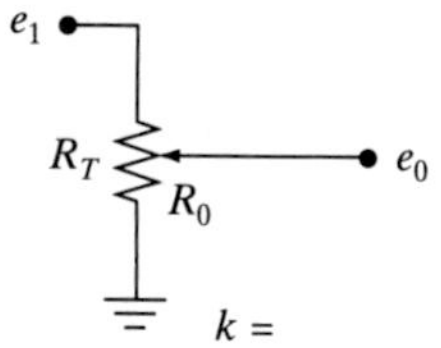 $e_1$, $R_T$, $R_0$, $e_0$<br>$k = \frac{R_0}{R_T}$ |
| Multiplication (by a constant > 1.0)<br>$e_0 = -10ke_1$ | 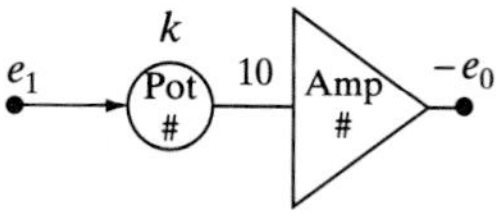 $e_1$, $k$, Pot #, 10, Amp #, $-e_0$ | 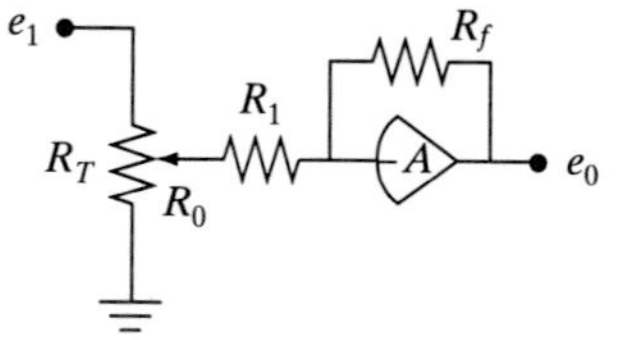 $e_1$, $R_T$, $R_0$, $R_1$, $R_f$, $A$, $e_0$<br>$k = \frac{R_0}{R_T}$ $10 = \frac{R_f}{R_1}$ |

This equation represents multiplication by a constant or simulates a coefficient in the differential equation. Most analog computers utilize fixed, or hard-wired, resistors in multiples of ten for scaling; thus, gains of 0.1, 1.0, 10.0, or 100.0 are possible from typical fixed-gain amplifiers for determining the order of magnitude of coefficients. The setting of precise values for the coefficients is done with potentiometers.

### I.2.3 Potentiometer

A variable electronic resistive voltage divider, or "pot," is a passive element that can be used to adjust the exact numerical value of the coefficients of the differential equation. Mathematically,

$$e_o = ke_i \quad \text{where} \quad 0 \le k \le 1.0 \tag{I.9}$$

Notice that the gain of a pot is less than unity, so that a fixed-gain amplifier with a gain of 10.0 or 100.0 is required in conjunction with a pot to obtain coefficients greater than unity. A bank of several pots is always supplied with analog computers to allow system coefficients to be adjusted.

### I.2.4 Summing Junction

Using more than one input resistor allows the addition, or summation, of several independent input signals. The computing equation reduces to a linear summation of the signals, with the possibility of different gains (usually in multiples of 10) for each signal:

$$e_o = -\frac{R_f}{R_1}e_1 - \frac{R_f}{R_2}e_2 - \frac{R_f}{R_3}e_3 \tag{I.10}$$

Summing junctions are ordinarily used to sum together the various terms required to form the differential equation.

### I.2.5 Integrator

The use of a capacitive feedback impedance with a resistive input impedance results in the integration function. The transfer gain can be stated, in operator notation, with $D = d/dt$, as

$$e_o = -\frac{1}{R_i C_f D} e_i \tag{I.11}$$

Writing this equation in integral form and considering the initial condition $e_o(0)$ yields

$$e_o(t) = -\left[\frac{1}{R_i C_f}\int_0^t e_i\,dt + e_o(0)\right] \tag{I.12}$$

The integration gain is $1/(R_i C_f)$. Note that the "dynamic gain" of an integrator has a different meaning than the "static gain" of a fixed-gain amplifier, since the integrator's dynamic gain specifies a frequency-dependent value, whereas the static gain of an amplifier is independent of frequency (over the range of utility of the device). The sign change across the amplifier requires that the initial condition be entered with a sign change. The initial conditions of integrators are usually automatically switched in and out with a relay. Notice that by using multiple input resistors, a summing integrator is possible (and is usually provided).

### I.2.6 Differentiator

If a resistive feedback impedance and a capacitive input impedance are used, the resulting equation implies the function of differentiation. Specifically,

$$e_o = -R_f C_i D e_i \tag{I.13}$$

or

$$e_o = -R_f C_i \frac{de_i}{dt} \tag{I.14}$$

The gain of the differentiator is a dynamic gain that increases with frequency, whereas the dynamic gain of an integrator decreases with frequency. Differentiators amplify high-frequency signals more than low-frequency signals and hence produce more noise on the output. For this reason, differentiators are seldom used in analog computation.

All of the preceding components are commonly available on most analog computers. Some machines contain multipliers for the multiplication of two signals, nonlinear diode function generators, and other nonlinear functions such as limiting, deadband, relay, hysteresis, and Boolean logic functions. These nonlinear functions are quite significant and should not be ignored. However, it is the purpose of this presentation to discuss the organization and scaling of problems for the analog computer. Little difficulty should be encountered in utilizing them once the scaling problem is solved, toward which end this appendix is directed.

Table I.1 illustrates the electronic circuit and the simplified analog computer block diagram symbol for each of the common linear operations just discussed. The block diagram symbols are used throughout the remainder of this appendix in drawing analog computer diagrams.

## I.3 CLASSICAL ANALOG COMPUTER DIAGRAMS

The goal of classical analog computation is to solve the classical differential equation for the highest-order derivative of an $n$th-order equation and integrate $n$ times to yield the solution. The equation that results for the $n$th derivative is formed at the input of the first integrator by summing the lower order derivatives with appropriate coefficients (gains).

For example, consider the second-order differential equation

$$\ddot{e} + 0.5\dot{e} + 2.0e = v(t) \tag{I.15}$$

with initial conditions

$$e(0) = e_{int} \tag{I.16}$$

and

$$\dot{e}(0) = \dot{e}_{int} \tag{I.17}$$

Solving for the highest order derivative yields

$$\ddot{e} = v(t) - 0.5\dot{e} - 2.0e \tag{I.18}$$

Assuming that this equation is formed at the input of an integrator, the output would be $-\dot{e}$, and the initial condition $\dot{e}(0)$ would have the opposite sign of $-\dot{e}_{int}$, as shown in Figure I.2. An integration of the $-\dot{e}$ signal yields $e$. The initial condition of $e$ would be entered with the opposite sign of $+e$.

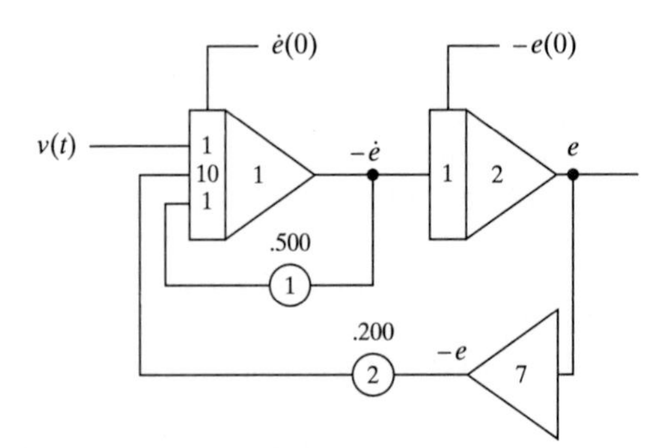


**Figure I.2** Classical analog computer diagram.

Now the equation for $\ddot{e}$ must be wired at the input to integrator number 1. Scaling $-\dot{e}$ with a pot set at 0.500 yields one term. The $e$ signal must be inverted, scaled by 0.200, and then entered into the summing integrator with a gain of 10 to obtain the proper sign and coefficient of the term. The system input $v(t)$ must be available from a function generator or some similar source, depending upon the input signal. Summation of these three terms with the gains shown yields Eq. (I.18) for $\ddot{e}$.

The classical approach to analog computer diagrams is not pursued further here, as the state-space approach appears more natural and therefore will be discussed in the remainder of the chapter.

## I.4 STATE-SPACE ANALOG COMPUTER DIAGRAMS

As mentioned earlier, the analog computer is designed to conveniently perform single integrations at a time; thus, having an $n$th-order dynamic system model of $n$ first-order equations in state-space format is advantageous, since single integrations of the derivative state vector $\dot{\mathbf{x}}$ yields the solution of the system. The equations for the derivatives of each of the state variables are formed or wired at the input of the integrators. This format is further attractive, since the coefficients in the **A** and **B** matrices will result in integrator gains when scaled in magnitude and time. Note that magnitude scaling can be handled with matrix multiplication.

Consider a state-space system (with a single input) that has not been magnitude scaled, namely,

$$\dot{\mathbf{x}} = \mathbf{Ax} + \mathbf{Bu} \tag{I.19}$$

or

$$\begin{bmatrix} \dot{x}_1 \\ \dot{x}_2 \\ \vdots \\ \dot{x}_n \end{bmatrix} = \begin{bmatrix} a_{11} & a_{12} & \cdots & a_{1n} \\ a_{21} & a_{22} & \cdots & a_{2n} \\ \vdots & & & \vdots \\ a_{n1} & a_{n2} & \cdots & a_{nn} \end{bmatrix} \begin{bmatrix} x_1 \\ x_2 \\ \vdots \\ x_n \end{bmatrix} + \begin{bmatrix} b_1 \\ b_2 \\ \vdots \\ b_n \end{bmatrix} [u_1] \tag{I.20}$$

with

$$\mathbf{x}(0) = \begin{bmatrix} x_1(0) \\ x_2(0) \\ \vdots \\ x_n(0) \end{bmatrix} \tag{I.21}$$

This system is simulated by the analog computer diagram shown in Figure I.3. Notice that the interconnecting lines have been omitted for simplicity and that all coefficients have been assumed to be positive. (In actuality, most of the coefficients will be zero.)

Now, the input to each integrator forms the equation of the derivative of a state variable, and the output will be that state variable with a sign change; an inverter often (but not always) will be required for computation or monitoring. Notice that the initial condition of an integrator must be entered with a sign opposite to that of the output, which in this case allows the initial conditions to be entered with the same sign as that of the physical problem.

For example, consider the second-order system of Eq. (I.15), which is used to illustrate the classical analog computer diagram. This system can be converted to state-space form, yielding

$$\begin{bmatrix} \dot{x}_1 \\ \dot{x}_2 \end{bmatrix} = \begin{bmatrix} 0 & 1.0 \\ -2.0 & -0.5 \end{bmatrix} \begin{bmatrix} x_1 \\ x_2 \end{bmatrix} + \begin{bmatrix} 0 \\ 1.0 \end{bmatrix} [u_1] \tag{I.22}$$

with

$$\mathbf{x}(0) = \begin{bmatrix} x_1(0) \\ x_2(0) \end{bmatrix} \tag{I.23}$$

The latter system is programmed on the analog computer as shown in Figure I.4. Notice that the solution for $x_1$ is the output of amp 7, and the solution for $x_2$ is the output of amp 8. The equation for $\dot{x}_2$ contains $-2.0x_1$, that was obtained with a pot set at .200 and an integrator gain of 10. With a slight rearrangement, Figure I.4 is observed to be the same as Figure I.2, which should obviously be the case.

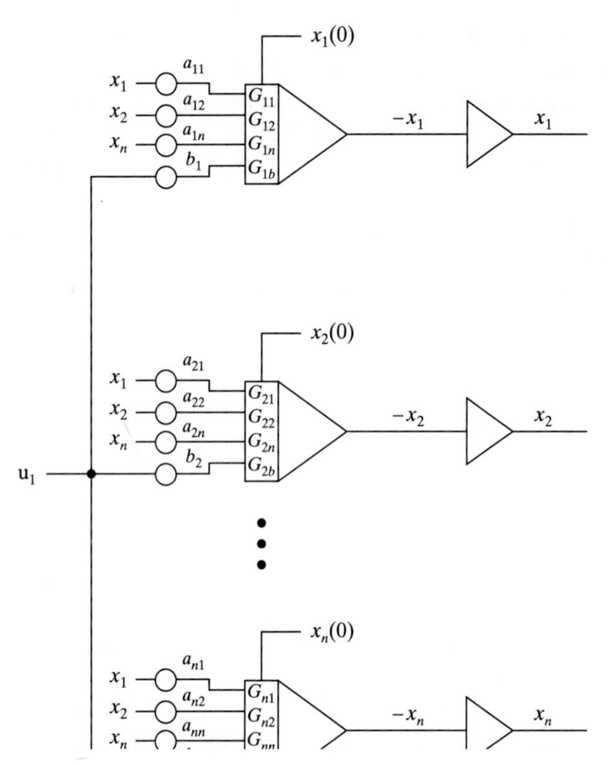


**Figure I.3** General state-space analog computer diagram.

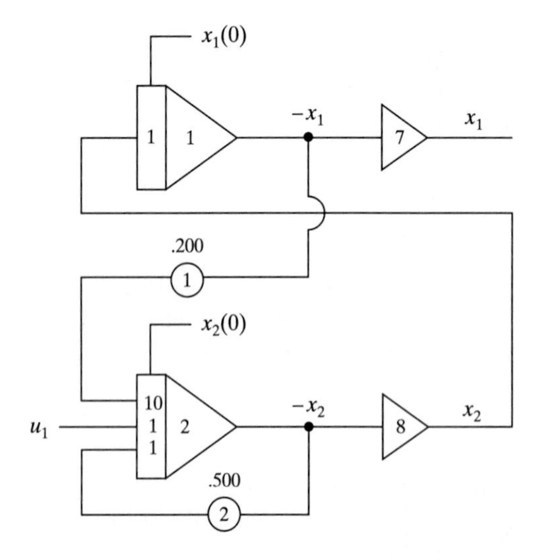


**Figure I.4** State-space analog computer diagram for a second-order system.

## I.5 MAGNITUDE AND TIME SCALING

The purpose of analog computation is to develop an analogy between the physical system variables and the voltages that represent those variables on the analog computer; thus, a conversion must be made from the physical variables to the analog computer variables. The analogy provides these conversion factors, or magnitude scale factors. A magnitude scale factor must be associated with each state variable. Note that this scaling must be done in either the classical or the state-space format. It is easily handled in the state-space format with scaling matrices.

### I.5.1 Magnitude Scaling

Consider the set of $n$th-order linear state-space equations with $m$ inputs, denoted

$$\dot{\mathbf{x}} = \mathbf{A}\mathbf{x} + \mathbf{B}\mathbf{u} \tag{I.24}$$

with initial conditions

$$\mathbf{x}(0) = \mathbf{x}_{int} \tag{I.25}$$

where $\mathbf{x}$ and $\mathbf{u}$ are the variables and input vectors representing the physical system. Any system response or output $\mathbf{y}$ may be found by the linear combinations

$$\mathbf{y} = \mathbf{C}\mathbf{x} + \mathbf{D}\mathbf{u} \tag{I.26}$$

The physical system state variables $\mathbf{x}$ can be scaled to analog computer state variables $\mathbf{e}$ by the transformations

$$\mathbf{x} = \boldsymbol{\alpha}\,\mathbf{e} \tag{I.27}$$

where

$$\boldsymbol{\alpha} = \begin{bmatrix} \alpha_1 & 0 & 0 & \cdots & 0 \\ 0 & \alpha_2 & 0 & \cdots & 0 \\ 0 & 0 & \alpha_3 & \cdots & 0 \\ \vdots & & & & \\ 0 & 0 & 0 & \cdots & \alpha_n \end{bmatrix} \tag{I.28}$$

is a constant $n$th-order diagonal scaling matrix (in physical units/volt).

In a similar fashion, the inputs $\mathbf{u}$ to the physical system can be scaled to inputs $\mathbf{v}$ to the analog computer system by the transformation

$$\mathbf{u} = \boldsymbol{\beta}\,\mathbf{v} \tag{I.29}$$

where

$$\boldsymbol{\beta} = \begin{bmatrix} \beta_1 & 0 & \cdots & 0 \\ 0 & \beta_2 & \cdots & 0 \\ \vdots & & & \\ 0 & 0 & \cdots & \beta_m \end{bmatrix} \tag{I.30}$$

is a constant $m$th-order diagonal scaling matrix. Note that usually $m = 1$.

Thus, observing that $\dot{\mathbf{x}} = \boldsymbol{\alpha}\,\dot{\mathbf{e}}$ and substituting the preceding expressions for $\mathbf{x}$ and $\mathbf{u}$ into Eq. (I.24), we obtain, after a bit of algebraic rearrangement,

$$\dot{\mathbf{e}} = \boldsymbol{\alpha}^{-1}\mathbf{A}\boldsymbol{\alpha}\,\mathbf{e} + \boldsymbol{\alpha}^{-1}\mathbf{B}\boldsymbol{\beta}\,\mathbf{v} \tag{I.31}$$

or

$$\dot{\mathbf{e}} = \tilde{\mathbf{A}}\mathbf{e} + \tilde{\mathbf{B}}\mathbf{v} \tag{I.32}$$

with initial conditions

$$\mathbf{e}(0) = \boldsymbol{\alpha}^{-1}\mathbf{x}(0) \tag{I.33}$$

and outputs

$$\mathbf{y} = \mathbf{C}\,\boldsymbol{\alpha}\,\mathbf{e} + \mathbf{D}\,\boldsymbol{\beta}\,\mathbf{v} \tag{I.34}$$

or

$$\mathbf{y} = \tilde{\mathbf{C}}\mathbf{e} + \tilde{\mathbf{D}}\mathbf{v} \tag{I.35}$$

Now notice that the inverse of a diagonal matrix is merely a diagonal matrix with the scalar inverse of the respective elements; that is,

$$\boldsymbol{\alpha}^{-1} = \begin{bmatrix} \frac{1}{\alpha_1} & 0 & 0 & \cdots & 0 \\ 0 & \frac{1}{\alpha_2} & 0 & \cdots & 0 \\ 0 & 0 & \frac{1}{\alpha_3} & \cdots & 0 \\ \vdots & & & & \\ 0 & 0 & 0 & \cdots & \frac{1}{\alpha_n} \end{bmatrix} \tag{I.36}$$

Thus, the tilde, matrices or analog computer, matrices become

$$\tilde{\mathbf{A}} = \boldsymbol{\alpha}^{-1}\mathbf{A}\boldsymbol{\alpha} = \begin{bmatrix} \frac{\alpha_1}{\alpha_1}a_{11} & \frac{\alpha_2}{\alpha_1}a_{12} & \frac{\alpha_3}{\alpha_1}a_{13} & \cdots & \frac{\alpha_n}{\alpha_1}a_{1n} \\ \frac{\alpha_1}{\alpha_2}a_{21} & \frac{\alpha_2}{\alpha_2}a_{22} & \frac{\alpha_3}{\alpha_2}a_{23} & \cdots & \frac{\alpha_n}{\alpha_2}a_{2n} \\ \vdots & & & & \\ \frac{\alpha_1}{\alpha_n}a_{n1} & \frac{\alpha_2}{\alpha_n}a_{n2} & \frac{\alpha_3}{\alpha_n}a_{n3} & \cdots & \frac{\alpha_n}{\alpha_n}a_{nn} \end{bmatrix} \tag{I.37}$$

in which the diagonal elements are unchanged, and

$$\tilde{\mathbf{B}} = \boldsymbol{\alpha}^{-1}\boldsymbol{B\beta} = \begin{bmatrix} \frac{\beta_1}{\alpha_1}b_{11} & \frac{\beta_2}{\alpha_1}b_{12} & \cdots & \frac{\beta_m}{\alpha_1}b_{1m} \\ \frac{\beta_1}{\alpha_2}b_{21} & \frac{\beta_2}{\alpha_2}b_{22} & \cdots & \frac{\beta_m}{\alpha_2}b_{2m} \\ \vdots & & & \\ \frac{\beta_1}{\alpha_n}b_{n1} & \frac{\beta_2}{\alpha_n}b_{n2} & \cdots & \frac{\beta_m}{\alpha_n}b_{nm} \end{bmatrix} \tag{I.38}$$

or, in general, the element in the $i$th row and $j$th column becomes

$$\tilde{b}_{ij} = \frac{\beta_j}{\alpha_i}b_{ij} \tag{I.39}$$

and similarily for $\tilde{a}_{ij}$.

For $k$ outputs,

$$\tilde{\mathbf{C}} = \mathbf{C}\boldsymbol{\alpha} = \begin{bmatrix} c_{11}\alpha_1 & c_{12}\alpha_2 & \cdots & c_{1n}\alpha_n \\ c_{21}\alpha_1 & c_{22}\alpha_2 & \cdots & c_{2n}\alpha_n \\ \vdots & & & \\ c_{k1}\alpha_1 & c_{k2}\alpha_2 & \cdots & c_{kn}\alpha_n \end{bmatrix} \tag{I.40}$$

and

$$\tilde{\mathbf{D}} = \mathbf{D}\boldsymbol{\beta} = \begin{bmatrix} d_{11}\beta_1 & d_{12}\beta_2 & \cdots & d_{1m}\beta_m \\ d_{21}\beta_1 & d_{22}\beta_2 & \cdots & d_{2m}\beta_m \\ \vdots & & & \\ d_{k1}\beta_1 & d_{k2}\beta_2 & \cdots & d_{km}\beta_m \end{bmatrix} \tag{I.41}$$

### I.5.2 Time Scaling

Often, the response of the physical system model proceeds at a slower or faster pace than is practical for recording or plotting purposes. That is, the response of the physical system may be complete in milliseconds or may take hours to arrive at steady state. For this reason, it is frequently desirable to slow down or speed up the simulated response on the computer. This can conveniently be done by making the transformation

$$t = \gamma \tau \tag{I.42}$$

where $\gamma$ is a time scaling factor, $t$ is the real time of the physical system, and $\tau$ is computer time ($\gamma > 1$ accelerates the solution of the system, while $\gamma < 1$ retards the solution). This transformation implies that

$$dt = \gamma\, d\tau \tag{I.43}$$

or that

$$\frac{d}{d\tau} = \gamma \frac{d}{dt} \tag{I.44}$$

Thus, the analog computer state-variable system becomes

$$\frac{d}{d\tau}\mathbf{e} = \gamma\,\tilde{\mathbf{A}}\mathbf{e} + \gamma\tilde{\mathbf{B}}\mathbf{v} \tag{I.45}$$

with initial conditions

$$\mathbf{e}(0) = \boldsymbol{\alpha}^{-1}\,\mathbf{x}(0) \tag{I.46}$$

and outputs

$$\mathbf{y} = \tilde{\mathbf{C}}\mathbf{e} + \tilde{\mathbf{D}}\mathbf{v} \tag{I.47}$$

Notice from the foregoing description of the system that time scaling affects only the $\tilde{\mathbf{A}}$ and $\tilde{\mathbf{B}}$ matrices and thus the integrator gains; it does not affect initial conditions or outputs. Time scaling affects only the time required to obtain a response. You will know if time scaling is necessary if all of the integrator gains are very large ($> 100$) or very small ($< 0.01$).

## I.6 SELECTION OF MAGNITUDE AND TIME SCALE FACTORS

Just as the most difficult problem facing digital simulation is the selection of a proper integration step size, the most difficult task in analog simulation is selecting the magnitude scale factors for each state variable. The scale factors $\boldsymbol{\alpha}$ establish the analogy between the physical variables and the analog computer variables. They are selected such that a maximum response in the physical system will result in a maximum response in the computer variables that will be within certain limits.

Most analog computers operate best with signals between $-10$ volts and $+10$ volts. (Some use signals between $-100$ volts and $+100$ volts.) It is desirable to have the analog computer variables be as large as possible, but never to exceed $\pm 10$ volts, in order to facilitate plotting and to achieve maximum accuracy with existing equipment. Further, since all amplifiers will produce a given level of noise, it is desirable to keep the signals as large as possible to maintain a high signal-to-noise ratio.

### I.6.1 Selection of Magnitude Scale Factors

The magnitude scale factors are selected in such a manner that the maximum (absolute) value of the physical response of a state variable $x_i$ will cause a maximum response of the computer variable $e_i$ to be about (but not greater than) 10 volts. Thus,

$$\alpha_i = \frac{|x_{imax}|}{|e_{imax}|} \tag{I.48}$$

If the computed value of $\alpha$ is not a convenient number, it should be rounded off higher (rather than lower) to prevent overloading ($e_i > \pm 10$ volts). For example, if $x_{imax}$ is predicted to be 1.37 cm, then we would select

$$\alpha_i = \frac{1.37 \text{ cm}}{10 \text{ volt}} = 0.137 \frac{\text{cm}}{\text{volt}} \tag{I.49}$$

which would be rounded off to 0.15 or 0.20 cm/volt. The output scale factors are selected in a similar manner as

$$\beta_j = \frac{|u_{jmax}|}{|v_{jmax}|} \tag{I.50}$$

The preceding discussion specifies how to select the scale factors, but the problem of predicting $x_{imax}$ and $u_{jmax}$ still faces us. Determining $u_{jmax}$ is relatively easy, since the system input is usually known (e.g., a step, a ramp, sinusoidal, etc.). While the determination of $x_{imax}$ is slightly more difficult, it is greatly simplified by the approach set out next.

In some systems, the largest value of a state variable occurs at its initial condition, and the response decays thereafter. In other instances, the maximum response occurs at or near the steady-state value, as shown in Figure I.5.

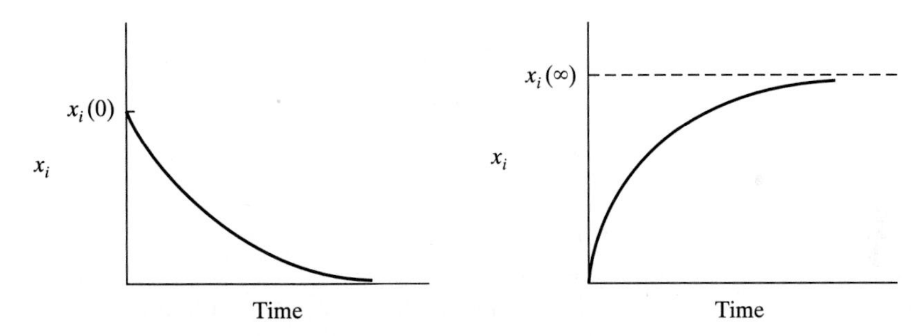


**Figure I.5** Typical responses illustrating maximum values.

The initial conditions are known, and the steady-state maxima can be estimated. To do the estimation, we observe that the steady-state values occur when $\dot{\mathbf{x}} = \mathbf{0}$, and thus,

$$\mathbf{A}\mathbf{x}(\infty) = -\mathbf{B}\mathbf{u} \tag{I.51}$$

This yields a set of algebraic equations in which the energy or effort supplied by the inputs is put into the state variables. If we solve each of the equations for successive state variables, then assuming that all of the input effort goes into these variables entirely, we can observe the ultimate steady-state maxima of the state variables. For example, consider a second-order system with one input; Equation (I.51) becomes

$$a_{11}x_1(\infty) + a_{12}x_2(\infty) = -b_1 u \tag{I.52}$$

$$a_{21}x_1(\infty) + a_{22}x_2(\infty) = -b_2 u_1 \tag{I.53}$$

Now, assuming that all of the input effort goes into $x_1$ and none goes into $x_2$, we obtain the ultimate (indicated by *) steady-state maxima. From Eq. (I.52) with $x_2(\infty)$ assumed to be zero, it follows that

$$x_1^*(\infty) = \frac{-b_1}{a_{11}} u_{1max} \tag{I.54}$$

From Eq. (I.53),

$$x_1^*(\infty) = \frac{-b_2}{a_{21}} u_{1max} \tag{I.55}$$

Whichever of these two equations yields the largest absolute value for $x_i$ will be used as the steady-state maximum. Thus,

$$x_1^*(\infty)_{max} = \text{MAX of} \left\{ \left| \frac{b_1}{a_{11}} u_{1\,max} \right| \text{ or } \left| \frac{b_2}{a_{21}} u_{1\,max} \right| \right\} \tag{I.56}$$

And similarly,

$$x_2^*(\infty)_{max} = \text{MAX of} \left\{ \left| \frac{b_1}{a_{12}} u_{1max} \right| \text{or} \left| \frac{b_2}{a_{22}} u_{1max} \right| \right\} \tag{I.57}$$

We are now ready to estimate the maximum value we expect from a state variable. The maximum will occur either at the initial condition or at the value estimated by the steady-state maximum method; that is

$$x_{imax} = \text{MAX of } \{ |x_i(0)| \text{ or } |x_i^*(\infty)_{max}| \} \tag{I.58}$$

The method just outlined should predict all of the maximum expected values of the state variables necessary to allow the selection of magnitude scale factors $\tilde{\boldsymbol{\alpha}}$ and $\boldsymbol{\beta}$. Then, having selected the scale factors, we can compute the matrices $\tilde{\mathbf{A}}$ and $\tilde{\mathbf{B}}$. The elements in those matrices represent the integrator gains $(1/RC)$ to be used in wiring the computer diagram and thus have units of 1/second or 1/time. It is desirable for these gains to be in the range 0.1 to 10.0 (or possibly 100.0). If all of the coefficients or gains are within this range, the solution of the system can be obtained directly. If all of the gains are in this range except for one or so, trade-offs can be made by raising one scale factor. Notice in Eq. (I.39) that increasing a scale factor will decrease the gains in one row and raise the gains in one column. Since the $\mathbf{A}$ matrix usually has numerous zero entries, the trade-off of lowering one gain while raising its symmetrical element can be effective.

### I.6.2 Selection of Time Scale Factors

If all of the gains are either too low or too high, time scaling will be necessary. The time scale factor $\gamma$ alters all gains by a common amount in the $\tilde{\mathbf{A}}$ and $\tilde{\mathbf{B}}$ matrices. The time scale factor is usually selected in multiples of 10.

For example, if the matrix

$$\tilde{\mathbf{A}} = \begin{bmatrix} 0 & 250 \\ 100 & 75 \end{bmatrix} \tag{I.59}$$

were obtained, a time scale factor of 1/100 would reduce the coefficients to a usable value. After time scaling:

$$\gamma\tilde{\mathbf{A}} = \begin{bmatrix} 0 & 2.5 \\ 1.0 & 0.75 \end{bmatrix} \tag{I.60}$$

**Example I.1 Fluid RLC System**

As an example of the scaling process, consider the series *RLC* fluid system with definitions of state variables

$$u_1 = \text{Pressure input (kPa)}$$

$$x_1 = \text{Pressure in capacitor (kPa)}$$

$$x_2 = \text{Flow in inductor (cm}^3\text{/s)}$$

state-space differential equations

$$\begin{bmatrix} \dot{x}_1 \\ \dot{x}_2 \end{bmatrix} = \begin{bmatrix} 0\,\dfrac{1}{\text{s}} & 5\,\dfrac{\text{kPa}}{\text{cm}^3} \\ -10\,\dfrac{\text{cm}^3}{\text{kPa s}^2} & -2.5\,\dfrac{1}{\text{s}} \end{bmatrix} \begin{bmatrix} x_1 \\ x_2 \end{bmatrix} + \begin{bmatrix} 0\,\dfrac{1}{\text{s}} \\ 10\,\dfrac{\text{cm}^3}{\text{kPa s}^2} \end{bmatrix} [u_1] \tag{I.61}$$

and initial conditions

$$\begin{bmatrix} x_1(0) \\ x_2(0) \end{bmatrix} = \begin{bmatrix} -20\text{ kPa} \\ 0\text{ cm}^3\text{/s} \end{bmatrix} \tag{I.62}$$

Let the system be subjected to a step input

$$u_1 = 10\text{ kPa} \quad \text{for} \quad t \geq 0 \tag{I.63}$$

The first step in solving the system is to select the magnitude scale factors. This requires knowledge of $x_{imax}$. The steady-state maximums can be predicted from Eq. (I.61) by setting $\mathbf{x} = \mathbf{0}$. The first equation yields no information, since the input has no direct influence on $\dot{x}_1$. The second equation results in

$$10x_1(\infty) + 2.5\,x_2(\infty) = 10\,u_1 \tag{I.64}$$

By setting $x_2(\infty) = 0$ and allowing all of the input effort to drive $x_1$, we find that

$$x_1^*(\infty) = \frac{10}{10}\,u_{1max} = u_{1max} \tag{I.65a}$$

$$x_1^*(\infty) = 10\text{ kPa} \tag{I.65b}$$

By setting $x_1(\infty) = 0$ and allowing all of the input to go into the second state variable, we find that

$$x_2^*(\infty) = \frac{10}{2.5}\,u_{1max} = 4\,u_{1max} \tag{I.66a}$$

$$x_2^*(\infty) = 40\text{ cm}^3\text{/s} \tag{I.66b}$$

The maximum values for **x** can now be estimated from the maximum of either the initial conditions or the steady-state maxima:

$$x_{1max} = \text{MAX of } \{|x_1(0)| \quad \text{or} \quad |x_1^*(\infty)_{max}|\} \tag{I.67}$$
$$= \text{MAX of } \{20 \quad \text{or} \quad 10\} = 20 \text{ kPa}$$

$$x_{2max} = \text{MAX of } \{|x_2(0)| \quad \text{or} \quad |x_2^*(\infty)_{max}|\} \tag{I.68}$$
$$= \text{MAX of } \{0 \quad \text{or} \quad 40\} = 40 \text{ cm}^3/\text{s}$$

The magnitude scale factors can now be computed:

$$\alpha_1 = \frac{x_{1max}}{e_{1max}} = \frac{20 \text{ kPa}}{10 \text{ volt}} = 2.0 \frac{\text{kPa}}{\text{volt}} \tag{I.69}$$

$$\alpha_2 = \frac{x_{2max}}{e_{2max}} = \frac{40 \text{ cm}^3/\text{s}}{10 \text{ volt}} = 4.0 \frac{\text{cm}^3/\text{s}}{\text{volt}} \tag{I.70}$$

$$\beta_1 = \frac{u_{1max}}{v_{1max}} = \frac{10 \text{ kPa}}{10 \text{ volt}} = 1.0 \frac{\text{kPa}}{\text{volt}} \tag{I.71}$$

Next, we can compute the $\widetilde{\mathbf{A}}$ and $\widetilde{\mathbf{B}}$ matrices:

$$\widetilde{\mathbf{A}} = \begin{bmatrix} a_{11} & \frac{\alpha_2}{\alpha_1} a_{12} \\ \frac{\alpha_1}{\alpha_2} a_{21} & a_{22} \end{bmatrix} = \begin{bmatrix} 0 & 10 \\ -5 & 2.5 \end{bmatrix} \tag{I.72}$$

$$\widetilde{\mathbf{B}} = \begin{bmatrix} \frac{\beta_1}{\alpha_1} b_1 \\ \frac{\beta_1}{\alpha_2} b_2 \end{bmatrix} = \begin{bmatrix} 0 \\ 2.5 \end{bmatrix} \tag{I.73}$$

The reader should verify that the elements in these matrices all have units of 1/second.

The initial conditions of **e** are computed from the following (notice the units result in volts):

$$\mathbf{e}(0) = \boldsymbol{\alpha}^{-1}\mathbf{x}(0) \tag{I.74}$$

or

$$e_1(0) = \frac{1}{\alpha_1} x_1(0) = \frac{\text{volt}}{2 \text{ kPa}}(-20 \text{ kPa}) = -10 \text{ volt} \tag{I.75}$$

$$e_2(0) = \frac{1}{\alpha_2} x_2(0) = \frac{\text{volt}}{4 \text{ cm}^3/\text{s}}(0 \text{ cm}^3/\text{s}) = 0 \text{ volt} \tag{I.76}$$

Recall that the initial condition of an integrator has a sign opposite to that of the integrator output. Thus, since the integrator output is $-e_1$, the initial condition is input as $+e_i(0)$, in other words, with its original sign.

The final simulation problem to be wired on the analog computer can be stated as

$$\begin{bmatrix} \dot{e}_1 \\ \dot{e}_2 \end{bmatrix} = \begin{bmatrix} 0 & 10 \\ -5 & -2.5 \end{bmatrix} \begin{bmatrix} e_1 \\ e_2 \end{bmatrix} + \begin{bmatrix} 0 \\ 2.5 \end{bmatrix} [v_1] \tag{I.77}$$

with initial conditions

$$\begin{bmatrix} e_1(0) \\ e_2(0) \end{bmatrix} = \begin{bmatrix} -10 \text{ volt} \\ 0 \end{bmatrix} \tag{I.78}$$

where $\boldsymbol{\alpha}$ is as previously defined, and the analogy is given by

$$\alpha_1 = 2.0 \frac{\text{kPa}}{\text{volt}} \tag{I.79}$$

$$\alpha_2 = 4.0 \frac{\text{cm}^3/\text{s}}{\text{volt}} \tag{I.80}$$

$$\beta_1 = 1.0 \frac{\text{kPa}}{\text{volt}} \tag{I.81}$$

The system input is a step input, or a constant for $t \geq 0$. The magnitude of this constant is

$$v_1 = \frac{u_1}{\beta_1} = 10 \text{ volt} \tag{I.82}$$

The elements in $\tilde{\mathbf{A}}$ and $\tilde{\mathbf{B}}$ represent integrator gains. The analog computer diagram for this system is shown in Figure I.6.

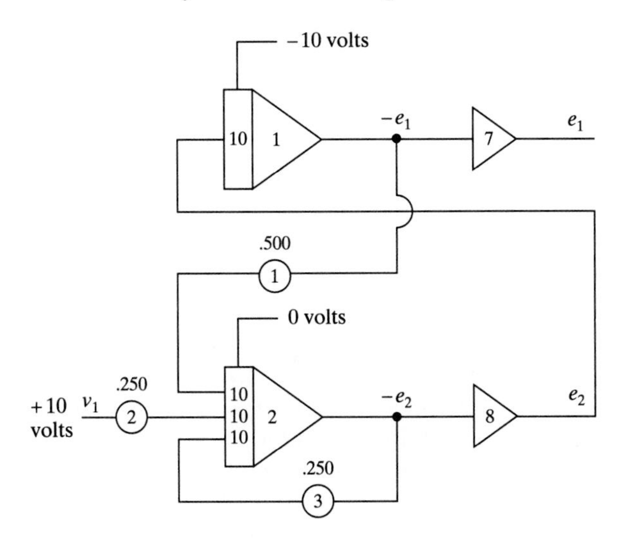


**Figure I.6** Analog computer diagram for Example I.1.

The preceding system was simulated on the analog computer, and the responses of Figure I.6 were obtained. The basic units of these plots are volts. The analog computer state variables $\mathbf{e}$ can be converted back into physical system state variables x by multiplying the voltage scales by the magnitude scale factors $\boldsymbol{\alpha}$. Two scales are shown in Figure I.7.

Any $\mathbf{y}$ output could be handled in a similar fashion. If an output is a function of only one computer state variable, the conversion $\tilde{c}_{ij}$ can best be handled with a scale factor applied to the voltage scale, rather than using a scaling pot to obtain the coefficient. If an output is a function of more than one computer state variable, it will probably be easier to let the computer perform the summation than perform hand addition of two curves; in this case, pots would be used to obtain the $\tilde{\mathbf{C}}$ coefficients.

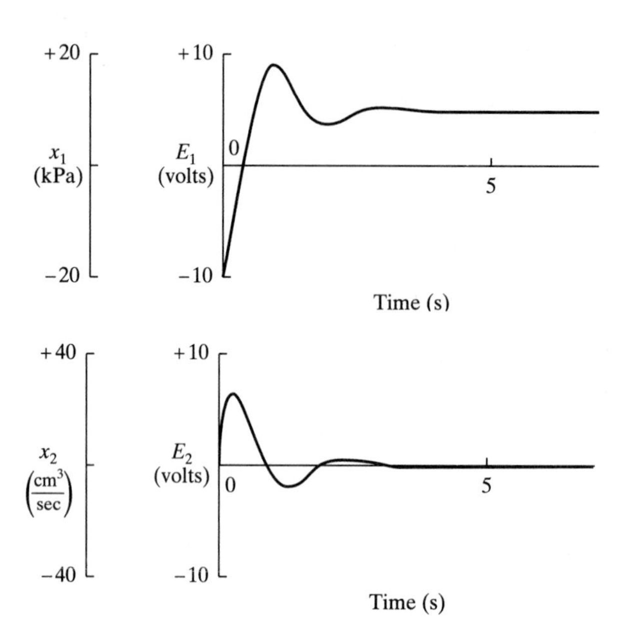


**Figure I.7** Responses from analog computer for Example I.1.

## I.7 SUMMARY

This appendix has introduced the fundamental concepts of operational amplifiers and how they can be used to obtain the various functions required in differential equations (i.e., multiplication by a constant, addition and subtraction, integration, etc.). A method was presented for taking these mathematical functions and forming a differential equation in classical or state-space form. The response of the electrical circuit is then a solution of the differential equation, which might be analogous to the differential equation of some other type of system. Details of magnitude scaling and the selection of maximum values of the state variable were given. The prediction of maximum values is of interest in general systems analysis beyond analog computer scaling. The appendix concluded with the presentation of a methodology for the selection of time scale factors, together with some examples of the application of the methodology.

## REFERENCES

I.1 Jenness, Roger R. *Analog Computation and Simulation: Laboratory Approach*. Allyn & Bacon, Boston, 1965.

I.2 Warfield, John N. *Introduction to Electronic Analog Computers*. Prentice Hall, Englewood Cliffs, NJ, 1959.

## PROBLEMS

**I.1** Using ciruit analysis with the standard assumptions on the amplifier gain and input impedance, *derive* the computing equation for the summing junction shown in Figure PI.1.

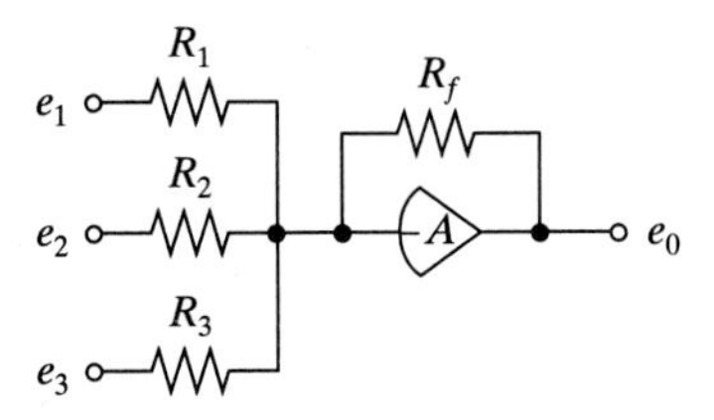


**Figure PI.1** Summing junction.

**I.2** *Obtain* the analog computer solution to the first-order differential equation

$$\dot{e}_1 = -2\,e_1 \qquad \text{with } e_1(0) = 0 \text{ volts}$$

which has already been scaled to computer units. *Submit* your computer block diagram and the plot of your results.

**I.3** A mechanical system is represented in state space as

$$\dot{x}_1 = x_2 \qquad u_1 = \text{position (mm)}$$

$$x_1 = \text{position (mm)}$$

$$\dot{x}_2 = -15\frac{1}{s^2}x_1 - 1.5\frac{1}{s}x_2 + 15\frac{1}{s^2}u_1 \qquad x_2 = \text{velocity (mm/s)}$$

with initial conditions

$$x_1(0) = 3.0 \text{ mm}$$
$$x_2(0) = 0.0 \text{ mm/s}$$

*Scale* this system and obtain an analog computer solution of it.

Plot the following responses with an input $u_1 = 2.0$ mm for $t \geq 0$:.

$$y_1 = x_1 \qquad y_1 = \text{position (mm)}$$

$$y_2 = 0.2\frac{\text{N}}{\text{mm/s}}x_2 \qquad y_2 = \text{force (N)}$$

Perform your work on the system with the following prelab, laboratory, and postlab procedures:

### Prelab:

**a.** *Scale* the system equations and initial conditions to a computer state-variable set of equations.

**b.** *Draw* the computer diagram and determine the patch panel layout.

### Laboratory:

**a.** *Wire* the system on the analog computer.

**b.** *Observe* the responses on the oscilloscope in the "REP OP" mode.

**c.** When you are sure that your scaling and responses are correct (rescaling might be necessary), *plot* your outputs on the *x-y* plotter.

**d.** *Vary* the pot settings for the $a_{21}$, $a_{22}$, $b_2$, and initial-condition coefficients, and observe the effect each of these has on your response.

**Postlab:**

**a.** *Convert* the computer units (volts) back into physical units on the plot, and label the axes properly.

**b.** *Discuss* the effects of varying the aforementioned coefficients.

**c.** *Submit* a summary of your work.

**I.4** *Obtain* the solution to Problem I.3 using digital simulation. *Compare* the analog and digital solutions to the analytical solution found in Appendix E. *Plot* all three solutions for $x_1$ on a single graph. *Discuss* the advantages and disadvantages of each method.

# APPENDIX J

# Questions About The Concepts Presented

1. *What* are the purposes of modeling, and *what* can we do with a model of a dynamic system?
2. *What* is the functional definition of linearity? *What* is a linear equation? *What* is a linear differential equation?
3. *What* are the so-called elementary inputs that are used as inputs to dynamic systems? *Are* there other typical inputs?
4. The solution to a nonhomogeneous differential equation is composed of two parts. *What* are they, and *how* are they found?
5. *Discuss* the concept of the time constant of a first-order differential equation. *How* can the time constant be determined from the linear differential equation? *How* can the time constant be determined from the time response of the system?
6. There are two performance factors for a first-order differential equation. *What* are they? *Explain* what each factor means in terms of the response of the system.
7. There are three performance factors for a second-order differential equation. *What* are they? *Explain* what each factor means in terms of the response of the system.
8. *What* are the characteristics of an undamped, an underdamped, a critically damped, and an overdamped system?
9. *Explain* what is meant by the systems approach in the treatment of dynamic systems.

10. *Discuss* the meaning of effort and flow variables for dynamic systems. *Mention* the effort and flow variables in electrical, mechanical, fluid, and thermal systems.
11. *Explain* the characteristics of the three types of fundamental components that are the building blocks of all systems. *What* are examples of each type of component in electrical, mechanical, fluid, and thermal systems?
12. *Discuss* the different types of friction in mechanical systems.
13. *What* is an op-amp, and *how* can it be used in electronic circuits?
14. *What* different types of fluid resistance are there, and *what* is the energy loss mechanism in each?
15. *What* types of fluid capacitance are there, and *what* causes each?
16. *What* is a common explanation for the effect of fluid inductance?
17. *What* are resistance and capacitance in a thermal system?
18. *What* is a good way to confirm that an equation has been derived using consistent terms?
19. *What* are the general forms of the Laplace transform of a first-, second-, and higher order derivative? *Discuss* the form of the Laplace transform of a differential equation.
20. *How* can the time domain solution of a differential equation be found by the Laplace transform technique?
21. *Discuss* what is meant by partial fractions and *why* partial fraction expansion is necessary in the solution of differential equations.
22. *What* is the transfer function of a dynamic system? *How* is it usually expressed?
23. *How* are state variables defined to convert a classical linear $n$th-order differential equation to a state-space equation without input derivatives?
24. *How* are state variables defined to convert a classical differential equation to state-space form with input derivatives? *How* does this affect the inputs to the derivatives of the state variables?
25. *How* can a set of linear state-space differential equations be converted to a classical differential equation?
26. *Is* the state-space representation of dynamic systems limited to linear systems? If a system is nonlinear, *can* it be represented in state-space format? *Can* a nonlinear system always be reduced to a classical differential equation?
27. *Describe* how a state-space differential equation representation can be derived directly from systems circuit modeling equations or from engineering modeling equations.
28. *Discuss* the general approach to solving, and the format of, the equations used in the numerical integration of a first-order differential equation on the digital computer. *Name* some of the integration methods that could be used.

**29.** *What* trade-offs are involved in selecting the step size for numerical integration on the digital computer? *Are* there limitations on how far the trade-offs can be taken?

**30.** *Discuss* the concepts and equations used to select an appropriate step size for Runge-Kutta integration. *How* would the final time and the print interval be selected?

**31.** *Discuss* and *compare* the time domain solutions of dynamic systems using analytical techniques and digital simulation techniques. Under *what* conditions would analytical solutions be more desirable than simulation solutions, and vice versa?

# ANSWERS TO SELECTED PROBLEMS

## CHAPTER 2

**2.3** Digital simulation employs numerical calculations at discrete points of time to describe the response of a system, while analog simulation finds the continuous response of a physical system by constructing a model whose behavior is analogous to the system in question.

**2.6** Classical, transfer function, and state space.

**2.9** System dynamic characteristics expressed in terms of time constants, natural frequencies, and damping ratios.

**2.12** The classical form consists of a single *nth*-order differential equation, while the state-space form is composed of $n$ first-order equations.

**2.15** $\dot{y}_1 = y_2, \quad \dot{y}_2 = y_3, \quad \dot{y}_3 = 3y_2 - 2y_3 + \sin\omega t$

**2.21** See Section 2.4.1.

**2.24** The two definitions are related to the so-called force-voltage or force-current analogies. We use the force-voltage analogy because a "flow" variable should be the rate of change of a physical variable with respect to time (in this case, the rate of change of position) and because this selection makes the concept of impedance more natural (i.e., a brick wall is a large mechanical impedance).

## CHAPTER 3

**3.3** $\dot{y} = (2k/3b)(\delta - y) + (2/3)\dot{\delta}$, $y$ is the position of the right end of the damper.

**3.6** $0.102 \le \mu \le 0.302$.

**3.9** $m\ddot{y} + \dot{m}\dot{y} + C\dot{y}^2 = \text{Thrust} - mg$

**3.12** Many solutions are possible. One is a beam 10.5 inches long with a 0.5" × 0.5" cross section.

**3.15** Assuming a uniform wheel disk, $\left(m_c + \frac{3}{2}m_w d^2\right)\dot{v} + bv = (2T_a)/d$.

**3.18** $m\ddot{z} = -k_1 z - b\dot{z} + k_2(x - z)$

**3.21** $J$ = inertia, $k$ = stiffness, $T$ = torque, $e$ = engine, $c$ = clutch plate, $p$ = propeller, $w$ = water, $l$ = left, $r$ = right.

$$J_e\ddot{\theta}_e + k_l\theta_e - k_l\theta_{cl} = T_e(t)$$

$$J_{cl}\ddot{\theta}_{cl} + k_l\theta_{cl} - k_l\theta_e = T_c$$

$$J_{cr}\ddot{\theta}_{cr} + k_r\theta_{cr} - k_r\theta_p = -T_c$$

$$J_p\ddot{\theta}_p + k_r\theta_p - k_r\theta_{cr} = -T_w$$

**3.24** The displacement of the bar $y$ is positive up, as is the displacement of the mass $z$. The rotation of the bar $\theta$ is positive ccw.

$$m\ddot{y} + 4ky + ka\theta - kz = 0$$
$$J\ddot{\theta} + kay + 3ka^2\theta = 0$$
$$m\ddot{z} + kz - ky = 0$$
$$J = (ma^2)/3$$

**3.27** $ma^2\ddot{\theta} + ka^2\theta + fa\,\text{sign}\,(\dot{\theta}) = 2aF(t),\ f =$ friction force

**3.30**

$$\ddot{\theta} + \frac{r^2M_2 + r^2M_1 + J_{cg1}}{rM_2\,L\cos\theta}\,\ddot{\phi} - \dot{\theta}^2\tan\theta = 0$$
$$\ddot{\theta} + \frac{LM_2r\cos\theta}{J_{cg2} + M_2L^2}\,\ddot{\phi} + \frac{LM_2g\sin\theta}{J_{cg2} + L^2M_2} = 0$$

**3.33** See Problem 3.24.

## CHAPTER 4

**4.3** 20 volts

**4.6** 12.06 f

**4.9**

$$R_{eq} = \frac{R_5\left[\dfrac{R_1R_2}{(R_1 + R_2)} + \dfrac{R_3R_4}{(R_3 + R_4)}\right]}{R_5 + \dfrac{R_1R_2}{(R_1 + R_2)} + \dfrac{R_3R_4}{(R_3 + R_4)}}$$

**4.12**

$$i_{R_1} = \frac{e_0 - e_1}{R_1},\quad i_{C_1} = C_1De_1 \text{ with } e_1(0),\ i_{R_1} = i_{C_1}$$
$$(R_1C_1D + 1)e_1 = e_0$$
$$\tau = 0.022 \text{ seconds},\quad G = 1.0$$

**4.15**

$$i_{R_1} = \frac{e_0 - e_1}{R_1},\ i_{C_1} = C_1De_1 \text{ with } e_1(0),\ i_L = \frac{e_1 - e_2}{LD} \text{ with } i_L(0),$$
$$i_{C_2} = C_2De_2 \text{ with } e_2(0),\ i_{R_2} = \frac{e_2}{R_2},\ i_{R_1} = i_{C_1} + i_L,\ i_L = i_{C_2} + i_{R_2}$$
$$u_1 = e_0,\ x_1 = e_1,\ x_2 = i_L,\ x_3 = e_2$$
$$\dot{x}_1 = -\frac{1}{R_1C_1}x_1 - \frac{1}{C_1}x_2 + \frac{1}{R_1C_1}u_1$$

$$\dot{x}_2 = \frac{1}{L}x_1 - \frac{1}{L}x_3$$

$$\dot{x}_3 = -\frac{1}{C_2}x_2 - \frac{1}{R_2C_2}x_3$$

**4.18**

$$i_{R_1} = \frac{e_0 - e_1}{R_1}, \quad i_L = \frac{e_0 - e_1}{R_2 + LD} \quad \text{with } i_L(0), \quad i_C = CDe_1 \text{ with } e_1(0),$$

$$i_{R_1} + i_L = i_C$$

$$\left[\left(\frac{R_1}{R_1 + R_2}\right) LCD^2 + \frac{L}{(R_1 + R_2)}D + 1\right]e_1 = \left[\frac{L}{(R_1 + R_2)}D + 1\right]e_0$$

$$e_1(0) = \text{known}, \quad \dot{e}_1 = \frac{1}{C}i_L(0) + \frac{e_0(0) - e_1(0)}{R_1C}$$

**4.21**

$$i_{R_1} = \frac{e_1 - e_4}{R_1}, \quad i_L = \frac{e_2 - e_4}{R_2 + LD} \quad \text{with} \quad i_L(0), \quad i_{R_4} = \frac{e_4}{R_4},$$

$$i_C = CDe_4 \quad \text{with } e_4(0), \quad i_{R_1} + i_L = i_{R_4} + i_C$$

$$u_1 = e_1, \quad u_2 = e_2, \quad x_1 = i_L, \quad x_2 = e_4$$

$$\dot{x} = -\frac{R_2}{L}x_1 - \frac{1}{L}x_2 + \frac{1}{L}u_2$$

$$\dot{x}_2 = \frac{1}{C}x_1 - \left(\frac{1}{R_1C} + \frac{1}{R_4C}\right)x_2 + \frac{1}{R_1C}u_1$$

$$x_1(0) = i_L(0), \quad x_2(0) = e_4(0)$$

**4.24**

$$e_o = \frac{-\frac{R_f}{R_i}}{R_fCD + 1}e_i, \quad i_v = \frac{e_o}{R_o + LD}$$

$$u_1 = e_i, \quad x_1 = e_o, \quad x_2 = i_v$$

$$\dot{x}_1 = -\frac{1}{R_fC}x_1 - \frac{1}{R_fC}u_1$$

$$\dot{x}_2 = \frac{1}{L}x_1 - \frac{R_o}{L}x_2$$

$$x_1(0) = e_o(0), \quad x_2(0) = i_v(0)$$

**4.27**

$$\frac{e_o}{e_i} = -\frac{R_f}{R_i}\frac{(R_iC_iD + 1)}{(R_fC_fD + 1)}$$

**4.30** If $R_{meter} = 100\ \text{k}\Omega$, then voltage reading will be degraded by 7.6% (i.e., reading = 92.4% of undisturbed voltage). If $R_{meter} = 1\ \text{M}\Omega$, then voltage reading will be degraded by 0.8%.

## CHAPTER 5

**5.3** Recall that $\rho = M/V$, and that $\partial\rho = \partial\rho/\partial V|_{P_o,T_o}\,\partial V$, and that $M =$ constant.

**5.6**

$$C_t = \frac{V}{\beta}\left[1 + \frac{3\beta}{kR}\right]$$

**5.9**

$$u(y) = u_{max}\left[1 - \left(\frac{y}{h/2}\right)^2\right], \quad \bar{u} = \frac{2}{3}u_{max}$$

$$\frac{\partial P}{\partial x} = \frac{2\mu u_{max}}{(h/2)^2}, \quad P_1 - P_2 = \frac{2\mu u_{max}}{(h/2)^2}\ell$$

$$Q = wh\bar{u}, \quad \delta P = \frac{12\,\mu\ell}{wh^3}Q, \quad R = \frac{12\,\mu\ell}{wh^3}$$

**5.15**

$$\frac{\partial}{\partial P_r}\left[P_r^{\frac{2}{k}} - P_r^{\frac{k+1}{k}}\right] = 0, \quad \frac{2}{k}P_r^{\frac{2-k}{k}} - \frac{(k+1)}{k}P_r^{\frac{1}{k}} = 0, \quad P_r = \left(\frac{2}{k+1}\right)^{\frac{k}{k-1}}$$

(using a positive exponent)

**5.18** Model of system using incompressible orifice equation and no mass:

$$\delta P_0 - \delta P_1 = K_o Q^2\,\text{sign}(Q), \quad K_o = \frac{\rho}{2C_d^2 A_o^2}, \quad A\delta P_1 - k_s z = 0$$

$$Q = \frac{V}{\beta}\delta\dot{P}_1 + \dot{V}, \quad V = V_0 + Az, \quad \dot{V} = A\dot{z}, \quad \beta = n(\delta P_1 + P_{atm})$$

$$\dot{z} = \frac{\sqrt{\dfrac{\left|\delta P_0 - \dfrac{k_s}{A}z\right|}{K_o A^2}}\,\text{sign}\left(\delta P_0 - \dfrac{k_s}{A}z\right)}{\left[\dfrac{V}{\beta}\dfrac{k_s}{A^2} + 1\right]}$$

state variables: $u_1 = \delta P_0, \;\; x_1 = z$

$$\dot{x}_1 = \frac{\sqrt{\dfrac{\left|u_1 - \dfrac{k_s}{A}x_1\right|}{K_o A^2}}\,\text{sign}\left(u_1 - \dfrac{k_s}{A}x_1\right)}{\left[\dfrac{(V_0 + A\,x_1)}{n\left(\dfrac{k_s}{A}x_1 + P_{atm}\right)}\dfrac{k_s}{A^2} + 1\right]}, \quad \text{outputs: } y_1 = z = x_1, \quad y_2 = \delta P_1 = \frac{k_s}{A}x_1$$

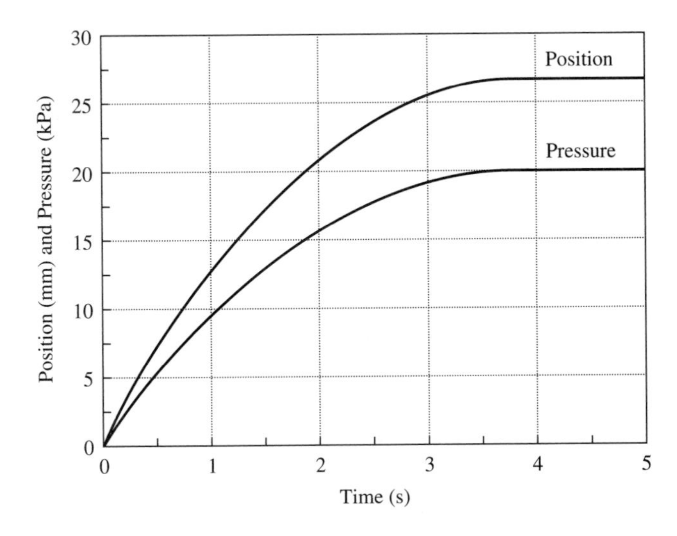


**Problem 5.18** Pneumatic actuator, incompressible orifice, no mass.

Model of system using incompressible orifice equation with mass:

change force balance to $A\delta P_1 - M\ddot{z} - k_s z = 0$:

$$[MD^2 + k_s]z = A\delta P_1, \quad \frac{V}{\beta}\delta\dot{P}_1 = \sqrt{\frac{|\delta P_0 - \delta P_1|}{K_o}}\,\text{sign}(\delta P_0 - \delta P) - A\dot{z}$$

state variables: $u_1 = \delta P_0, \;\; x_1 = z, \;\; x_2 = \dot{z}, \;\; x_3 = \delta P_1$

$$\dot{x}_1 = x_2, \quad \dot{x}_2 = -\frac{k_s}{M}x_1 + \frac{A}{M}x_2$$

$$\dot{x}_3 = \frac{n(x_3 + P_{atm})}{(V_0 + Ax_1)}\left\{\sqrt{\frac{|u_1 - x_3|}{K_o}}\,\text{sign}(u_1 - x_3) - Ax_2\right\}$$

outputs: $y_1 = z = x_1, \quad y_2 = \delta P_1 = x_3$

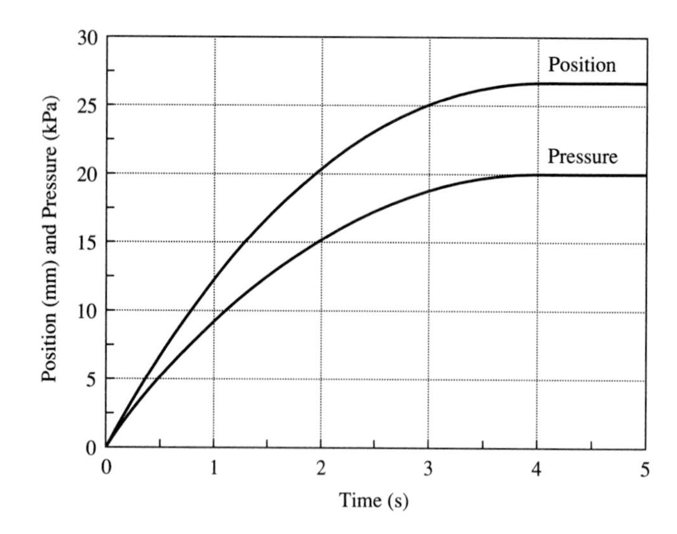


**Problem 5.18** Pneumatic actuator, incompressible orifice, with mass.

Model of system using compressible orifice equation (incompressible approximation) with mass:

change flow equation to (neglecting temperature variations):

$$\dot{m}_{in} = C_d A_o \sqrt{\frac{2}{RT^*}} (\delta P_0 + P_{atm}) \sqrt{|P_r - P_r^2|} \operatorname{sign}(P_r - P_r^2), \quad C_d A_o = \sqrt{\frac{\rho_o}{2K_o}}$$

where $P_r = \dfrac{\delta P_1 + P_{atm}}{\delta P_0 + P_{atm}}$ and if $P_r \leq 0.528$, then $P_r = 0.528$

change continuity equation to:

$$\delta \dot{P}_1 = \frac{\beta}{V}\left[\frac{\dot{m}}{\rho_1} - A\dot{z}\right], \quad \rho_1 = \frac{\delta P_1}{RT_1}$$

state variables: $u_1 = \delta P_0, \quad x_1 = z, \quad x_2 = \dot{z}, \quad x_3 = \delta P_1$

$$\dot{x}_1 = x_2, \quad \dot{x}_2 = -\frac{k_s}{M} x_1 + \frac{A}{M} x_2$$

$$P_r = \frac{x_3 + P_{atm}}{u_1 + P_{atm}}, \text{ and if } P_r \leq 0.528, \text{ then } P_r = 0.528$$

$$\dot{x}_3 = \frac{n(x_3 + P_{atm})}{(V_0 + Ax_1)} \left\{ C_d A_o \sqrt{2RT_1} \frac{(\delta P_0 + P_{atm})}{(\delta P_1 + P_{atm})} \sqrt{|P_r - P_r^2|} \operatorname{sign}(P_r - P_r^2) - Ax_2 \right\}$$

outputs: $y_1 = z = x_1, \quad y_2 = \delta P_1 = x_3$

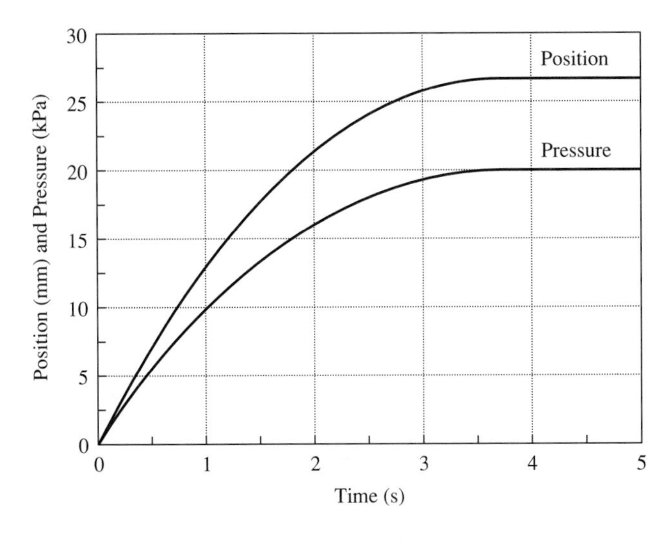


**Problem 5.18** Pneumatic actuator, compressible orifice, with mass.

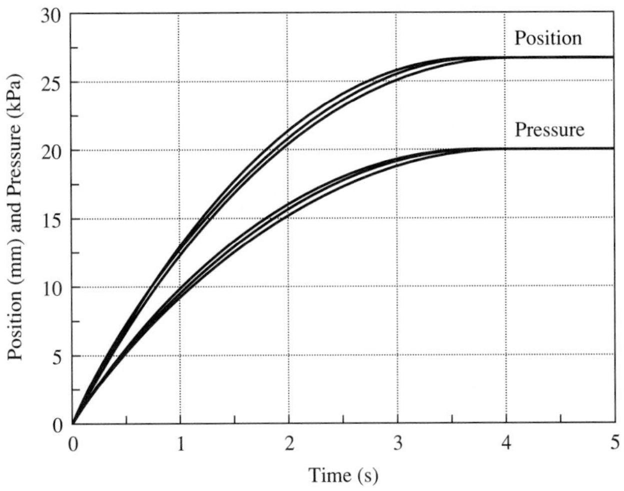


**Problem 5.18** Pneumatic actuator, all three models.

## CHAPTER 6

**6.3** If $Q_h$ and $k_t$ are constant (with respect to $r$), then

$$\int_{r_i}^{r_o} \frac{dr}{r} = -\frac{2\pi k_t L}{Q_h} \int_{T_i}^{T_o} dT, \ldots \text{continue.}$$

**6.6** $Q_h = 4455$ watts

**6.9** $Q_h = 292$ watts

**6.12** Hint:

$$h_{eq} = \frac{\sigma(T_H^4 - T_L^4)}{(T_H - T_L)} = \frac{\sigma[(T_H - T_L)(T_H + T_L)(T_H^2 + T_L^2)]}{(T_H - T_L)} \text{ and } T_L = T_H - \Delta T$$

**6.15**

$$q_h - q_{conv} = MC_p\dot{T}_B, \quad q_{conv} = hA(T_B - T_\infty)$$

$$\left[\frac{MC_p}{hA}D + 1\right]T_B = T_\infty + \frac{q_h}{hA}, \quad \tau = \frac{MC_p}{hA}, \quad T_B(\infty) = T_\infty + \frac{q_h}{hA}$$

**6.18**

$$\left[\frac{MC_p}{h_{side}A_{side}}D + 1\right]T_c = T_\infty$$

Note that the top and bottom of the can might not experience convection. If the can is sitting on a glass shelf, then the bottom of the can is sealed, and the liquid does not rise all the way to the top; thus, the area for convection might be just the sides.

For the wall of the can,

$$N_b = \frac{ht}{k} = \frac{2\dfrac{\text{Btu}}{\text{hr ft}^2\,°\text{F}}\dfrac{0.004\text{ in}}{12\text{ in/ft}}}{137\dfrac{\text{Btu}}{\text{hr ft }°\text{F}}} = 4.9 \times 10^{-6}$$

therefore, neglect conduction.

For the liquid inside of the can,

$$N_b = \frac{hd_c}{4k} = \frac{2\dfrac{\text{Btu}}{\text{hr ft}^2\,°\text{F}}\dfrac{2.6\text{ in}}{12\text{ in/ft}}}{4(137)\dfrac{\text{Btu}}{\text{hr ft }°\text{F}}} = 0.00079$$

therefore, a single-lump model is good.

$$h_{side} = 2\frac{\text{Btu}}{\text{hr ft}^2\,°\text{F}}, \quad \tau = \frac{MC_p}{h_{side}A_{side}} = \frac{0.767\text{ lbm }1.0\dfrac{\text{Btu}}{\text{lbm }°\text{F}}}{2.0\dfrac{\text{Btu}}{\text{hr ft}^2\,°\text{F}}\,32.7\dfrac{\text{in}^2}{144\text{ in}^2/\text{ft}^2}} = 1.69\text{ hr}$$

$$\tau_{settle} = 4\tau = 6.76\text{ hr}$$

$$\text{time to 50 °F(in 40 °F air)} = 2.1\text{ hr}$$
$$\text{time to 50 °F(in 10 °F air)} = 0.8\text{ hr}$$

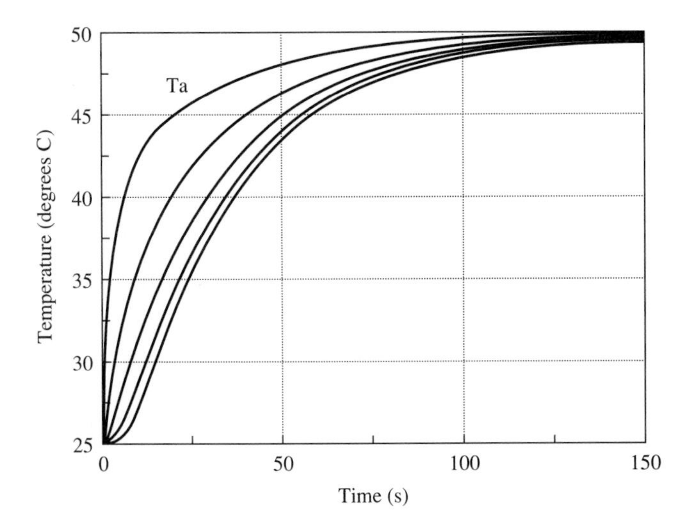


**Problem 6.24** Temperature response.

The settling time of the system is about 150 seconds, compared to $4 \times (1.51) = 6.04$ seconds for the time constant of an individual node.

## CHAPTER 7

**7.3**

$$\omega_m = \frac{\dfrac{e_o}{k_m}}{\left[\dfrac{J R_o}{k_m^2} s + 1\right]}, \quad \omega_m = 1000 \text{ rpm } [1 - e^{-\frac{t}{\tau}}], \quad \tau = 9.92 \text{ ms}$$

## CHAPTER 8

**8.3** Amplitude $= 4.55 \times 10^{-3}$ units, phase $= -89.7$ degrees, it lags

**8.6** $\omega/\omega_n > 20$

**8.9** $\omega_n = 2.83$ rad/s $= 0.45$ Hz, $\zeta = 0.086$, $\omega_{nd} = 0.448$ Hz, Amplitude $= 0.0523$ mm

**8.12** **a.** $\dfrac{15}{2s^2 + 8s + 32}$

**b.** $\dfrac{1}{0.01s^2 + 0.02s + 1}$

**c.** $\dfrac{134}{s^2 + 20s + 25}$

**d.** $\dfrac{1}{10s^2 + 600s + 1000}$

**e.** $\dfrac{1}{100s^2 + 400s + 1}$

**f.** $\dfrac{2.5}{7s^3 + 5s^2 + 2s + 3}$

**g.** $\dfrac{8s^2 + 24s + 96}{5s^4 + 17s^3 + 62s^2 + 3s - 36}$

**h.** $\dfrac{4}{s^3 + 2s^2 + 4s + 5}$

**8.15**

$$\text{TF} = \frac{s\Theta_2(s)}{T(s)} = \frac{k}{J_1J_2s^3 + (J_1b_2 + J_2b_1)s^2 + (J_1k + b_1b_2 + J_2k)s + (b_1 + b_2)}$$

Characteristic equation roots are $-3.9978 \pm j\,28.9194$, and $-7.7298$.

**8.18**

| | Eigenvalues | Damping ratio | Frequency, rad/s | Time constant |
|---|---|---|---|---|
| **a.** | $-1.125 \pm j\,0.4841$ | 0.9816 | 1.2247 | |
| **b.** | $0.1338 \pm j\,0.9424$ | (unstable) | | |
| | $-1.5176$ | | | 0.6589 |
| **c.** | $0.0534 \pm j\,0.8297$ | 0.0642 | 0.8314 | |
| | $-2.8933$ | | | 0.3456 |
| **d.** | $0.0826 \pm j\,1.2488$ | (unstable) | | |
| | $-1.9152$ | | | 0.5221 |
| **e.** | $0$ | | | |
| | $-1 \pm j\,2$ | 0.4472 | 2.236 | |
| **f.** | $0 \pm j\,2$ | 0 | 2 | |
| | $-1$ | | | 1.0 |
| **g.** | $-0.269 \pm j\,1.769$ | 0.1503 | 1.7893 | |
| | $-0.481 \pm j\,0.7413$ | 0.5433 | 0.8837 | |
| **h.** | $-0.3029 \pm j\,2.3077$ | 0.1301 | 2.3275 | |
| | $-0.6971 \pm j\,0.5024$ | 0.8113 | 0.8593 | |

**8.21 a.** $\dfrac{G}{s + 8}$

**b.** $\dfrac{Gs}{2s + 3}$

**c.** $\dfrac{G\omega}{(0.25s + 7.5)(s^2 + \omega)}$

**d.** $\dfrac{s}{s^2 + 5s + 4}$

**e.** $\dfrac{G}{2.5s^2 + 2s + 3.75}$

**f.** $\dfrac{Gs}{(s^2 + 0.25s + 0.5)(s^2 + \omega^2)}$

**8.24** One solution is $R = 2.5\ \Omega$, $C = 20\ \mu\text{f}$.

## CHAPTER 9

**9.3**

$$[G\tau D + 1]e_1 = G[\tau D + 1]e_0, \quad \text{where } \tau = R_1 C, \quad \text{and } G = \frac{1}{\left(1 + \dfrac{R_1}{R_2}\right)}$$

$$e_1(t) = Ge_0\left(1 - e^{-\frac{t}{G\tau}}\right) + \left(e_0 - e_0(0^-)\right)e^{-\frac{t}{G\tau}} + e_1(0^-)e^{-\frac{t}{G\tau}}$$

$$e_1(\infty) = Ge_0$$

**9.6** See figure on next page.

$$x_1 = e_1, \qquad \dot{x}_1 = -\frac{x_1}{R_1 C_1} - \frac{x_2}{C_1} + \frac{u_1}{R_1 C_1}$$

$$x_2 = i_L, \qquad \dot{x}_2 = \frac{x_1}{L} - \frac{x_3}{L}$$

$$x_3 = e_2, \qquad \dot{x}_3 = \frac{x_2}{C_2} - \frac{x_3}{R_2 C_2}$$

**9.15 a.** Roots: $\lambda_1 = -1$, $\lambda_2 = -4$ System is overdamped.

$z(t) = C_1 e^{-t} + C_2 e^{-4t} + z_p(t) \quad z_p(t) = 0.25, \quad z(t = 3) = 0.233$

**b.** Roots $= -0.4 \pm j\,1.1576$, $\omega_n = 1.2247$ rad/s, $\zeta = 0.3266$.
System is underdamped. See Table E.3. $\tau = 5.1\text{s}$, $z(t = 15) = 0.268$

**c.** Roots $= -0.125 \pm j\,0.696$, $\omega_n = 0.7071$ rad/s, $\zeta = 0.1768$
System is underdamped. See Table E.3. $\tau = 8.88$ s, $z(t = 27) = 1.94$

**9.18 a.** $z(t = 1) = 0.081 \qquad \dot{z}(t = 1) = 0.0134$

**b.** $z(t = 5) = -0.047 \qquad \dot{z}(t = 5) = -0.1434$

**c.** $z(t = 8) = 0.4431 \qquad \dot{z}(t = 8) = -0.8057$

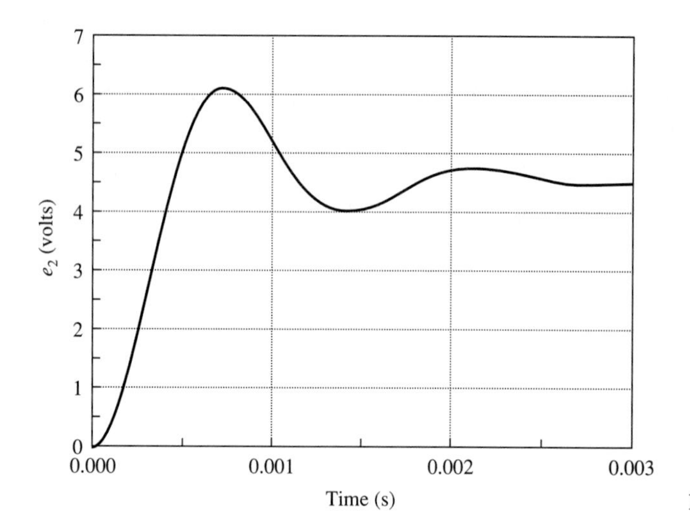


**Problem 9.6** Voltage response.

**9.21**

$$\frac{e_1}{e_2} = \frac{\left(\dfrac{R_2}{R_1 + R_2}\right)}{\left(\dfrac{R_2}{R_1 + R_2}\right) R_1CD + 1} = \frac{0.706}{35.3 \times 10^{-6} D + 1}$$

$e_1(\infty) = 3.53$ volts

**9.24** Steady state value of displacement is **a.** 0.618 cm and **b.** 1.0 cm. Input amplitude is doubled, but response output is not necessarily doubled for nonlinear system.

**9.27** Roots $= -1.0 \pm j\,5.0,\ -1.0,\ T_1 = 1$ s.

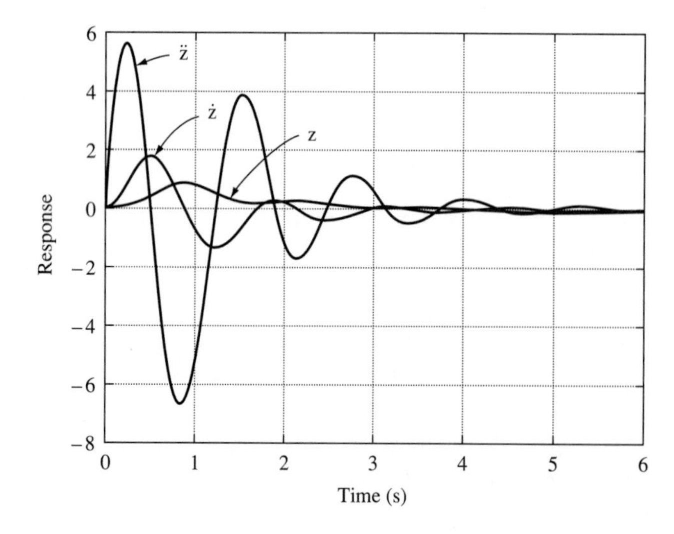


**Problem 9.27** Transient response.

**9.33** Equation of motion: $mL^2\ddot{\phi} + (1/2)\rho A C_d L^3 \dot{\phi}|\dot{\phi}| + mgL\sin\phi = 0$, $C_d = 0.7$, $\rho = 1.061 \times 10^{-7}$ lbf s/in$^4$, $A = 1.767$ in$^2$, $m = 4.58 \times 10^{-4}$ lbf s$^2$/in
Successive amplitudes of pendulum over three cycles: 1.57, 1.53, 1.49, 1.46.

**9.36** There is little temperature gradient through the plate due to its low thermal resistance. Variation of conductivity with temperature has little effect in this case.

## CHAPTER 10

**10.3**

$$z = \frac{\frac{b}{a}u - \frac{F_L}{G_p A a'}}{\frac{M}{G_p A a'}D^2 + \frac{R_o A^2}{G_p A a'} + 1}$$

$$G_s = \frac{b}{a}, \quad k_s = G_p A a', \quad \omega_n = \sqrt{\frac{k_s}{M}}, \quad \zeta = \frac{R_o A^2}{2\sqrt{k_s M}}$$

For each valve, there is only one piston diameter that meets all three of the specifications simutaneously. If the cost of the valve is not a factor, then valve #1 has the best performance; however, it might cost more.

| Valve # | Piston diameter (in) | Stiffness (lbf/in) | Natural frequency (Hz) | Damping ratio |
|---|---|---|---|---|
| 1 | 0.75 | 17,671 | 41.6 | 0.75 |
| 2 | 0.75 | 16,364 | 40.0 | 0.52 |
| 3 | 0.75 | 14,800 | 38.0 | 0.41 |

$$\tan(5°) = \frac{1.5''}{(a+b)}, \quad a = b = 8.57''$$

## APPENDIX C

**C.3**

$$\mathbf{e} = \begin{bmatrix} -70 \\ -10 \\ 35 \\ -65 \\ -20 \end{bmatrix}$$

**C.6**

$$3.9$$

**C.9**

$$\mathbf{c} - 2\mathbf{d} = \begin{bmatrix} 10 \\ -8 \\ 23 \\ -6 \end{bmatrix}$$

**C.12**

$$\sqrt{98}$$

**C.15**

$$\mathbf{C} = \begin{bmatrix} 11 & -22 & -11 & -10 \\ 42 & -6 & 36 & 2 \\ -42 & 72 & 30 & 32 \end{bmatrix}$$

**C.18**

$$\mathbf{d} = \begin{bmatrix} 98 \\ 66 \end{bmatrix}$$

**C.21**

$$\begin{bmatrix} -3 \\ -2 \end{bmatrix}$$

**C.24**

$$[\mathbf{ABC}]^T = \begin{bmatrix} 165 & 180 & 45 \\ 435 & 380 & 115 \\ 765 & 740 & 205 \\ 45 & 68 & 13 \end{bmatrix}$$

**C.27**

$$332$$

**C.30**

$$|\mathbf{B}| = 144$$

**C.33**

$$|\mathbf{B}| = -2$$

**C.39**

$$a_{13} = 4$$

**C.42**

$$\mathbf{A}^{-1} = \begin{bmatrix} 7 & -3 & -3 \\ -1 & 0 & 1 \\ -1 & 1 & 0 \end{bmatrix}$$

**C.45**

$$\mathbf{B} = \begin{bmatrix} -0.3750 & 0.2500 & 0.7500 \\ 0.3750 & -0.2500 & 0.2500 \\ 0.1250 & 0.2500 & -0.2500 \end{bmatrix}$$

**C.48**

$$\mathbf{B} = \begin{bmatrix} 1.0000 & -3.0000 & 2.0000 \\ -3.0000 & 3.0000 & -1.0000 \\ 2.0000 & -1.0000 & 0.0000 \end{bmatrix}$$

**C.51**

$$a_{12} = -0.866, \quad a_{21} = 0.866$$

## APPENDIX D

**D.3**

$$\mathbf{X} = \begin{bmatrix} 1 \\ 2 \\ -1 \end{bmatrix}$$

**D.6**

$$\mathbf{X} = \begin{bmatrix} 1.9444 \\ 1.6111 \\ 0.2778 \end{bmatrix}$$

**D.9**

$$\mathbf{X} = \begin{bmatrix} 0.0000 \\ 0.5000 \\ 0.0000 \end{bmatrix}$$

**D.12**

$$\mathbf{X} = \begin{bmatrix} 0.0000 \\ 0.0000 \\ 0.0000 \end{bmatrix}$$

**D.15**

$$\mathbf{X} = \begin{bmatrix} 0.0000 \\ 0.0000 \\ 0.0000 \end{bmatrix}$$

**D.18**

$$r(\mathbf{A}) = 3$$

**D.21**

$$r(\mathbf{A}) = 3, \;\; r(\mathbf{A}^b) = 3$$

**D.24**

$$\lambda_1 = 1.7639 \quad \lambda_2 = 6.2361 \quad \text{vector} = \begin{bmatrix} 0.618 & 1.0 \\ 1.0 & -0.618 \end{bmatrix}$$

**D.27**

$$\lambda_1 = 2 + j2.45 \quad \lambda_2 = 2 - j2.45 \quad \text{vector} = \begin{bmatrix} -0.9258 & -0.9258 \\ 0 + j0.3780 & 0 - j0.3780 \end{bmatrix}$$

**D.30**

$$\lambda_1 = 6.5826 \quad \lambda_2 = -2.5826 \quad \text{vector} = \begin{bmatrix} 0.8367 & -0.8367 \\ 0.5477 & 0.5477 \end{bmatrix}$$

**D.33**

$$\lambda_1 = 2 + j4.5826 \quad \lambda_2 = 2 - j4.5826 \quad \text{vector} = \begin{bmatrix} -0.8367 & -0.8367 \\ 0 + j0.5477 & 0 - j0.5477 \end{bmatrix}$$

## APPENDIX E

**E.3**

$$z(t) = z_h + z_p, \quad z_h = Ae^{-\frac{t}{\tau}}, \quad z_p = B \sin \omega t + C \cos \omega t$$

$$z(t) = z_0 e^{-\frac{t}{\tau}} + \frac{Gu_0}{1 + (\omega\tau)^2}\left[\sin \omega t + \omega\tau\left(e^{-\frac{t}{\tau}} - \cos \omega t\right)\right]$$

## APPENDIX F

**F.3**

$$\omega_n = \sqrt{a^2 + b^2}, \ \zeta = \frac{-a}{\sqrt{a^2 + b^2}}$$

**F.6 (a)**

$$\frac{1}{s^2 + 4s + 1} = \frac{-0.2887}{s + 3.7321} + \frac{0.2887}{s + 0.2679}$$

**(b)**

$$\frac{s + 20}{s^4 + 17s^3 + 85s^2 + 175s + 250} = \frac{-0.0235}{s + 10} + \frac{0.15}{s + 5}$$

$$+ \frac{-0.0623 - j0.0971}{s + 1 - j2} + \frac{-0.0623 + j0.0971}{s + 1 + j2}$$

**F.9**

$$Z(s) = \frac{20\,U(s) + 4z_o + (sz_o + \dot{z}_o)}{s^2 + 4s + 10}$$

$$Z(s) = \frac{(\beta_1 s + \beta_0)U(s)}{(s^3 + \alpha_2 s^2 + \alpha_1 s + \alpha_0)}$$

## Appendix G

**G.3**

$$u_1 = v, \ x_1 = z, \ x_2 = Dz, \ x_3 = D^2z, \ x_4 = D^3z$$

$$\dot{x}_1 = x_2 \qquad x_1(0) = 20$$

$$\dot{x}_2 = x_3 \qquad x_2(0) = 0$$

$$\dot{x}_3 = x_4 \qquad x_3(0) = 50$$

$$\dot{x}_4 = -6x_1 - 5x_2 - 4x_3 - 3x_4 + 10u_1 \quad x_4(0) = 0$$

**G.6**

$$u_1 = w, \quad u_2 = v, \quad x_1 = g, \quad x_2 = \dot{g}$$

$$\dot{x}_1 = x_2$$

$$\dot{x}_2 = -20x_1 - 2x_2 + 10u_1 + u_2$$

**G.9**

$$u_1 = u(t), \quad x_1 + z, \quad x_2 = \dot{z} - 2u(t)$$

$$\dot{x}_1 = x_2 + 2u_1 \qquad x_1(0) = -1$$

$$\dot{x}_2 = -0.5x_1 - 0.1x_2 + (1 - 2 \times 0.1)u \qquad x_2(0) = 10 - 2u(0^-) = 10$$

**G.12**

$$u_1 = v, \quad x_1 = e, \quad x_2 = \dot{e} - 5v, \quad x_3 = \ddot{e} - 5\dot{v} - 7v$$

$$\dot{x}_1 = x_2 + 5u_1$$

$$\dot{x}_2 = x_3 + 7u_1$$

$$\dot{x}_3 = -0.8x_1 - 0.3x_2 - 0.2x_3 + (9 - 5 \times 0.3 - 7 \times 0.2)u_1$$

$$x_1(0) = 0$$

$$x_2(0) = 0$$

$$x_3(0) = 0 - 5\omega$$

## APPENDIX I

**I.3**

$$x_1^*(\infty) = u_{1max} = 2\text{ mm}, \quad x_2^*(\infty) = \frac{15}{1.5}u_{1max} = 20\ \frac{\text{mm}}{\text{s}}$$

$$x_{1max} = \text{max of } \{|x_1(0)| \text{ or } |x_1^*(\infty)|\} = \text{max of } \{3 \text{ or } 2\} = 3\text{ mm}$$

$$x_{2max} = \text{max of } \{|x_2(0)| \text{ or } |x_2^*(\infty)|\} = \text{max of } \{0 \text{ or } 20\} = 20\ \frac{\text{mm}}{\text{s}}$$

$$\alpha_1 = \frac{3\text{ mm}}{10\text{ v}} = 0.3\ \frac{\text{mm}}{\text{v}}$$

$$\alpha_2 = \frac{20\ \frac{\text{mm}}{\text{s}}}{10\text{ v}} = 2.0\ \frac{\frac{\text{mm}}{\text{s}}}{\text{v}}$$

$$\beta_1 = \frac{2\text{ mm}}{10\text{ v}} = 0.2\ \frac{\text{mm}}{\text{v}}$$

$$\dot{e}_1 = 6.667e_2 \qquad e_1(0) = 10\text{ volts}$$

$$\dot{e}_2 = -2.25e_1 - 1.5e_2 - 1.5v \qquad e_2(0) = 0\text{ volts}$$

# Index

## A



## B



## C

## D



## E

## F



## G



## H



## I

## O



## P



## R

## T



## U



## V



## W



## Y



## Z